QM
23.2
.M66

100572

Montgomery, Royce L.
Basic Anatomy for the Allied
Health Professions.

**East Texas Baptist College
Library**
Marshall, Texas 75670

Basic Anatomy
for the
Allied Health Professions

Basic Anatomy for the Allied Health Professions

ROYCE L. MONTGOMERY, Ph.D.
Department of Anatomy
University of North Carolina
School of Medicine
Chapel Hill

Urban & Schwarzenberg
Baltimore—Munich 1981

URBAN & SCHWARZENBERG, INC.
7 E. Redwood Street
Baltimore, Maryland 21202
USA

URBAN & SCHWARZENBERG
Pettenkoferstrasse 18
D-8000 München 2
Germany

© Urban & Schwarzenberg 1980

All rights, including that of translation, reserved. No part of this publication may be reproduced, stored in a retrieval system, or transmitted in any other form or by any means, electronic, mechanical, recording, or otherwise without the prior written permission of the publisher.

The illustrations in this atlas originally appeared in the following Urban & Schwarzenberg publications:

Lippert: Anatomie
Benninghoff/Goerttler: Lehrbuch der Anatomie des Menschen, edited by Helmut Ferner and Jochen Staubesand
Sobotta/Figge: Atlas of Human Anatomy
Toldt/Hochstetter: Anatomischer Atlas

excepting those cited below.

Figures 5-14 and 5-17A are reproduced from *Grant's* Atlas of Anatomy with permission from The Williams & Wilkins Company.

Figures 1-1, 1-4, 1-5F, 5-17B, 5-17C, 9-7, 9-9A, 9-20A, 9-21A, 9-21C, 9-33A, 11-4B, 11-5A, 12-1A, 12-4C, 13-1B, 13-2B, 13-6B, 14-2D and 14-4B have been drawn for this atlas by *Christine D. Young.*

Cover illustration by *Christine D. Young.*

Library of Congress Cataloging in Publication Data

Montgomery, Royce L.
 Basic anatomy for the allied health professions.
  1. Anatomy, Human. 2. Allied health personnel.
 I. Title. [DNLM: 1. Anatomy—Atlases. QS17 M788b]
 QM23.2.M66   611   79-19131
 ISBN 0-8067-1231-7

ISBN 0-8067-1231-7 (Baltimore)
ISBN 3-541-71231-7 (Munich)

Printed in the United States of America

Preface

The aim of this anatomy "text atlas" is to provide a useful, appropriately-illustrated learning tool for students in introductory courses in human anatomy. The broad scope and high cost of the well-illustrated textbooks of anatomy used by medical students make such books inappropriate for students of the allied health professions. The answer, therefore, seemed to lie in a compilation of anatomical illustrations from medical textbooks, supplemented with easy-to-comprehend explanations.

This anatomy "text atlas" is specifically written for students seeking careers in nursing and allied health professions (i.e., medical assistant, physician's assistant, surgical assistant, medical laboratory technologist, mortician, medical record keeper, physical therapist, dental hygienist, radiologic technologist and respiratory therapist). This book is also understandable and affordable by students in the liberal arts and physical education.

The basic content of this "text atlas" is anatomy with emphasis on anatomical plates and illustrations supplemented with clinical applications.

I am indebted to Ms. Kathrinn Plemmons for typing the manuscript and to Ms. Genice Chin for assisting with the labelling of illustrations. And also, I am particularly grateful to Mary Hsieh, Carola Sautter and Braxton Mitchell for their help and advice not only in encouraging the early stages of the book but also in seeing the work through to its completion.

Autumn, 1980   ROYCE L. MONTGOMERY, Ph.D.

Dedication
To my students—past, present and future;
my wife, Jane; my children, Todd, Scott and Jill;
and my mother and father, Jimmie and E.L.

Contents

1. Introduction 3
2. The Integumentary System 23
3. The Skeletal System 31
4. Articulations and Ligaments 70
5. The Muscular System 84
6. The Nervous System 122
7. The Special Senses 214
8. The Digestive System 239
9. The Cardiovascular System 275
10. The Lymphatic System 346
11. The Respiratory System 367
12. The Urinary System 388
13. The Reproductive System 397
14. The Endocrine Glands 426

Answers 440

Index 443

1. Introduction

> STUDENT OBJECTIVES
> After you have read this chapter, you should be able to:
> 1. Define anatomy.
> 2. Define the anatomical position.
> 3. Define directional terms used in association with the body.
> 4. Describe the common anatomical planes of the body.
> 5. Discuss the levels of structural organization that comprise the human body.
> 6. Compare systemic anatomy with regional anatomy.
> 7. Explain applied anatomy.

DEFINITION OF ANATOMY

The term *anatomy* is of Greek origin and means to cut apart or dismember. Anatomy deals with *structure* and the relationships among structures.

ANATOMICAL POSITION

In describing the *human* body, the directional terms always refer to the body in the *anatomical position* (fig. 1-1) in which the body is standing erect with the arms and hands by the sides, the palms facing forward with the thumbs turned outward, and the feet together.

DIRECTIONAL TERMS

It is important that a student understand the proper directional terms to be employed in human anatomy, for these are used constantly (fig. 1-2).

Anterior and *posterior*: front and back, respectively.
Dorsal and *ventral*: the proper terms for back and front, respectively, of the trunk and upper arm; permissible for the lower limbs except the foot.
Dorsal and *volar* (or *palmar*): the proper term for the back and the palm of the hand and the corresponding aspects of the forearm.
Dorsal and *plantar*: the proper terms, respectively, for the upper aspect and the sole of the foot.
Superior and *inferior*: superior is toward the head and inferior is in the direction of the feet.
Cranial and *caudal*: seldom used but are the same as superior and inferior.
Superficial and *deep*: superficial is nearer the surface from any aspect, as opposed to deep.
Medial and *lateral*: medial is toward the median plane, and lateral is toward the side.
Proximal and *distal*: proximal, toward the median plane; distal, away from the center; used mostly in connection with the limbs.
External and *internal*: used chiefly to indicate respective surfaces of hollow organs or of the body as a whole.
Radial and *ulnar*: the proper terms for denoting the sides of the forearm, wrist, and hand; toward the radius or toward the ulna, respectively.
Tibial and *fibular*: the proper terms for denoting the sides of the leg, ankle, and foot; toward the tibia or toward the fibula, respectively.

PLANES OF THE BODY

As the body is three-dimensional, so are there three planes of the body (fig. 1-2).

Sagittal plane: the dorsoventral longitudinal plane; through or parallel to the sagittal suture.

Coronal plane: the longitudinal plane, from side to side, at a right angle to the sagittal suture; through or parallel to the coronal suture.

Transverse plane: the horizontal plane, at any level, at right angles to both sagittal and coronal planes.

STRUCTURAL ORGANIZATION

Those individuals fortunate enough to study the anatomy of the human body should recognize several levels of structural organization (fig. 1-3). The *chemical level*, which includes all chemical substances essential for maintaining life, is the lowest level of structural organization. The chemical makeup of the human body involves atoms joined together in various ways to form the next higher level of organization, known as the *cellular level.* Cells form the basic structural and functional units of the human body. The next higher level of structural organization is the *tissue level*, which includes groups of similar cells performing specific functions. Different kinds of tissues are joined together to form higher levels of organization known as *organ levels.* Organs include several different tissues that perform specific functions. The next higher level of structural organization in the human body is the *system level,* which includes organs with a common function.

SCOPE

The study of human anatomy can be approached in several ways: by regions; by systems; or by a combination of the two, which is called *practical* or *applied anatomy.*

The *primary regions* of the body (fig. 1-4) are the *head, neck, trunk,* and *upper* and *lower limbs.*

The *primary systems* of the body (fig. 1-5) are: the *integumentary (A), skeletal (B), articular (C), nervous (D), muscular (E), special senses (F), digestive (G), vascular (H), lymphatic (I), respiratory (J), urinary (K), reproductive (L),* and *endocrine (M).*

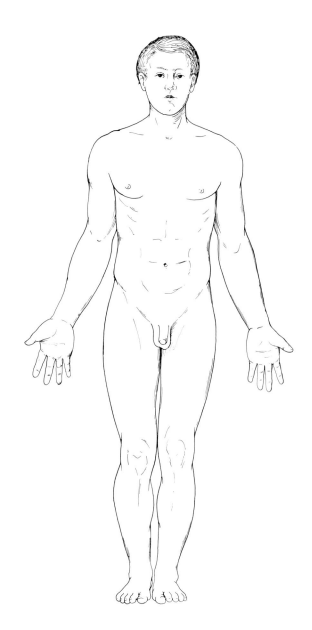

FIG. 1-1. *Anatomical position*

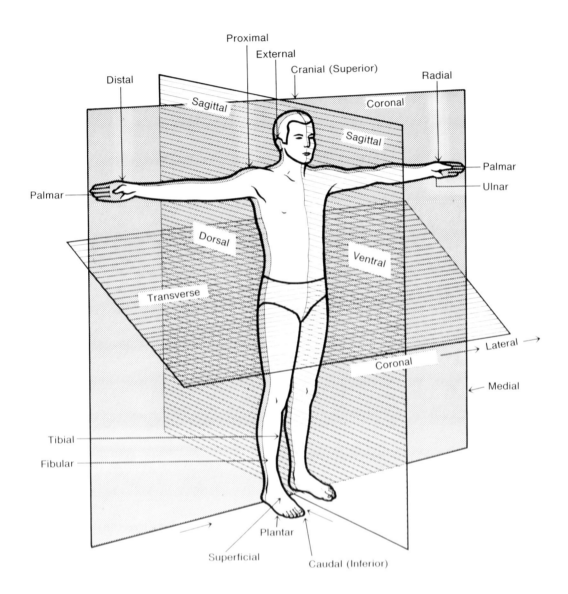

FIG. 1-2. *Directional terms*

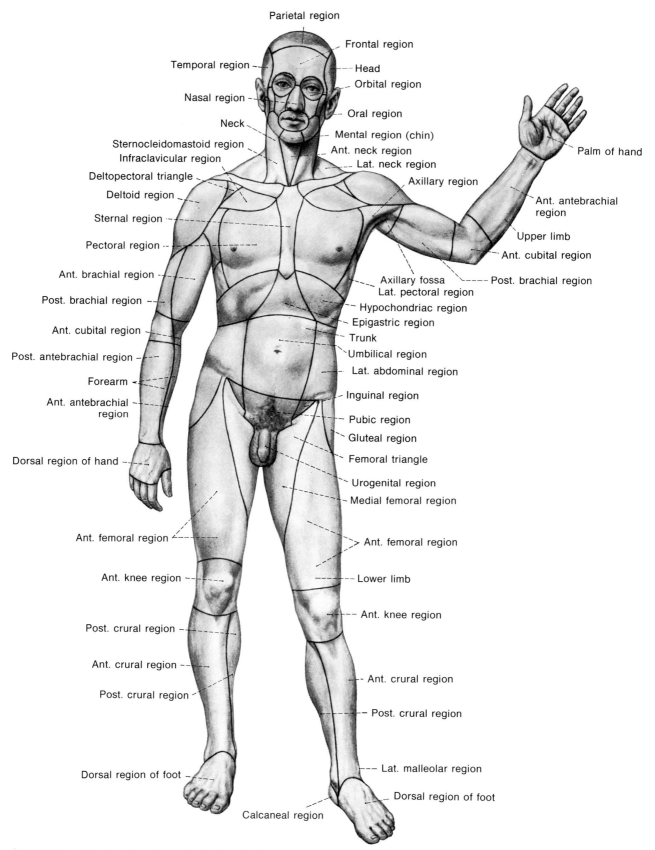

FIG. 1-3A. *Regional anatomy (anterior view)*

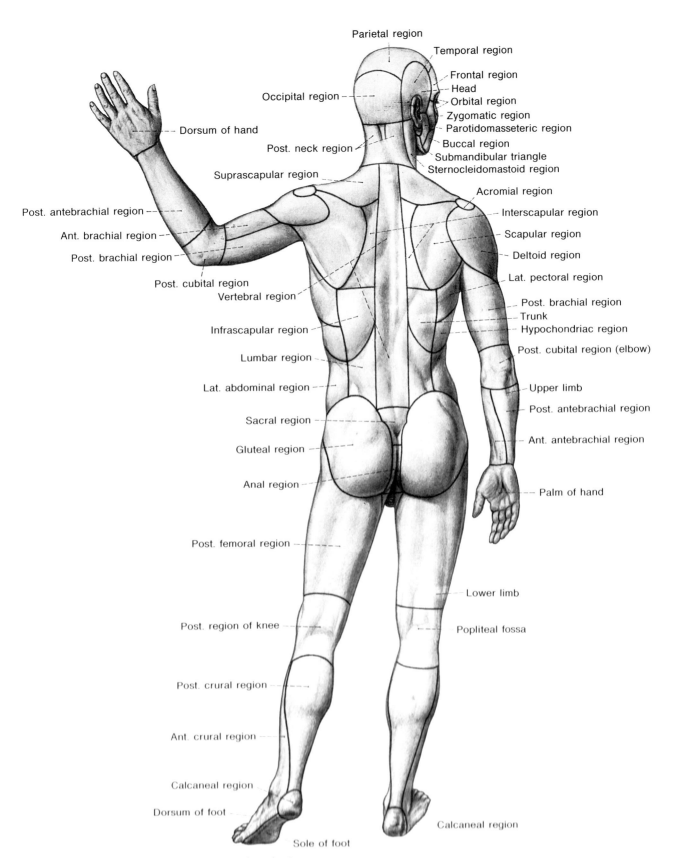

FIG. 1-3B. *Regional anatomy (posterior view)*

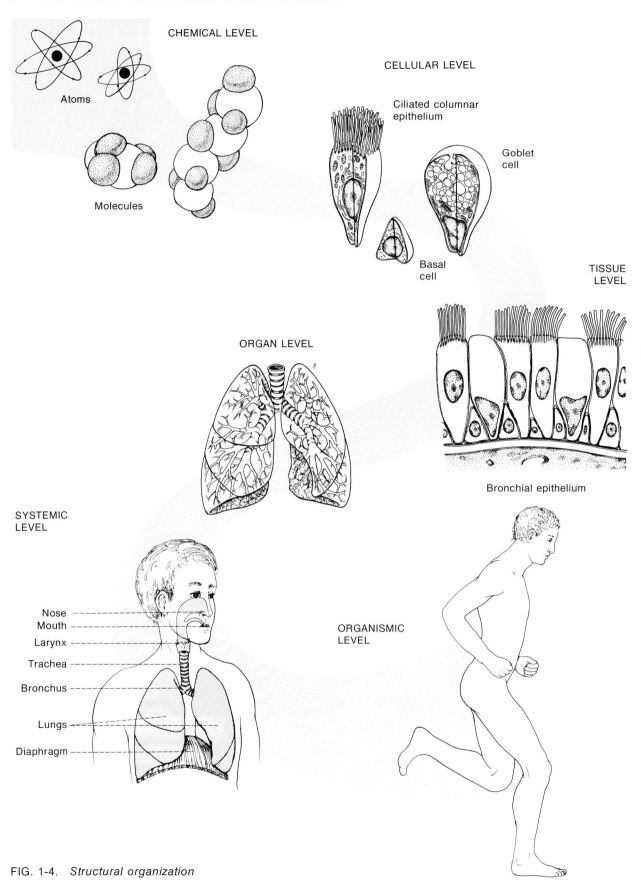

FIG. 1-4. *Structural organization*

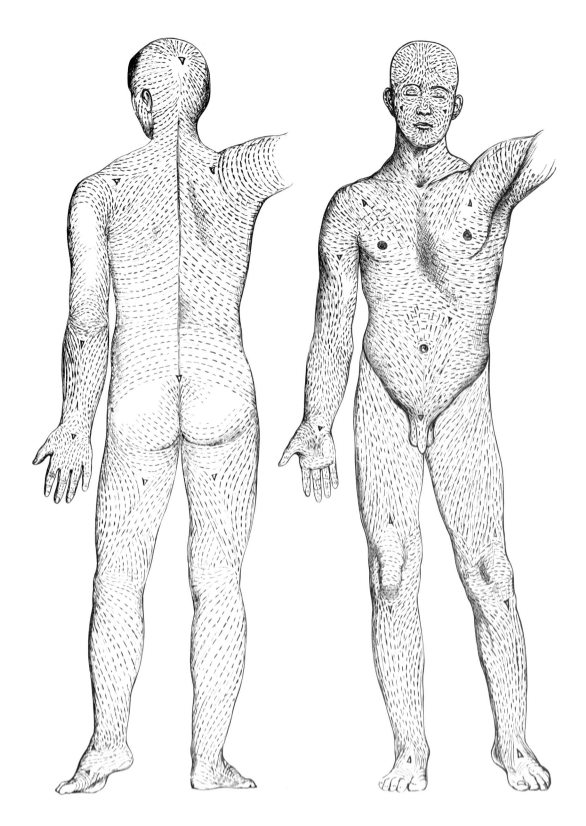

FIG. 1-5A. *Systemic anatomy: integumentary system (skin)*

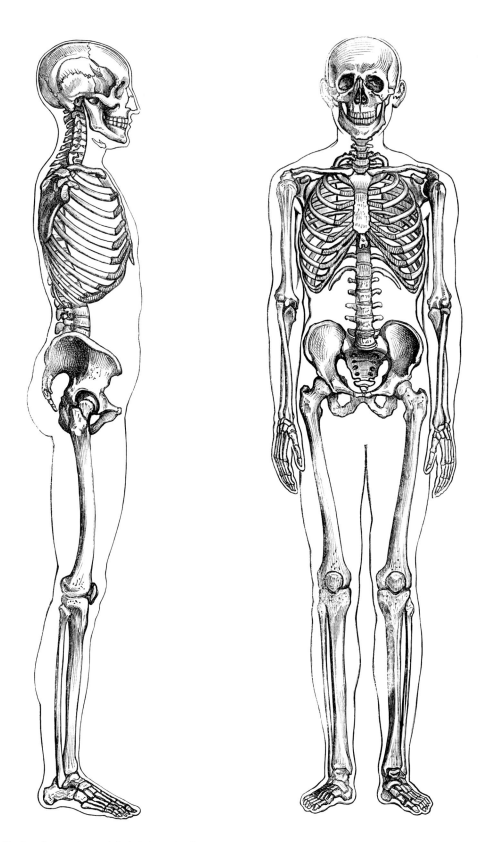

FIG. 1-5B. *Systemic anatomy (skeletal system)*

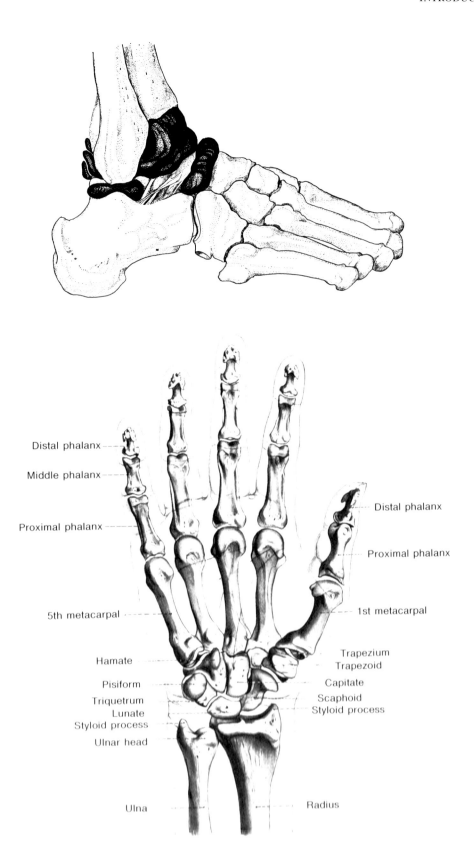

FIG. 1-5C. *Systemic anatomy (articular and ligamental system)*

12 ■ BASIC ANATOMY FOR THE ALLIED HEALTH PROFESSIONS

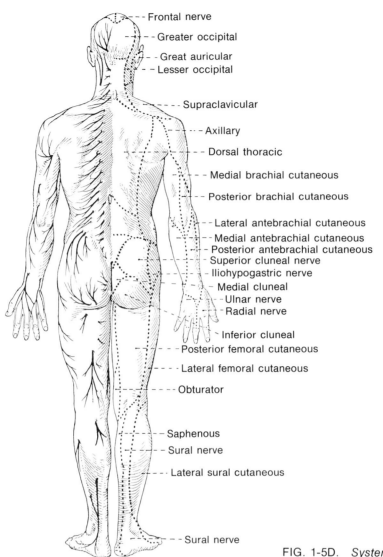

FIG. 1-5D. *Systemic anatomy (nervous system)*

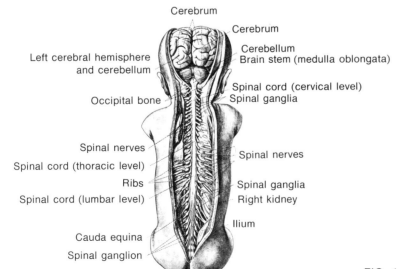

FIG. 1-5D. *Systemic anatomy (nervous system)*

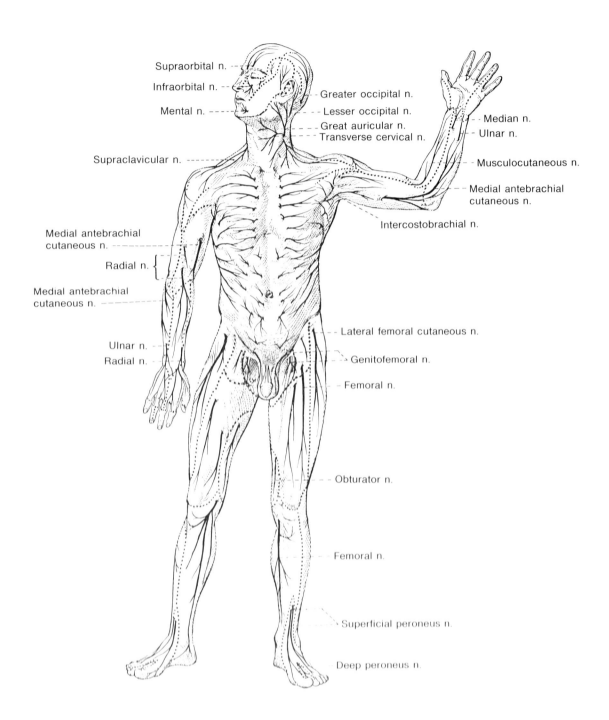

FIG. 1-5D. *Systemic anatomy (nervous system)*

14 ■ BASIC ANATOMY FOR THE ALLIED HEALTH PROFESSIONS

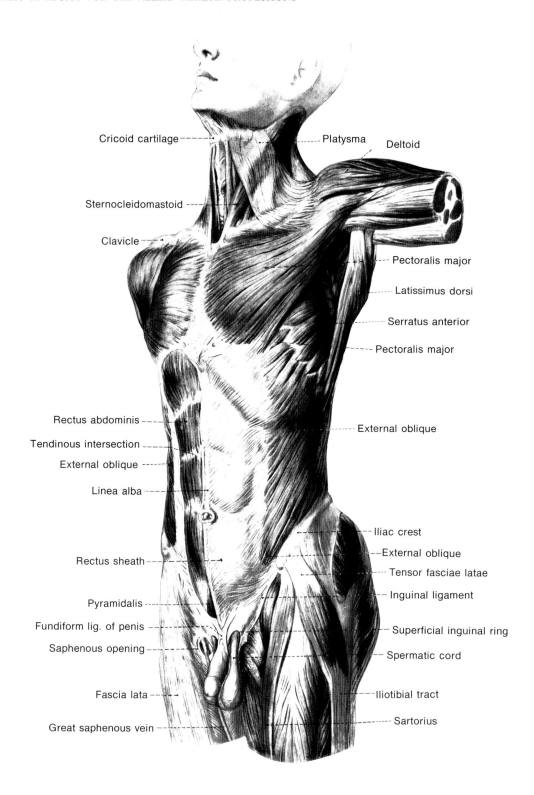

FIG. 1-5E. *Systemic anatomy (muscular system)*

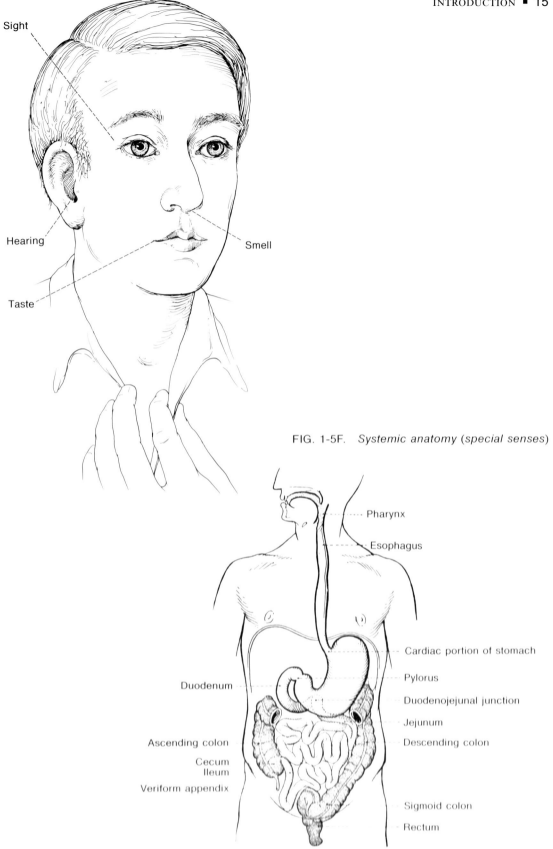

FIG. 1-5F. *Systemic anatomy (special senses)*

FIG. 1-5G. *Systemic anatomy (digestive system)*

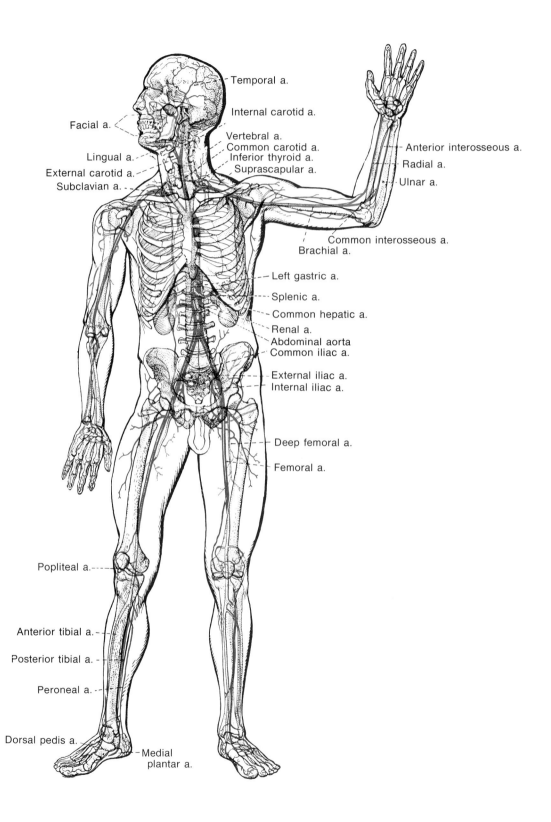

FIG. 1-5H. *Systemic anatomy (cardiovascular system)*

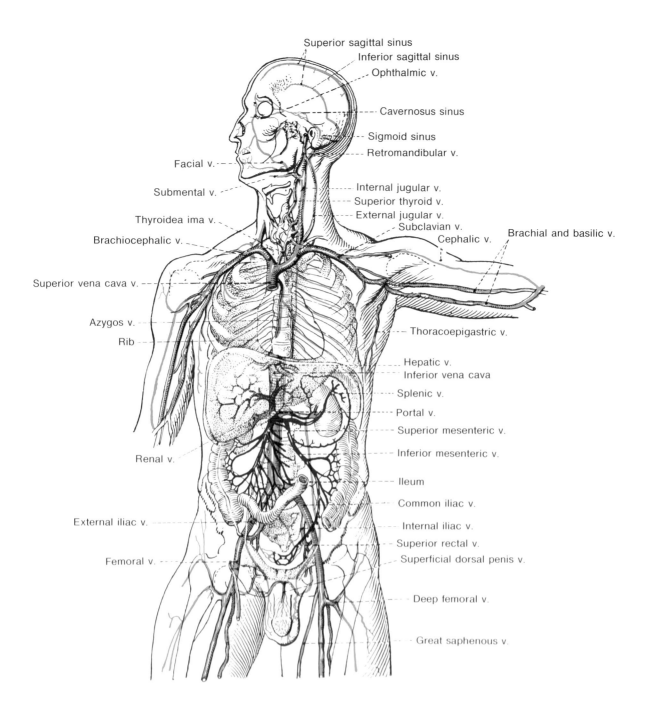

FIG. 1-5H. *Systemic anatomy (cardiovascular system)*

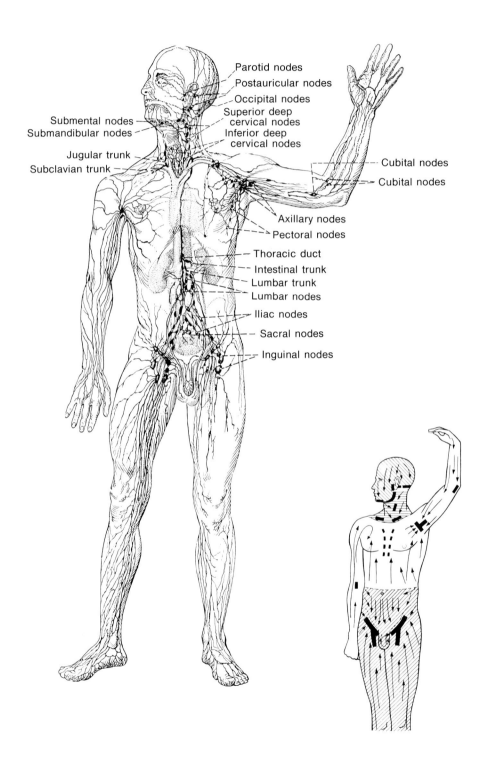

FIG. 1-5l. *Systemic anatomy (lymphatic system)*

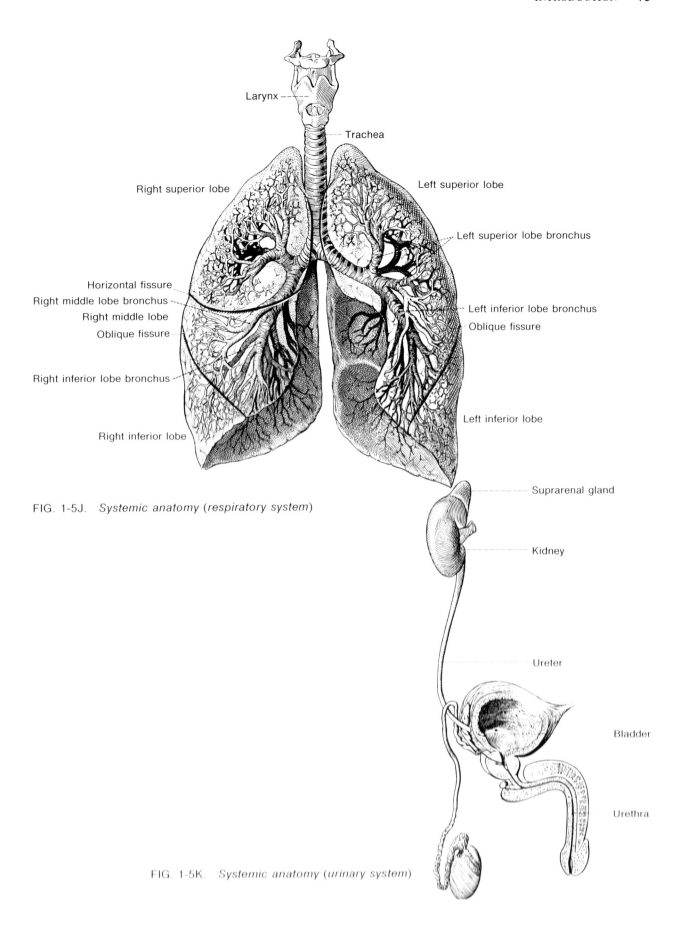

FIG. 1-5J. *Systemic anatomy (respiratory system)*

FIG. 1-5K. *Systemic anatomy (urinary system)*

20 ■ BASIC ANATOMY FOR THE ALLIED HEALTH PROFESSIONS

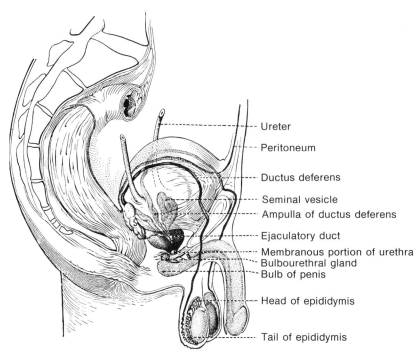

FIG. 1-5L. *Systemic anatomy (reproductive system)*

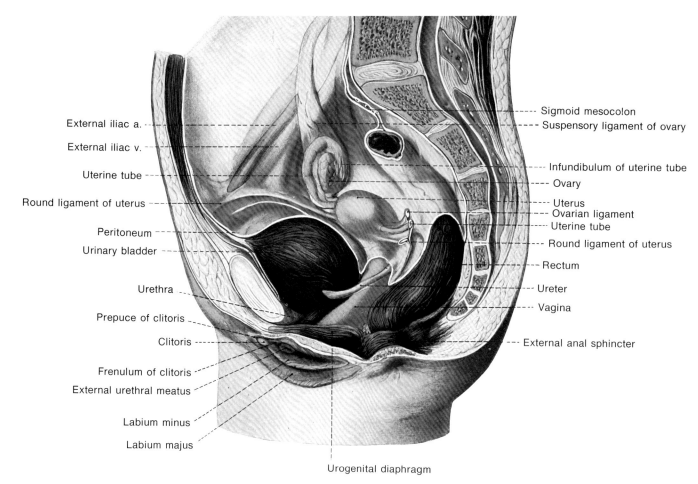

FIG. 1-5L. *Systemic anatomy (reproductive system)*

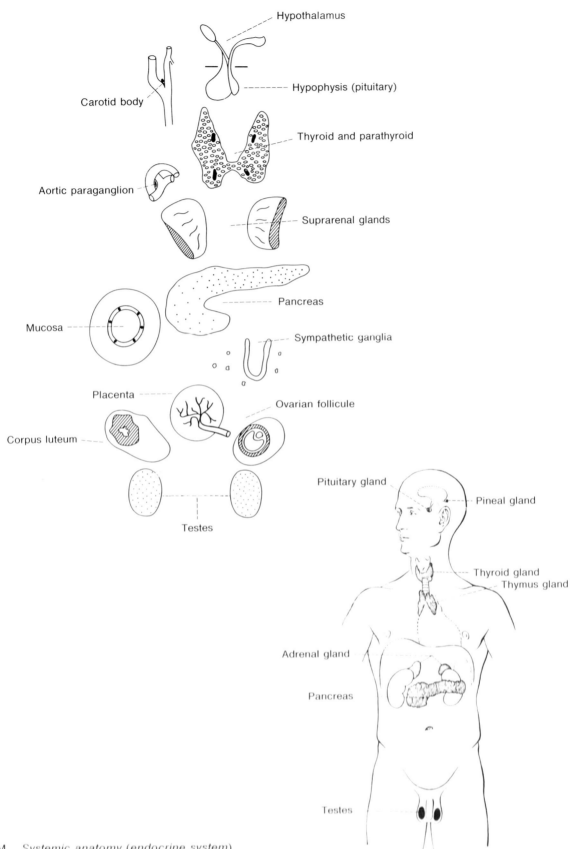

FIG. 1-5M. *Systemic anatomy (endocrine system)*

REVIEW QUESTIONS:

1. Anatomy deals with _____.
2. The palms face _____ in the anatomical position.
3. The proper term used for the sole of the foot is: _____.
4. The thumb is _____ to the shoulder.
5. The radial side of the forearm is _____ to the ulnar side.
6. The coronal plane is at a _____ angle to the sagittal plane.
7. The _____ plane is at right angles to both the _____ and _____ planes.
8. The _____ level is the lowest level of structural organization.
9. The _____ level includes groups of similar cells performing specific functions.
10. Organs include several different _____ that perform specific functions.
11. Applied anatomy involves a combination of _____ and _____ anatomy.
12. In dissection of a cadaver, which method of study would be least suited? _____

2. The Integumentary System

> STUDENT OBJECTIVES
> After you have read this chapter, you should be able to:
> 1. Identify the layers of skin.
> 2. List the structural layers of the epidermis.
> 3. List the composition of the dermis.
> 4. Discuss the epidermal derivatives of the skin.
> 5. Explain the composition and functions of superficial fascia.
> 6. Discuss classifications of burns.

The *integumentary system* includes the skin and its associated appendages: nails, hair, and glands.

The *skin*, or *integument*, is the largest organ of the body. It covers the body and protects the underlying tissues from mechanical stress, temperature, bacteria, radiation, excessive loss of fluids, and organic and inorganic materials. It also receives important stimuli from the environment and thus provides for a network of alerting signals by means of sensory organs.

SKIN AND ITS STRUCTURE

The *skin* is composed of two distinct layers: the more superficial and thinner *epidermis*, and the deeper *dermis*.

Epidermis

The *epidermis* (fig. 2-1) is stratified squamous epithelium and is made up of five strata. These are, from superficial to deep, the *stratum corneum*, the *stratum lucidum*, the *stratum granulosum*, the *stratum spinosum*, and the *stratum basale*. The stratum corneum consists of several layers of dead scalelike cells which form the keratinized layer. The most external layers are constantly being sloughed off to be replaced by new cells formed in the basale layer. The stratum lucidum contains several layers of translucent cells. The cells of the stratum granulosum contain dark-staining granules. The stratum spinosum consists of several layers of irregularly placed cells with spinelike processes. The stratum basale contains a single layer of columnar cells that undergo mitosis. The new cells produced in the basale layer are pushed upward into each of the above mentioned strata. Eventually, the cells die and slough off. Certain cells located in the basale layer produce pigment when exposed to ultraviolet radiation.

Dermis

The *dermis* (fig. 2-1) contains peglike *papillae* (stratum papillare) (fig. 2-2) that tend to interlock the epidermis and dermis. The dermis also contains interwoven connective tissue (stratum reticulare), blood vessels, lymph vessels, nerves, and glandular tissue.

On the palm of the hand and the sole of the foot there is a complicated *pattern of cristae* (fig. 2-3) conforming to the *papillae of the dermis*, which is best developed on the subterminal sensory pads of the digits. The orifices of the sweat (sudoriferous) glands of the palms and soles open along the cristae. The individual dissimilarity of patterns is well known.

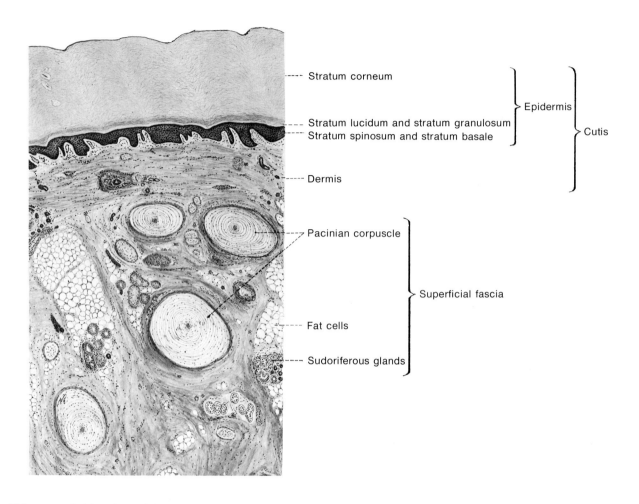

FIG. 2-1. *Epidermis and dermis*

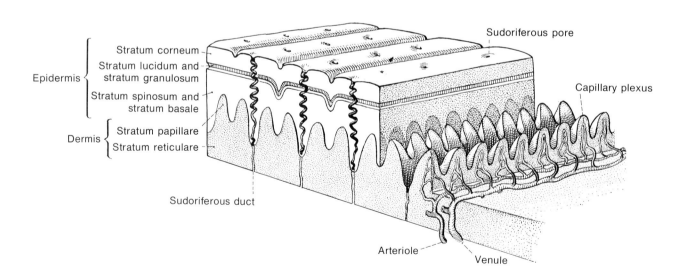

FIG. 2-2. *Dermal papillae*

THE INTEGUMENTARY SYSTEM ■ 25

FIG. 2-3A. *Actual size*

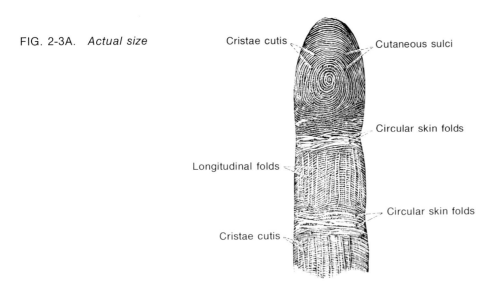

FIG. 2-3B. ×20

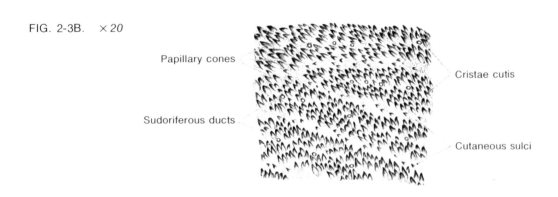

FIG. 2-3C. ×13

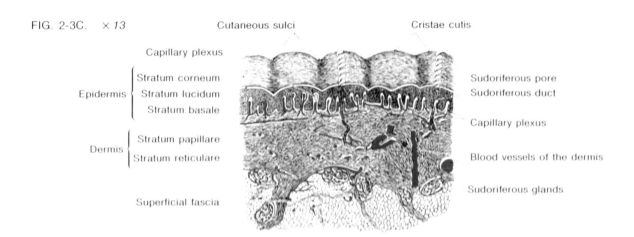

FIG. 2-3. *Pattern of cristae*

26 ■ BASIC ANATOMY FOR THE ALLIED HEALTH PROFESSIONS

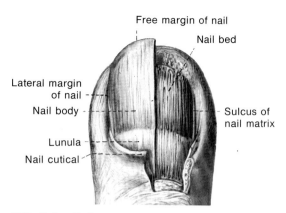

FIG. 2-4. *Nails*

APPENDAGES OF THE SKIN

These *appendages of the skin* include the nails, hairs, and glands of the skin, including the mammary glands.

Nails

The *nails* (fig. 2-4) rest upon a nail bed of dermis. They grow from the matrix beneath the root, where the nail bed is less vascular and less adherent to the dermis. This zone is whitish in color and is known as the *lunula*. Growth of the nail occurs in the light-colored proximal end of the nail bed known as the *crescent*.

Hair

The *hairs* (fig. 2-5) occur over most of the body, being absent from the lips, palms of the hands, soles of the feet, terminal phalanges of the digits, and all mucous membranes. Hairs may be coarse or fine, round or flattened, luxuriant or sparse, and may occur in many colors.

Aging causes most males, but few females, to lose a portion of the hair on the scalp. The age at which this hair loss occurs is variable. Graying hair is also associated with advancing age. Heredity plays an important role in both the graying and loss of hair.

The *hair shaft* is surrounded by its *inner* and *outer sheaths*. The *bulb* from which the hair grows is located near the root of the hair shaft and is surrounded by the *hair follicle*. The *arrector pili muscle*, composed of smooth fibers, extends from the superficial surface of the dermis to the sheath near the base of the hair follicle, from the direction in which the hair slopes. Contraction of the arrector pili muscles erects the hair and discharges sebaceous gland secretion.

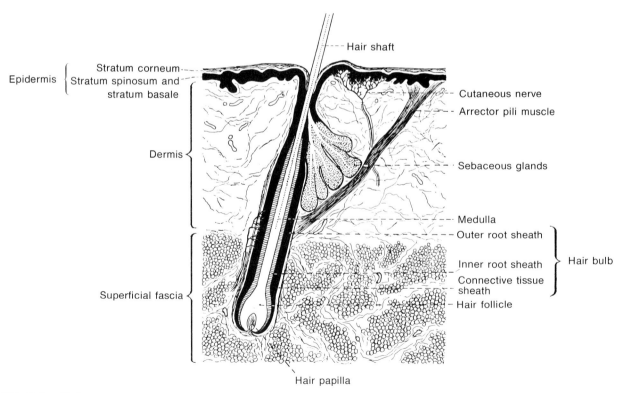

FIG. 2-5. *Hair*

THE INTEGUMENTARY SYSTEM ▪ 27

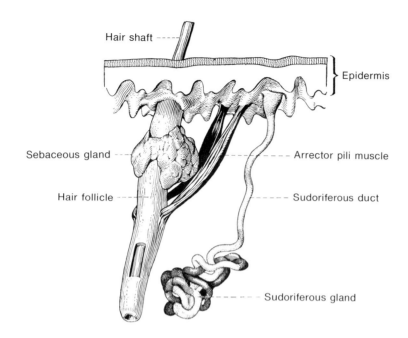

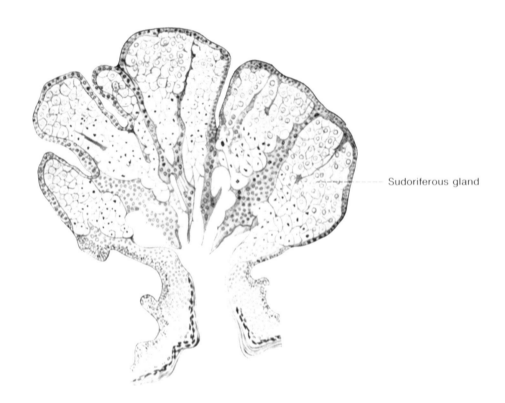

FIG. 2-6. *Sebaceous and sudoriferous glands. above: In relation to skin structures. below: Detail of sebaceous gland*

Glands

The *glands of the skin* are the *sebaceous* and *sweat glands*.

Sebaceous glands (fig. 2-6) are located in the dermis and may open directly into hair follicles or may be independent of them. The sebaceous glands on the tarsus of the eyelid are especially well developed.

Sudoriferous or *sweat glands* (fig. 2-6) occur on most areas of the body and are particularly plentiful on the upper lip, forehead, axillae, palms of the hands and soles of the feet. They are located upon the deep surface of the dermis with coiled tubes extending to the surface, independent of the hair follicles. Modified sweat glands include the *ciliary glands* of the eyelid, the *ceruminous glands* of the external auditory meatus, and the *mammary glands*.

MAMMARY GLANDS The *mammary glands* (fig. 2-7) consist of lobular glandular tissue connected by fibrous tissue in which there is considerable fat. *Lactiferous ducts* pass from the lobes to the nipple. The mammary gland is composed of 12–15 separate glands, each having its own lactiferous ducts which radiate out to converge at the nipple.

The *areola area* is pigmented and contains sebaceous and sudoriferous glands that produce evaginations of the soft alveolar skin.

Helical network The muscle fibers of the nipple are arranged into a *helical network* (fig. 2-8). Stimula-

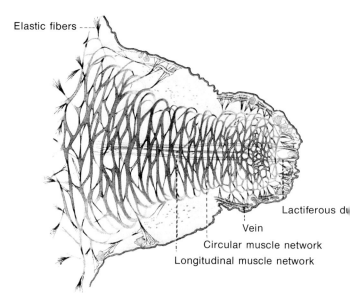

FIG. 2-8. *Helical network*

tion of the nipple results in contraction of the helical muscles to produce the erection reflex of the nipple.

Developmental stages The milk glands (mammary glands) (fig. 2-9A–E) remain in a dormant stage in children until puberty (fig. 2-9C). Then the tissue beneath the areola begins to enlarge, raising this area above the remaining breast, which is still flat. Next, the lobules of the mammary gland begin to grow at individually different rates. Glandular tubules enlarge and form branching networks (fig. 2-9A, B, and D).

During pregnancy there is further development in which the solid glandular tubules develop *lumens*. Microscopic sections through the breast of a pregnant woman indicate the mammary lobes are fully developed and clearly separated from adjacent lobes by intervening connective tissue. The lactiferous ducts course in a convergent path to the nipple.

Following the nursing stage, the mammary gland regresses to a glandular condition similar to the stage prior to pregnancy (fig. 2-9E).

SUPERFICIAL FASCIA

The *superficial fascia* (fig. 2-1) is composed of loose areolar connective tissue containing fat, blood vessels, lymphatics, nerves, receptors, and glands. This tissue nourishes, protects, and acts as a food reserve for the skin. The superficial fascia also attaches the skin to the deep fascia surrounding the muscles.

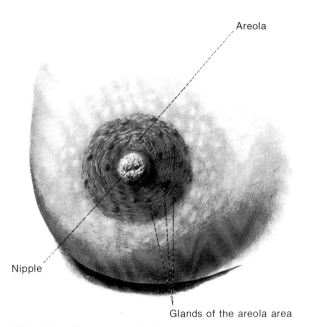

FIG. 2-7. *Mammary glands*

THE INTEGUMENTARY SYSTEM ■ 29

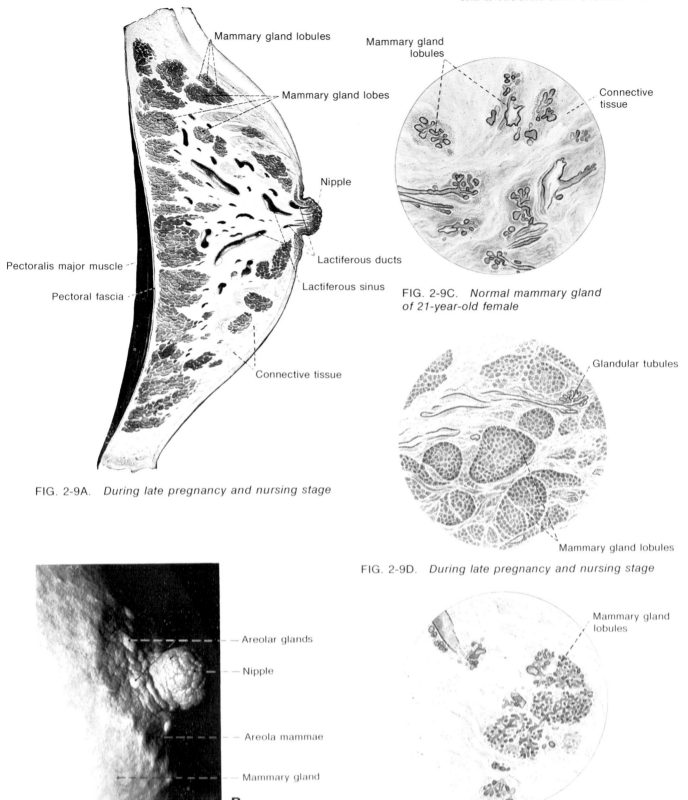

FIG. 2-9A. During late pregnancy and nursing stage

FIG. 2-9B. Nipple and areola of female breast

FIG. 2-9C. Normal mammary gland of 21-year-old female

FIG. 2-9D. During late pregnancy and nursing stage

FIG. 2-9E. Dormant stage following pregnancy

FIG. 2-9. Developmental stages of mammary gland

CLINICAL APPLICATIONS

Skin may be damaged by thermal, electrical, radioactive, or chemical agents. Each of these agents damages the skin in the form of *burns*. Burns may be placed into one of the following three categories:

Classification of Burns	Tissue Damage	Results
first degree	epidermis only	reddened skin
second degree	epidermis and dermis damaged	blistered skin
third degree	epidermis and dermis destroyed	charred or white skin

Today, burns covering 50% or less of the body are not usually fatal.

A few of the major bacterial invaders of the skin include staphylococcus (boils, abscesses, and carbuncles), and streptococcus (cellulitis and scarlet fever). Some of the more common *viral diseases* involving the skin include rubeola (measles) and rubella (German measles), varicella (chicken pox), and herpes (cold sores). *Fungal infections* of the skin are referred to as dermatomycoses, the more common of which are named according to the affected area of skin: tinea pedis (athlete's foot) and tinea unguium (infection of the fingernails).

Skin cancer outnumbers all other cancers and is considered to be caused by overexposure to sunshine and its ultraviolet rays. The three most common types of skin cancer are *basal cell*, *squamous cell*, and *melanoma*.

The most serious breast disease is cancer, which arises most frequently in the upper quadrant of the breast. *Breast cancer* shows less favorable prognosis with longer duration of symptoms, larger size of tumor, and greater involvement of the lymph nodes.

REVIEW QUESTIONS:

1. The _____ or _____ is the largest organ of the body.
2. The skin is composed of two distinct layers, the _____ and the _____.
3. The stratum _____ contains several layers of translucent cells.
4. The stratum _____ consists of several layers of dead scalelike cells which form the keratinized layer.
5. Certain cells within the stratum _____ produce pigment when exposed to ultraviolet radiation.
6. The dermis contains peglike _____ that tend to interlock the epidermis and dermis.
7. Growth of the nail occurs in the light-colored proximal end of the nail bed known as the _____.
8. Contraction of the _____ muscle erects the hair and discharges sebaceous gland secretion.
9. Sweat glands are also called _____ glands.
10. Modified sweat glands include the _____ glands of the eyelid, the _____ glands of the external auditory meatus, and the _____ glands.
11. The _____ fascia is deep to the dermis.
12. The dermis and epidermis are destroyed in a _____ degree burn.

3. The Skeletal System

> STUDENT OBJECTIVES
> After you have read this chapter, you should be able to:
> 1. Describe the functions of the skeletal system.
> 2. Contrast spongy (cancellous) with compact (dense) bone.
> 3. Describe the components of the axial and appendicular skeleton.
> 4. Identify the bones of the skull.
> 5. Identify the sutures and fontanels of the skull.
> 6. Describe the paranasal sinuses.
> 7. Identify the principal foramina of the skull.
> 8. Identify the bones of the vertebral column.
> 9. Identify the bones of the thorax.
> 10. Identify the bones of the shoulder girdle.
> 11. Identify the bones of the upper limb.
> 12. Identify the bones of the pelvic girdle.
> 13. Identify the bones of the lower limb.

The human skeleton, which consists of approximately 206 bones, provides a protective envelope for vital parts that otherwise would be particularly prone to injury; supplies a system of levers so that muscle action can operate to best advantage; acts as a prop to keep the organism from collapsing onto the ground; stores calcium; and provides bone marrow for the formation of blood cells. There are several ways of classifying bones: (1) by method of origin, which includes either intramembranous or endochondral ossification; or (2) by structure, which refers to spongy or compact; or (3) by shape, which relates to long, short, irregular, or flat types.

The following are definitions of a few of the general terms used in reference to the skeletal system:

Ala: a wing-shaped projection
Alveolus: a small cavity; used to define a tooth socket
Canal: a tubular channel through bone
Condyle: an articular eminence
Crest: a bony ridge
Diaphysis: the shaft of a long bone
Epicondyle: a ridge above a condyle
Epiphysis: a terminal piece of bone separated from a diaphysis by cartilage during early life
Foramen: a perforation through bone; the end of a canal
Fossa: a shallow depression
Lamina: a thin plane of bone
Linea: a line, marked either by a low crest or a depression
Meatus: a passageway; a large foramen; used particularly for an external opening of a canal
Process: a bony projection
Sinus: an air space in bone
Spine: a sharp process
Sulcus: a groove or furrow
Suture: the line of junction of neighboring bones
Symphysis: essentially the same as suture, but used only for particular junctions
Trochanter: each of two processes below the neck of the femur
Trochlea: a pulley-shaped structure
Tubercle: a nodular process

Bones develop from either membrane or cartilage. In the former the bone is laid down by osteoblasts among the fibers of a mesodermal membrane sheet, as in the roof of the skull. Cartilage bone is formed in cartilage. The actual ossification is the same in both cases.

In the transformation of cartilage into bone, one or more centers of ossification appear from which the bony tissue spreads. In a long bone this regularly appears in the region of the shaft (diaphysis). The ends (epiphyses) ossify later, usually long after birth. The centers of the long bones become hollow, largely by an eroding process, and become filled with bone marrow.

Bone is compact or spongy, the two differing only in the size of the interstitial cavities. As a rule the former occurs at the surface and the latter occurs deeper in the substance of a bone. The interstices of spongy bone are lined with delicate *endosteum* and filled with *red marrow*, which manufactures red blood corpuscles. The medullary cavity of the long bones is also lined with endosteum, which is connected through the nutrient canals with the periosteum. These canals carry vessels and nerves. Within the substance of the bone the Haversian canals, some of microscopic size, are surrounded by bony lamellae, this complex forming the *Haversian systems*. As already stated, the medullary cavities of the long bones are filled with *yellow bone marrow*, but cavities in other bones may contain only air (such as the paranasal sinuses).

The skeletal system is divided into axial and appendicular divisions. The axial skeleton includes the skull, vertebral column, ribs, sternum, and hyoid. The appendicular division contains the bones of the upper and lower limbs.

AXIAL DIVISION

Skull

ANTERIOR VIEW   The anterior aspect (fig. 3-1) may be divided into the following regions: *frontal, orbital, nasal, zygomatic, maxillary,* and *mandibular.*

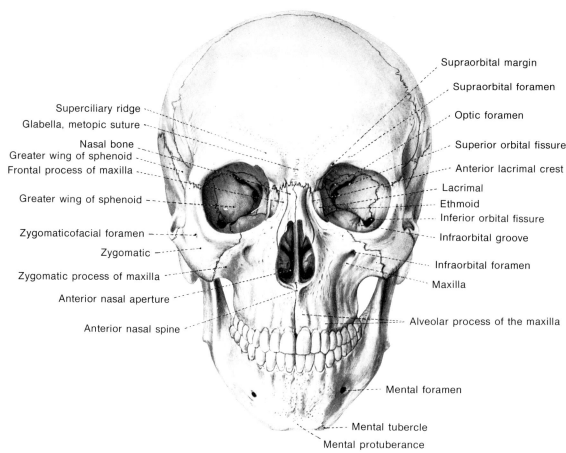

FIG. 3-1. *Anterior view of the skull*

Frontal Region The smooth dome of the forehead is formed by the *eminences* of the frontal bone, which ends inferiorly in the *supraorbital margins* and at the root of the nose. *Superciliary ridges* are present above the orbits. Near the medial termination of the supraorbital margin is the *supraorbital notch* or *foramen*, through which pass the supraorbital nerve and vessels. A slight elevation, the *glabella*, lies between the superciliary ridges in which the *frontal* or *metopic suture* may be present.

Orbital Region The orbit is pyramid-shaped with its walls formed by maxillary, zygomatic, sphenoid, frontal, palatine, ethmoid, and lacrimal bones. Near the apex of the cone are the *optic foramen* and *superior orbital fissure* (from the middle cranial fossa); inferolaterally the *inferior orbital fissure* (from the infratemporal fossa), anterior from which is the *infraorbital groove* and *canal*, passing to the *infraorbital foramen*. In the anterior portion of the medial wall is the *lacrimal fossa* for the lacrimal sac.

Maxillary Region Maxillary and zygomatic bones contribute to the inferior and lateral margins of the orbit. The frontal process of the maxilla extends to the root of the nose. The maxilla ends inferiorly in the *alveolar border*, with slight ridges marking the roots of the anterior teeth. The *zygomatic bone* extends posterolaterally to form the *zygomatic arch*. In the frontal aspect of the maxilla is the *infraorbital foramen* and, in the zygomatic bone, the *zygomaticofacial foramen*.

Nasal Region The *anterior nasal aperture* is the triangular area formed by the *maxillae* and *nasal bones*. Note the *anterior nasal spine*, *nasal conchae*, and *nasal septum*.

Mandibular Region At the junction of the two halves of the body of the mandible is the *mental protuberance* and, in the midline, the *symphysis menti*. The *mental tubercle* is the inferior and lateral parts of the protuberance. The superior part of the mandibular body supports the mandibular teeth, which are located in *alveoli*. The *mental foramen* lies inferior to the second premolar halfway between the upper border of the alveolar process and the lower border of the base.

LATERAL VIEW A major aspect of the lateral view of the skull (fig. 3-2) is the zygomaticotemporal region, which comprises both the dorsal part of the cranial vault and the part covered by the temporal muscle. It includes also the zygomatic arch, composed of the zygomatic process of the temporal bone, the zygomatic bone, and the zygomatic process of the maxilla. Near the middle of the temporal region is the curved *superior temporal line* and a short distance below it the *inferior temporal line*, marking the limit of the origin of the temporal muscle fibers. The two *parietal* bones form the greater portion of the sides and roof of the cranial cavity.

The *infratemporal fossa* lies inferior to the infratemporal crest, medial to the mandible and its superior processes, and lateral to the pterygoid plates. The prominence of the cheek is formed by the orbital surface of the *zygomatic bone*. On its lateral surface the zygomatic bone sends a *frontal process* upward along the lateral wall of the orbit to articulate with the *zygomatic process* of the frontal bone. The zygomatic bone sends a *temporal process* backward to articulate with the *zygomatic process* of the temporal bone. The two processes form the *zygomatic arch*.

Inferior to the zygomatic bone, extending both anteriorly and posteriorly, is the *maxilla*. A zygomatic process extends upward to articulate with the *maxillary process* of the zygomatic bone. Posteriorly, the maxilla shows a small process known as the *maxillary tuberosity*. Hidden behind the *coronoid process* of the mandible is the thin lamina of the *lateral pterygoid plate*.

The *greater wing of the sphenoid bone*, lying behind the zygomatic bone, articulates with the *frontal*, *parietal*, *temporal*, and *zygomatic bone*.

The *squamous portion of the temporal bone* forms the inferior lateral aspect of the skull. Posterior and inferior to the zygomatic process is the *external auditory meatus*. The *mastoid process* is the large bony process posterior to the external auditory meatus. The *styloid process* is anteromedial to the mastoid process.

The calvaria viewed from the lateral aspect includes the *frontal*, *parietal*, and *occipital bones*. The *coronal suture* is found between the frontal and parietal bones. The suture between the parietal and occipital bones is the *lambdoid*.

The lateral view of the mandible presents the *condyle*, *ramus*, *coronoid process*, *mandibular notch*, *angle*, *body*, and *alveolar process*.

BASAL VIEW (fig. 3-3) The prominent structure of the anterior region is the *superior alveolar arch*. Within the arch is the *hard palate*, formed by the palatine processes of the maxillary and the palatine bones. Anteriorly in the midline is the *incisive fossa* containing the *incisive foramina*. Piercing either palatine bone, close to the last molar, are two foramina; the larger and more anterior is the *greater palatine foramen*, which communicates with the pterygopalatine fossa, and the smaller is the *lesser palatine foramen*.

Opening superior to the palatal shelf are the *posterior nasal apertures*, with the *vomer bone* forming the medial septum. On either side of the choanae are the

34 ■ BASIC ANATOMY FOR THE ALLIED HEALTH PROFESSIONS

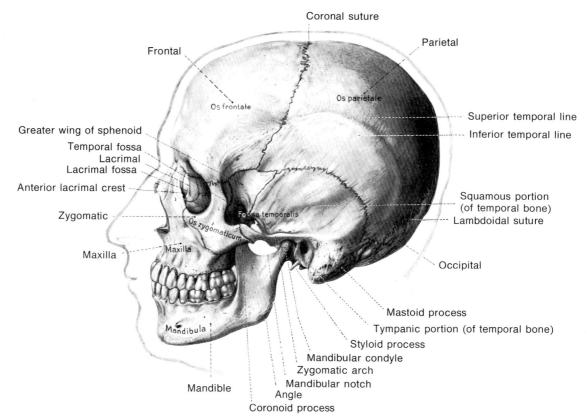

FIG. 3-2. *Lateral view of the skull*

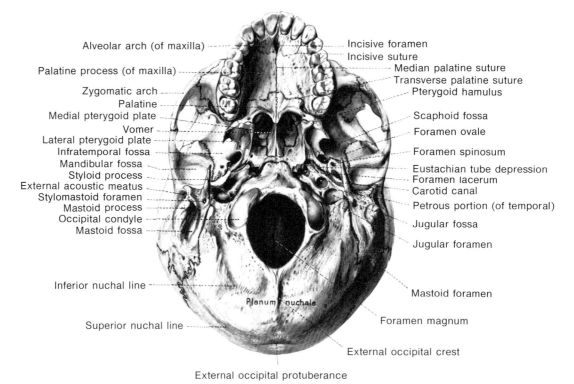

FIG. 3-3. *Basal view of the skull*

lateral and *medial pterygoid plates* with the *pterygoid fossa* between them, and the *pterygoid hamulus* inferiorly. Lateral to these is the roof of the infratemporal fossa. Posterolateral to the plates appears the large *foramen ovale*. Posterolateral to the foramen ovale at the base of the spine of the sphenoid is the *foramen spinosum*.

The zygomatic arch is a prominent feature of the anterior aspect. At its caudal end is the *mandibular fossa*, which receives the condyle of the mandible. This is bounded anteriorly by the *articular eminence*, and posteriorly by the *postglenoid tubercle*.

Adjoining the narrowest part of the basioccipital is the *foramen lacerum*, and piercing the anterior wall of its canal is the *pterygoid canal*. Piercing the substance of the petrous temporal bone is the *carotid canal*. Between the petrous temporal and occipital bones at the level of the condyle is the large *jugular foramen*, with the *jugular fossa* on its lateral wall. The orifice of the Eustachian tube is found between the petrous temporal bone and the spine of the sphenoid. Just lateral to the jugular foramen is the *styloid process* and at its base, almost medial to the mastoid process, the *stylomastoid foramen*.

The *occipital condyles* are located along the anterior margin of the *foramen magnum*. At the lateral base of each condyle is the *hypoglossal canal* and at the posterior base is the *condylar fossa*.

CRANIAL CAVITY

Roof of the Cranial Cavity The exterior of the skull or *calvaria* (fig. 3-4) is formed by the following bones: the *frontal*, two *parietal*, and the *occipital*. The *sagittal suture* separates the two parietal bones. The suture between the frontal and parietal bones is the *coronal suture*. The *lambdoid suture* separates the occipital and the parietal bones.

The posterior aspect of the skull shows a projection known as the *external occipital protuberance*. Extending laterally is the *superior nuchal line*.

The inner surface (fig. 3-5) has a shallow sagittal groove in the midline, marking the *superior sagittal venous sinus* of the dura, and branching *vascular grooves* indicate the position of *meningeal vessels*. Bordering the sagittal groove may be *granular pits*, which increase with age and occasionally are of sufficient depth to pass through the *diploë* to the outer table. They lodge, and are eroded by the arachnoid granulations. In addition, there are numbers of minute nutrient foramina.

Through this tissue course the diploic veins and the cells are largely filled with venous blood. The thickness of the vault varies individually. The *inner* and *outer layers* of the skull are composed of dense bone; between them is cancellous tissue known as the *diploë*.

Floor of the Cranial Cavity The floor of the cranium (fig. 3-6) is divisible into three *cerebral fossae: anterior, middle,* and *posterior*.

The anterior cerebral fossa contains the frontal lobes of the brain and is bounded posteriorly by the *lesser wings of the sphenoid*. In the midline are the *cribriform plates* (for the filaments of the olfactory

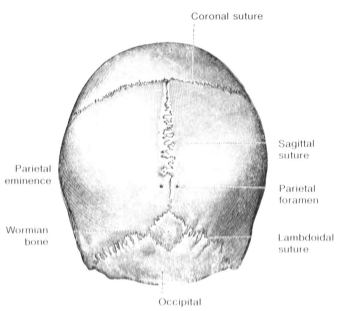
FIG. 3-4. *Exterior view of the skull (calvaria)*

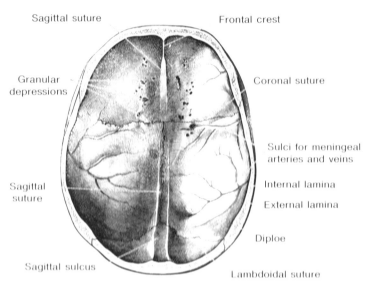
FIG. 3-5. *Interior view of the skull (calvaria)*

nerve) separated by the *crista galli*, both of the ethmoid bones, directly anterior to which is the *foramen cecum*. Hardly apparent at the lateral margin of the cribriform plate is the small *anterior ethmoid canal* (for an ethmoid vessel and nerve) and, posteriorly between the ethmoid and sphenoid bones, are the *posterior ethmoid canals*.

The middle cerebral fossa comprises a central raised portion and two lateral depressions. The former is saddle-shaped and is the *sella turcica* for the accommodation of the hypophysis or pituitary gland. Anteriorly is a ridge, the *tuberculum sellae*, on either side of which is an *anterior clinoid process*. Just anterior to this process is the optic foramen at the end of the *optic groove*. The posterior part of the sella turcica is formed from the crest of the *dorsum sellae*, ending laterally in the *posterior clinoid process*.

On either side of the sella is the *carotid groove*, extending from the medial part of the *foramen lacerum* at the base of the dorsum sellae to within the curve of the anterior clinoid process.

The lateral depressions of the middle cerebral fossa contain the temporal lobes of the brain. They are bounded posteriorly by the *eminence of the petrous portion of the temporal bone*. The *foramen lacerum, foramen ovale,* and *foramen spinosum* may be observed in this fossa. The *carotid canal* extends posterolaterally through the petrous temporal bone. Anterior to the ovale is the smaller *foramen rotundum* and, anterior to that, is the *superior orbital fissure*. The highest part of the surface of the petrous temporal bone is the *arcuate eminence,* marking the position of the superior semicircular canal.

The posterior cerebral fossa includes the cerebellum, pons, and medulla oblongata. Its most prominent feature is the *foramen magnum,* near the anterior

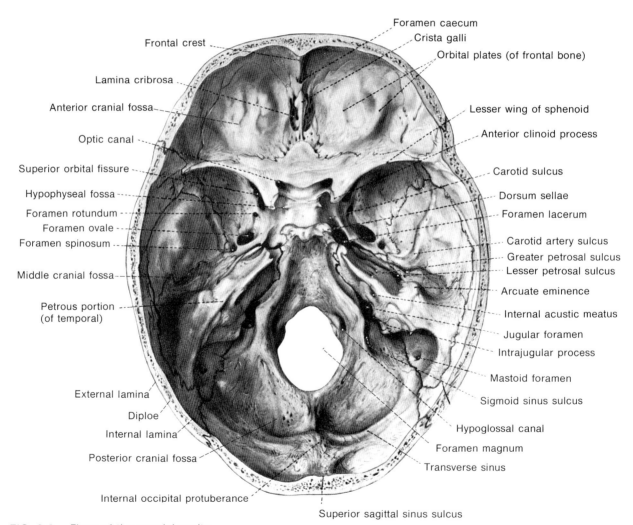

FIG. 3-6. *Floor of the cranial cavity*

border of which is the horizontally directed *hypoglossal canal.* Between the occipital and petrous temporal bones is the large *jugular foramen* with a laterally directed sulcus, the *sigmoid groove,* for the sigmoid venous sinus. This extends superolaterally and then posteromedially for the *transverse venous sinus.*

The *tentorium cerebelli,* a portion of the dura mater between the cerebrum and cerebellum, is attached along a line from the posterior clinoid process to the *transverse groove,* following the prominence of the petrous portion of the temporal bone.

ORBITAL CAVITY The pyramid-shaped *orbit* (fig. 3-7) has four walls, a base, and an apex. The *roof of the orbit* includes the orbital plate of the *frontal bone* and the lesser wings of the *sphenoid bone.* The *medial wall* includes the frontal process of the *maxilla,* the *lacrimal bone,* the orbital lamina of the *ethmoid bone,* and a small portion of the *sphenoid bone.* The *lateral wall* is formed by the *zygomatic bone* and the greater wing of the *sphenoid bone.* The *floor* is formed by the orbital surface of the *maxilla,* the *zygomatic bone,* and the *palatine bone.*

The *optic canal* is an aperture in the lesser wing of the sphenoid. The *superior orbital fissure* is bounded by the lesser and greater wings of the sphenoid bone. The *inferior orbital fissure* extends from the apex toward the base of the orbit.

NASAL CAVITY The *cavity of the nose* (fig. 3-8) extends from the *anterior* to the *posterior nasal apertures.* It is divided into two parts by the nasal septum, composed of the *ethmoid* and *vomer bones.* The *nasal conchae* are located on the lateral walls of the nasal cavity. The *inferior nasal concha* is a part of the maxillary bone. The *superior* and *middle nasal conchae* are part of the ethmoid bone.

The *inferior nasal meatus* is the airway between the bony palate and the inferior concha, while between the inferior concha and the middle concha is the *middle meatus.* Above the middle concha is the *superior meatus.* The nasal atrium is the space anterior to the middle concha. The *sphenopalatine foramen* opens into the nasal cavity directly posterior to the middle meatus. The middle nasal meatus is displaced by the *bulla ethmoidalis.* The deep groove anterior and inferior to the bulla is the *hiatus semilunaris.*

PARANASAL AIR SINUSES Air sinuses communicating with the nasal cavity occur in frontal, maxillary, ethmoid, and sphenoid bones (fig. 3-9).

All the paranasal sinuses are lined with mucous membrane, the cilia of which sweep minor discharges through the passages and into the nasal cavity. Inflammation and infection, however, may clog the restricted passages and allow the sinuses to fill with mucus.

The *maxillary sinus* is the largest of the paranasal sinuses and is situated in the maxilla between the

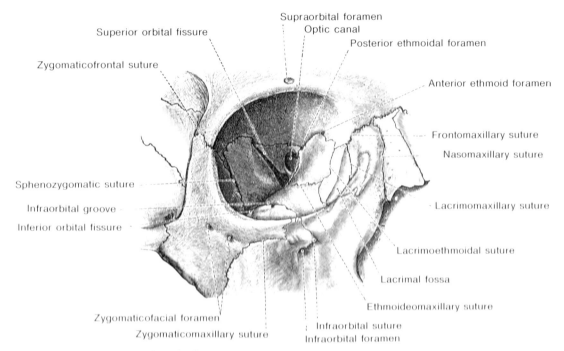

FIG. 3-7A. *Orbital cavity (anterior view)*

38 ■ BASIC ANATOMY FOR THE ALLIED HEALTH PROFESSIONS

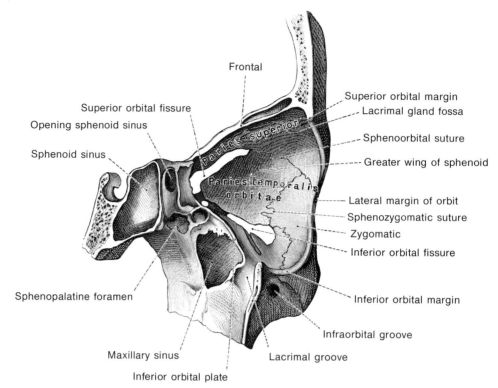

FIG. 3-7B. *Orbital cavity* (*lateral portion*)

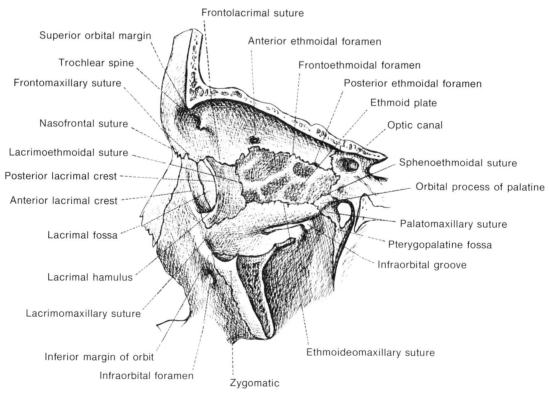

FIG. 3-7C. *Orbital cavity* (*medial view*)

orbit and upper alveolar arch, lateral to the nasal cavity. It opens beneath the middle concha into the deep anterior portion of the semilunar hiatus. Its floor is close to the roots of the upper molars.

The *frontal sinus* and anterior ethmoidal sinuses are sometimes units of a common sinus. The former is in the frontal bone between the anterior part of the orbit and the anterior cerebral fossa. It may be separated from its fellow of the opposite side by a septum. Its size is extremely variable. It may open into the nasal cavity directly or, more frequently, it may share a duct, the frontonasal, with the anterior ethmoid sinuses, opening beneath the middle concha.

The *ethmoidal sinuses* are the smallest of the paranasal sinuses and are variable in number, occupying the ethmoidal labyrinth between nasal cavity and orbit. They are considered to be divisible into two groups. The *anterior ethmoidal sinuses* drain into the infundibulum, usually shared by the frontal sinus, beneath the middle concha, and the *posterior ethmoidal sinuses* open beneath the superior concha.

The *sphenoidal sinus* is a large sinus in the sphenoid bone, of variable size, divided by a complete bony septum into two halves that are usually asymmetrical. The location is posterior to the superior concha and superior to the nasopharynx. The opening is anterior to the *sphenoethmoidal recess* of the nasal cavity.

MANDIBLE   The *mandible* (fig. 3-10) comprises a horizontal *body,* supporting the lower teeth, and on either side a mandibular *ramus* posterior to the teeth. The ramus has an *angle* providing attachment for the masseter and medial pterygoid muscles, and two superior processes: an anterior *coronoid process,* for insertion of the temporal muscle, and a posterior *condylar* or articular *process.* The latter has a *head* and *neck.*

PRINCIPAL FORAMINA AND CANALS OF THE SKULL Following is a list of the *principal foramina and canals of the skull,* with their locations and the structures passing through them. It should be realized that a foramen is often the termination of a canal, in which case the same name is usually applicable to both. In addition to the structures mentioned, foramina frequently contain connective tissue and lymphatics, which are not mentioned in the following list. The various foramina should be associated with specific bones (fig. 3-11).

Internal auditory meatus: the large foramen on the posteromedial aspect of the petrous portion of the temporal bone; transmits facial and vestibulocochlear nerves

Foramen cecum: between frontal bone and crista

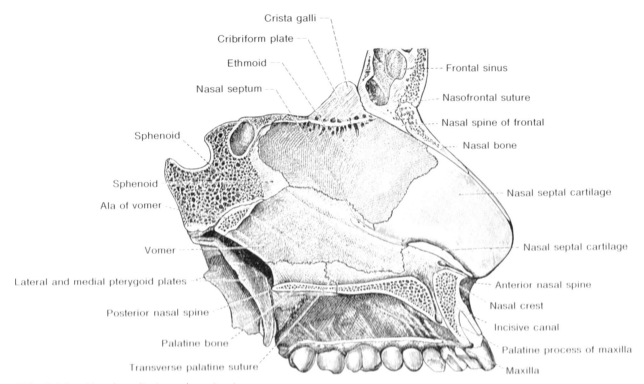

FIG. 3-8A. *Nasal cavity (nasal septum)*

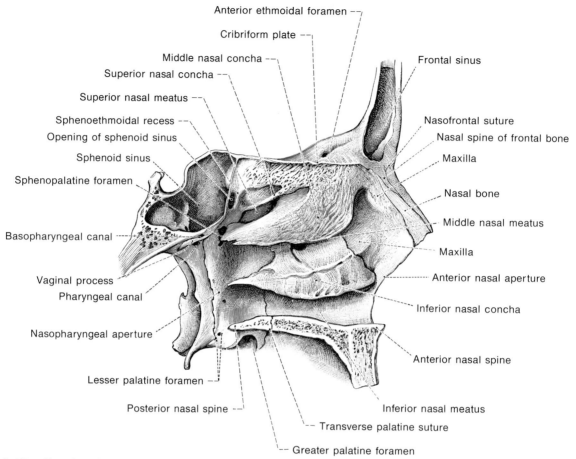

FIG. 3-8B. *Nasal cavity (nasal conchae)*

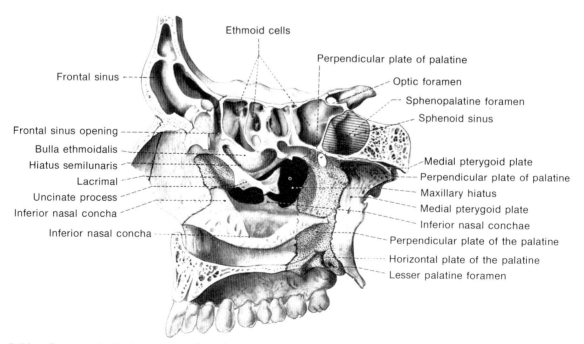

FIG. 3-9A. *Paranasal air sinuses (median view)*

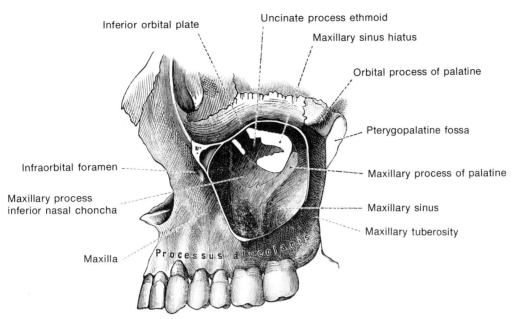

FIG. 3-9B. *Paranasal air sinuses (maxillary sinus)*

galli; may end blindly or transmit a vein from the nose to the superior sagittal sinus.

Carotid canal: through the petrous portion of the temporal bone; internal carotid artery and its sympathetic plexus

Condylar fossa: variable; posterior to the condyle; the vein from the sigmoid venous sinus to the suboccipital plexus

Anterior and posterior ethmoid foramina: medial wall of orbit; anterior and posterior ethmoidal nerves and vessels

Hypoglossal canal: lateral to the condyle; hypoglossal nerve and a meningeal artery

Incisive foramen: incisive fossa of anterior palate; branch of the palatine artery and usually the nasopalatine nerve

Infraorbital foramen: anterior end of the infraorbital canal, from the orbit piercing the maxilla below the orbit; infraorbital nerve and vessels

Jugular foramen: between the occipital and the petrous temporal bones; formation of internal jugular vein, glossopharyngeal, vagus, and accessory nerves

Foramen lacerum: at the junction of the basioccipital, petrous temporal, and sphenoid; contains cartilage

Foramen magnum: spinal cord junction with the brain stem and its meninges, accessory nerve, vertebral arteries, and spinal arteries and veins

Mandibular foramen: medially on the ramus of the mandible; inferior alveolar nerve and vessels

Mental foramen: anterior mandible; mental branch of the mandibular nerve and mental vessels

Optic foramen: from anterior cranial fossa to orbit; the optic nerve and ophthalmic artery

Inferior orbital fissure: infraorbital vessels and nerves

Superior orbital fissure: from the middle cranial fossa to the orbit; ophthalmic vein, the oculomotor, trochlear and abducent nerves, and branches of the ophthalmic division of the trigeminal nerve

Foramen ovale: through the sphenoid bone in the middle cranial fossa; mandibular division of the trigeminal

Greater palatine foramen: through the posterior palate to the pterygopalatine fossa; greater palatine branches of pterygopalatine ganglion and palatine branches of the maxillary vessels

Lesser palatine foramen: through the posterior palate; lesser palatine branches of greater palatine vessels and nerves

Pterygoid canal: from the foramen lacerum to the pterygopalatine fossa; nerve of the pterygoid canal

Pterygopalatine fossa: laterally between maxilla and sphenoid; pterygopalatine ganglion and branches of the maxillary artery

Foramen rotundum: from the middle cranial fossa to the pterygopalatine fossa; maxillary division of the trigeminal nerve

Sphenopalatine foramen: from pterygopalatine fossa to middle meatus of the nasal cavity; sphenopalatine nerves and vessels

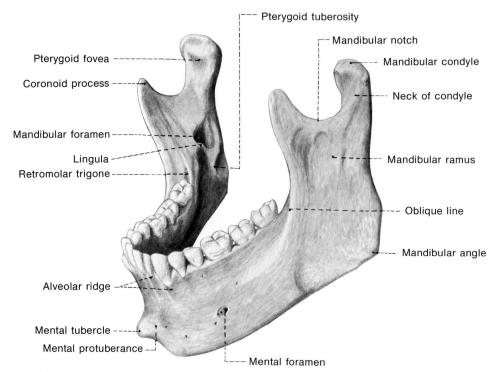

FIG. 3-10. *Mandible*

Foramen spinosum: through the sphenoid bone at the base of its spine; middle meningeal vessels
Stylomastoid foramen: posterior to the styloid process; facial nerve
Supraorbital notch: occasionally a foramen; in the frontal bone; supraorbital nerve and vessels

The *cranium,* enclosing the brain and its membranes, contains eight bones. The *bones of the face* total fourteen (fig. 3-12).

FONTANELLES In the skull of the newborn the bones of the cranial vault do not always meet at the points where three or four of them will come in contact at a later stage. This is particularly true at the juncture of frontal and parietal bones—the "soft spot." These areas are termed *fontanelles* (fig. 3-13) and are covered by a membrane.

Vertebral Column

The *vertebral column* (fig. 3-14) is composed of bony segments, the vertebrae, interconnected with *intervertebral disks* of fibrocartilage, which impart considerable elasticity to the column as a whole. There are 33 vertebrae (with some variation in the coccygeal region) comprising 7 cervical, 12 thoracic, 5 lumbar, 5 sacral (fused in the adult into one complex), and 4 coccygeal. Each series has particular characteristic details.

A typical vertebra (fig. 3-15) has a *body,* posterior to which is the *vertebral foramen.* On either side of the foramen is a *pedicle,* surmounted by lateral projections comprising *superior* and *inferior articular processes,* and the *transverse process.* In the dorsal midline is the *spinous process,* which is connected with the transverse processes by the *laminae.* Between the pedicles of adjacent vertebrae are spaces, the *intervertebral foramina,* for the passage of spinal nerves and vessels.

CERVICAL LEVEL The seven *cervical vertebrae* constitute the portion of the column superior to the first rib. The first (atlas) and the second (axis) cervical vertebrae are considerably different from the other cervical vertebrae.

The chief characteristic of the cervical vertebrae (fig. 3-16) is the presence of a *transverse foramen* through each transverse process. The pedicles of the transverse process bound the foramen. The spinous processes (which do not occur on the atlas) are usu-

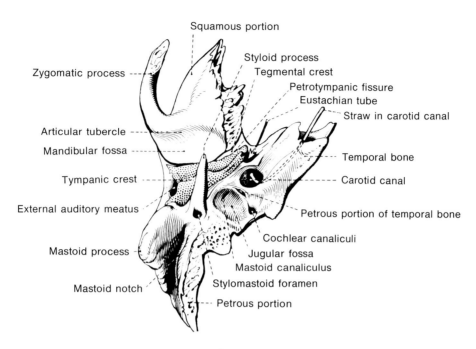

FIG. 3-11A. *Principal foramina and canals of specific bones of the skull (foramina of the temporal bone)*

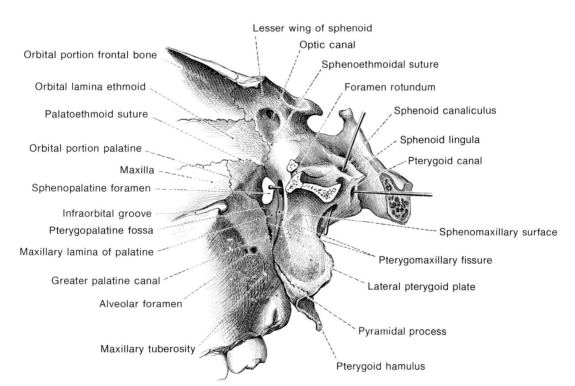

FIG. 3-11B. *Principal foramina and canals of specific bones of the skull (foramina associated with the pterygopalatine fossa)*

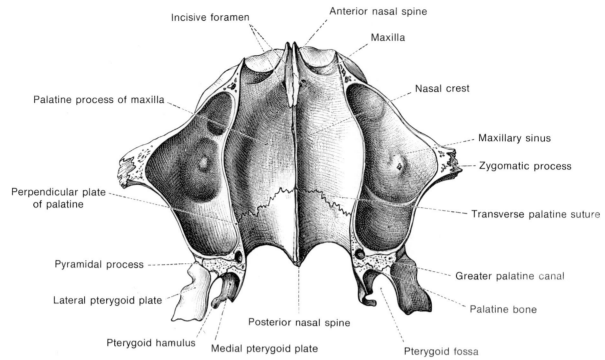

FIG. 3-11C. *Principal foramina and canals of specific bones of the skull (foramina associated with maxillary and palatine bones)*

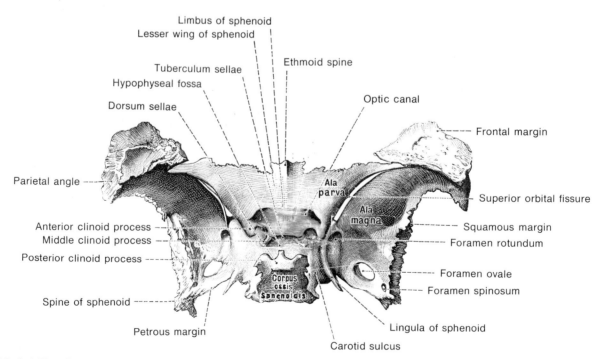

FIG. 3-11D. *Principal foramina and canals of specific bones of the skull (sphenoid bone viewed posterosuperiorly)*

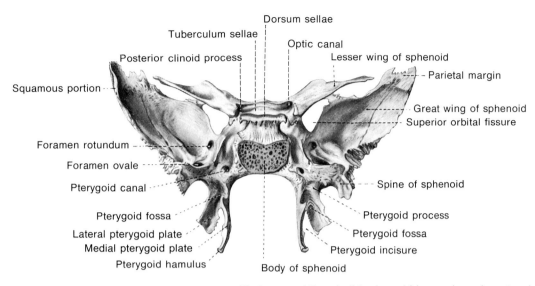

FIG. 3-11E. *Principal foramina and canals of specific bones of the skull (sphenoid bone viewed posteroinferiorly)*

ally bifid except on the seventh vertebra, in which it is longer than on the other vertebrae, and is called the *vertebra prominens.*

The *atlas* (fig. 3-17) is considerably modified for the purpose of supporting and permitting sufficient movement of the head. It differs from the other cervical vertebrae in being devoid of body and spine; hence, it is largely a ring of bone with heavy transverse processes. The superior and inferior articular facets differ in form from the articular processes of the other cervical vertebrae.

The *axis* (fig. 3-18) is the second cervical vertebra. It differs from the succeeding cervical vertebrae chiefly in that from the body projects the cranially directed odontoid process (dens), representing the body of the atlas that has fused with the axis.

The third to seventh units of the series conform to the general pattern already described for the cervical vertebrae, except for the spine of the seventh, which is of the thoracic type.

THORACIC LEVEL The twelve *thoracic vertebrae* (fig. 3-19) are characterized by the presence of facets for the articulation of ribs. They increase in size (chiefly of the body), in caudal sequence, and, from the middle of the series, the spines increase in length in both directions. The slope of all spines is caudal.

LUMBAR LEVEL The five *lumbar vertebrae* (fig. 3-20) are the largest of the column. They are characterized by the length of their transverse processes and by the craniocaudal width of their spinous processes.

SACRAL LEVEL Various aspects of the five *sacral vertebrae* (fig. 3-21) fuse giving, at puberty, a solid sacrum. The pedicles, laminae, and transverse process elements form the plates of the lateral masses, each pierced by four sacral foramina for the passage of nerves and vessels. The sacrum is of an irregular, curved, triangular shape with body directed cranially and apex caudally. The lateral auricular surface, for articulation with the ilium, is concerned with the first two vertebrae and part of the third. The female sacrum is slightly flatter cranially and more curved caudally, and is more tilted than in the male.

COCCYGEAL LEVEL The *coccyx* (fig. 3-14) functions as an anchorage for pelvic structures and it may thus have reached its minimum reduction in the human. It usually comprises three to five vertebrae; the first, being most often separate until after middle life, is the largest, while the others are successively smaller, the terminal unit being nothing but a bony nub. The vertebral canal is absent from this series.

INTERVERTEBRAL DISKS Between the vertebrae are fibrocartilaginous *intervertebral disks* (fig. 3-22). They form strong joints and permit various movements. The outer fibrous portion is called the *anulus fibrosus* and the inner more pliable portion is called the *nucleus pulposus.*

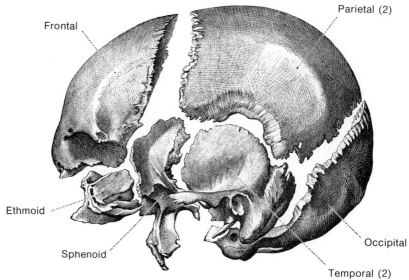

FIG. 3-12A. *The eight bones of the skull*

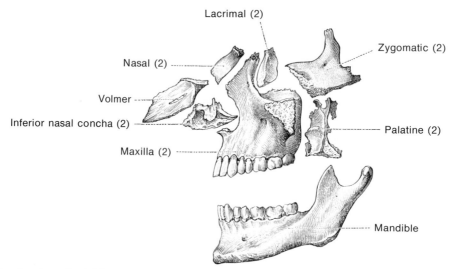

FIG. 3-12B. *The fourteen facial bones*

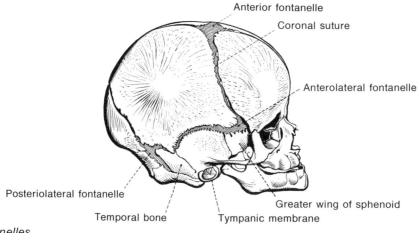

FIG. 3-13. *Fontanelles*

THE SKELETAL SYSTEM ■ 47

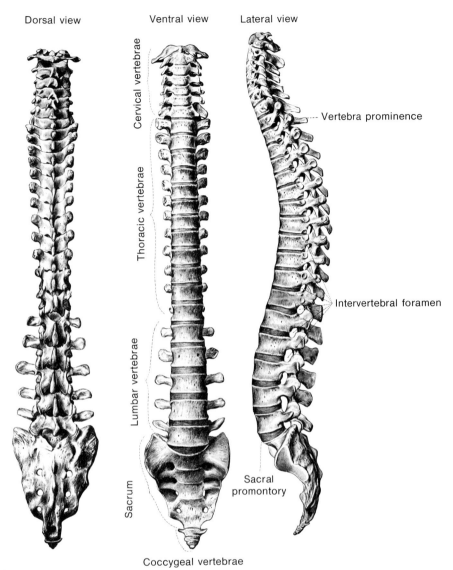

FIG. 3-14. *Vertebral column*

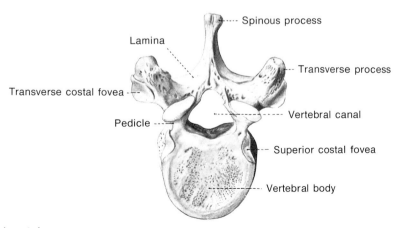

FIG. 3-15. *A typical vertebra*

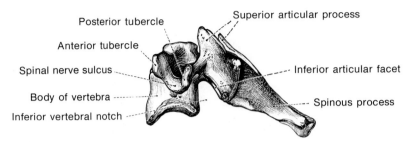

FIG. 3-16A. *Lateral view*

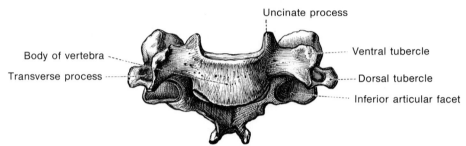

FIG. 3-16B. *Ventral view*

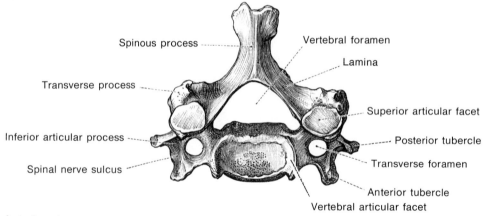

FIG. 3-16C. *Anterior view*

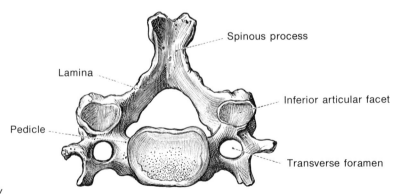

FIG. 3-16D. *Posterior view*

FIG. 3-16. *Cervical vertebrae*

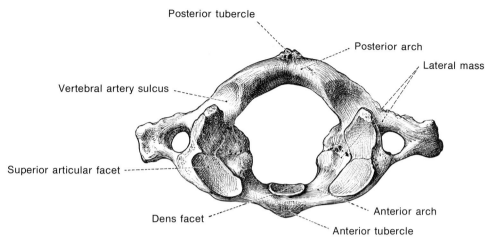

FIG. 3-17. *Atlas*

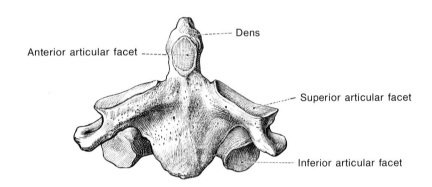

FIG. 3-18. *Axis*

Ribs

The *ribs* (fig. 3-23) enable the thoracic cavity to act as an elastic box or bellows. The chief function of the ribs is to prevent the collapse of the lungs.

Normally, 12 *pairs of ribs* are present, of which the first 7 pairs end in costal cartilages that are attached to the sternum. These are termed *true ribs*. The remaining five pairs are called *false ribs* and they do not reach the sternum. The *costal cartilages* of the 8th, 9th, and 10th pairs are each joined to the one above but the short cartilages of the last two pairs of ribs are free and are known as *floating ribs*.

The ribs are curved around the lateral thorax from vertebral column to costal cartilages, and are attached so that raising them increases the capacity of the chest cavity. The arc of their curvature increases in successive ribs from cranial to caudal direction.

The head of a typical rib articulates with two vertebrae, so there are two terminal *articular facets* (*superior* and *inferior*). Lateral to the head is the *neck*, and then the *tubercle*, with the *tubercular facet* marking the articulation with the transverse process of the vertebra. Somewhat lateral to this is a marked bend, the *angle*, followed ventrally by the *shaft* of the rib. The lower inner side of the shaft is concave and this is termed the *costal groove*.

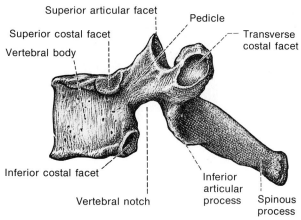

FIG. 3-19A. *Lateral view*

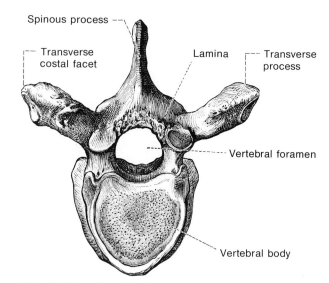

FIG. 3-19B. *Cranial view*

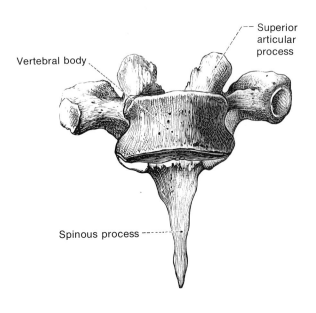

FIG. 3-19C. *Ventral view*

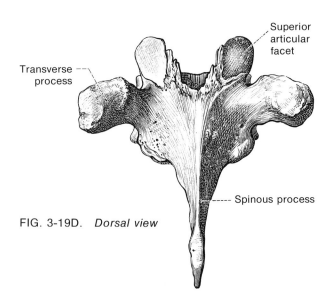

FIG. 3-19D. *Dorsal view*

FIG. 3-19. *Thoracic vertebrae*

THE SKELETAL SYSTEM ■ 51

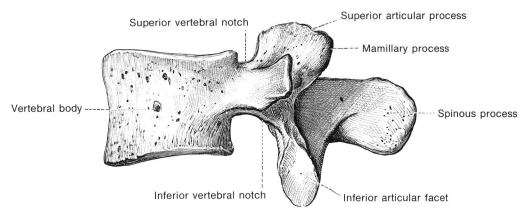

FIG. 3-20A. *Lateral view*

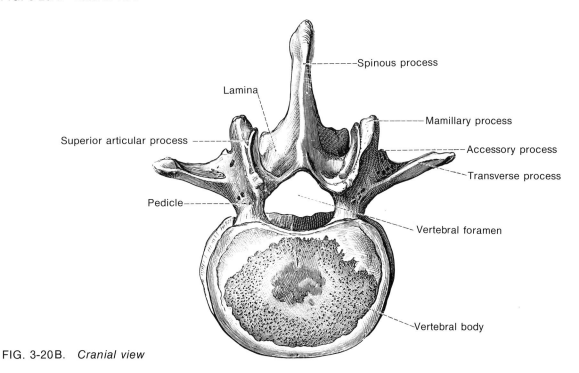

FIG. 3-20B. *Cranial view*

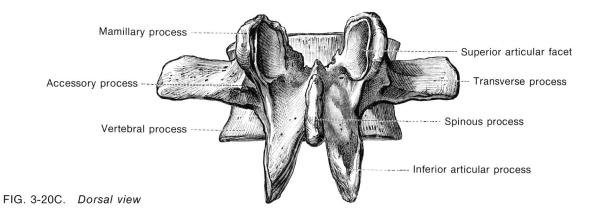

FIG. 3-20C. *Dorsal view*

FIG. 3-20. *Lumbar vertebrae*

52 ■ BASIC ANATOMY FOR THE ALLIED HEALTH PROFESSIONS

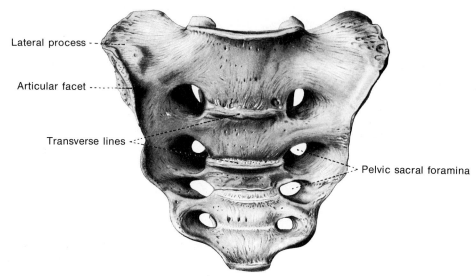

FIG. 3-21A. *Ventral view*

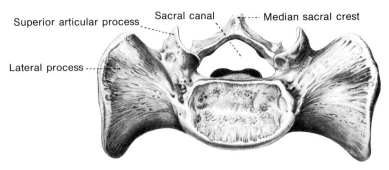

FIG. 3-21B. *Cranial view*

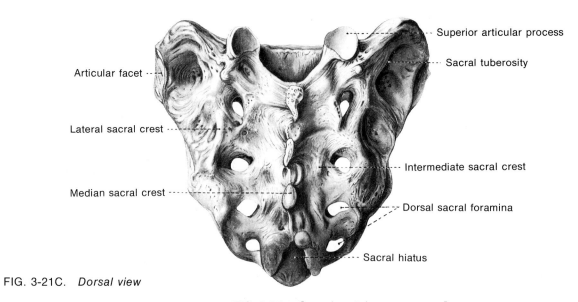

FIG. 3-21C. *Dorsal view*

FIG. 3-21. *Sacral vertebrae*

The first rib is the shortest of the true ribs. Its inner border may be grooved by the subclavian vessels, and between these grooves is the *scalene tubercle* for the attachment of the anterior scalene muscle. The second rib is much longer and is characterized by a central tubercle marking the attachment of the serratus posterior muscle.

The last three ribs each articulates with but one vertebra, so the head has only a single facet; the last two have no neck, and no tubercle as they do not articulate with the transverse processes of the vertebrae; and the last is short and slender and devoid of an angle.

Sternum

The *sternum* (fig. 3-24) comprises the *manubrium*, the *body*, and a terminal *xiphoid process*.

The manubrium has a *jugular notch, clavicular facets*, and *lateral facets* for the first costal cartilages. The second costal cartilages articulate at the junction of the manubrium with the body of the sternum (*sternal angle*).

Hyoid

The *hyoid* (fig. 3-25) is U-shaped and contains a *body* and *greater* and *lesser horns*.

APPENDICULAR DIVISION

The *skeleton of the appendages* comprises the *clavicles, scapulae, coxal bones*, and the *bones of the upper* and *lower limbs*.

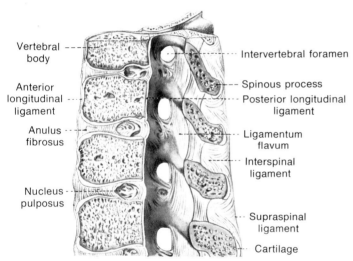

FIG. 3-22A. *Sagittal section*

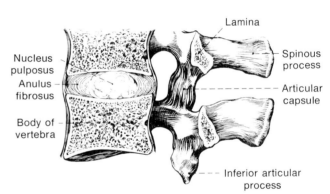

FIG. 3-22C. *Sagittal section*

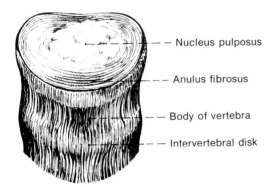

FIG. 3-22B. *Cross section*

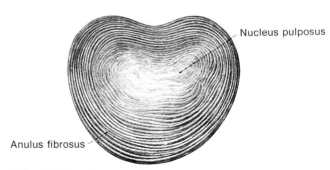

FIG. 3-22D. *Cross section*

FIG. 3-22. *Intervertebral disk*

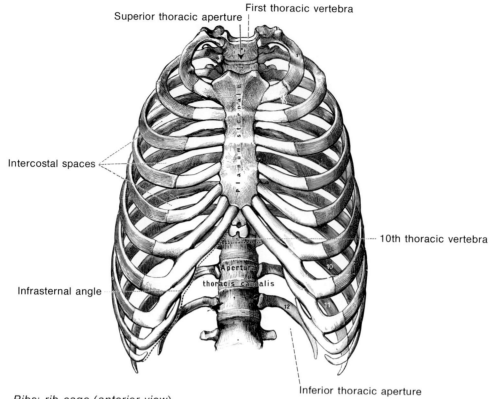

FIG. 3-23A. *Ribs: rib cage (anterior view)*

FIG. 3-23B. *Ribs: rib cage (posterior view)*

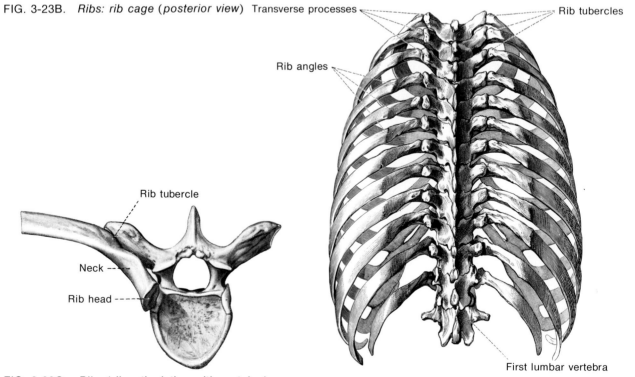

FIG. 3-23C. *Ribs (rib articulating with vertebra)*

Upper Limbs

The *upper limbs* (fig. 3-26) comprise the bones of the *shoulder, clavicle, arm, forearm, wrist,* and *hand.* The anatomical position of the human arm is with the palms presented ventrally.

Scapula

The *scapula* (fig. 3-27) has a *body* separated by a *neck* from a *head,* which bears the *glenoid cavity* for

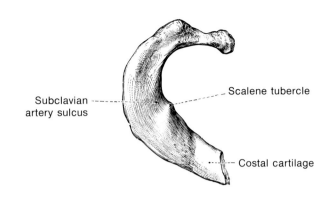

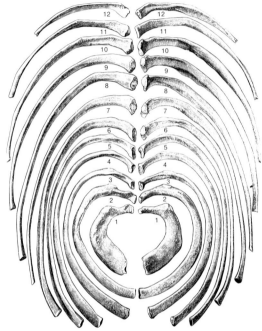

FIG. 3-23D. *Ribs (the 12 pairs of ribs)*

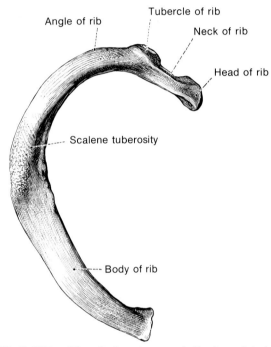

FIG. 3-23E. *Ribs: first and second ribs (cranial view)*

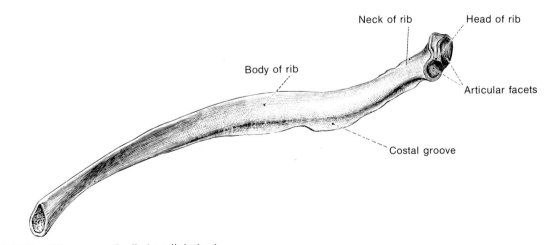

FIG. 3-23F. *Ribs: seventh rib (medial view)*

56 ■ BASIC ANATOMY FOR THE ALLIED HEALTH PROFESSIONS

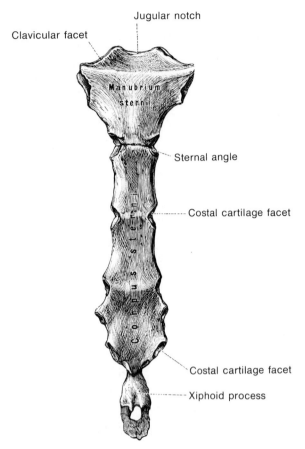

FIG. 3-24A. *Sternum (anterior view)*

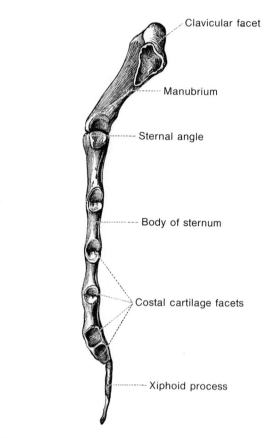

FIG. 3-24B. *Sternum (lateral view)*

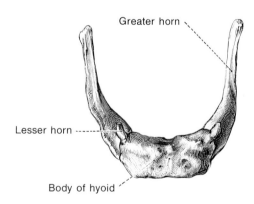

FIG. 3-25A. *Hyoid: cranial view*

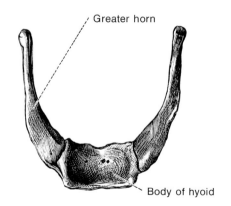

FIG. 3-25B. *Hyoid: caudal view*

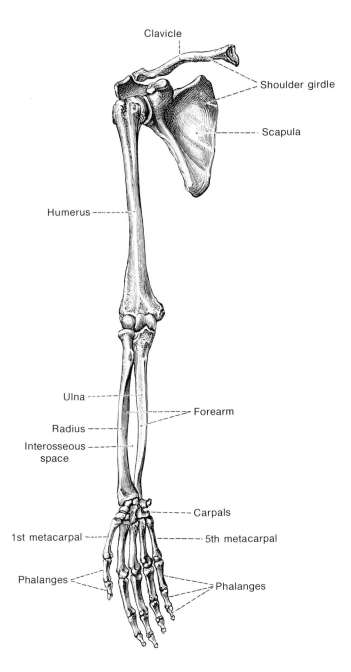

FIG. 3-26. *Upper limb*

articulation with the humerus. The end view of the scapula appears triradiate, with three *fossae: subscapular* ventrally, *supraspinous,* and *infraspinous* dorsally. The latter two fossae are separated by the *spine of the scapula.* The body has *lateral, medial,* and *superior borders* and a *superior* and *inferior angle.* The spine ends ventrally in the *acromion* or *acromial process,* bearing a facet for articulation with the clavicle. Near the neck is an indentation in the superior border, the *suprascapular notch,* through which passes the suprascapular nerve, and at the ventral termination of the superior border is the *coracoid process.* Near the ventral termination of the inferior border is the *infraglenoid tubercle,* for attachment of the long head of the triceps muscle. Near the coracoid base, upon the cranial margin of the glenoid, is the *supraglenoid tubercle,* for attachment of the long head of the biceps brachii muscle.

CLAVICLE The *clavicle* (fig. 3-28) is a bone with a double curvature, articulating with the manubrium at one end and the acromial process of the scapula at the other end.

HUMERUS The *humerus* (fig. 3-29) is the bone of the arm. It has a *shaft,* surmounted proximally by a *head* with a spherical articular surface, adjoining the more ventral *greater tubercle* and more medial *lesser tubercle,* between which is the *intertubercular groove.* Distal to the tubercles is the *surgical neck* of the humerus. Distally is the *deltoid tuberosity.*

At the distal end of the humerus are two pulley-shaped articular surfaces, the more medial *trochlea* (for the ulna), and the more lateral and smaller *capitulum* (for the radius). Directly above the former, from the anterior aspect, is the *coronoid fossa* (for the coronoid process of the ulna), and above the capitulum is the *radial fossa.* Upon the medial margin of the distal end of the humerus is the large ridge of the *medial epicondyle,* and upon the lateral margin, the smaller *lateral epicondyle.* Upon the posterior or dorsal aspect of the distal end of the humerus is the *olecranon fossa,* into which fits the *olecranon process* when the forearm is extended.

BONES OF THE FOREARM There are two major bones in the forearm (fig. 3-30), the *radius* and the *ulna.*

The *radius* has a *head,* with a concave terminal facet for articulation with the capitulum of the humerus, and a smaller facet on the ulnar side, for articulation with the radial notch of the ulna. Distal to the head is the *neck,* and immediately below it, is the *tuberosity* of the radius, a pronounced elevation upon the ulnar side.

The distal end of the radius becomes rather abruptly the broadest part of the bone, with a distal articular surface and ulnar facet, for articulation with the carpal bones and ulna, respectively. The distal tip, upon the radial side, is the *styloid process.*

The *ulna* is the longer of the two bones of the forearm. It has a *shaft* and a *head,* which is the distal extremity rather than the proximal one as in the radius. At the proximal end is a large concavity, the *trochlear notch,* for articulation with the trochlea of the humerus. The *olecranon process* extends back of this and

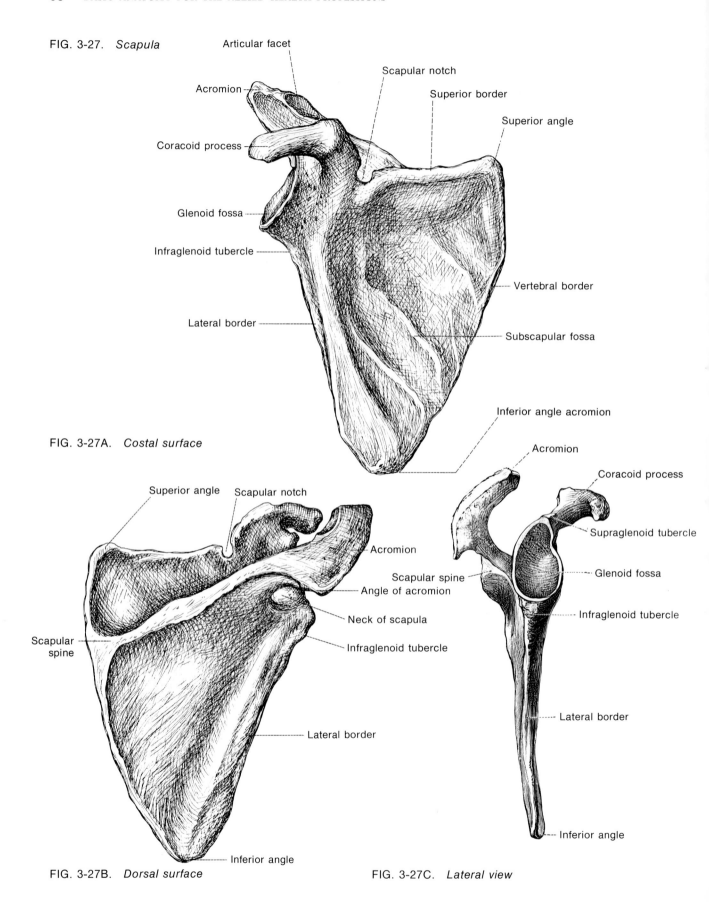

FIG. 3-27. *Scapula*

FIG. 3-27A. *Costal surface*

FIG. 3-27B. *Dorsal surface*

FIG. 3-27C. *Lateral view*

THE SKELETAL SYSTEM ▪ 59

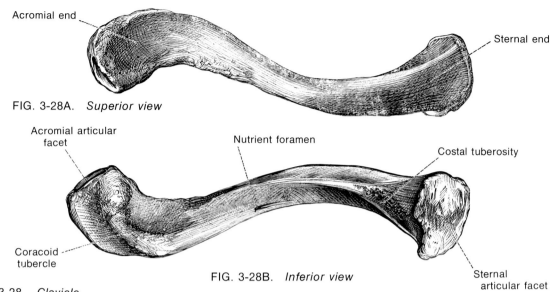

FIG. 3-28A. *Superior view*

FIG. 3-28B. *Inferior view*

FIG. 3-28. *Clavicle*

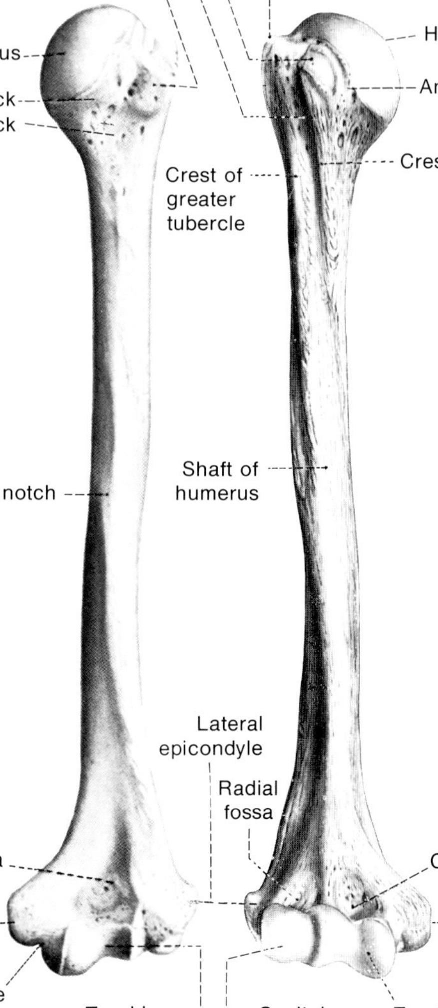

FIG. 3-29. *Humerus*   FIG. 3-29A. *Posterior view*   FIG. 3-29B. *Anterior view*

the *coronoid process* extends frontwards. Upon the lateral side of this process is the *radial notch*, with a facet for articulation with the head of the radius. Projecting beyond the head on the ulnar side is the *styloid process*.

BONES OF THE HAND   The bones of the hand (fig. 3-31) are the *carpals, metacarpals,* and the *phalanges,* which form the digits.

Carpals   Beginning on the radial side of the proximal row, the sequence of the *carpal bones* is as follows: *scaphoid, lunate, triangular,* and *pisiform;* the distal row sequence, beginning on the radial side, is: the *trapezium, trapezoid, capitate,* and *hamate.* The pisiform is considered to be a sesamoid bone developed in the tendon of the flexor carpi ulnaris. Scaphoid and lunate articulate with the radius, while lunate and triangular articulate with the cartilaginous disk that is interposed between these bones and the head of the ulna. All four distal carpal bones articulate with the metacarpals.

Metacarpals   A *metacarpal bone* is said to have a *body,* a proximal *base,* and a distal *head.*

Digits   The *five digits* are referred to by number, beginning with the thumb or pollex. The second is often termed the index, the third the middle, the fourth the ring, and the fifth the minimus. The thumb has two phalanges and the other digits, three.

Lower Limbs

The *lower limbs* (fig. 3-32) comprise bones of the *bony pelvis, thigh, leg, ankle,* and *foot.*

INNOMINATE (COXAL) BONE   The *bony pelvis* is composed of the two *coxal bones* and the *fixed (sacral and coccygeal) vertebrae* (fig. 3-33). It is thus a bony ring, with an aperture partly separated into two parts by a slight constriction, the *pelvic brim,* from the superior margin of the symphysis to the sacroiliac joint. The part of the aperture superior to this is the *false*

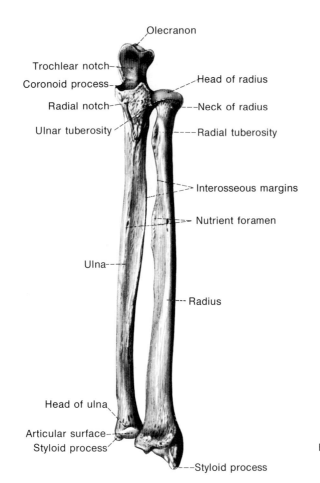

FIG. 3-30A.   *Bones of the forearm (anterior view)*

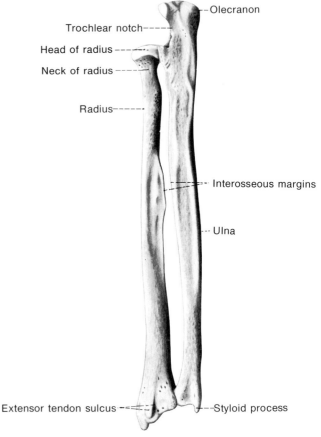

FIG. 3-30B.   *Bones of the forearm (posterior view)*

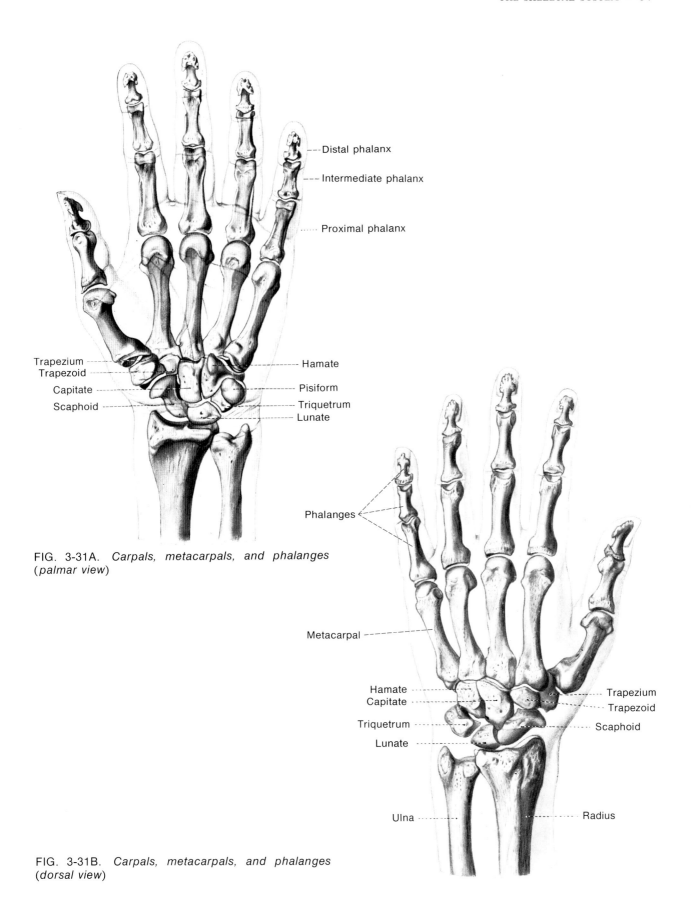

FIG. 3-31A. *Carpals, metacarpals, and phalanges* (*palmar view*)

FIG. 3-31B. *Carpals, metacarpals, and phalanges* (*dorsal view*)

pelvis, and the part inferior is the *true pelvis.* The pelvic inlet is then surrounded by the pelvic brim, and the pelvic outlet by the irregular ring of the inferior pelvic margin. The *pubic arch* is formed by the inferior rami of the pubes with the apex at the symphysis, forming an inverted V.

Slight, average sexual differences are seen in many parts of the skeleton, but those of the pelvis are most trustworthy. Even in the pelvis, however, they are individually variable and not always reliable. The chief sexual differences are that in the female the pelvis is more capacious, for childbirth, and is less robust. The ilia are less sloping (flatter), the anterior iliac spines are farther apart, and therefore the false pelvis is shallower as well as broader. The aperture enclosed by the pelvic brim is larger, broader, and more circular. The pubic arch is broader and its apical angle less acute, and the pelvic outlet—and therefore the true pelvis—is larger.

The *pelvic girdle* is the term used to denote the two hip bones together, without the fixed vertebrae. The hip bone is formed by the union of three elements: the *ilium, ischium,* and *pubis.* The point of junction of the three bones is near the center of the *acetabulum,* and the ischium also joins the pubis near the inferior ramus. In the adult the sutures are practically obliterated, but the lines of juncture may be detected by close scrutiny of the slight differences in the texture of the bone.

The ilium has a narrow *body,* a broad *blade* with a lateral *gluteal fossa,* and a medial *iliac fossa.* The long cranial border constitutes the *iliac crest.* At its dorsal termination is first the *posterior superior iliac spine* and below the articular surface is the *posterior inferior spine.* At the opposite end of the crest, ventrally, is the *anterior superior iliac spine.* Dorsally at the narrowest part of the ilium is the greater sciatic notch, for the accommodation of the sciatic nerve trunk.

The *acetabulum* is the large concavity with notched margin upon the lateral aspect of the hip bone. Dorsomedial to this is the *ischial spine* (for the sacrospinous ligament), superior to which is the *greater sciatic notch,* and inferior, the *lesser sciatic notch.* The sacrospinous ligament, aided by the sacrotuberous ligament, converts these notches, respectively, into the *greater* and *lesser sciatic foramina.* Caudal to the lesser sciatic notch is the large *ischial tuberosity,* to which is attached the sacrotuberous ligament and the hamstring muscles. Ventral to it is the large *obturator foramen.* Inferior to it is the *ramus of the ischium,* which extends ventrally to join the *inferior ramus of the pubis.* The broad part of this bone is the body of the pubis, to which are attached the interpubic fibers

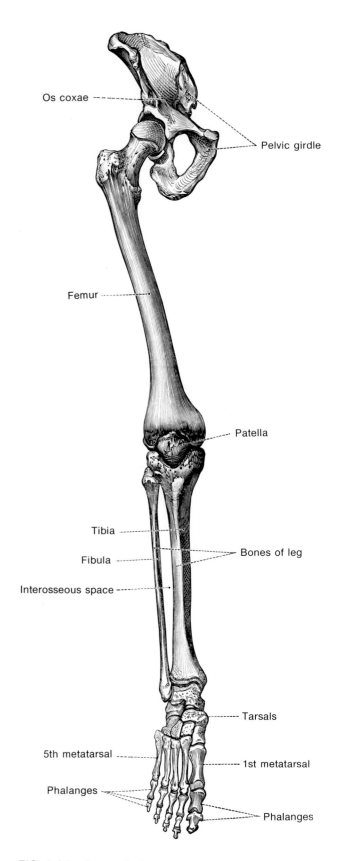

FIG. 3-32. *Lower limb*

THE SKELETAL SYSTEM ■ 63

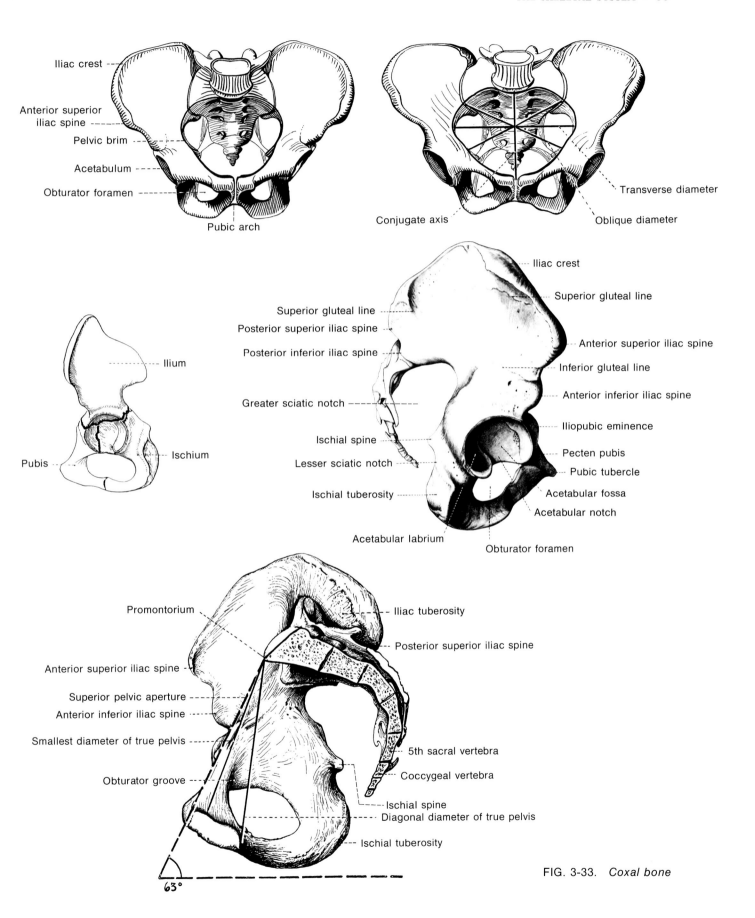

FIG. 3-33. *Coxal bone*

of the *symphysis pubis*. The bone continues toward the acetabulum as the *superior ramus of the pubis*.

FEMUR The *femur* or thigh bone (fig. 3-34) has a *shaft* and two ends. At the proximal end is the globular *head* with a *neck* and two *trochanters*. The more proximal of these is the *greater trochanter*, rising proximal to the neck. At its medial base, posteriorly, is the *trochanteric fossa;* more distally and dorsomedially is the *lesser trochanter;* and between the two is the *trochanteric crest*. Paralleling this crest on the anterior side of the neck is the *trochanteric line*. Posteriorly along the shaft extends the *linea aspera*. As the linea approaches the distal end it splits into two faint crests, the *lateral* and the *medial supracondylar lines*, between which is the *popliteal surface*.

At the distal end of the femur are two articular surfaces, the *lateral* and *medial condyles*, the latter extending slightly farther distally, between which is the *intercondylar notch*. Ventrally they join in a depression, the *patellar surface*, to accommodate the patella bone. Upon the sides of the extremity are the *lateral epicondyle* and the *medial epicondyle*, upon the proximal part of which is the *adductor tubercle*.

The *patella* (fig. 3-35) is the largest sesamoid bone of the body and it articulates with the patellar surface of the femur.

BONES OF THE LEG The *leg* is the segment of the lower limb between knee and ankle. The morphology is largely comparable to that of the forearm bones.

Tibia The *tibia* (fig. 3-36) is the most medial of the two bones. It alone bears the weight of the body. It has a *shaft* and two ends. The proximal end is as broad as the distal end of the femur. It bears a *lateral* and *medial condyle*, with corresponding separate facets for articulation with the femur, separated by the *intercondylar area*, upon which is located the *intercondylar eminence* and several depressions for the semilunar cartilages.

At the distal end of the bone is a *lateral fibular notch* or sulcus, in which lies the fibula. Upon the distal extremity is the articular surface, for the ankle joint, and a process extending beyond it, the *medial malleolus*.

Fibula The *fibula* (fig. 3-36) is the lateral bone of the leg. It is slender and has no mechanical function in the support of the body. It comprises a *shaft* and two ends, the more proximal of which is termed the *head*. Upon the medial side of the head is a facet for articulation with the tibia. It has a distal projection, the *lateral malleolus*. Distal to the head is a slender *neck*.

BONES OF THE FOOT The *foot* is composed of *tarsals* (ankle bones), *metatarsals*, and *digits* or toes.

Tarsals The following is a list of the *tarsal* bones (fig. 3-37).

Calcaneus: this is the bone of the heel and is the largest of the tarsal bones. In the medial aspect there is a sharp projection, the *sustentaculum tali*.

Talus: this bone is interposed between the leg and the calcaneus.

Cuboid: this bone has facets for articulation with the calcaneus.

Navicular: this bone articulates with the talus posteriorly, the three cuneiforms anteriorly, and sometimes with the cuboid laterally.

Cuneiform bones: these articulate posteriorly with the navicular.

Metatarsals The five *metatarsals* (fig. 3-37) are referred to by number, the first being that of the great toe. Each has a *shaft*, a proximal *base*, and a distal *head*.

Digits Each *phalanx* has a *shaft*, a proximal *base*, and a distal *head* (fig. 3-37). The *hallux* or great toe has two phalanges, and each of the remaining digits has three.

CLINICAL APPLICATIONS

Osteoporosis is a process of decalcification or demineralization of bone that frequently occurs in elderly individuals. It is therefore characteristic of the skeletal system to become more vulnerable to injuries as one approaches later life. The chief complications of senile osteoporosis include bone pain, kyphosis, fracture, and invalidism.

Rickets and *osteomalacia* are associated with skeletal defects that are caused by a deficiency of vitamin D. Rickets may be seen in children with inadequate vitamin intake. The skeletal deformities (bending and bowing of the leg bones) are associated with defective calcification of bone and hypertrophy of epiphyseal cartilage. Osteomalacia is the adult counterpart of rickets and is characterized by a soft and flexible skeletal system.

Deformities of the spinal column may involve exaggerations of the normal anteroposterior curvatures. A spine may project abnormally backward forming a *kyphosis* or abnormally forward forming a *lordosis*. A complex distortion of the spine in a lateral direction is known as *scoliosis*.

A common response of bone to injury is *fracture*. Classification of fractures include: *simple fractures*, where the skin is intact; *compound fractures*, where the skin is broken and the bone communicates with the outside; *compression fractures*, which involve can-

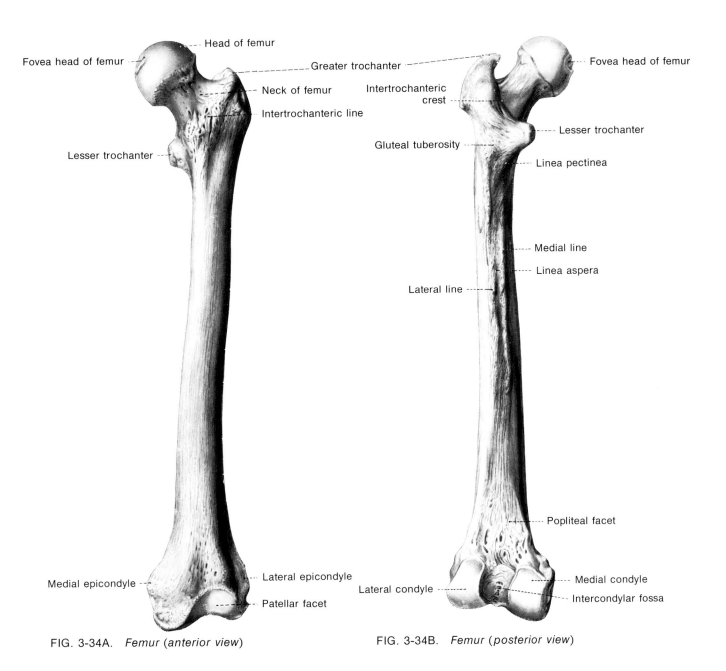

FIG. 3-34A. *Femur (anterior view)*

FIG. 3-34B. *Femur (posterior view)*

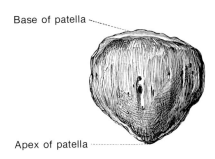

FIG. 3-35A. *Patella (anterior view)*

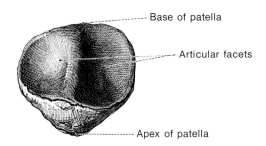

FIG. 3-35B. *Patella (posterior view)*

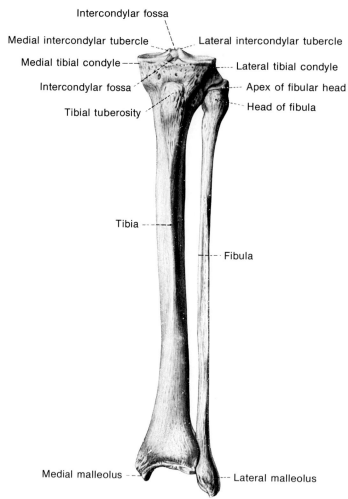

FIG. 3-36. *Fibula and tibia*

cellous bone such as the vertebrae and pelvis; and, *comminuted fractures*, which involve fragmentation of bone. Fractures in children frequently break like a green willow stick, which shows a fracture to both the periosteum and bone on the convex side, and mere bending on the concave side with the periosteum intact. This type of injury is referred to as a *green stick fracture.*

Fractures of the cranium usually cause greater damage to the inner table of the bone than to the outer. This is primarily because of the brittle nature of the former and also because blows to the head tend to flatten the convexity of the skull, causing a diffuse force on the inside. Fractures of the skull therefore frequently prove to be more extensive than they at first appear.

The *middle cranial fossa* is one of the more vulnerable portions of the skull, largely because of the pituitary fossa, the numerous foramina, and the possible transmission of violence through the lower jaw.

THE SKELETAL SYSTEM ■ 67

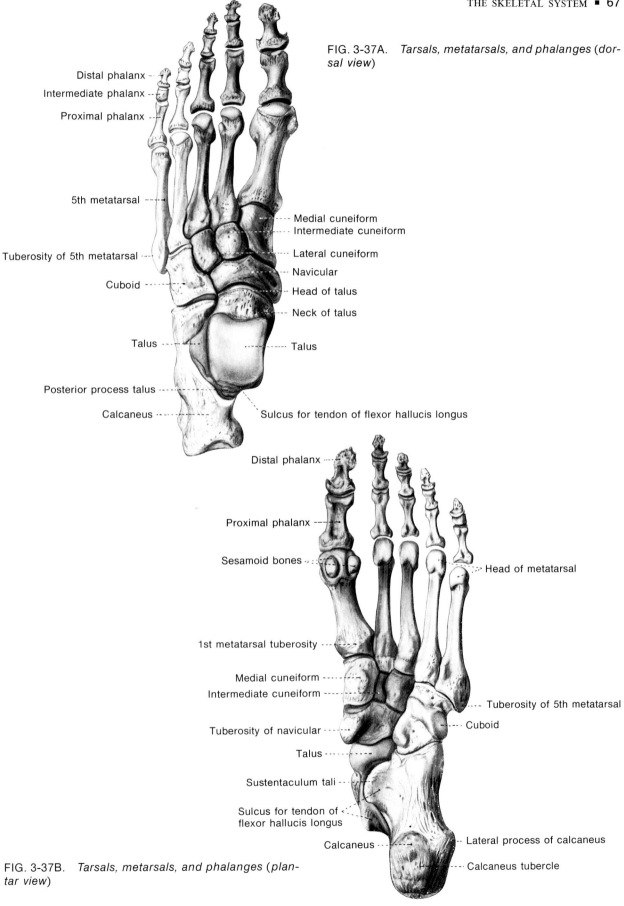

FIG. 3-37A. *Tarsals, metatarsals, and phalanges (dorsal view)*

FIG. 3-37B. *Tarsals, metarsals, and phalanges (plantar view)*

REVIEW QUESTIONS:

1. The skeletal system is divided into the _____ and _____ divisions.
2. Near the apex of the orbital cavity are located the _____ foramen and the _____ fissure.
3. The zygomatic bone extends posterolaterally to form the _____.
4. The anterior nasal aperture is the triangular area formed by the _____ and _____ bones.
5. The _____ foramen lies inferior to the second mandibular premolar halfway between the upper border of the alveolar process and the lower border of the base.
6. The _____ suture is located between the parietal and occipital bones.
7. The greater palatine foramen communicates with the _____ fossa.
8. The _____ fossa is bounded anteriorly by the articular eminence and posteriorly by the postglenoid tubercle.
9. The orifice of the _____ tube is found between the petrous temporal bone and the spine of the sphenoid.
10. The _____ suture separates the two parietal bones.
11. The floor of the cranium is divisible into three cerebral fossae:
    a.
    b.
    c.
12. The posterior part of the sella turcica is formed from the crest of the _____, ending laterally in the _____.
13. The _____ marks the position of the superior semicircular canal.
14. The _____ is the most prominent feature of the posterior cerebral fossa.
15. The _____ is an aperture in the lesser wing of the sphenoid.
16. The _____ is bounded by the lesser and greater wings of the sphenoid.
17. The inferior nasal concha is a part of the _____ bone.
18. The superior and middle nasal conchae are part of the _____ bone.
19. The _____ is the deep groove anterior and inferior to the bulla ethmoidalis.
20. The _____ sinus is the largest of the paranasal sinuses.
21. List the cranial nerves that pass through the jugular foramen.
22. The _____ are spaces located between the pedicles of adjacent vertebrae for the passage of spinal nerves and vessels.
23. The chief characteristic of the cervical vetebrae is the presence of a _____ through each transverse process.
24. The _____ is the second cervical vertebra.
25. The _____ vertebrae are characterized by the presence of facets for the articulation of ribs.
26. The outer fibrous portion of an intervertebral disk is known as the _____.
27. The first seven pairs of ribs, commonly known as _____, end in costal cartilages that are attached to the sternum.
28. The inferior inner side of the rib shaft is concave and is termed the _____.
29. The sternum comprises the _____, the _____ and a terminal _____.
30. List three fossae of the scapula.
31. The scapular spine ends ventrally in the _____.

32. The clavicle attaches to the _____ and _____ process.
33. On the posterior aspect of the distal end of the humerus is the _____ fossa.
34. The _____ process is located on the distal end of the radius.
35. List, beginning on the radial side of the proximal row, the sequence of the carpal bones.
36. The _____ is the term used to denote the two hip bones together, without the fixed vertebrae.
37. The hip bone is formed by the union of the _____, the _____ and the _____ bones.
38. The _____ is the large concavity with notched margins upon the lateral aspect of the hip bone.
39. The adductor tubercle is located on the proximal part of the _____ epicondyle of the _____.
40. The _____ is the largest sesamoid bone of the body.
41. The _____ is located on the distal end of the tibia.
42. The _____ is the bone of the heel and is the largest of the tarsal bones.
43. The _____ fossa, viewed anteriorly, is proximal to the capitulum.
44. Directly above the trochlea, viewed anteriorly, is the _____ fossa.
45. The head of a typical rib articulates with _____.
46. The _____ differs from the other cervical vertebrae in being devoid of body and spine.
47. The vertebra prominens is located on the _____ vertebra.
48. The sphenopalatine foramen connects the _____ fossa with the _____ cavity.
49. The foramen rotundum connects the _____ fossa with the _____ fossa.
50. The _____ canal extends from the foramen lacerum to the pterygopalatine fossa.

4. Articulations and Ligaments

> STUDENT OBJECTIVES
> After you have read this chapter, you should be able to:
> 1. Define an articulation.
> 2. Describe a synovial joint.
> 3. Classify joints on the basis of function and structure.
> 4. Discuss and compare the movements at various joints.
> 5. Contrast the structure, type of movement, and location of fibrous, cartilaginous, and synovial joints.

Articulations are of three types: *synarthrosis* (immovable joints), *amphiarthrosis* (slightly movable joints), and *diarthrosis* (freely movable joints).

A. Synarthrosis
   1. Suture (coronal, sagittal, lambdoidal)
   2. Synchondrosis (temporary joint between the diaphysis and epiphysis of a long bone)
B. Amphiarthrosis
   1. Symphysis (intervertebral, pubic symphysis)
   2. Syndesmosis (distal ends of the tibia and fibula)
C. Diarthrosis
   1. Gliding (intercarpal, intertarsal)
   2. Hinge (elbow, knee, ankle, interphalangeal)
   3. Ellipsoidal (radiocarpal)
   4. Pivot (atlas-axis, radioulnar)
   5. Saddle (carpometacarpal)
   6. Ball-and-socket (hip, shoulder)

Synarthrodial and amphiarthrodial joints are simple. In contrast, diarthrodial or synovial joints have: a hyaline covering on the articular parts of the bones; a joint cavity in between; at times, one or more intervening cartilaginous disks; synovial fluid; an encapsulating synovial membrane; and a capsular ligamentous sheath for holding the articular surfaces in apposition.

The *ligaments* surrounding a joint may be simply sleevelike or may adopt a most complicated pattern for the resistance of particular stresses. They not only strengthen the joint but confine the degree of movement to within proper limits. An articular cartilaginous disk may divide the cavity into two distinct compartments, as in the temporomandibular joint.

Bursae, in some respects similar to diarthroses, are sacs of synovial membrane containing a small amount of fluid, which occur in many parts of the body. They may be present wherever one structure frequently moves back and forth over another as between two muscle tendons, where a tendon passes over a bony projection, or between the skin and a bony prominence. *Synovial sheaths* are in the same category as bursae; they surround tendons that run in grooves, as in the case of the tendons of the wrist and ankle.

ARTICULATIONS OF THE SKULL

Temporomandibular Articulation

The temporomandibular joint (fig. 4-1) is between the articular (mandibular) fossa of the temporal bone and the articular (condylar) process of the mandible. An articular disk and a double synovial cavity are surrounded by a fibrous ligamentous capsule. The *articular disk* is an oval plate consisting of fibrous tis-

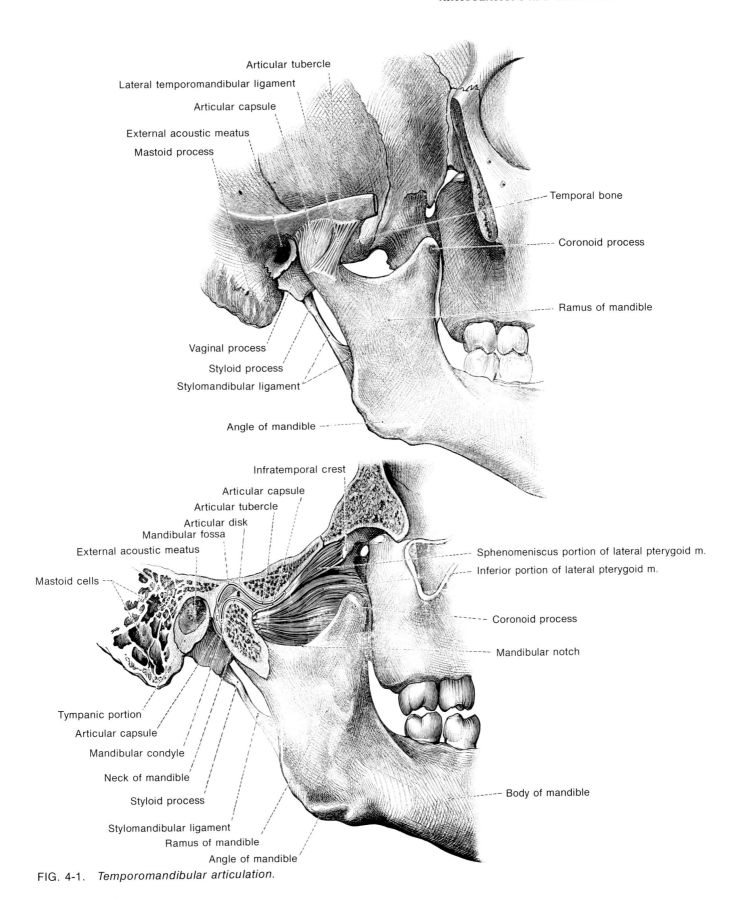
FIG. 4-1. *Temporomandibular articulation.*

sue, which completely divides the joint into superior and inferior compartments. The upper surface of the disk is saddle-shaped from anterior to posterior, accommodating the form of the articular fossa and articular tubercle; its inferior surface, in contact with the head of the mandible, is concave. The articular disk blends medially and laterally with the fibrous capsule to attach to the medial and lateral poles of the mandibular condyle. Anteriorly, the disk attaches to the anterior lip of the articular eminence and to the front of the articular margin of the condyle. Posteriorly, the disk attaches to the posterior wall of the mandibular fossa and the back of the articular margin of the condyle. Anteriorly, the disk is attached to the tendon of the lateral pterygoid muscle. When the mouth is opened the condyle of the mandible rotates around a horizontal axis; this movement is then combined with gliding of the condyle forward and downward in contact with the lower surface of the articular disk. At the same time, the disk slides forward and downward on the temporal bone. This movement results from the attachments of the disk to the medial and lateral poles of the condyle of the mandible and from the contraction of the *lateral pterygoid*, which carries the lead with its articular disk onto the *articular tubercle*. Some hinging and gliding forward of the head of the mandible occurs until it articulates with the most anterior part of the disk and the mouth is opened fully. Associated ligaments are as follows: the *sphenomandibular ligament*, which extends from the spine of the sphenoid to the lingula upon the medial surface of the mandibular ramus; the *stylomandibular ligament*, which extends from the styloid process to the angle of the mandible. The *lateral temporomandibular ligament* is attached above to the tubercle on the root of the zygoma, and below to the lateral surface and posterior border of the neck of the mandible.

Atlantooccipital Articulation

The atlantooccipital joint (fig. 4-2) is located between the occipital condyles and the superior articular facets of the atlas, there being a synovial cavity on either side, each surrounded by a capsular ligament and by an *anterior* and a *posterior atlantooccipital membrane*. The anterior membrane extends from the anterior border of the foramen magnum to the ventral arch of the atlas, and the posterior membrane from the posterior border of the foramen to the dorsal arch of the atlas.

INTERVERTEBRAL ARTICULATIONS

The intervertebral joints (fig. 4-3) are located between the bodies of the vertebrae and between the articular processes. There are also ligamentous connections between other parts of adjacent vertebrae.

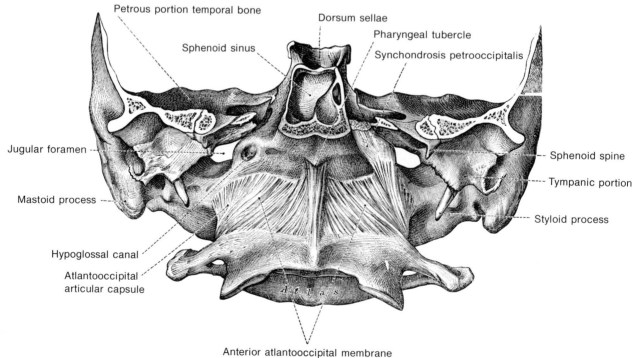

FIG. 4-2A. *Atlantooccipital articulation* (*anterior view*)

Intervertebral disks (23 in number), consisting of particularly dense fibrocartilage, are interposed between the vertebral bodies and these impart a certain degree of elasticity to the column. The outer margin of the disk contains concentric fibrous lamellae known as the *anulus fibrosus*. The inner portion contains the *nucleus pulposus*. Those disks between the sacral and the coccygeal vertebrae undergo ossification. Attached to the dorsal surfaces of the vertebral bodies is the *posterior longitudinal ligament*, composed of fibers of various lengths, and to the ventral surfaces, the *anterior longitudinal ligament*, which runs the full length of the column, with some alteration in the sacral region.

COSTAL AND STERNAL ARTICULATIONS

See figure 4-4.

Costosternal Articulations

The first rib unites with the sternum by cartilage. The remainder have small synovial cavities for articulations with adjoining sternebra with an *intraarticular ligament*.

Sternoclavicular Articulations

The articulation of clavicle with sternum has two synovial cavities separated by an articular disk of fibrocartilage.

ARTICULATIONS OF THE SHOULDER

There are two *joints of the shoulder,* the *acromioclavicular* and the *shoulder joint proper* or scapulohumeral articulation (fig. 4-5).

Acromioclavicular Articulation

A synovial cavity and a variable articular disk of cartilage are present in the *acromioclavicular articu-*

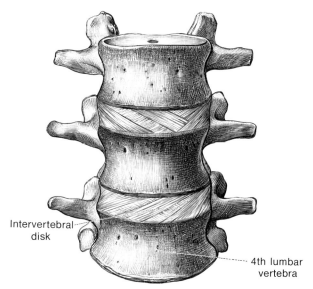

FIG. 4-3A. *Intervertebral articulation (anterior view)*

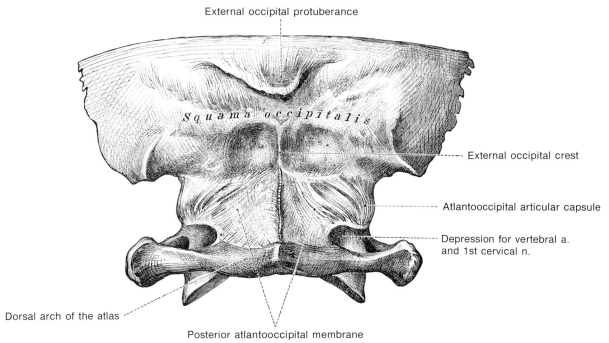

FIG. 4-2B. *Atlantooccipital articulation (posterior view)*

74 ■ BASIC ANATOMY FOR THE ALLIED HEALTH PROFESSIONS

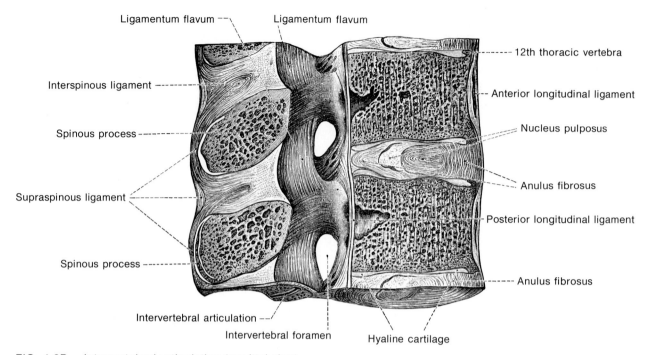

FIG. 4-3B. *Intervertebral articulation (sagittal view)*

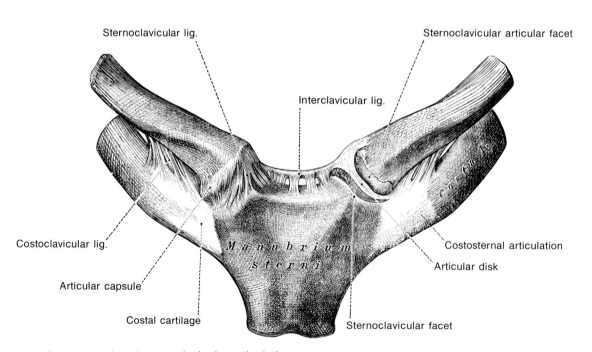

FIG. 4-4. *Costosternal and sternoclavicular articulations*

lation. Associated with the joint is the powerful *coracoclavicular ligament,* divided into two parts: the *conoid ligament,* from the bend of the *coracoid process* to the conoid tubercle of the clavicle; and the *trapezoid ligament,* from the coracoid to a point on the clavicle nearer the lateral end.

Shoulder Joint Articulation

The *shoulder joint* has great mobility; hence the synovial cavity is extensive and its capsular ligament is lax. Accordingly, the joint is inherently weak and subject to dislocations, which the muscles of fixation are only partially successful in preventing. Extending from the margin of the glenoid cavity is a rim of fibrocartilage, the glenoid labrum, into which fits the head of the humerus. Fused with the superior border of the labrum is the tendon of the long head of the biceps brachii muscle, which passes through the synovial cavity. Three thickenings of the capsular ligament are named *superior, middle,* and *inferior glenohumeral ligaments.*

ARTICULATION OF THE ELBOW

The *elbow joint* (fig. 4-6) comprises a humeroradial and a humeroulnar joint within a single synovial cavity. A fibrous capsule that is relatively weak in front and behind completely invests the joint. The capsule is especially strong in the medial and lateral sides, forming the following ligaments: the *radial collateral ligament,* from the lateral epicondyle of the humerus to the annular ligament of the radius and the border of the radial notch of the ulna; and the *ulnar collateral ligament,* which is divisible into three parts, an *anterior band* from the medial epicondyle to the coronoid process of the ulna, a *posterior band* from the medial epicondyle to the olecranon, and a *transverse band* that extends in an oblique direction from the olecranon to the coronoid process of the ulna.

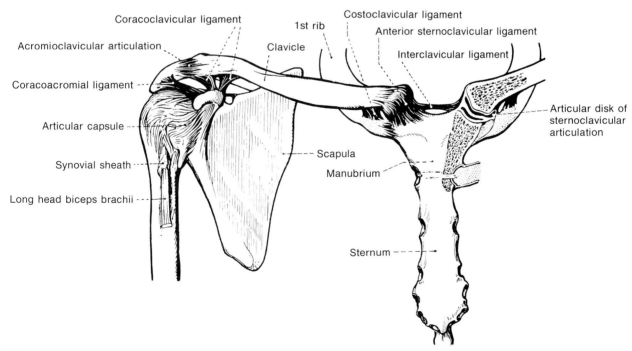

FIG. 4-5. *Acromioclavicular and shoulder joint articulations*

Superior Radioulnar Articulation

In the *superior radioulnar articulation*, the head of the radius can rotate within the radial notch of the ulna, and the former bone is bound in position by the *annular ligament of the radius*.

ARTICULATIONS OF THE WRIST

See figure 4-7.

Radiocarpal Articulation

The *radiocarpal articulation* is between the radius and ulnocarpal articular disk, as well as the scaphoid, lunate, and the triangular bones.

Intercarpal Articulations

Intercarpal articulations consist of joints between the individual carpal bones. A synovial cavity extends between the bones, which are anchored together by a system of short interosseous ligaments between each bone.

Carpometacarpal Articulations

The *carpometacarpal joints* of the second and third digits have synovial cavities at the ends, and to some extent between their bases, that are continuous with the common carpal synovial cavity. This is also true for the fourth and fifth digits; however, in some cases these two metacarpals may have no communication with the other. The first metacarpal usually has a small synovial cavity of its own.

METACARPOPHALANGEAL AND INTER-PHALANGEAL ARTICULATIONS

Each metacarpophalangeal and interphalangeal articulation (fig. 4-7) has a small synovial cavity and a capsular ligament with a strengthening *collateral ligament* on either side.

ARTICULATIONS OF THE HIP BONE

Articulations of the hip bone are the *sacroiliac, symphysis pubis,* and *hip joints* (fig. 4-8).

Sacroiliac Articulation

The capsular ligament of the *sacroiliac articulation* is divisible primarily into an *anterior sacroiliac ligament* and a *posterior interosseous ligament*. Both *short* and *long posterior sacroiliac ligaments* lie superficial to the posterior interosseous ligament.

Associated topographically, but hardly in relationship with this joint, are two ligaments of much importance. The *sacrotuberous ligament* is partly a caudal continuation of the long posterior sacroiliac ligament, which arises from the posterior iliac spines, and partly extends from the sacrum and coccyx. The

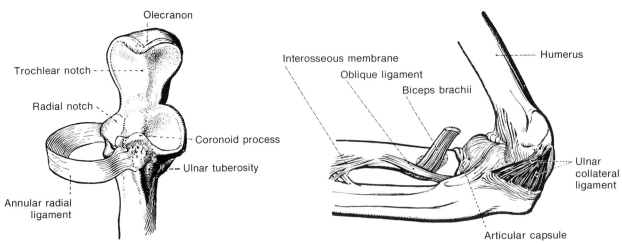

FIG. 4-6. *Articulation of the elbow*

ARTICULATIONS AND LIGAMENTS ■ 77

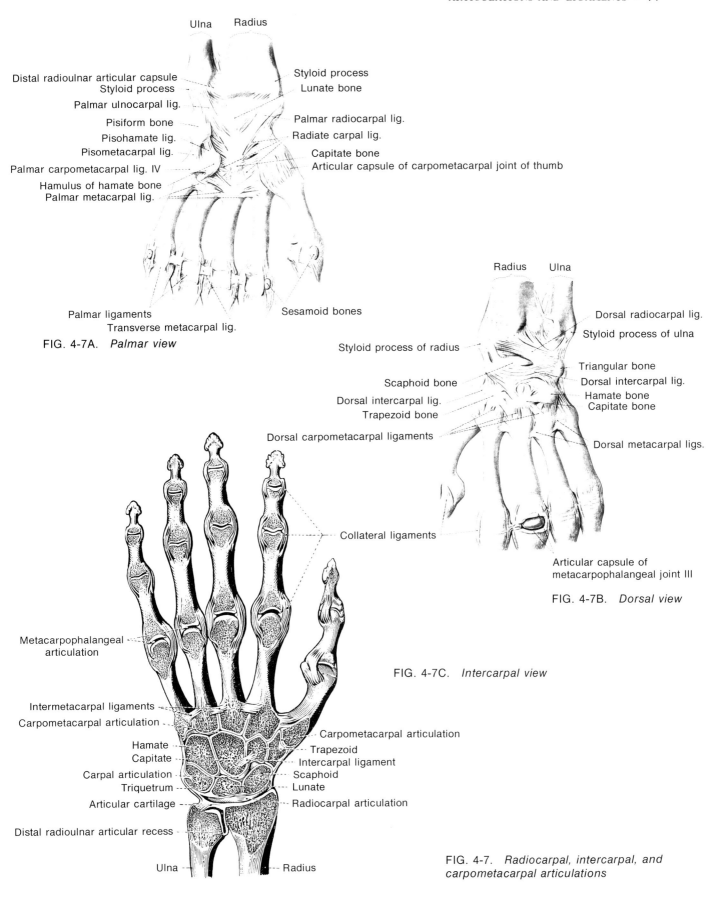

FIG. 4-7A. *Palmar view*

FIG. 4-7B. *Dorsal view*

FIG. 4-7C. *Intercarpal view*

FIG. 4-7. *Radiocarpal, intercarpal, and carpometacarpal articulations*

sacrospinous ligament extends from the inferior sacrum and adjoining coccyx to the ischial spine.

Symphysis Pubis Articulation

The *symphysis pubis articulation* is the midline joint between the two pubic bones. The articular surface of each pubis is covered by thin hyaline cartilage, and between these surfaces is an *interpubic disk* of fibrocartilage, in the center of which appears a rudimentary cavity without synovial membrane.

Hip Joint Articulation

At the hip, the head of the femur fits into the socket of the acetabulum, to which the three coxal bones contribute. This joint is considerably stronger than the shoulder joint because the pelvis is solidly anchored to the body and this articulation must be capable of supporting the entire weight of the body (as when standing on one leg) as well as transferring the body weight stably, as when walking and running.

The articular surface of the acetabulum is C-shaped, the hiatus being continuous with the acetabular notch, across which extends the *transverse acetabular ligament*. Between the latter and the notch pass the vessels and nerves of the joint. Around the acetabular margin is a cartilaginous lip, the *glenoid labrum*.

The capsular ligament is sleevelike. Some of the deeper fibers encircle the joint. The more superficial part is not of uniform thickness and the fibers gather into tracts, attached to the three bones of the hip, and to the trochanteric line and the crest of the femur. These fibrous thickenings are the *iliofemoral, ischiofemoral,* and *pubofemoral ligaments*. Movement occurs in all directions, including rotation.

ARTICULATION OF THE KNEE

The *knee joint* (fig. 4-9) is the largest, most complicated, and in some ways the most remarkable joint in the body. Often it is subjected to heavy lateral stresses from which only its ligaments can protect it, and yet dislocation of the knee is infrequent.

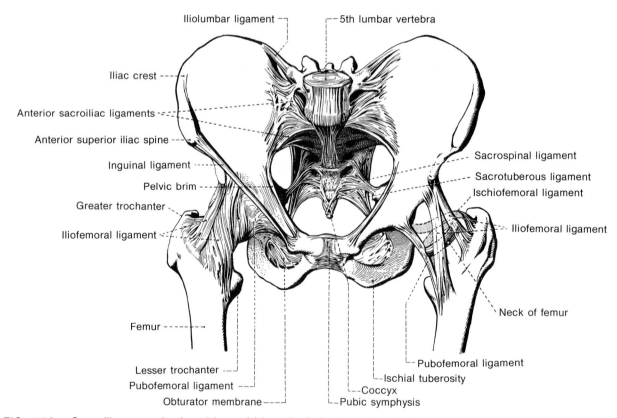

FIG. 4-8A. *Sacroiliac, symphysis pubis, and hip articulation (anterior view)*

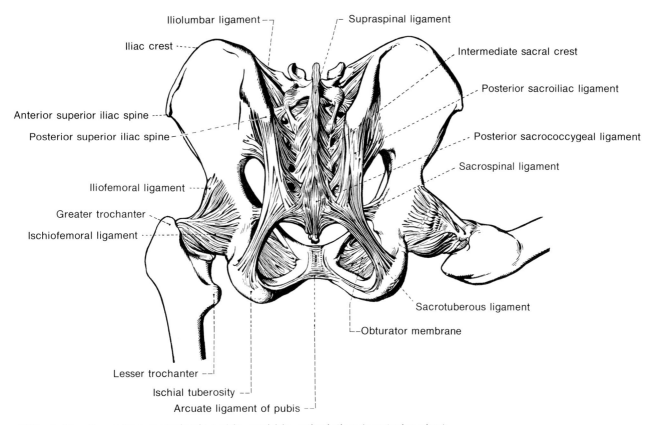

FIG. 4-8B. *Sacroiliac, symphysis pubis, and hip articulation (posterior view)*

Bones comprising the knee joint consist of the *femur,* the *tibia,* and the *patella.* The patella enters into the formation of the joint by fitting upon the patellar surface of the femur, with the synovial membrane intervening. The synovial cavity is also the largest in the body. The *menisci* (semilunar cartilages), interposed between the femoral and tibial condyles, are outside of the synovial membrane but indent it so that each indentation partly divides the cavity into two parts. The two menisci, *lateral* and *medial,* are C-shaped, each with an anterior and a posterior horn cartilage with ligamentous continuation directed toward the intercondylar notch, and joined ventrally by the *transverse ligament* of the knee. The menisci thus act as ligamentous washers between the bones forming the joint. There are also two central tendons from the tibia, crossed, as their names imply: the *anterior cruciate ligament* attached to the lateral condyle of the femur, and the *posterior cruciate ligament* attached to the intercondylar aspect of the medial condyle.

Capsular fibers occur chiefly as sharply defined *fibular* and *tibial collateral ligaments,* from femur to leg bones. Capsular in appearance is also the *oblique popliteal ligament,* covering the dorsal part of the joint.

ARTICULATION OF THE ANKLE

The *talocrural articulation* (fig. 4-10) must bear the full weight of the body, and must be ready to accept a tremendous excess load, as occurs in jumping. The direct load is borne through the tibia and talus, the medial malleolus of the tibia and the lateral malleolus of the fibula being applied to the sides of the articular surface of the talus; hence the fibula bears no weight.

The synovial cavity is simple but sends a process between the distal ends of the tibia and fibula. The capsular ligament is weak fore and aft and particularly strong at the sides, as is usual in hinge joints. The *lateral ligament* constitutes the *anterior* and *posterior talofibular* and *calcaneofibular ligaments.* The *medial ligament* comprises chiefly the *deltoid ligament,* which extends from the medial malleolus to the calcaneus, navicular bone and talus.

INTERTARSAL AND TARSOMETATARSAL ARTICULATIONS

The *intertarsal joints* (fig. 4-10) are even more remarkable than the other articulations of the lower limbs because the longitudinal arches of the foot,

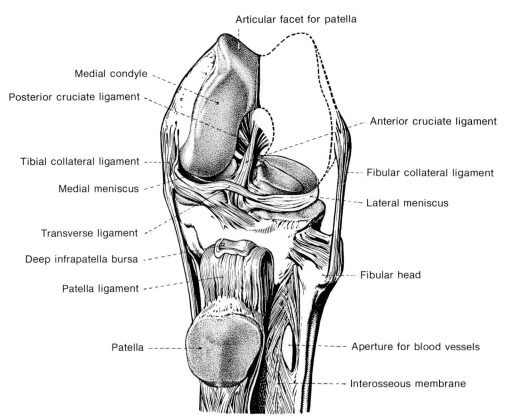

FIG. 4-9A. *Articulation of the knee (anterior view)*

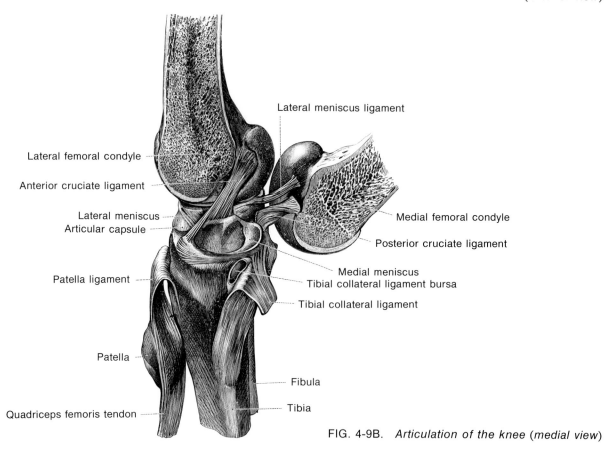

FIG. 4-9B. *Articulation of the knee (medial view)*

which bear as much weight as the other joints, must rely solely on the strength of ligaments, whereas in the more proximal joints the weight is applied chiefly to the bone. The intertarsal joints are numerous and complicated.

The *tarsometatarsal joints* (fig. 4-10) include a system of interosseous ligaments between the tarsal and metatarsal bones, and a capsular sheath forming ligamentous bands of extreme complexity. The most important of these is the *long plantar ligament,* stretching from the plantar surface of the calcaneus to the cuboid bone and the four lateral metatarsals.

METATARSOPHALANGEAL AND INTERPHALANGEAL ARTICULATIONS

Each metatarsophalangeal and interphalangeal articulation (fig. 4-10) has a small synovial cavity and a capsular ligament, with a small *collateral ligament* on either side, and a thickened *accessory plantar ligament.* Worthy of note is the *transverse metatarsal ligament* connecting the heads of all the metatarsal bones.

CLINICAL APPLICATIONS

Dislocation of the temporomandibular joint may occur whenever the articular disk and condyle of the mandible are moved anterior and forward of the summit of the articular tubercle. The strong lateral temporomandibular ligament in addition to the articular disk prevents the fracturing of the thin tympanic plate during abrupt posterior movements.

Dislocation of the sternal or acromial ends of the clavicle are uncommon. The ligaments are stronger than the clavicle and therefore fractures rather than dislocations usually occur.

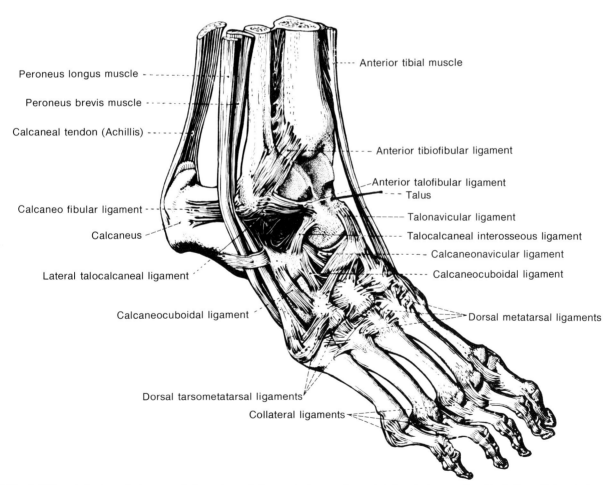

FIG. 4-10A. *Intertarsal, tarsometatarsal, metatarsophalangeal, and interphalangeal articulation (lateral dorsal view)*

82 ■ BASIC ANATOMY FOR THE ALLIED HEALTH PROFESSIONS

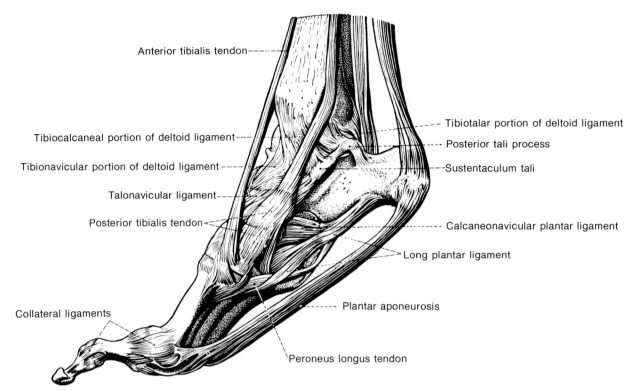

FIG. 4-10B. *Intertarsal, tarsometatarsal, metatarsophalangeal, and interphalangeal articulation* (*medial plantar view*)

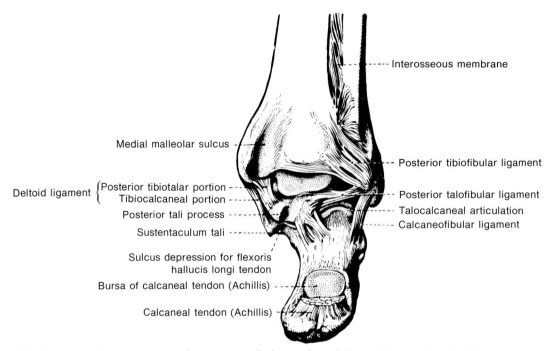

FIG. 4-10C. *Intertarsal, tarsometatarsal, metatarsophalangeal, and interphalangeal articulation* (*posterior view*)

Anterior dislocation of the shoulder usually involves the head of the humerus passing through the anterior portion of the capsule to slip beneath the coracoid process. Posterior dislocation of the shoulder always shows the head of the humerus beneath the spine of the scapula or acromial process.

Injury to the intervertebral disk most commonly involves the cervicothoracic and lumbosacral junctions. The posterior portion of the anulus fibrosus ruptures and the nucleus pulposus is forced posteriorly. This herniation of the nucleus pulposus may extend into the midline beneath the posterior longitudinal ligament or laterally toward the intervertebral foramen. Central herniation may involve the spinal cord, whereas lateral protrusions tend to compress spinal nerves.

Dislocations of the hip are either anterior or posterior. The attachment of the iliofemoral ligament immediately above the acetabulum and of the ischiofemoral ligament below it favor movement of the head of the femur only in an anterior or posterior direction.

The anterior cruciate ligament prevents anterior displacement of the tibia while the posterior cruciate ligament limits the posterior tibial displacement. The knee is seldom dislocated without damage to the popliteal vessels. The tibia is displaced anteriorly more frequently than posteriorly. Damage to the semilunar cartilages usually produces a torn piece of cartilage, which gets caught between the articular surfaces. Until this torn cartilage is removed from between the tibia and femur the joint is useless.

REVIEW QUESTIONS:

1. Define synarthrosis, amphiarthrosis, and diarthrosis.
2. Diarthrodial or synovial joints have _____ on the articular parts of the bone.
3. List the associated ligaments of the temporomandibular joint:
   a.
   b.
   c.
4. The _____ joint is located between the occipital condyles and the superior articular facets of the atlas.
5. The outer margin of the intervertebral disk contains concentric fibrous lamellae known as the _____.
6. The _____ is attached to the dorsal surfaces of the vertebral bodies.
7. Both the _____ and the _____ articulations have synovial cavities separated by an articular disk of fibrocartilage.
8. The coracoclavicular ligament is divided into _____ and _____ ligaments.
9. The deltoid ligament extends from the _____ to the _____, _____, _____ bone.
10. The anterior cruciate ligament prevents _____ displacement of the tibia while the posterior cruciate ligament limits _____ tibial displacement.

5. The Muscular System

STUDENT OBJECTIVES

After you have read this chapter, you should be able to:
1. Define the criteria employed in naming skeletal muscles.
2. Identify principal skeletal muscles by name, origin, insertion and innervation when they are involved in the following actions:
    a. facial expression
    b. mastication
    c. movement of the eyeball
    d. movement of the tongue
    e. movement of the head
    f. fixation of the hyoid
    g. movement of the vertebral column
    h. movement of the anterior abdominal wall
    i. respiration
    j. movement of the shoulder girdle
    k. movement of the arm
    l. movement at the elbow
    m. movement at the wrist and fingers
    n. movement of the hip
    o. movement at the knee
    p. movement at the ankle and toes

The skeletal muscles of the human body (fig. 5-1), approximately 400 in number, are named on the basis of one or more of the following characteristics: (1) size (maximus, minimus); (2) action (flexor, extensor); (3) direction of muscle fibers (rectus, oblique); (4) location (pterygoid, temporalis); (5) shape (deltoid, trapezius); (6) number of origins (biceps, triceps); and (7) origins and insertions (sternothyroid, sternohyoid).

This chapter identifies the principal skeletal muscles and briefly describes the origins, insertions, actions and innervations of each major muscle group. Clinical applications related to the muscular system are also considered.

IDENTIFICATION OF THE PRINCIPAL SKELETAL MUSCLES

Muscles of Facial Expression

The *muscles of facial expression* (fig. 5-2) arise either from fascia or from the surface of the skull and insert into the skin. The individual muscles are arranged in sheets or thin bands that blend with adjacent muscles. The variety of facial expressions are primarily caused by the action of muscles that are associated with the orifices of the orbits, nose, and mouth. See table 5-1.

Muscles of Mastication

The *muscles of mastication* (fig. 5-3) extend from the skull to the mandible and act upon the temporomandibular joint. These muscles protrude and retract the mandible, and move it from side to side. See table 5-2.

Muscles That Move the Eyeball

The *muscles of the eyeball* may be classified as either extrinsic (outside of the eyeball) or intrinsic (inside the eyeball). Those that move the eyeball are extrinsic. See figure 5-4 and table 5-3.

THE MUSCULAR SYSTEM ■ 85

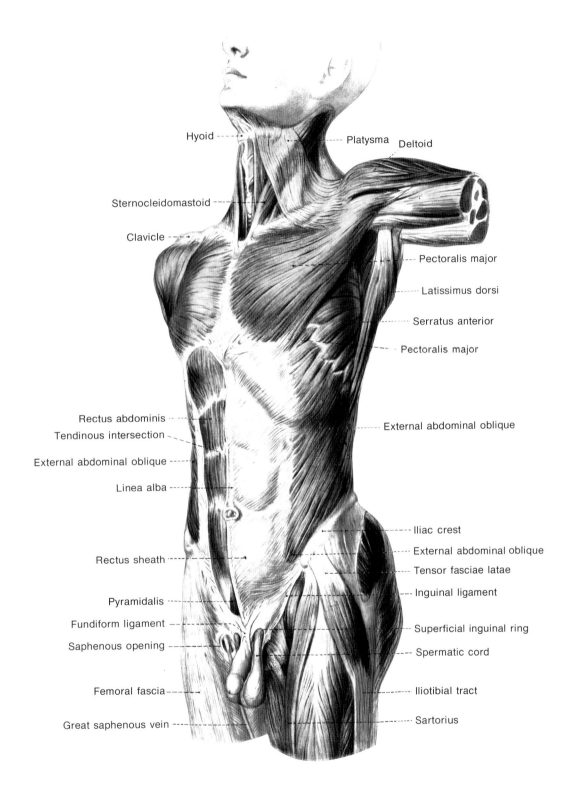

FIG. 5-1A. *The muscular system (anterior view)*

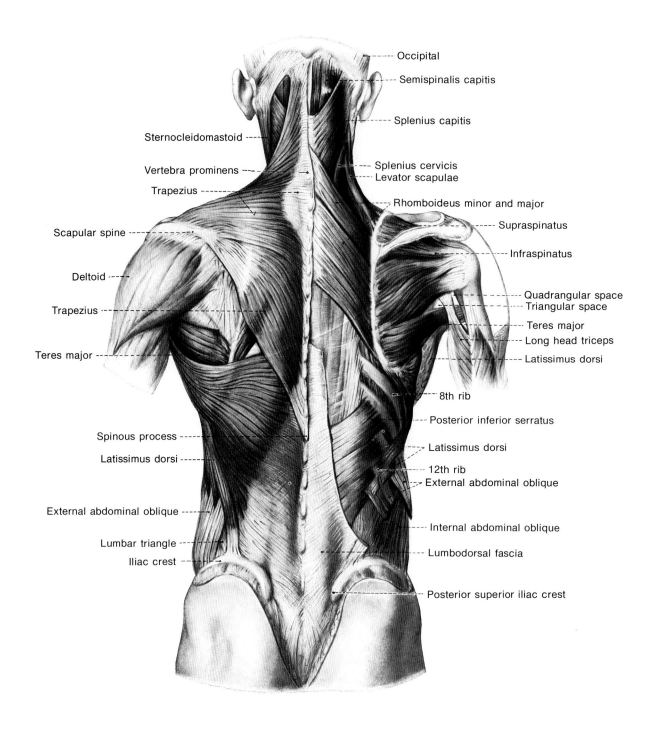

FIG. 5-1B. *The muscular system (posterior view)*

TABLE 5-1: Muscles of Facial Expression

MUSCLE	ORIGIN	INSERTION	ACTION	INNERVATION
Epicranius				
Frontalis	Galea aponeurotica	Fibers of the orbicularis oculi	Draws scalp forward, raises eyebrows	Facial
Occipitalis	Occipital bone and mastoid process	Galea aponeurotica	Draws scalp backward	Facial
Orbicularis oris	Muscle fibers encircling the mouth	Skin surrounding the mouth	Closes and compresses lips	Facial
Zygomaticus major	Zygomatic bone	Skin at angle of mouth and orbicularis oris	Draws angle of mouth upward and backward as in smiling or laughing	Facial
Levator labii superioris	Lower margin of orbit	Upper lip	Elevates upper lip	Facial
Depressor labii inferioris	Mandible	Skin of lower lip	Depresses lower lip	Facial
Buccinator	Alveolar processes of maxilla and mandible	Orbicularis oris	Compresses cheeks	Facial
Mentalis	Mandible	Skin of chin	Elevates skin of chin	Facial
Platysma	Fascia over pectoral and cervical region	Mandible, muscles around angle of mouth	Draws lip downward	Facial
Risorius	Fascia over masseter muscle	Skin at angle of mouth	Draws angle of mouth laterally	Facial
Orbicularis oculi	Medial wall of orbit	Circular path around orbit	Closes eyelids	Facial
Levator labii superioris alaeque nasi	Nasal process of the maxilla	Ala of the nose and upper lip	Elevates upper lip, dilates nostrils	Facial
Dilator naris	Maxilla	Skin of nostril	Dilates nostrils	Facial
Compressor nasi	Maxilla	Fibrocartilage of nose	Dilates nostrils	Facial
Levator anguli oris	Maxilla	Angle of mouth	Elevates angle of mouth	Facial
Depressor anguli oris	Mandible	Angle of mouth	Depresses angle of mouth	Facial

TABLE 5-2: Muscles of Mastication

MUSCLE	ORIGIN	INSERTION	ACTION	INNERVATION
Masseter	Zygomatic arch and malar process of maxilla	Angle and ramus of mandible	Elevates mandible	Mandibular branch of trigeminal nerve
Temporalis	Temporal fossa	Coronoid process of mandible	Elevates and retracts mandible	Mandibular division of trigeminal nerve
Medial pterygoid	Medial surface of lateral pterygoid plate of sphenoid	Angle and inner surface of ramus of mandible	Protracts and elevates mandible; moves mandible from side to side	Mandibular branch of trigeminal nerve
Lateral pterygoid	Lateral surface of lateral pterygoid plate of sphenoid	Neck of condyle of mandible and articular disc	Protracts and depresses mandible; moves mandible from side to side	Mandibular branch of trigeminal nerve

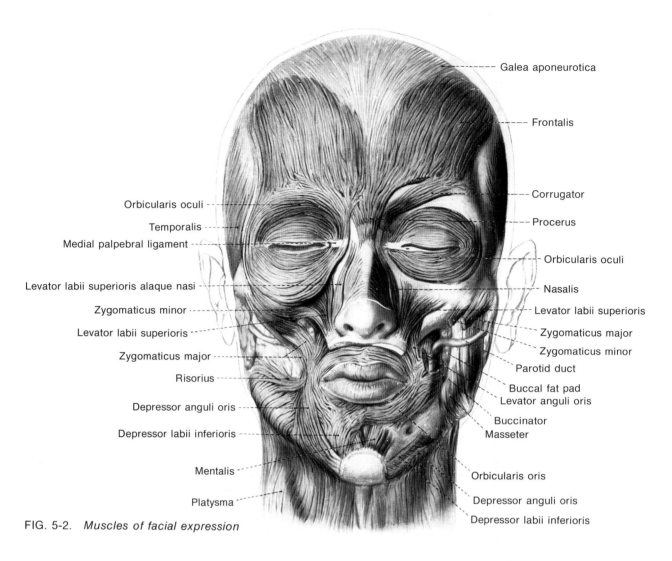

FIG. 5-2. *Muscles of facial expression*

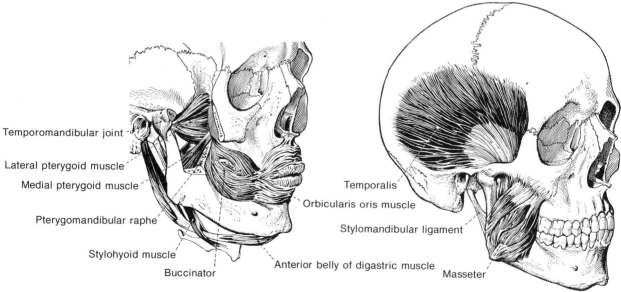

FIG. 5-3. *Muscles of mastication*

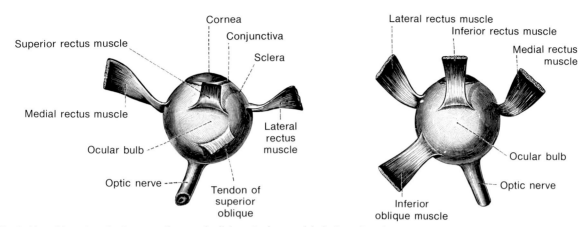

FIG. 5-4A. *Muscles that move the eyeball (posterior and inferior views)*

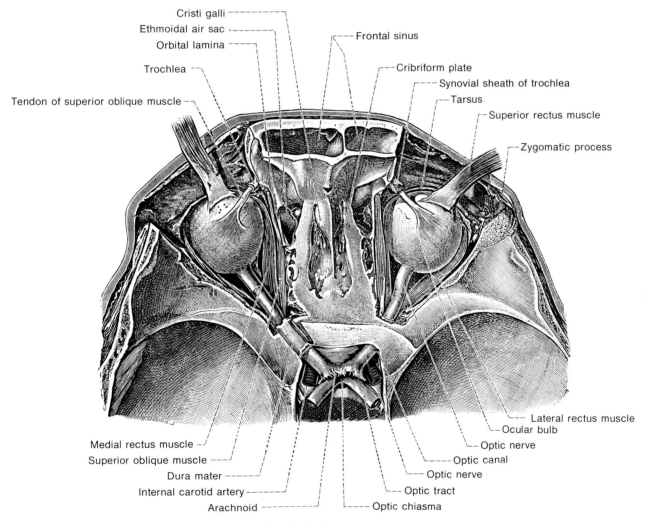

FIG. 5-4B. *Muscles that move the eyeball (dorsal view)*

Muscles That Move the Tongue

The form and shape of the tongue depends on both its intrinsic and extrinsic muscles. The position of the tongue depends on its extrinsic muscles. See figure 5-5 and table 5-4.

Muscles That Move the Head

The atlantooccipital articulation is the joint involved in movements of flexion and extension of the head. See figure 5-6A and table 5-5.

Muscles That Fix the Hyoid

The *infrahyoid muscles* are four straplike muscles that anchor the hyoid bone to the sternum, the clavicle, and the scapula.

The *suprahyoid muscles* connect the hyoid bone to the skull. See figure 5-6B and table 5-6.

Muscles That Move the Vertebral Column

The *muscles of the vertebral column* (fig. 5-7) are arranged in two main groups, *anterior* and *posterior*.

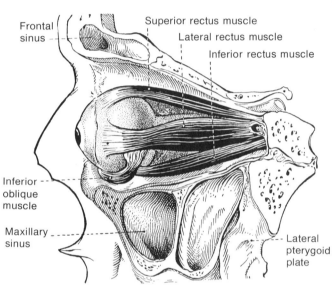

FIG. 5-4C. *Muscles that move the eyeball (lateral view)*

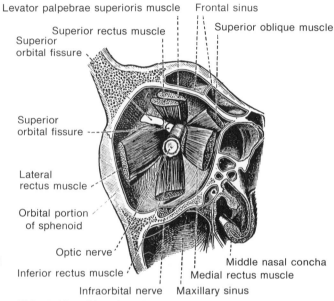

FIG. 5-4D. *Muscles that move the eyeball (anterior view)*

TABLE 5-3: Muscles that Move the Eyeball—the Extrinsic Muscles

MUSCLE	ORIGIN	INSERTION	ACTION	INNERVATION
Superior rectus	Common tendon	Superior and central part of eyeball	Rolls eyeball upward	Oculomotor
Inferior rectus	Common tendon	Inferior and central part of eyeball	Rolls eyeball downward	Oculomotor
Lateral rectus	Common tendon	Lateral side of eyeball	Rolls eyeball laterally	Abducens
Medial rectus	Common tendon	Medial side of eyeball	Rolls eyeball medially	Oculomotor
Superior oblique	Common tendon and trochlea	Eyeball between superior and lateral recti	Rotates eyeball on its anteroposterior axis; moves cornea downward and laterally—note that it moves through a ring of fibrocartilagenous tissue called the trochlea	Trochlear
Inferior oblique	Orbital plate of maxilla	Eyeball between superior and lateral recti	Rotates eyeball on its anteroposterior axis; moves cornea upward and laterally	Oculomotor

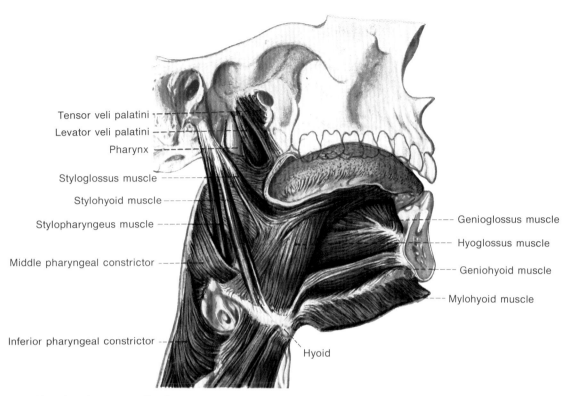

FIG. 5-5. *Muscles that move the tongue*

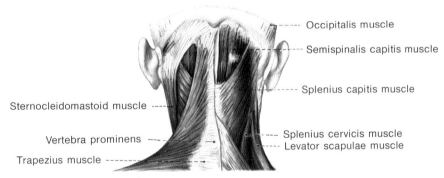

FIG. 5-6A. *Muscles that move the head*

TABLE 5-4: Muscles that Move the Tongue

MUSCLE	ORIGIN	INSERTION	ACTION	INNERVATION
Genioglossus	Mandible	Hyoid and undersurface of tongue	Depresses and thrusts tongue forward (protraction)	Hypoglossal
Styloglossus	Styloid process	Side and undersurface of tongue	Elevates tongue and draws it backward (retraction)	Hypoglossal
Hypoglossus	Body of hyoid bone	Side of tongue	Depresses tongue and draws down its sides	Hypoglossal
Chondro-glossus	Hyoid	Muscles of tongue	Depresses sides of tongue	Hypoglossal

TABLE 5-5: Muscles that Move the Head

MUSCLE	ORIGIN	INSERTION	ACTION	INNERVATION
Sternocleido-mastoid	Sternum and clavicle	Mastoid process	Contraction of both muscles flexes the head on the chest; contraction of one muscle rotates face toward side opposite contracting muscle	Accessory nerve
Semispinalis capitis	Articular processes of the lower 4 cervical vertebrae; transverse processes of the 7th cervical and upper 6 or 7 thoracic vertebrae	Occipital bone	Both muscles extend head; contraction of one muscle rotates face toward same side as contracting muscle	Dorsal rami of spinal nerves
Splenius capitis	Ligamentum nuchae and spines of 7th cervical and first 4 thoracic vertebrae	Occipital bone and mastoid process	Both muscles extend head; contraction of one rotates it to same side as contracting muscle	Dorsal rami of middle and lower cervical nerves
Longissimus capitis	Transverse processes of the upper 3 thoracic vertebrae; articular processes of the 1st–3rd cervical vertebrae	Mastoid process	Extends head and rotates face toward side opposite contracting muscle	Dorsal rami of middle and lower cervical nerves

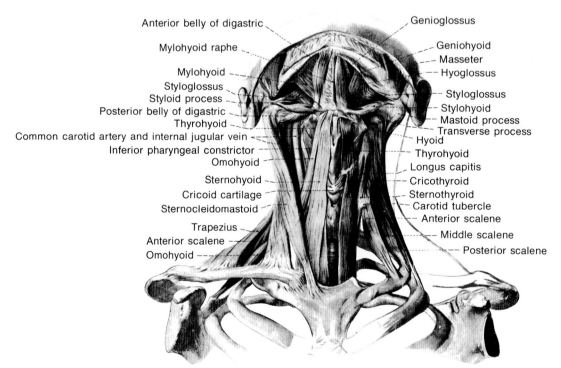

FIG. 5-6B. *Muscles that fix the hyoid*

TABLE 5-6: Muscles that Fix the Hyoid

MUSCLE	ORIGIN	INSERTION	ACTION	INNERVATION
Infrahyoid muscles				
Sternohyoid	Sternum	Hyoid	Draws hyoid downward	Cervical plexus
Omohyoid				
Inferior belly	Scapula	Intermediate tendon	Draws hyoid downward	Cervical plexus
Superior belly	Intermediate tendon	Hyoid	Draws hyoid downward	Cervical plexus
Sternothyroid	Sternum	Thyroid cartilage	Draws thyroid and hyoid downward	Cervical plexus
Thyrohyoid	Thyroid cartilage	Hyoid	Draws hyoid downward	Cervical plexus
Suprahyoid muscles				
Digastric				
Anterior belly	Mandible	Intermediate tendon	Depresses mandible	Mandibular division of trigeminal nerve
Posterior belly	Mastoid notch	Intermediate tendon	Elevates hyoid	Facial
Stylohyoid	Styloid process	Hyoid	Elevates hyoid	Facial
Mylohyoid	Mandible	Hyoid	Elevates floor of mouth	Mandibular division of trigeminal nerve
Geniohyoid	Mandible	Hyoid	Elevates floor of mouth	First cervical

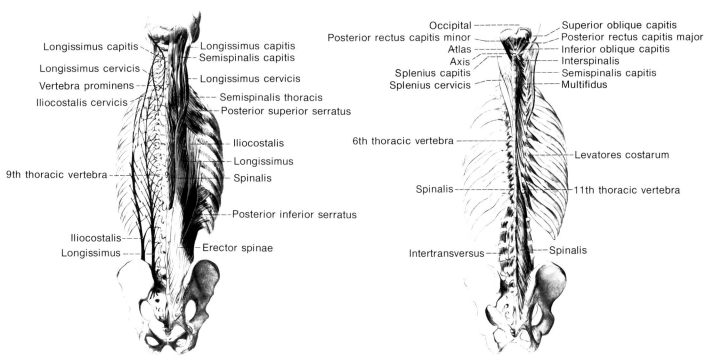

FIG. 5-7A. *Muscles that move the vertebral column (superficial)*

FIG. 5-7B. *Muscles that move the vertebral column (deep)*

The anterior group includes muscles in the neck and in the abdomen. The posterior aspect of the column includes the deep muscles of the back. These muscles can be grouped according to the direction and attachments of their component bundles.

The sacrospinalis (erector spinae) muscle consists of three posterior groupings: iliocostalis, longissimus, and spinalis. These groups in turn consist of a series of overlapping muscles. The iliocostalis group is laterally placed, the longissimus group is intermediate in placement, and the spinalis is medially placed. See table 5-7.

Muscles That Act on the Anterior Abdominal Wall

The anterolateral abdominal wall is muscular and aponeurotic. There are three pairs of broad and flat muscles that form the major portion of the anterolateral abdominal wall. The rectus abdominis is situated on each side of the midline. The aponeuroses of the muscles meet at the midline to form the *linea alba*. See figure 5-8.

Contributions of Anterior Abdominal Wall Muscles and Fasciae to Inguinal Region

The fascia and aponeurosis of the *external oblique* give rise to the: (1) *inguinal ligament;* (2) *lacunar ligament;* (3) *reflected inguinal ligament;* (4) *intercrural fibers;* (5) *medial and lateral crura;* (6) *superficial inguinal ring;* and (7) *external spermatic fascia*.

The *internal oblique* and its fascia give rise to the *cremaster muscle* and *fascia*.

The combined fibers of the *transverse* and *internal oblique* form the *conjoined tendon*.

The *transversalis fascia* gives rise to the *internal spermatic fascia* and the *deep inguinal ring*. See figure 5-8 and table 5-8.

Muscles of Respiration

During inspiration, the diaphragm contracts, pulling down its central tendon to enlarge the vertical diameter of the thorax. Intercostal muscles are active in deep and forced respiration. See figure 5-9 and table 5-9.

TABLE 5-7: Muscles that Move the Vertebral Column

MUSCLE	ORIGIN	INSERTION	ACTION	INNERVATION
Quadratus lumborum	Iliac crest, transverse processes of lower 4 lumbar vertebrae	12th rib and upper 4 lumbar vertebrae	Flexes vertebral column laterally	T12–L1
Sacrospinalis (erector spinae)				
Lateral				
Iliocostalis lumborum	Iliac crest	Lower 6 or 7 ribs	Extends lumbar region of vertebral column	Dorsal rami of lumbar nerves
Iliocostalis thoracis	Lower 6 ribs	Upper 6 ribs	Maintains erect position of spine	Dorsal rami of thoracic nerves
Iliocostalis cervicis	First 4 or 5 ribs	Transverse processes of the 4th–6th cervical vertebrae	Extends cervical region of vertebral column	Dorsal rami of cervical nerves
Intermediate				
Longissimus thoracis	Transverse processes of lumbar vertebrae	Transverse processes of all thoracic and upper lumbar vertebrae, and the 9th and 10th ribs	Extends thoracic region of vertebral column	Dorsal rami of spinal nerves
Longissimus cervicis	Transverse processes of upper 6 thoracic vertebrae	Transverse processes of 2nd–6th cervical vertebrae	Extends cervical region of vertebral column	Dorsal rami of spinal nerves
Longissimus capitis	Transverse processes of upper 5 or 6 thoracic vertebrae	Mastoid process	Extends head and rotates it to opposite side	Dorsal rami of middle and lower cervical nerves
Medial				
Spinalis thoracis	Spines of upper lumbar and lower thoracic vertebrae	Spines of upper thoracic vertebrae	Extends vertebral column	Dorsal rami of spinal nerves

THE MUSCULAR SYSTEM • 95

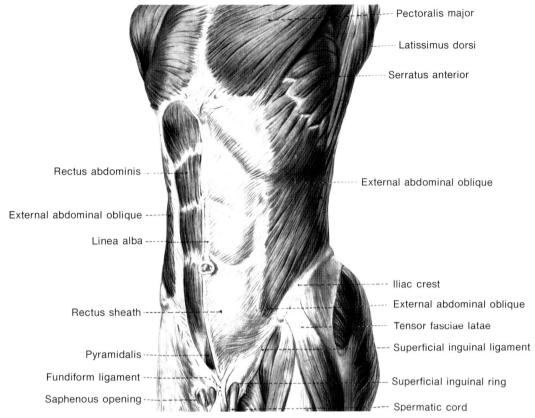

FIG. 5-8A. *Anterior abdominal wall (anteriolateral view)*

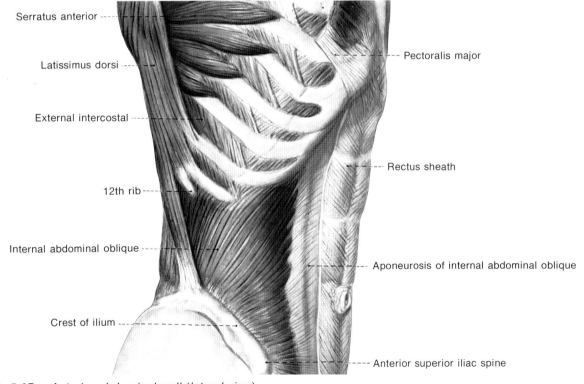

FIG. 5-8B. *Anterior abdominal wall (lateral view)*

96 ■ BASIC ANATOMY FOR THE ALLIED HEALTH PROFESSIONS

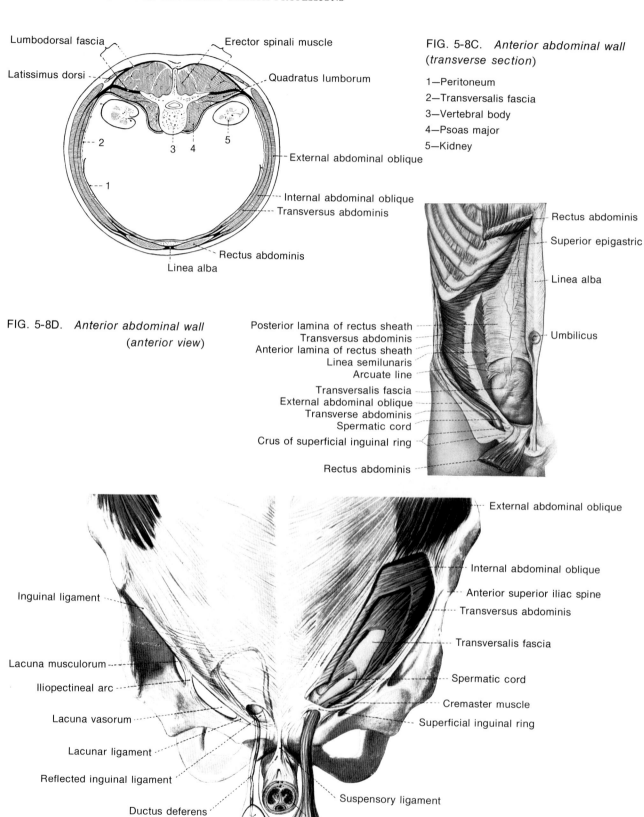

FIG. 5-8C. *Anterior abdominal wall (transverse section)*

1—Peritoneum
2—Transversalis fascia
3—Vertebral body
4—Psoas major
5—Kidney

FIG. 5-8D. *Anterior abdominal wall (anterior view)*

FIG. 5-8E. *Anterior abdominal wall (inguinal region)*

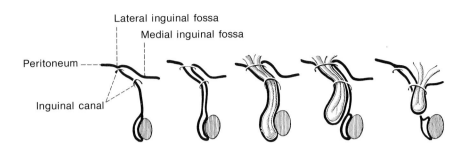

FIG. 5-8F. *Inguinal hernia*

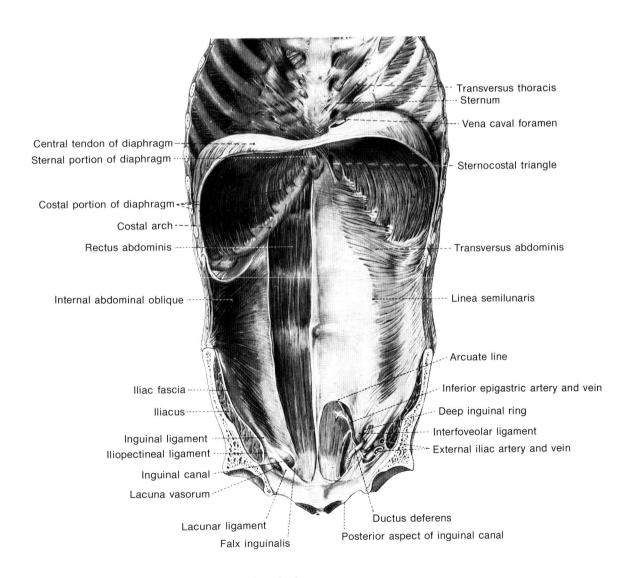

FIG. 5-8G. *Anterior abdominal wall (posterior view)*

TABLE 5-8: Muscles that Act on the Anterior Abdominal Wall

MUSCLE	ORIGIN	INSERTION	ACTION	INNERVATION
Rectus abdominis	Pubic crest and symphysis pubis	Cartilage of 5th–7th ribs and xiphoid process	Flexes thorax on the pelvis	Branches of 7th–12th intercostal nerves
External oblique	Lower 8 ribs	Iliac crest, linea alba (aponeurosis)	One side alone flexes vertebral column laterally; compresses viscera	Branches of 8th–12th intercostal nerves, iliohypogastric, and ilioinguinal nerves
Internal oblique	Iliac crest, inguinal ligament, and lumbodorsal fascia	Cartilage of lower 4 ribs, linea alba	Compresses abdomen; one side alone bends vertebral column laterally	Branches of 8th–12th intercostal nerves, iliohypogastric, and ilioinguinal nerves
Transversus abdominis	Iliac crest, inguinal ligament, lumbar fascia, and cartilages of last 6 ribs	Xiphoid process, linea alba, and pubis	Compresses abdomen; flexes vertebral column laterally	Branches of 7th–12th intercostal nerves, iliohypogastric, and ilioinguinal nerves

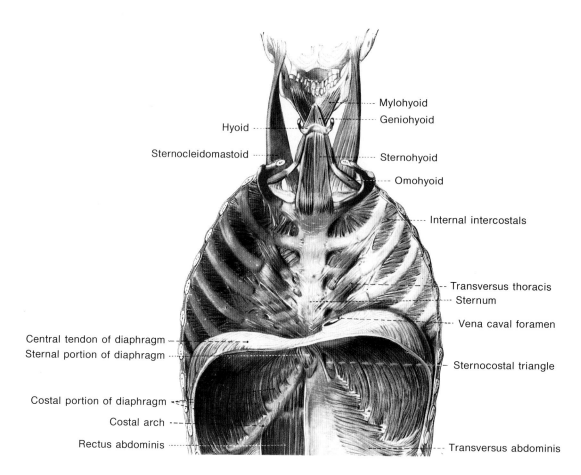

FIG. 5-9. *Muscles of respiration*

THE MUSCULAR SYSTEM ■ 99

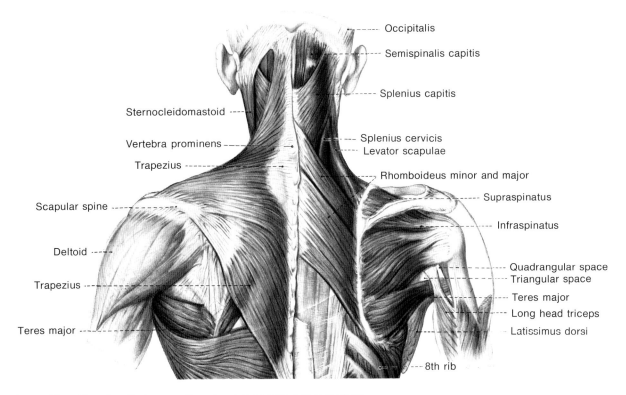

FIG. 5-10A. *Muscles that move the shoulder girdle (dorsal view)*

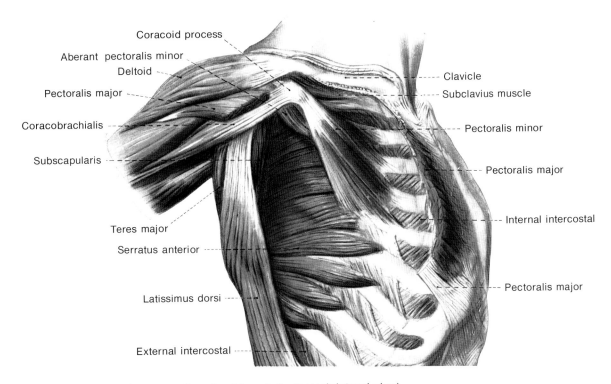

FIG. 5-10B. *Muscles that move the shoulder girdle (anteriolateral view)*

TABLE 5-9: Muscles of Respiration

MUSCLE	ORIGIN	INSERTION	ACTION	INNERVATION
Diaphragm	Xiphoid process, costal cartilages of lower 6 ribs, and lumbar vertebrae	Central tendon	Increases vertical length of thorax during respiration	Phrenic nerve
External intercostals	Each from the lower border of each rib	Upper border of the next rib below	Elevate ribs to increase lateral and anteroposterior dimensions of the thorax	Intercostal nerves
Internal intercostals	Each from the ridge on the inner surface of each rib	Upper border of the rib below	Draw adjacent ribs together during expiration to decrease the lateral and anteroposterior dimensions of the thorax	Intercostal nerves

Muscles That Move the Shoulder Girdle

The scapula, positioned by its associated muscles, extends the mobility and articulation of the upper limb. See figure 5-10 and table 5-10.

Muscles That Move the Humerus

The muscles that extend to the proximal end of the humerus from the thorax and pectoral region are individually capable of producing several different movements of the shoulder joint. See figure 5-11 and table 5-11.

Muscles of the Arm That Move the Forearm

The *muscles of the arm* (fig. 5-12) are classified as preaxial and postaxial. The *preaxial* are innervated by the musculocutaneous nerve and include the *coracobrachialis, biceps,* and the *brachialis.* The *postaxial* muscles are innervated by the radial nerve and include the *triceps* and the *anconeus.* See table 5-12.

TABLE 5-10: Muscles that Move the Shoulder Girdle

MUSCLE	ORIGIN	INSERTION	ACTION	INNERVATION
Subclavius	1st rib	Clavicle	Depresses clavicle	Brachial plexus (C5–C6)
Pectoralis minor	3rd–5th ribs	Coracoid process of scapula	Depresses scapula; elevates ribs in forced inspiration	Medial pectoral nerve
Serratus anterior	Upper 8 or 9 ribs	Vertebral border and inferior angle of scapula	Moves scapula forward and rotates scapula upward	Long thoracic nerve
Trapezius	Occipital bone and spines of 7th cervical and all thoracic vertebrae	Acromion process, clavicle, and spine of scapula	Adducts, elevates, or depresses scapula	Accessory nerve XI and C3–C4
Levator scapulae	Upper 4 or 5 cervical vertebrae	Vertebral border of scapula	Elevates scapula	Dorsal scapular nerve
Rhomboideus major	Spines of 2nd–5th thoracic vertebrae	Vertebral border of scapula	Moves scapula backward and upward	Dorsal scapular nerve
Rhomboideus minor	Ligamentum nuchae, spines of 7th cervical, and 1st thoracic vertebrae	Root of the spine of the scapula	Adducts scapula	Dorsal scapular nerve

Muscles of the Forearm That Move the Wrist and Fingers

The *muscles of the forearm* (fig. 5-13) are organized into anterior and posterior compartments. The *anterior compartment* contains *preaxial muscles* (*flexors*) and the *posterior compartment* contains *postaxial muscles* (*extensors*). The preaxial muscles are innervated by the median or ulnar nerves. All the postaxial muscles are innervated by the radial nerve. (For a complete diagram of the upper limbs, see figure 5-14.)

Muscles That Act on the Femur and Knee Joint

The *thigh* and *hip muscles* may be separated into four *compartments:* the *anterior femoral,* the *posterior femoral,* the *lateral femoral,* and the *medial femoral.* The muscles within each compartment are classified by locations because all have common origins, actions, and innervations. (See figure 5-15 and table 5-15.)

The *anterior femoral compartment* contains muscles that extend the leg at the knee. The anterior femoral

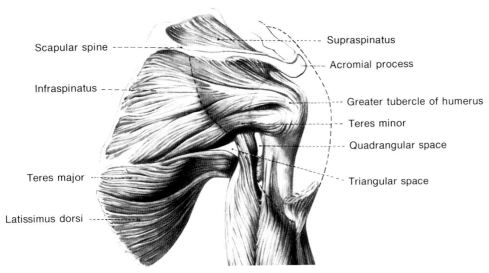

FIG. 5-11A. *Muscles that move the humerus (posterior view)*

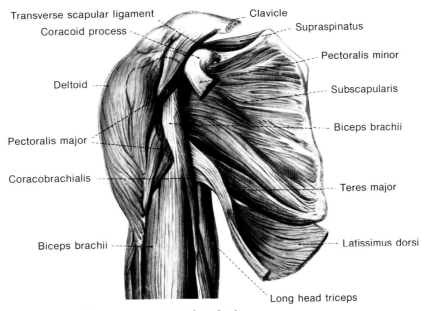

FIG. 5-11B. *Muscles that move the humerus (anterior view)*

TABLE 5-11: Muscles that Move the Humerus

MUSCLE	ORIGIN	INSERTION	ACTION	INNERVATION
Pectoralis major	Clavicle, sternum, cartilage of 2nd–6th ribs	Greater tubercle of humerus	Flexes and adducts arm	Medial and lateral pectoral nerve
Latissimus dorsi	Spines of lower 6 thoracic vertebrae, lumbar and sacral vertebrae, ilium and lower 4 ribs	Intertubercular groove of humerus	Draws arm downward and backward; extends and adducts arm medially	Thoracodorsal nerve
Deltoid	Acromion process and spine of scapula and clavicle	Lateral surface of body of humerus	Abducts arm	Axillary nerve
Supraspinatus	Fossa superior to spine of scapula	Greater tubercle of humerus	Assists deltoid muscle in abducting arm	Suprascapular nerve
Infraspinatus	Fossa inferior to spine of scapula	Greater tubercle of humerus	Rotates humerus outward	Suprascapular nerve
Teres major	Inferior angle of scapula	Lesser tubercle of humerus	Assists in adduction and medial rotation of arm	Lower subscapular nerve
Teres minor	Axillary border of scapula	Greater tubercle of humerus	Rotates humerus outward	Axillary nerve

TABLE 5-12: Muscles of the Arm that Move the Forearm

MUSCLE	ORIGIN	INSERTION	ACTION	INNERVATION
Biceps brachii	Long head from tuberosity above glenoid cavity; short head from coracoid process of scapula	Radial tuberosity	Flexes and supinates forearm	Musculocutaneous nerve
Brachialis	Anterior surface of humerus	Coronoid process of ulna	Flexes forearm	Musculocutaneous and radial nerves
Brachioradialis	Supracondyloid ridge of humerus	Styloid process of radius	Flexes forearm	Radial nerve
Triceps brachii	Long head from infraglenoid tuberosity of scapula; lateral head from lateral and posterior surface of humerus superior to radial groove; medial head from posterior surface of humerus inferior to radial groove	Olecranon process of ulna	Extends forearm	Radial nerve
Supinator	Lateral epicondyle of humerus, ridge on ulna	Oblique line of radius	Supinates forearm	Deep radial nerve
Pronator teres	Medial epicondyle of humerus, coronoid process of ulna	Midlateral surface of radius	Pronates forearm	Median nerve

THE MUSCULAR SYSTEM ■ 103

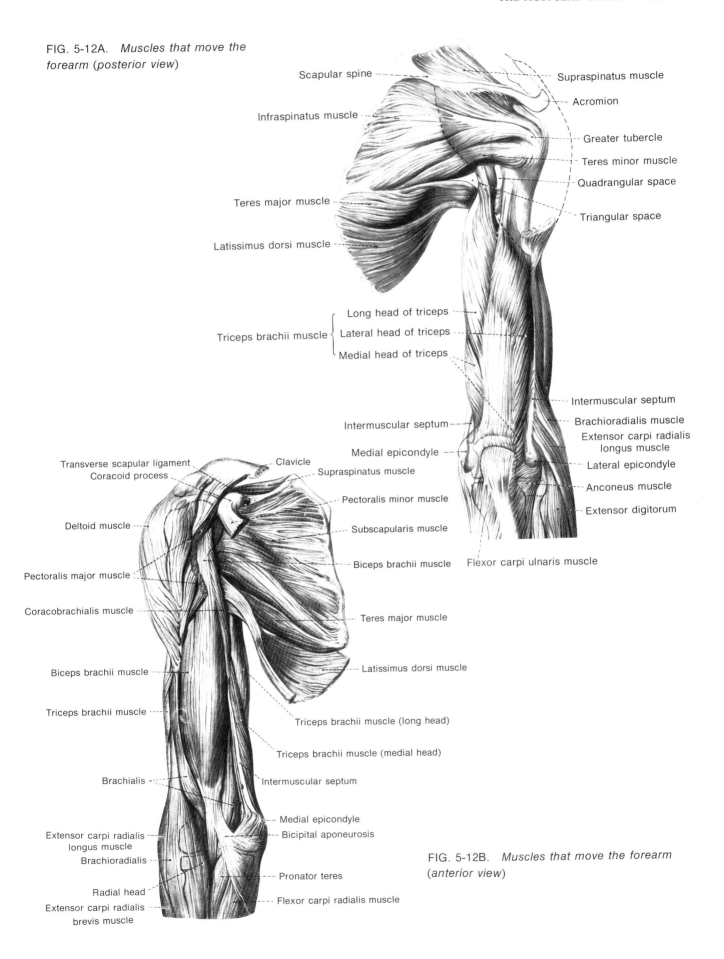

FIG. 5-12A. *Muscles that move the forearm (posterior view)*

FIG. 5-12B. *Muscles that move the forearm (anterior view)*

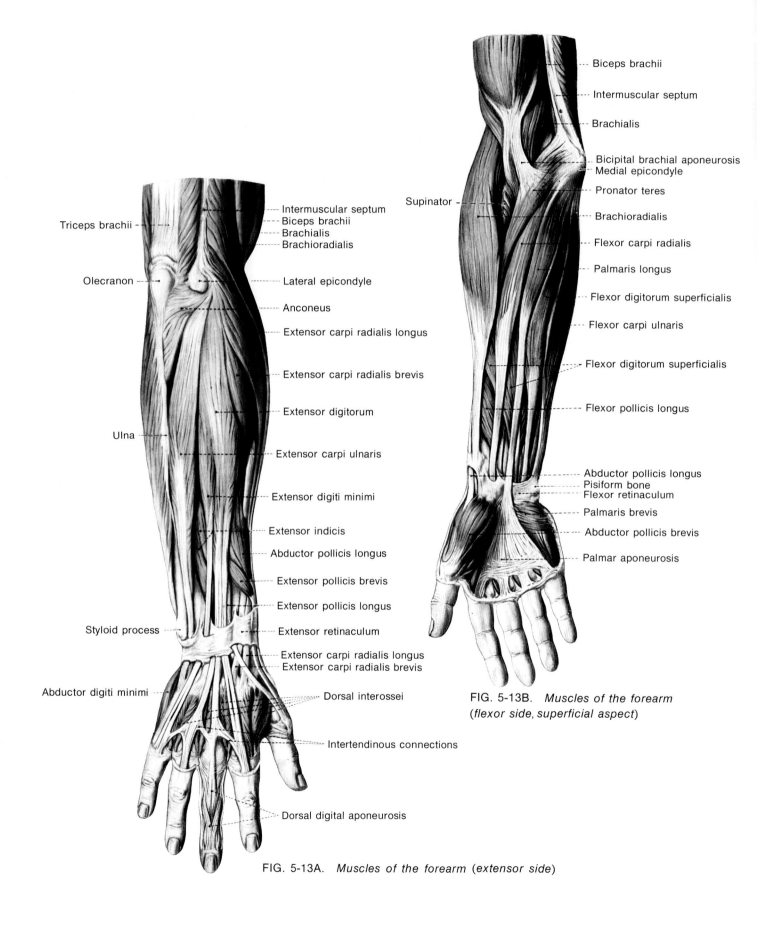

FIG. 5-13A. *Muscles of the forearm (extensor side)*

FIG. 5-13B. *Muscles of the forearm (flexor side, superficial aspect)*

THE MUSCULAR SYSTEM ■ 105

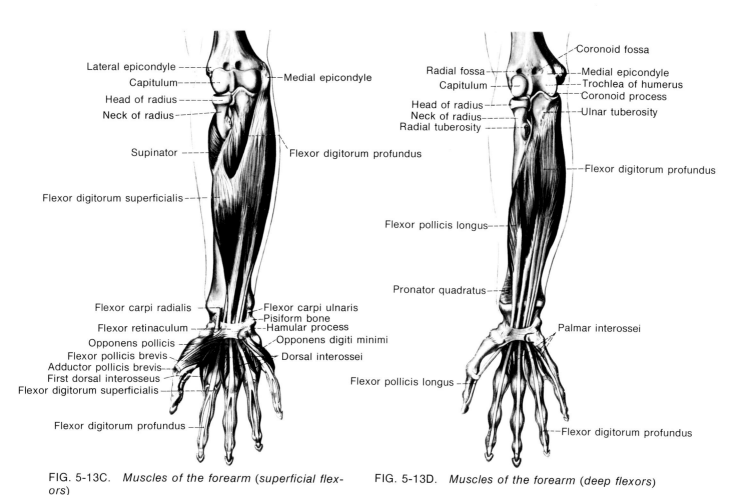

FIG. 5-13C. Muscles of the forearm (superficial flexors)

FIG. 5-13D. Muscles of the forearm (deep flexors)

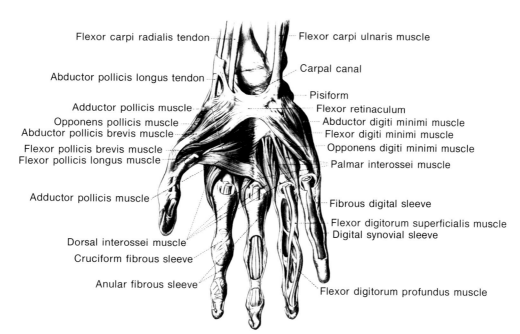

FIG. 5-13E. Muscles of the hand (palmar side)

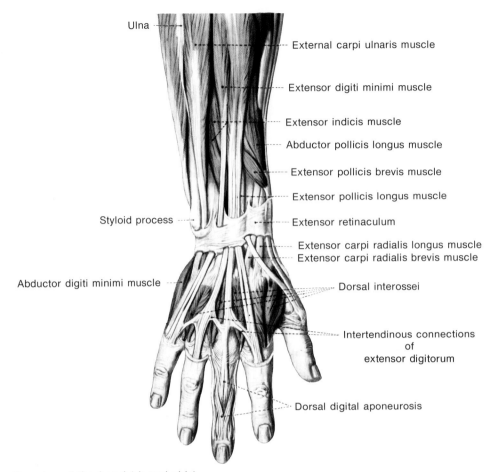

FIG. 5-13F. *Muscles of the hand (dorsal side)*

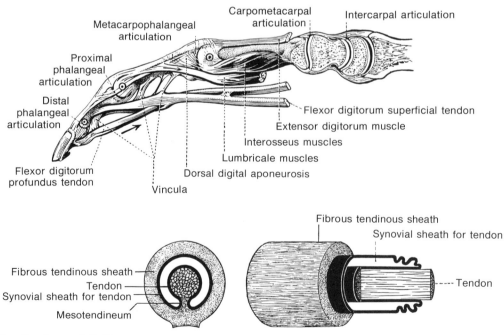

FIG. 5-13G. *Tendons of digits*

THE MUSCULAR SYSTEM ■ 107

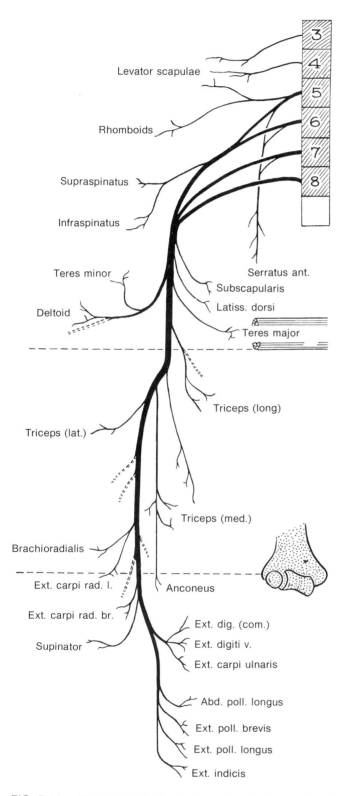

FIG. 5-14. *Diagrammatic illustration of motor innervation for upper limb* (Reproduced by permission from J.C.B. Grant's *"An Atlas of Anatomy,"* 7th Ed., copyright © 1978, The Williams & Wilkins Company.)

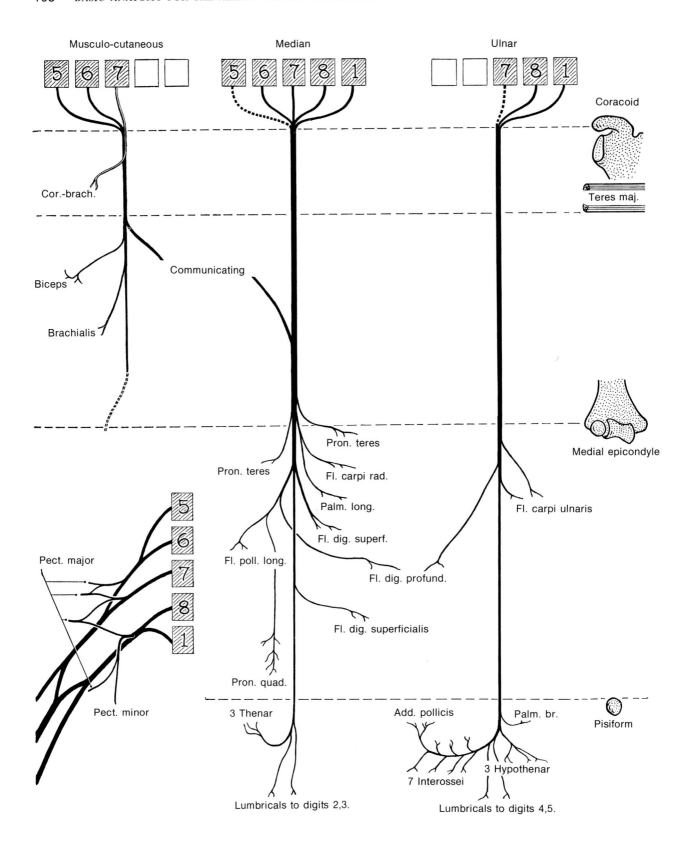

FIG. 5-14 (continued) *Diagrammatic illustration of motor innervation for upper limb* (Reproduced by permission from J. C. B. Grant's *"An Atlas of Anatomy,"* 7th ed. copyright © 1978, The Williams & Wilkins Company.)

TABLE 5-13: Muscles of the Forearm that Move the Wrist and Fingers

MUSCLE	ORIGIN	INSERTION	ACTION	INNERVATION
Flexor carpi radialis	Medial epicondyle of humerus	2nd and 3rd metacarpals	Flexes and abducts wrist	Median nerve
Flexor carpi ulnaris	Medial epicondyle of humerus	Pisiform, hamate, and 5th metacarpal	Flexes and adducts wrist	Ulnar nerve
Flexor digitorum superficialis	Medial epicondyle of humerus	Middle phalanges by tendons that are split for passage of deep flexor tendons	Flexes middle phalanges of each finger	Median nerve
Flexor digitorum profundus	Medial epicondyle, interosseous membranes	Distal phalanges	Flexes distal phalanges of each finger	Median and ulnar nerves
Extensor carpi radialis longus	Lateral epicondyle of humerus	2nd metacarpal	Extends and abducts wrist	Radial nerve
Extensor digitorum communis	Lateral epicondyle of humerus	Middle and distal phalanges of each digit	Extends phalanges	Radial nerve
Extensor indicis	Dorsal surface of ulna	Tendon of extensor digitorum into 2nd digit	Extends 2nd digit	Radial nerve
Extensor carpi ulnaris	Lateral epicondyle of humerus	5th metacarpal	Extends and adducts wrist	Radial nerve
Extensor pollicis brevis	Radius	Proximal phalanx of 1st digit	Extends 1st digit	Radial nerve
Extensor pollicis longus	Ulna	Distal phalanx of 1st digit	Extends 1st digit	Radial nerve
Abductor pollicis brevis	Scaphoid and lunate bones	Proximal phalanx of 1st digit	Abducts 1st digit	Median nerve
Opponens pollicis	Lunate	1st metacarpal	Moves 1st metacarpal medially	Median nerve
Flexor pollicis brevis	Lunate	Proximal phalanx of 1st digit	Flexes proximal phalanx of 1st digit	Median nerve
Adductor pollicis	Lunate, capitate, 2nd and 3rd metacarpals	Proximal phalanx of 1st digit	Adducts 1st digit	Ulnar nerve
Palmaris brevis	Palmar aponeurosis	Skin on ulnar side of palm	Deepens the hollow of the hand	Ulnar nerve
Abductor digiti minimi	Pisiform	Proximal phalanx of 5th digit	Abducts 5th digit	Ulnar nerve
Flexor digiti minimi brevis	Hamulus	Phalanx of 5th digit	Flexes 5th digit	Ulnar nerve
Opponens digiti minimi	Hamulus	5th metacarpal	Moves 5th metacarpal	Ulnar nerve
Lumbricales manus	Tendons of flexor digitorum profundus	Phalanges of 2nd–5th digits	Flex digits at metacarpophalangeal joint	Median and ulnar nerve
Dorsal interossei	Metacarpals	Proximal phalanges of 2nd–4th digits	Adduct 2nd–4th digits	Ulnar nerve
Palmar interrossei	2nd, 4th, and 5th metacarpals	Proximal phalanges of 2nd, 4th, and 5th digits	Adduct digits	Ulnar nerve

TABLE 5-14: Muscles that Act on the Femur and Knee Joint

MUSCLE	ORIGIN	INSERTION	ACTION	INNERVATION
Psoas major	Transverse processes and bodies of all lumbar vertebrae	Lesser trochanter of femur	Flexes the thigh and vertebral column	L2–L3
Iliacus	Iliac fossa	Tendon of psoas major	Flexes and rotates femur laterally	Femoral nerve
Gluteus maximus	Iliac crest, sacrum, coccyx, and sacrospinal ligament	Fascia lata and gluteal tuberosity of femur	Extends, abducts, and rotates femur laterally	Inferior gluteal nerve
Gluteus medius	Ilium	Greater trochanter of femur	Abducts thigh	Superior gluteal nerve
Gluteus minimus	Ilium	Greater trochanter of femur	Abducts and rotates femur medially	Superior gluteal nerve
Tensor fasciae latae	Iliac crest	Iliotibial tract of fascia lata	Flexes and abducts femur	Superior gluteal nerve
Adductor longus	Crest and symphysis pubis	Linea aspera of femur	Adducts femur	Obturator nerve
Adductor brevis	Inferior ramus of pubis	Linea aspera of femur	Adducts femur	Obturator nerve
Adductor magnus	Inferior ramus of pubis; ischium to ischial tuberosity	Linea aspera of femur	Adducts femur	Obturator nerve
Piriformis	Sacrum, sacrotuberous ligament	Greater trochanter of femur	Rotates femur laterally	S1–S2
Obturator internus	Margin of obturator foramen	Greater trochanter of femur	Rotates thigh laterally	Obturator nerve
Pectineus	Iliopectineal line	Linea aspera of femur	Adducts thigh	Femoral nerve
Quadriceps femoris				
Rectus femoris	Anterior inferior iliac spine	Patella	Extends leg	Femoral nerve
Vastus lateralis	Greater trochanter	Patella and tibial tuberosity	Extends leg	Femoral nerve
Vastus medialis	Linea aspera of femur	Patella and tibial tuberosity	Extends leg	Femoral nerve
Vastus intermedius	Anterior and lateral surfaces of femur	Patella and tibial tuberosity	Extends leg	Femoral nerve
Hamstrings				
Biceps femoris	Long head from ischial tuberosity; short head from femur	Head of fibula	Flexes leg	Tibial nerve from sciatic nerve
Semitendinosus	Ischial tuberosity	Proximal portion of tibia	Flexes leg	Tibial nerve from sciatic nerve
Semimembranosus	Ischial tuberosity	Medial condyle of tibia	Flexes leg	Tibial nerve from sciatic nerve
Gracilis	Pubic symphysis and arch	Medial surface of tibia	Flexes knee and adducts thigh	Obturator nerve
Sartorius	Anterior superior spine of ilium	Medial surface of tibia	Flexes leg	Femoral nerve

THE MUSCULAR SYSTEM ▪ 111

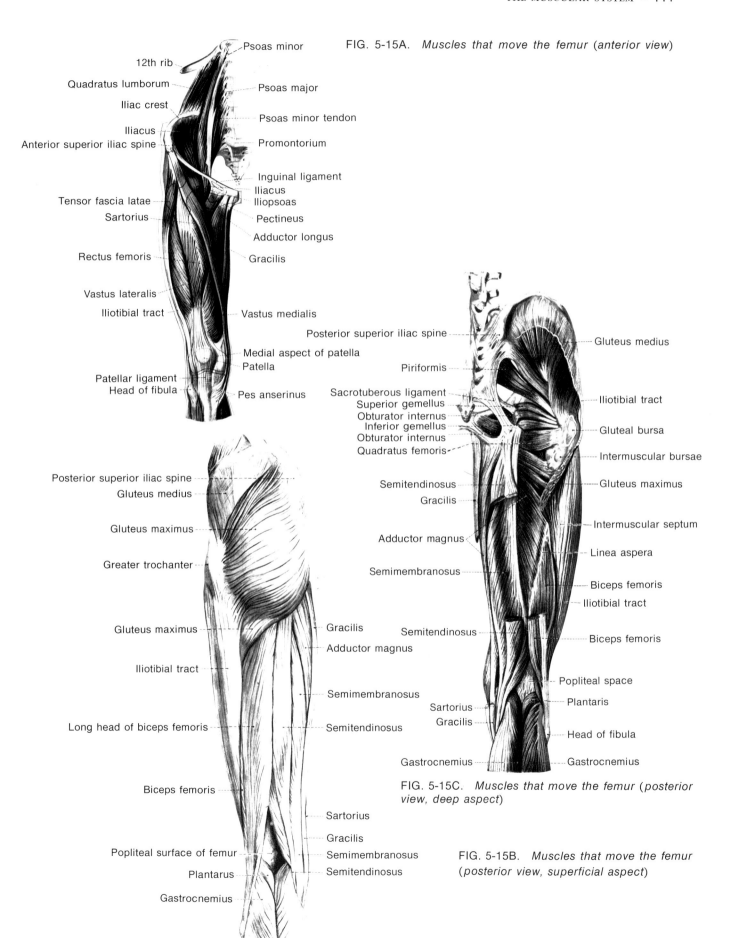

FIG. 5-15A. *Muscles that move the femur (anterior view)*

FIG. 5-15C. *Muscles that move the femur (posterior view, deep aspect)*

FIG. 5-15B. *Muscles that move the femur (posterior view, superficial aspect)*

TABLE 5-15: Muscles that Move the Foot and Toes

MUSCLE	ORIGIN	INSERTION	ACTION	INNERVATION
Gastrocnemius	Condyles of femur	Calcaneus	Flexes planta	Tibial nerve
Soleus	Head of fibula and medial border of tibia	Calcaneus	Flexes planta	Tibial nerve
Peroneus longus	Head of fibula and lateral aspect of tibia	1st metatarsal and 1st cuneiform bone	Flexes planta and everts foot	Superficial peroneal nerve
Peroneus brevis	Fibula	5th metatarsal	Flexes planta and everts foot	Superficial peroneal nerve
Peroneus tertius	Fibula	5th metatarsal	Dorsally flexes and everts foot	Deep peroneal nerve
Tibialis anterior	Lateral condyle of tibia	1st metatarsal and 1st cuneiform bone	Dorsally flexes and inverts foot	Deep peroneal nerve
Tibialis posterior	Interosseous membrane between tibia and fibula	2nd–4th metatarsals; navicular; 3rd cuneiform; cuboid	Flexes planta and inverts foot	Tibial nerve
Flexor digitorum longus	Tibia	Distal phalanges of 2nd–5th digits	Flexes toes and planta	Tibial nerve
Extensor digitorum longus	Tibia and fibula	Middle and distal phalanges of 2nd–5th digits	Extends toes and dorsally flexes foot	Deep peroneal nerve
Extensor digitorum brevis	Calcaneus	Proximal phalanx of 1st–4th digits	Extends phalanges of 1st–4th digits	Deep peroneal nerve
Abductor hallucis	Calcaneus	Proximal phalanx of great toe	Abducts great toe	Medial plantar nerve
Flexor digitorum brevis	Calcaneus	Middle phalanx of 1st–4th digits	Flexes 1st–4th digits	Medial plantar nerve
Abductor digiti minimi	Calcaneus	Proximal phalanx of 5th digit	Abducts 5th digit	Lateral plantar nerve
Quadratus plantae	Calcaneus	Tendons of flexor digitorum longus	Flexes distal phalanges of 2nd–5th digits	Lateral plantar nerve
Lumbricales pedis	Tendons of flexor digitorum longus	Distal phalanges of 2nd–5th digits	Flex toes at metatarso phalangeal joints	Medial plantar nerve
Flexor hallucis brevis	Cuboid and cuneiform bones	Proximal phalanx of 1st digit	Flexes 1st digit	Medial plantar nerve
Adductor hallucis	2nd–5th metatarsals	Proximal phalanx of 1st digit	Adducts 1st digit	Lateral plantar nerve
Flexor digiti minimi brevis	5th metatarsal	Proximal phalanx of 5th digit	Flexes 5th digit	Lateral plantar nerve
Dorsal interossei	Metatarsals	Proximal phalanx of 2nd–4th digits	Abduct 2nd–4th digits	Lateral plantar nerve
Plantar interossei	3rd–5th metatarsals	Proximal phalanges of 3rd–5th digits	Adduct 3rd–5th digits	Lateral plantar nerve

group is classified as postaxial and includes the following muscles: *sartorius, quadriceps femoris* (*rectus femoris, vastus lateralis, vastus medialis, vastus intermedius*), and the *articularis genus*. This group of muscles originates from either the ilium or femur and is innervated by the femoral nerve.

The *posterior femoral compartment* contains the "hamstring" group, which includes the *semitendinosus, semimembranosus,* and *biceps femoris*. These muscles are classified as preaxial and arise from the ischial tuberosity. They are innervated by the tibial division of the sciatic nerve.

The *lateral femoral compartment* includes both preaxial muscles (*obturator internus, superior* and *inferior gemelli,* and *quadratus femoris*) and postaxial muscles (*gluteus maximus, gluteus medius, gluteus*

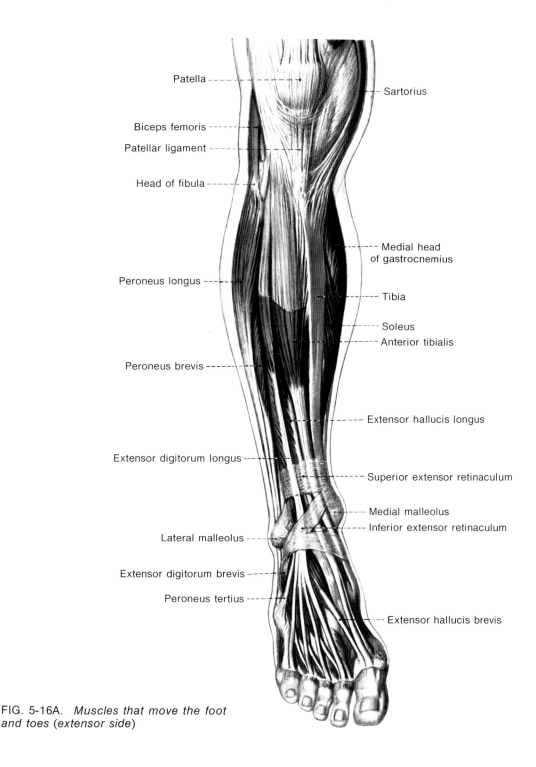

FIG. 5-16A. *Muscles that move the foot and toes* (*extensor side*)

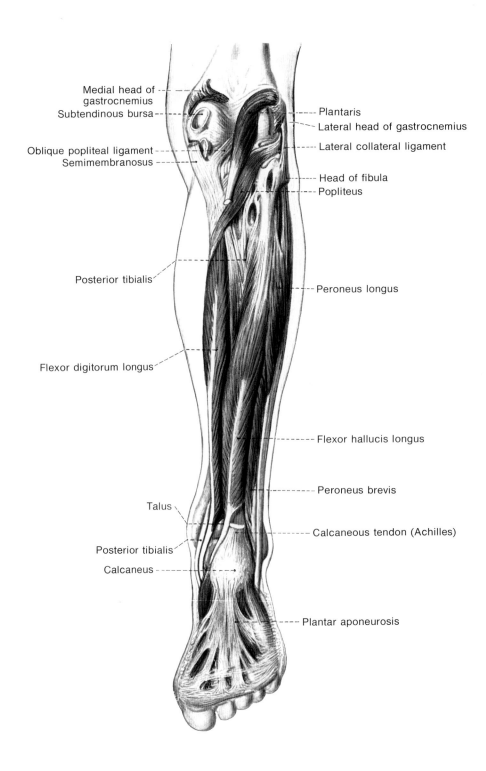

FIG. 5-16B. *Muscles that move the foot and toes (posterior aspect, deep flexors)*

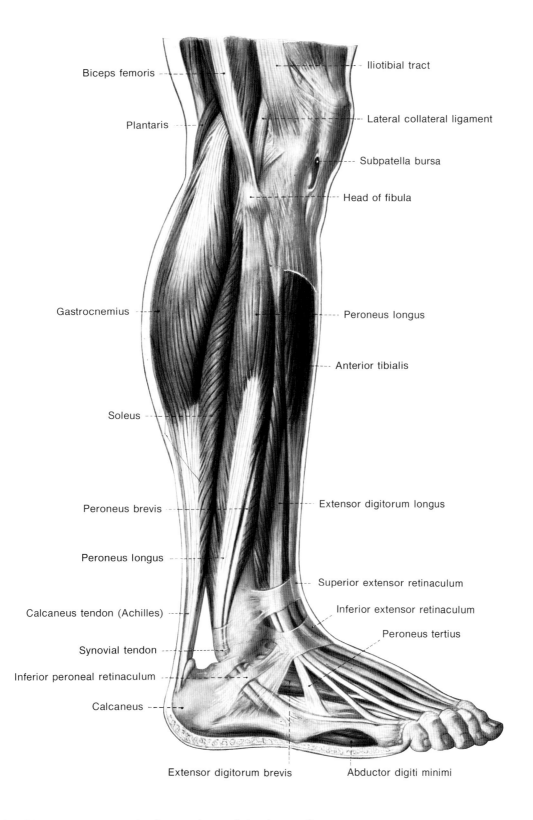

FIG. 5-16C. *Muscles that move the foot and toes (lateral aspect)*

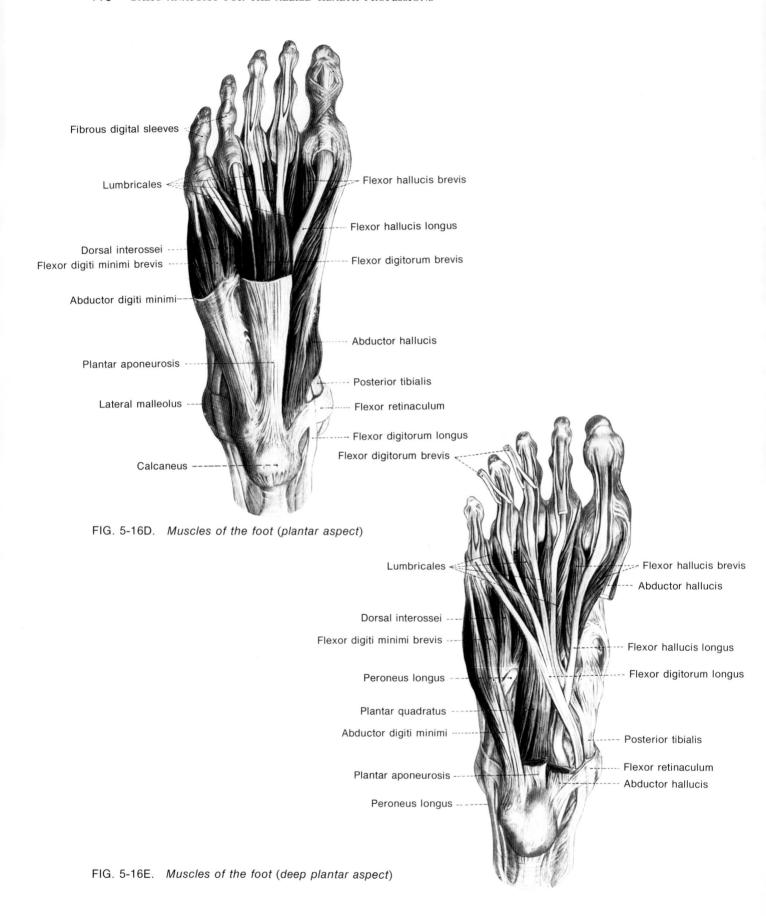

FIG. 5-16D. *Muscles of the foot (plantar aspect)*

FIG. 5-16E. *Muscles of the foot (deep plantar aspect)*

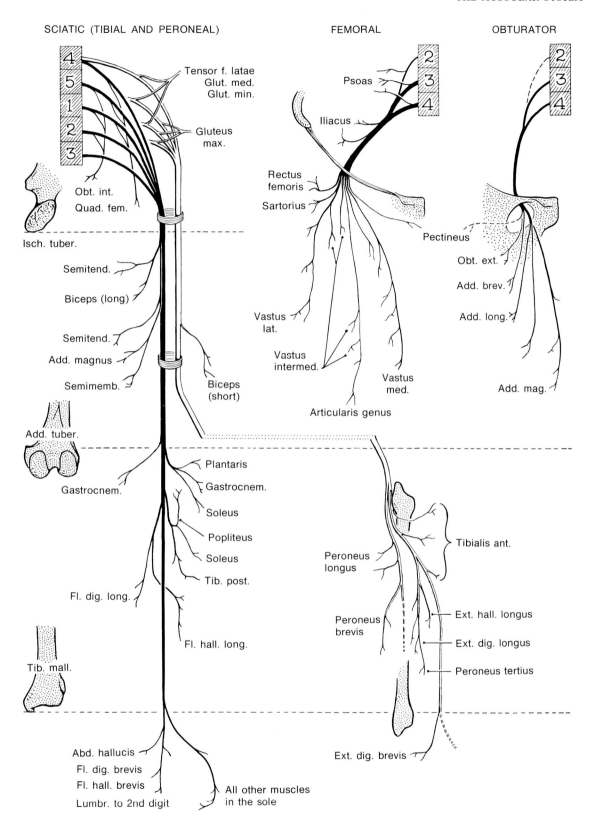

FIG. 5-17A. *Diagrammatic illustration of motor innervation for lower limb* (Reproduced by permission from J.C.B. Grant's *"An Atlas of Anatomy,"* 7th Ed., copyright © 1978, The Williams & Wilkins Company.)

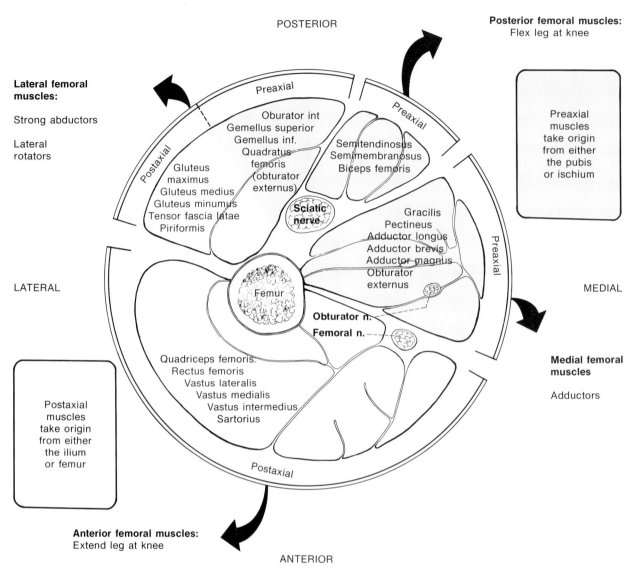

FIG. 5-17B. Diagrammatic representation of the four nerve-muscle groups of the hip and thigh

minimus, tensor fasciae latae, and *piriformis*). The preaxial group of muscles takes origin from the pubis or ischium and is innervated by either the obturator or the tibial nerve. The postaxial muscles arise from the ilium or the femur and are supplied by the femoral or the common peroneal nerve. Abduction and lateral rotation are the primary actions of the muscles of the lateral femoral compartment.

The *medial femoral compartment* includes the *adductor longus, adductor magnus, adductor brevis, gracilis, pectineus,* and *obturator externus.* All are preaxial adductors arising from the ramus of the pubis and ischium, and are innervated by the obturator nerve.

Muscles That Move the Foot and Toes

The anterior aspect of the leg contains *anterior* and *lateral muscular compartments* separated by the anterior intermuscular septum. The posterior aspect of the leg has *superficial* and *deep posterior compartments* separated by the transverse intermuscular septum. (See figure 5-16.)

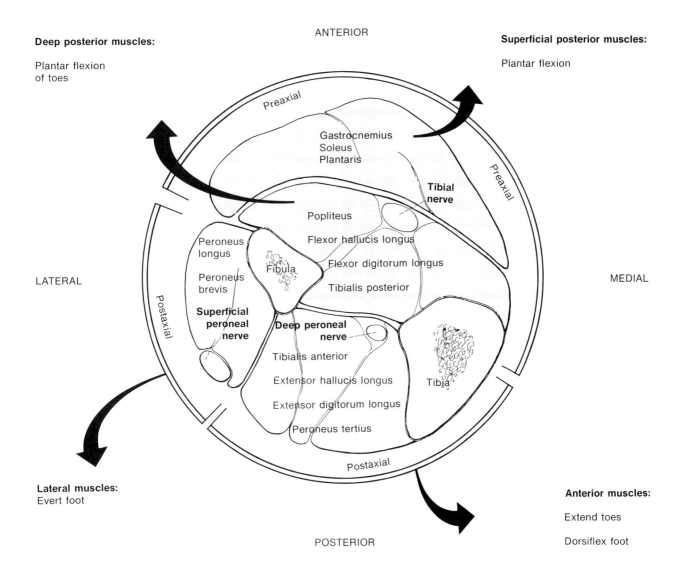

FIG. 5-17C. *Diagrammatic representation of the nerve-muscle groups of the leg*

The anterior and lateral compartments are postaxial in classification. They contain the *extensors* and are innervated by the common peroneal nerve. The two posterior compartments are preaxial in classification. They contain the *flexor muscles* of the foot and toes, and are innervated by the tibial nerve. (Lower limb innervation is diagrammed in figure 5-17.)

Muscles of the *anterior compartment* include the *tibialis anterior, extensor hallucis longus, extensor digitorum longus*, and the *peroneus tertius*. Muscles of the *lateral compartment* include the *peroneus longus* and *peroneus brevis*. The muscles of the *superficial posterior compartment* include the *gastrocnemius, soleus*, and *plantaris*. The muscles of the *deep posterior compartment* include the *popliteus, flexor hallucis longus, flexor digitorum longus*, and the *tibialis posterior*. See table 5-15.

CLINICAL APPLICATIONS

Myopathy refers to muscular weakness and wasting without evidence of neural involvement. The most common type of myopathy includes the *muscular dys-*

trophies, a class of hereditary diseases associated with degeneration of the muscle fibers.

Muscular atrophy refers to muscular weakness and wasting associated with lesions of the central or peripheral nervous system.

Distortion of facial muscles may be associated with injuries to the facial nerve.

Upper lesions of the brachial plexus (Erb-Duchenne palsy) may cause the limb to hang limply and medially rotated. It resembles the arm position of someone seeking a tip.

Lower lesions of the brachial plexus (Klumpke palsy) may result in the claw hand appearance.

Injury to the long thoracic nerve usually results in a *winged scapula* in which the vertebral border and inferior angle of the scapula show excessive dorsal protrusion.

Wrist-drop or flexion of the wrist is commonly associated with a radial nerve injury.

Ulnar nerve injuries usually cause wasting of the hypothenar eminence with a hand deformity characteristic of a *claw.*

Lesions of the sciatic nerve commonly involve hamstring muscles and all muscles below the knee, causing the foot to assume the plantar flexed position or *foot-drop.*

Muscles of the anterior and lateral compartments of the leg are sometimes paralyzed with injuries to the common peroneal nerve, causing the foot to be plantar flexed and inverted (*equinovarus*).

Muscles of the posterior compartment of the leg and the sole of the foot are paralyzed with injuries to the tibial nerve, causing dorsiflexion and eversion of the foot (*calcaneovalgocavus*).

Abdominal herniae include the following. (1) The *indirect inguinal hernia* enters the inguinal canal through the deep inguinal ring lateral to the inferior epigastric vessels. This hernia, as all abdominal herniae, consists of the hernial sac (peritoneum), hernial contents (viscera of the peritoneal cavity), and hernial coverings (layers of the abdominal wall). The indirect inguinal hernia is the most common form of hernia. (2) The *direct inguinal hernia* enters the inguinal canal medial to the inferior epigastric vessels. It comprises approximately 15% of all inguinal herniae. (3) The *umbilical hernia* may be acquired or congenital. (4) The *epigastric hernia* occurs through the widest part of the linea alba superior to the umbilicus.

REVIEW QUESTIONS:

1. All muscles of facial expression are innervated by the _____ nerve.
2. The lateral pterygoid muscle _____ the mandible.
3. The temporalis muscle is innervated by the mandibular division of the _____ nerve.
4. The lateral rectus muscle is innervated by the _____ nerve.
5. The _____ nerve innervates the medial rectus muscle.
6. The _____ nerve innervates the superior oblique muscle.
7. All muscles of the tongue are innervated by the _____ nerve.
8. The _____ muscle depresses and thrusts the tongue forward.
9. The sternocleidomastoid muscle is innervated by the _____ nerve.
10. List four muscles that turn the head:
    a.
    b.
    c.
    d.
11. The infrahyoid muscles are innervated by the _____ plexus.
12. The omohyoid muscle takes its origin from the _____.
13. The thyrohyoid muscle inserts onto the _____.
14. The posterior belly of the digastric muscle is innervated by the _____ nerve.

15. The geniohyoid takes origin from the _____ and is innervated by the _____ nerve.
16. List the three major groups of muscle fibers forming the sacrospinalis muscle:
    a.
    b.
    c.
17. List the muscles that form the anterior abdominal wall:
    a.
    b.
    c.
    d.

6. The Nervous System

> STUDENT OBJECTIVES
>
> After you have read this chapter, you should be able to:
> 1. Classify the nervous system into central and peripheral divisions.
> 2. Describe the external features of the spinal cord.
> 3. Identify the membranes that form the meninges.
> 4. Describe the structure of the spinal cord in cross section.
> 5. List the principal ascending and descending tracts of the spinal cord.
> 6. Define a spinal nerve.
> 7. Identify the names of the 31 pairs of spinal nerves.
> 8. Define the composition and distribution of the cervical, brachial, lumbar, sacral and coccygeal plexuses.
> 9. Define a dermatome.
> 10. Define white matter, gray matter, nerve, ganglion, spinal tract, nucleus, horn, and commissure.
> 11. Identify the principal sensory areas of the brain.
> 12. Identify the principal motor areas of the brain.
> 13. Identify the components of the brain stem.
> 14. Identify the structural features of the cerebrum.
> 15. Describe the anatomical features of the cerebellum.
> 16. Describe the formation and circulation of cerebrospinal fluid.
> 17. Identify the 12 pairs of cranial nerves by name, number, location and function.
> 18. Define the autonomic nervous system.
> 19. Compare the sympathetic and parasympathetic divisions of the autonomic nervous system in terms of structure.
> 20. Compare the structural and functional differences between the somatic and autonomic nervous systems.

BASIC SUBDIVISIONS OF THE NERVOUS SYSTEM

The *nervous system* (fig. 6-1) is divided into *peripheral* and *central*. The *peripheral nervous system* includes 31 pairs of spinal nerves, 12 pairs of cranial nerves, and all ganglia and plexuses associated with the *autonomic innervation*.

The *central nervous system* includes the *cerebrum, brain stem, cerebellum,* and *spinal cord*. The *brain stem* is further separated into the *diencephalon, mesencephalon, pons,* and *medulla oblongata.*

CENTRAL NERVOUS SYSTEM

External Aspect of the Cerebral Hemispheres

The overall convolutional pattern of *gyri* in human brains of normal dimensions is basically the same (fig.

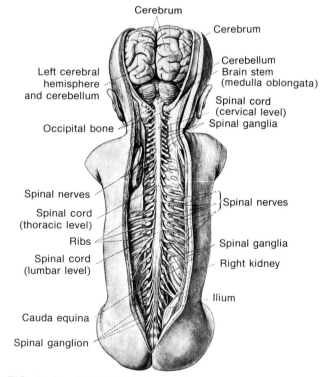

FIG. 6-1A. *Central nervous system*

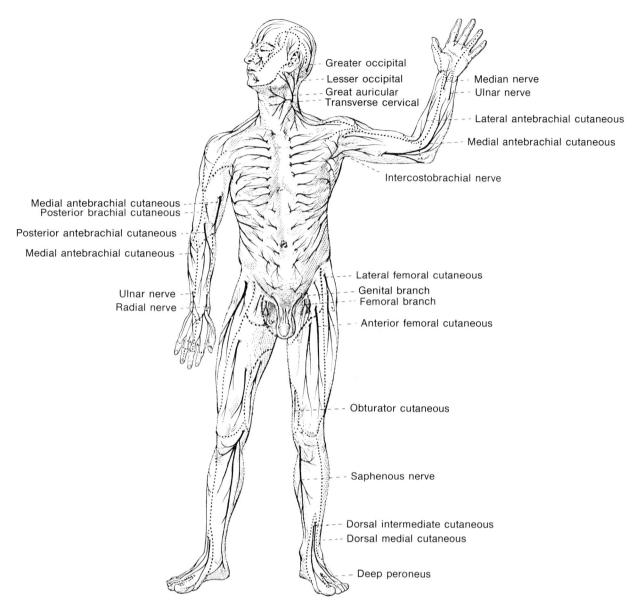

FIG. 6-1B. *Peripheral nervous system (anterior view)*

6-2). The deep furrows are referred to as *fissures* and those less pronounced are called *sulci*. The brain is subdivided into five lobes by means of the *fissures* and *sulci: frontal, parietal, occipital, temporal,* and *insular*. Histological and functional differences exist in each of the lobes.

The *longitudinal fissure* separates the cerebrum into right and left hemispheres. The corpus callosum is located at the bottom of the longitudinal fissure and connects the two cerebral hemispheres. Each hemisphere has a *dorsal, lateral, medial,* and *ventral surface*.

LATERAL SURFACE OF THE CEREBRAL HEMISPHERES The *central sulcus* (fissure of Rolando) and the *lateral fissure* (Sylvian) are prominent boundaries on the lateral aspect of the cerebral hemispheres (fig. 6-3). The *central sulcus* extends obliquely across the cerebral hemisphere to provide a convenient boundary between the frontal and parietal lobes. The *lateral fissure* separates the frontal and temporal lobes of the cerebral hemisphere.

The *frontal lobe* is situated anterior to the *central sulcus* and dorsal to the *lateral fissure*, which is subdivided into *precentral, superior, middle,* and *inferior*

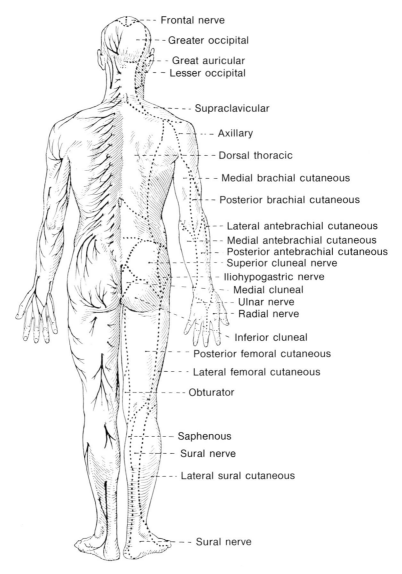

FIG. 6-1C. *Peripheral nervous system (posterior view)*

gyri. The *precentral gyrus* lies between the precentral sulcus and central sulcus. The remainder of the frontal lobe is composed of three gyri, the *superior, middle,* and *inferior frontal gyri.* The inferior frontal gyrus may be subdivided into *orbital, triangular,* and *opercular* parts.

The *parietal lobe* is located posterior to the central sulcus and is subdivided into *postcentral, superior,* and *inferior gyri.* The *postcentral sulcus* lies adjacent to the central sulcus with the postcentral gyrus lying between the two sulci. The superior and inferior parietal gyri are separated by the *interparietal sulcus.* The lateral fissure extends posteriorly to the *supramarginal gyrus,* which is part of the inferior parietal gyrus.

The lateral aspect of the *temporal lobe* presents two sulci that run somewhat parallel to the lateral fissure. The *superior temporal sulcus* begins at the temporal pole and extends posteriorly into the parietal lobe, where it connects with the *angular gyrus.* The *middle temporal sulcus* is ventral and somewhat parallel to the preceding sulcus. The *inferior temporal sulcus* is located on the ventral aspect of the brain. The *superior temporal gyrus* lies between the lateral fissure and the superior temporal sulcus. The *middle temporal gyrus* is situated between the superior and middle temporal sulci. The *inferior temporal gyrus* is located between the middle and inferior temporal sulci.

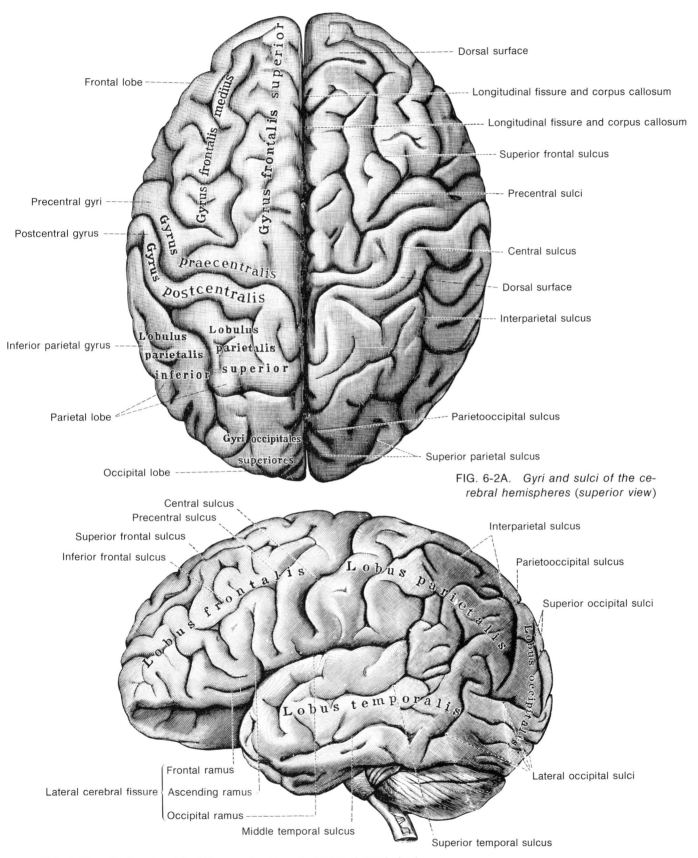

FIG. 6-2A. *Gyri and sulci of the cerebral hemispheres (superior view)*

FIG. 6-2B. *Gyri and sulci of the cerebral hemispheres (lateral view)*

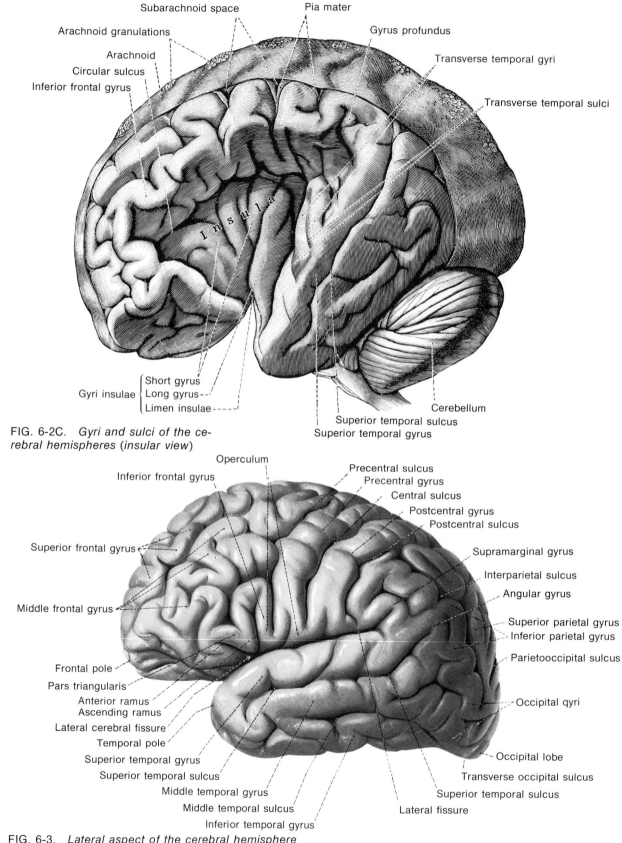

FIG. 6-2C. *Gyri and sulci of the cerebral hemispheres (insular view)*

FIG. 6-3. *Lateral aspect of the cerebral hemisphere*

The *occipital lobe* is a triangular area bounded by an imaginary line joining the calcarine fissure. The *insula* (Island of Reil) is a somewhat triangular, convex prominence that is concealed when the lateral cerebral fissure is closed.

VENTRAL SURFACE OF THE CEREBRAL HEMISPHERES The ventral surface of the cerebral hemispheres (fig. 6-4) is divided into several areas. The *orbital surface* of the frontal lobe rests upon the floor of the anterior cranial fossa. The *olfactory sulcus* contains the olfactory tract and the olfactory bulb. The *gyrus rectus* is located on the medial aspect of the olfactory sulcus. The inferior surface of the temporal and occipital lobes rests upon the tentorium cerebelli and the floor of the middle cranial fossa.

The *inferior temporal gyrus* is situated between the *middle* and *inferior temporal sulci*. The *fusiform gyrus* is located between the *inferior temporal sulcus* and the *collateral fissure*. The *parahippocampal gyrus* lies on the medial aspect of the *fusiform gyrus*. The *uncus* is the thickened anterior end of the hippocampal gyrus, which bends medially around the *hippocampal fissure*. The *rhinal fissure* separates the anterior portion of the fusiform gyrus from the hippocampal gyrus.

The *olfactory tract* passes caudally into the anterior border of the *anterior perforated substance*. Here the olfactory tract is enlarged to form a triangular area known as the *olfactory trigone*. The base of the olfactory trigone diverges to form the *medial* and *lateral olfactory striae*. The medial olfactory stria terminates in the subcallosal gyrus and the lateral olfactory stria extends laterally to the uncus and parahippocampal gyrus.

The *optic nerves* join posterior to the anterior perforated substance to form the *optic chiasma*, which continues caudally as the *optic tracts*. The *tuber cinereum* is located posterior to the optic chiasma and appears as a funnel-shaped structure that is continuous with the 3rd ventricle. The pituitary floor is connected with the floor of the hypothalamus by way of the *infundibulum*. The *mammillary bodies* are situated posterior to the tuber cinereum.

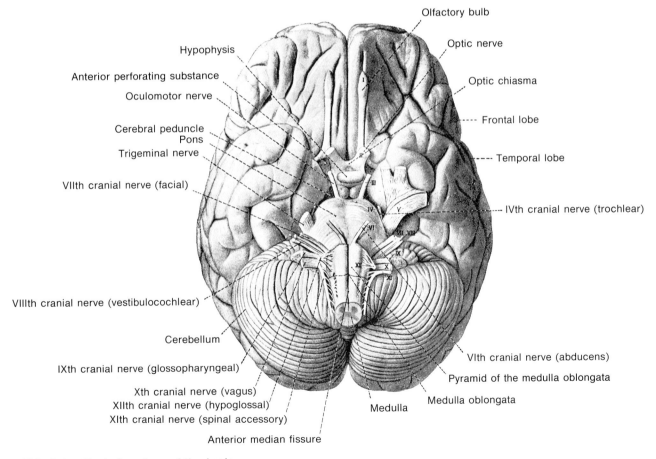

FIG. 6-4. *Ventral surface of the brain*

MEDIAL SURFACE OF THE CEREBRAL HEMISPHERES   The *corpus callosum* is the most convenient structure for orientation in the study of the medial surface of the cerebral hemisphere (fig. 6-5). Anteriorly, the corpus callosum turns inferiorly and then posteriorly to form the *genu* of the corpus callosum. The *rostrum* of the corpus callosum is a thin area, somewhat inferior to the genu, that extends downward as the *lamina terminalis*. Posteriorly, the corpus callosum is folded on itself to form the *splenium*. The *sulcus of the corpus callosum* separates the corpus callosum from the overlying *cingulate gyrus*. The *cingulate sulcus* parallels the course of the sulcus of the corpus callosum and separates the *superior frontal gyrus* from the cingulate gyrus.

The parietal and occipital lobes are separated by the *parietooccipital fissure*, which may be observed on the medial aspect of the cerebral hemisphere as a deep cleft which runs toward the splenium of the corpus callosum to join the *calcarine fissure*. The *cuneus* is the wedge-shaped area located between the parietooccipital and calcarine fissures. The *lingual gyrus* is situated inferior to the calcarine fissure.

The *cingulate* and *parahippocampal gyri* connected by an *isthmus* beneath the splenium of the corpus callosum form the *limbic lobe,* which is part of the *limbic system.* The *subcallosal gyrus* is located inferior to the rostrum of the corpus callosum. The subcallosal gyrus begins near the *anterior perforated substance* and continues as the *medial longitudinal stria.* The *paraolfactory area* is situated immediately anterior to the subcallosal gyrus. The *anterior commissure* is located in the lamina terminalis, which is anterior to the columns of the fornix.

The *septum pellucidum* is a thin membrane attached to the overlying corpus callosum, genu, and the underlying fornix. This membrane forms the medial wall of the lateral ventricles.

The inferior border of the septum pellucidum contains an arching band of longitudinal fibers known as the *fornix.* The columns of the fornix curve inferiorly, rostral to the *interventricular foramen* and disappear into the substance of the brain in the wall of the third ventricle.

The *thalamus* appears on the medial surface of the brain as a large mass of gray matter. The lateral limits of the third ventricle are formed by the two opposing contralateral thalami. Each medial surface of the thalamus shows an oval area, the *interthalamic adhesion,* which connects the two thalami.

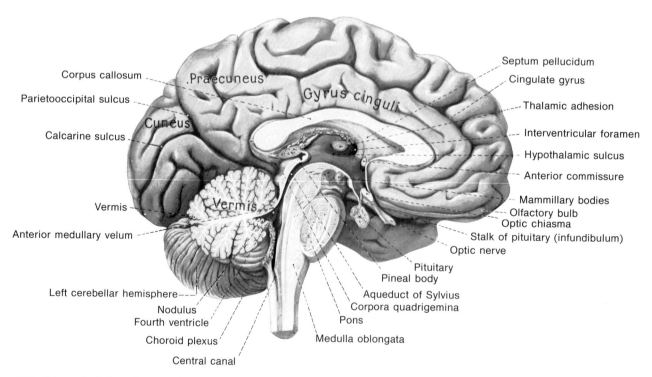

FIG. 6-5. *Medial surface of the brain*

The *hypothalamic groove* is a conspicuous groove, located below the interthalamic adhesions and separating the thalamus from the hypothalamus.

INTERNAL ASPECT OF THE CEREBRUM The prominent structures of the cerebrum (fig. 6-6) include the *caudate nucleus, internal capsule, lentiform nucleus, external capsule, claustrum, extreme capsule,* and *cerebral cortex of the insula.*

The arc structures of the cerebrum, *caudate nucleus, choroid plexus,* and *fornix,* parallel the ventricular system of the cerebral hemispheres.

The *lentiform nucleus* includes a lateral mass known as the *putamen* and a lighter-colored medial mass known as the *globus pallidus.* The lentiform nucleus sits between the *insula* and the *thalamus.* The medial aspect of the globus pallidus is closely associated with the *internal capsule.* The lateral surface of the lentiform nucleus is separated from the insula by the *external capsule, claustrum,* and *extreme capsule.* The *claustrum* is a thin plate of gray substance embedded in the white matter separating the *insula* from the *putamen.* The *external capsule* is a thin lamina of white matter separating the *claustrum* from the *putamen.* The *claustrum* is separated from the *cortex* of the *insula* by the *extreme capsule.* The *internal capsule* is a band of white substance that separates the *lentiform nucleus* laterally from the medially placed *thalamus* and *caudate nucleus.* The *caudate* and *lentiform nucleus* are separated by the *anterior limb of the internal capsule.* The *posterior limb of the internal capsule* separates the *lentiform nucleus* from the *thalamus.* The *genu* is located at the junction of the two limbs of the internal capsule. The *amygdala* is situated at the terminal part of the roof of the inferior horn of the lateral ventricle. It is continuous with the temporal lobe of the cortex.

The *ventricular system* (fig. 6-7) is lined with ependymal cells and contains cerebrospinal fluid. The *lateral ventricles* consist of *anterior, posterior,* and *inferior horns.* The anterior horns of the lateral ventricles lie inferior to the *corpus callosum.* The medial wall is formed by the *septum pellucidum,* which extends between the corpus callosum and the *fornix.* The head of the caudate nucleus protrudes into the ventricles from the ventrolateral side. The anterior horns of the lateral ventricle extend from the *interventricular foramen* (Monro's) to the splenium of the corpus callosum, where the ventricle divides into *posterior* and *inferior horns.* The *posterior horns* extend into the occipital lobe and taper to a point. The *inferior horns* curve ventrally and then rostrally into the temporal lobes. The *amygdaloid nucleus* extends

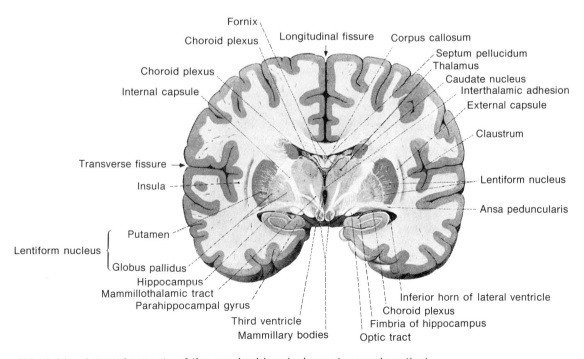

FIG. 6-6A. *Internal aspects of the cerebral hemispheres (coronal section)*

into the terminal portion of the inferior horn. The floor and medial wall of the inferior horns are formed by the *fimbria* and *hippocampus.*

The ventricles contain a *choroid plexus,* which is a rich network of blood vessels of the pia mater that is in contact with the ependymal lining of the ventricles. The *choroid plexus* of the anterior horn of the lateral ventricle continues into the third ventricle by way of the interventricular foramen. The third ventricle is continuous with the fourth ventricle by means of the *cerebral aqueduct.* The fourth ventricle joins the central canal of the spinal cord slightly rostral to the junction of the medulla and spinal cord.

The *head* of the *caudate nucleus* bulges into the anterior horn of the lateral ventricle, while the *tail* curves around to the roof of the inferior horn and extends rostrally as far as the *amygdaloid nucleus.*

Brain Stem

The *brain stem* (fig. 6-8) is separated into the *diencephalon, mesencephalon, pons,* and *medulla.*

DIENCEPHALON The *diencephalon* (fig. 6-9) includes: the (1) *epithalamus;* (2) *thalamus;* (3) *subthalamus;* and (4) *hypothalamus.* The dorsal and lateral surfaces of the diencephalon are covered by the cerebral hemispheres. Caudally, the medial portion of the diencephalon is continuous with the midbrain, while the lateral portion projects over the lateral part of the midbrain on each side. Each bilateral half of the diencephalon forms a lateral wall of the corresponding side of the third ventricle.

The *third ventricle* communicates with the lateral ventricles of the cerebral hemispheres by the *interventricular foramen.* The rostral limits of the third ventricle are formed by the *anterior commissure* and the *lamina terminalis.* The floor of the ventricle is formed by the *tuber cinereum,* which continues caudally as a funnel-like opening into the *infundibular stalk.* The ventricle extends caudally to the *posterior commissure,* below which it continues with the *cerebral aqueduct.* Above the *posterior commissure,* a small *pineal recess* of the ventricle projects into the base of the pineal body. The lateral ventricular wall above the *hypothalamic sulcus* is formed by the *thala-*

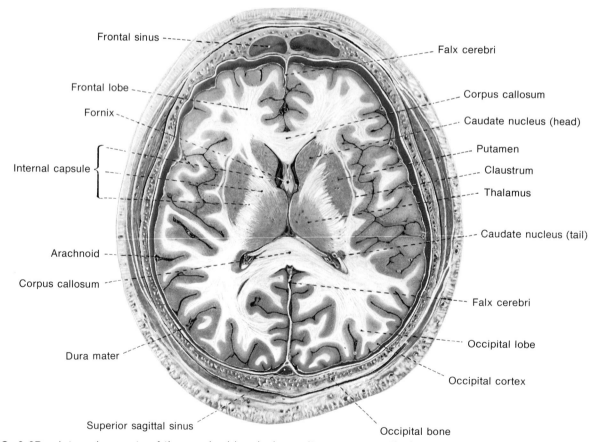

FIG. 6-6B. *Internal aspects of the cerebral hemispheres (transverse section)*

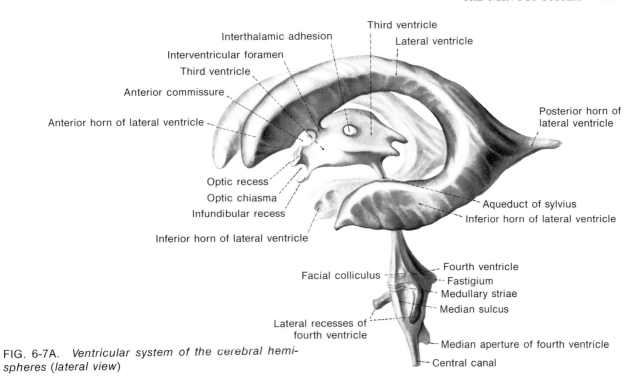

FIG. 6-7A. *Ventricular system of the cerebral hemispheres (lateral view)*

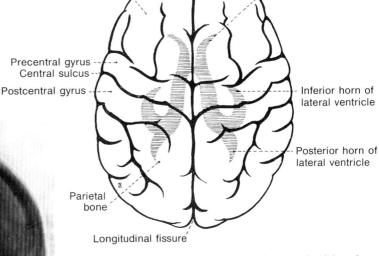

FIG. 6-7B. *Ventricular system of the cerebral hemispheres (diagrammatic projections of lateral ventricles on the area of the cerebral hemispheres)*

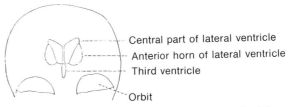

FIG. 6-7D. *Ventricular system of the cerebral hemispheres (diagrammatic outline of Fig. 6-7C.)*

FIG. 6-7C. *Ventricular system of the cerebral hemispheres (roentgenogram after injection of air into the ventricles)*

mus; below the sulcus the hypothalamus constitutes the wall on each side. The cavity of the third ventricle is frequently bridged by a bilateral medial extension of the gray substance of the thalami. These extensions fuse somewhat rostral to the middle portion of the ventricle and are variable in size as well as occurrence. The entire surface of the ventricle is covered with ependyma.

The *epithalamus* forms the upper and medial part of the diencephalon. It includes the *pineal body,* the *habenular trigone,* and part of the *posterior commissure.* The *pineal body* is an unpaired, cone-shaped structure that projects over the midbrain and lies in a groove between the superior colliculi.

The *habenular trigone,* as seen from above, is a depressed triangular area rostral to the superior colliculus on each side. It includes a fusiform mass of cells, the *habenular nuclei,* that are connected from one side to the other by the *habenular commissure.*

The *thalamus* is an oval mass of gray substance forming the largest subdivision of the diencephalon. Laterally, the thalamus rests against the occipital part of the internal capsule. Ventrally, it is continuous with the subthalamus and the hypothalamus. Caudally, the thalamus forms a somewhat angular prominence, the *pulvinar.* Beneath the pulvinar and separated from it by the brachium of the superior colliculus is an oval swelling, the *medial geniculate body,* which is the thalamic termination of the auditory pathway. On the lower surface of the lateral portion of the pulvinar another oval swelling constitutes the *lateral geniculate body,* in which the majority of the optic fibers terminate.

The subthalamus consists of a number of nuclei and fiber tracts that lie between the thalamus and the tegmentum of the midbrain.

The most ventral subdivision of the diencephalon constitutes the *hypothalamus.* Its total weight is only about 4 g, constituting approximately 0.3% of the total central nervous system weight. It forms most of the floor and the ventrolateral walls of the third ventricle and extends from the region just caudal to the mammillary bodies to the rostral level of the optic chiasma. Included in it are the *tuber cinereum, infundibulum,* and *mammillary bodies.* The pituitary is attached to the tuber cinereum by the infundibulum.

MESENCEPHALON The *mesencephalon,* or *midbrain* (fig. 6-10), is the portion of the brain stem that extends from the ventrocaudal portion of the diencephalon to the cephalic end of the pons.

The dorsal surface of the midbrain is formed by the *quadrigeminal plate* (tectum) which is the portion of the midbrain dorsal to the cerebral aqueduct. Externally, it shows four rounded elevations, the upper two constituting the *superior colliculi,* and the lower two constituting the *inferior colliculi.* The *brachium of the superior colliculus* extends rostrolaterally to the *lateral geniculate body* and the lateral root of the optic tract. The *brachium of the inferior colliculus* extends forward from the rostrolateral portion of the inferior colliculus to the *medial geniculate body.* Caudally, the quadrigeminal plate is continuous with the *anterior medullary velum* and the *superior cerebellar peduncles.* Immediately below the inferior colliculi, the *trochlear nerve* exits from the midbrain.

The ventral surface of the midbrain is formed by the *cerebral peduncles.* These are two bundles of fibers that lie close to each other at the rostral border of the pons, but which gradually diverge rostrally and dis-

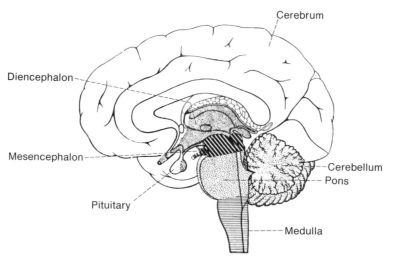

FIG. 6-8A. *Brain stem (orientation view)*

THE NERVOUS SYSTEM ▪ 133

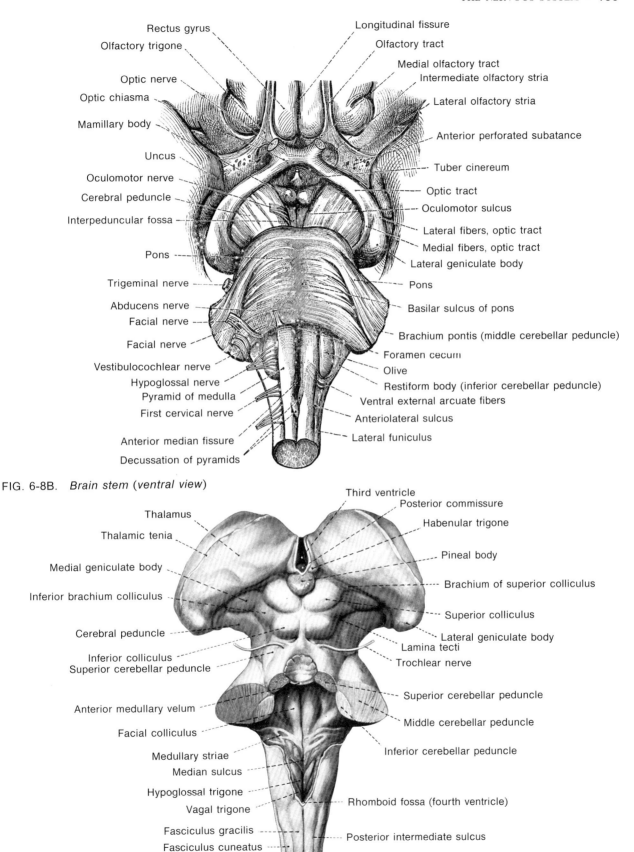

FIG. 6-8B. *Brain stem (ventral view)*

FIG. 6-8C. *Brain stem (dorsal view)*

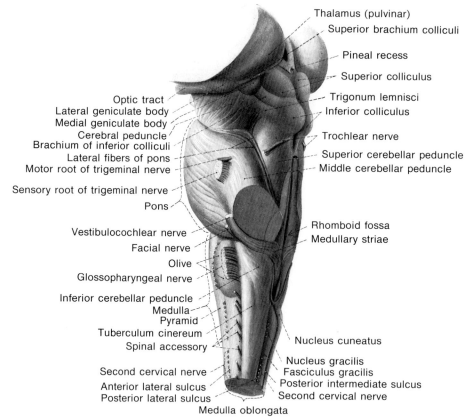

FIG. 6-8D. *Brain stem (lateral view)*

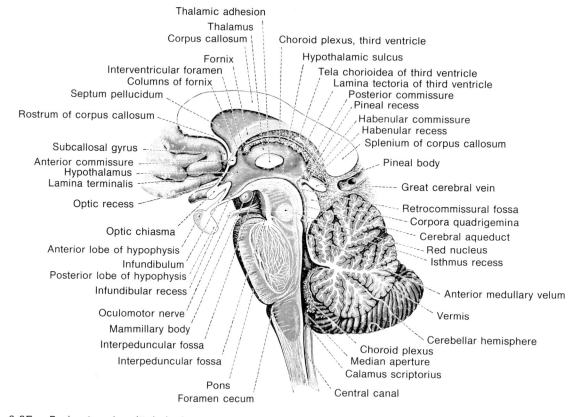

FIG. 6-8E. *Brain stem (sagittal view)*

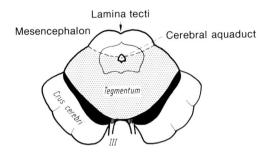

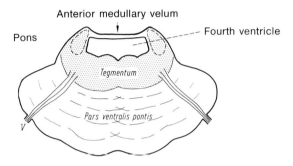

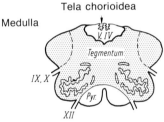

FIG. 6-8F. *Brain stem transverse sections (diagrammatic views)*

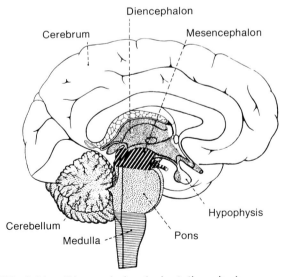

FIG. 6-9A. *Diencephalon (orientation view)*

appear beneath the *optic tract.* Their divergence results in the *interpeduncular fossa,* which lies between them. The *oculomotor nerves* leave the brain stem from the interpeduncular fossa on the medial side of the cerebral peduncle.

PONS   The *pons* (fig. 6-11) is a rounded structure located between the visible portion of the cerebral peduncle above and the medulla oblongata below. The pons is bilaterally symmetrical, with a *median sulcus* in which lies the *basilar artery.* The pons consists of two distinct parts, known as the *basal* and *tegmental portions.* The basal portion of the pons is distinctive of this part of the brain stem because fibers from the cerebral cortex terminate ipsilaterally and then cross to the opposite side to enter the middle cerebellar peduncle. The dorsal portion, or tegmentum, of the pons is similar to the medulla and midbrain, in that it contains ascending and descending tracts and nuclei of cranial nerves. The dorsal surface of the tegmentum contributes to the floor of the fourth ventricle. The pons merges laterally into the *middle cerebellar peduncle,* the attachment of the *trigeminal nerve* making the transition between the pons and the peduncle.

MEDULLA OBLONGATA   The *medulla oblongata* (fig. 6-12) is the caudal portion of the brain stem extending from the spinal cord to the inferior margin of the pons. It is continuous with the spinal cord at a transverse plane passing between the lower part of the decussation of the pyramids and the level of the highest rootlets of the first cervical nerve. The lower level of the medulla oblongata has the same diameter as the cord, but as it extends upward in a nearly vertical direction it expands and assumes an irregular conelike form. Between the lower and middle limits of the medulla, the central canal opens into the fourth ventricle. At this level the lateral margins of the fourth ventricle spread laterally, resulting in an open part of the medulla oblongata.

The *anterior median fissure* is the continuation of a fissure bearing the same name in the spinal cord. It is interrupted just above the *foramen magnum* by oblique strands of fibers that cross the median plane and form the *decussation of the pyramids.* Above the decussation the fissure continues to the inferior border of the pons, where it ends in the *foramen cecum.*

The *posterior median sulcus* is continuous with the posterior sulcus of the spinal cord. It becomes less obvious as it ascends, and at the tip of the fourth ventricle forms the lateral boundaries of the *obex.*

The *anterolateral sulcus* is a shallow furrow on the anterolateral surface of the medulla oblongata, continuous with the sulcus of the upper part of the cord

136 ■ BASIC ANATOMY FOR THE ALLIED HEALTH PROFESSIONS

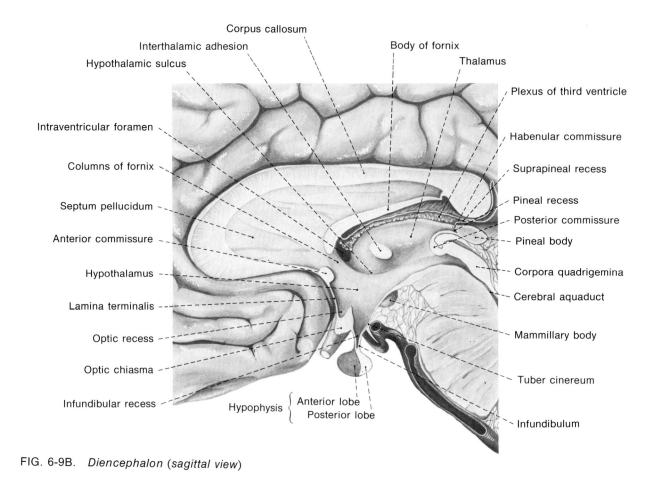

FIG. 6-9B. *Diencephalon (sagittal view)*

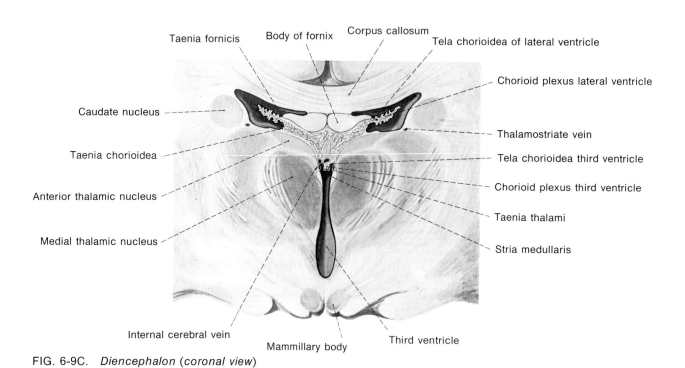

FIG. 6-9C. *Diencephalon (coronal view)*

THE NERVOUS SYSTEM ▪ 137

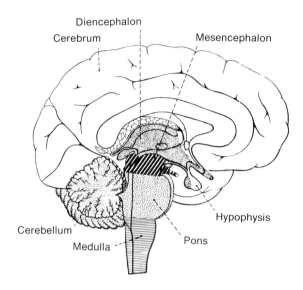

FIG. 6-10A. *Mesencephalon (orientation view)*

bearing the same name. It lies lateral to the pyramid and, at the levels where the olive is present, runs between the pyramid and the olive extending to the pons. Through it emerge the roots of the *hypoglossal nerve,* and in the same line at the inferior border of the pons emerge the roots of the *abducens nerve.*

The *pyramids* occupy the ventral region. They consist of longitudinally directed strands of nerve fibers that emerge from beneath the caudal margin of the pons and continue to the caudal end of the medulla oblongata, where they form the *decussation of the pyramids.*

The *glossopharyngeal, vagus,* and *accessory nerves* exit from the brain stem just dorsal to the olive.

In the closed part of the medulla oblongata the dorsal region consists of an upper continuation and expansion of the *fasciculus gracilis* and the *fasciculus cuneatus* of the spinal cord. The *fasciculus gracilis* gradually terminates in a triangular elevation, the *clava,* which is the superficial representation of the *nucleus gracilis.* The *fasciculus cuneatus,* beginning at a more rostral level, likewise terminates in a similar enlargement, the *cuneate tubercle:* the *tuberculum cinereum* is the rostral expansion of a zone that lies ventrolateral to the fasciculus cuneatus and is the surface expression of the descending root and nucleus of the trigeminal nerve.

The *inferior cerebellar peduncle* (restiform body) forms the greater part of the rostrodorsal region of the medulla oblongata on each side as it diverges laterally with the spreading of the fourth ventricle.

The *fourth ventricle* is rhomboid, being considerably widened at the level of the *middle cerebellar peduncle* and pointed at each end. The fourth ventricle is continuous with the central canal of the spinal cord

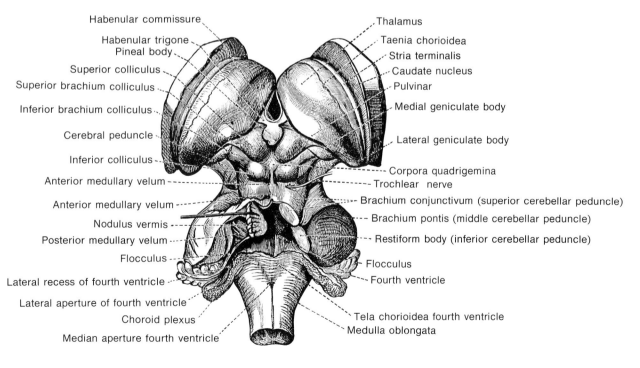

FIG. 6-10B. *Mesencephalon (dorsal view)*

138 ■ BASIC ANATOMY FOR THE ALLIED HEALTH PROFESSIONS

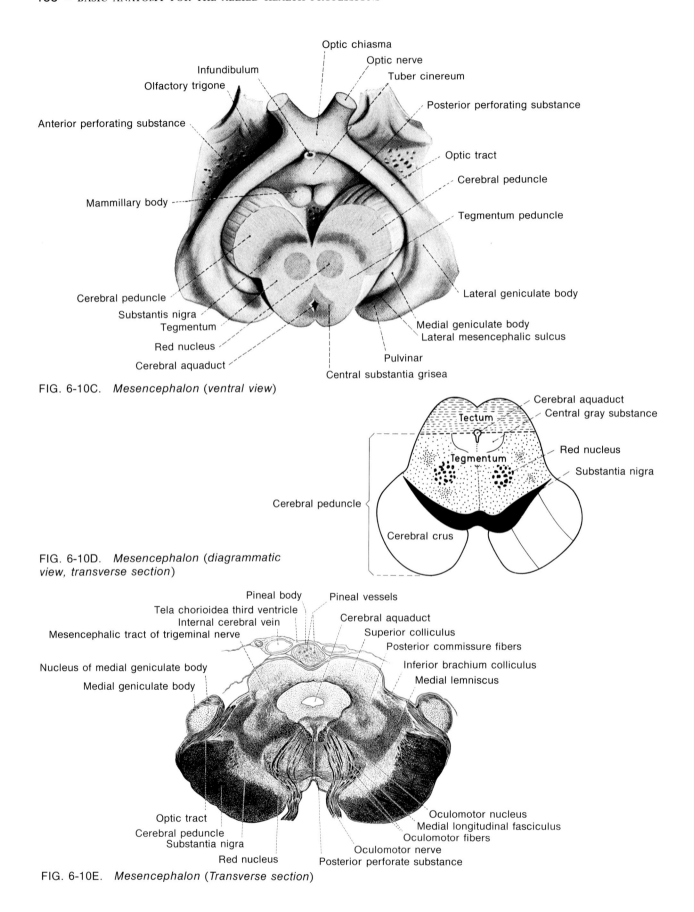

FIG. 6-10C. *Mesencephalon (ventral view)*

FIG. 6-10D. *Mesencephalon (diagrammatic view, transverse section)*

FIG. 6-10E. *Mesencephalon (Transverse section)*

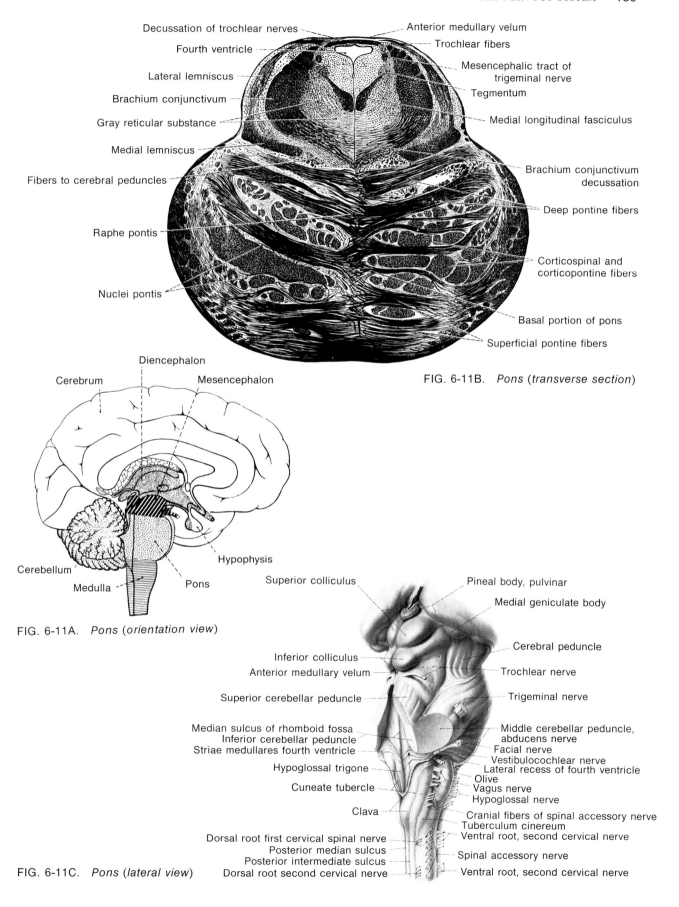

FIG. 6-11B. Pons (transverse section)

FIG. 6-11A. Pons (orientation view)

FIG. 6-11C. Pons (lateral view)

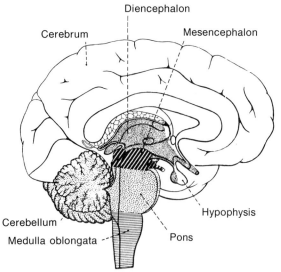

FIG. 6-12A. *Medulla oblongata (orientation view)*

below and with the cerebral aqueduct above. The entire cavity of the ventricle is lined with ependyma.

The roof of the rostral portion of the ventricle consists of a thin layer of white substance that is the *anterior medullary velum*. The roof of the inferior portion of the ventricle consists of a layer of ependymal epithelium that is covered with choroid tela of the fourth ventricle. The ventricular cavity extends laterally on each side into a *lateral recess*. At the lateral tip of each recess there is an opening, the *foramen of Luschka*. In the midplane of the lower part of the choroid tela there is a more or less well-marked opening, the *foramen of Magendie*.

The *choroid plexuses of the fourth ventricle* consist of highly vascular, lobular, villuslike processes formed of epithelium and the pia mater of the choroid tela. The floor of the fourth ventricle is a depressed, rhomboid area of grayish color that is divided bilaterally by a well-marked *median sulcus*. The lower part of the fossa is somewhat triangular and, at its lower extremity, opens into the central canal of the spinal cord. The inferior apex of the ventricle is known as the *calamus scriptorius*.

On each side of the median sulcus throughout the length of the rhomboid fossa, a continuous ridge, the *medial eminence*, is bounded laterally by the *limiting sulcus*. In the lower part of the rhomboid fossa the medial eminence forms a triangular area, the *hypoglossal trigone*.

Cerebellum

The surface of the *cerebellum* (fig. 6-13) is characterized by numerous folia separated by narrow furrows, the sulci, that show variation in depth. As was observed with the cerebral hemispheres, a few of the furrows are much broader and deeper than the majority and are known as fissures. These *fissures* (*primary* and *horizontal*) divide the cerebellum into lobes, of which folia are surface folds. The folia are more or less parallel in the various lobes. The three pairs of peduncles attaching the cerebellum to the brain stem consist of fibers to and from the organ. The *inferior cerebellar peduncle* (restiform body) connects the cerebellum with the medulla oblongata. The

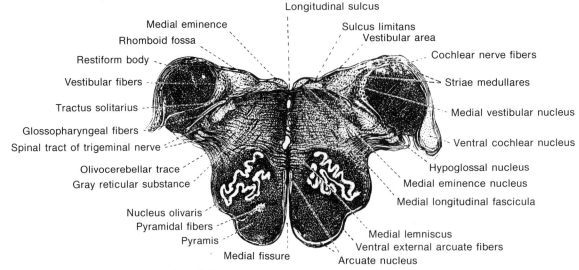

FIG. 6-12B. *Medulla oblongata (dorsal view)*

middle cerebellar peduncle (brachium pontis) connects the pons with the cerebellum. The *superior cerebellar peduncle* (brachium conjunctivum) connects the cerebellum with the midbrain and the diencephalon.

The cerebellum shows a large lateral hemisphere on each side and a narrow medial portion, the *vermis.* On the rostral surface there is no sharp boundary between the hemispheres and the vermis.

The *vermis* is subdivided into a number of lobules best seen in the sagittal section of the cerebellum. Beginning rostrally, these are the *lingula, central lobule, culmen, declive, folium, tuber, pyramid, uvula,* and *nodulus.*

External Aspect of the Spinal Cord

The *spinal cord* (fig. 6-14) occupies the upper two-thirds of the vertebral canal. It extends from the level of the foramen magnum to the upper border of the second lumbar vertebra. The distal end, *conus medullaris,* tapers to a point. The spinal cord is subdivided into four regions, as follows: (1) *cervical,* with 8 pairs of *cervical nerves;* (2) *thoracic,* with 12 pairs of *thoracic nerves;* (3) *lumbar,* with 5 pairs of *lumbar nerves;* and (4) *sacral,* with 5 pairs of *sacral nerves* and 1 pair of *coccygeal nerves.* Each spinal nerve has a *dorsal root* that contains afferent fibers, and a *ventral root* consisting of efferent fibers. Each dorsal root is characterized by an oval swelling, the *dorsal root ganglion,* which contains cell bodies of the incoming sensory fibers. The dorsal root fibers enter the spinal cord along the dorsolateral surface and the ventral root fibers emerge from the ventrolateral aspect. Each nerve root subdivides into numerous filaments as it approaches the cord. These filaments are called the *fila radicularia.*

The position of the caudal end of the spinal cord varies in adults, from the level of the 12th thoracic vertebra to the upper border of the 2nd lumbar. Most spinal cord segments do not correspond to their overlying vertebrae because of the difference in growth rate between the spinal cord and the vertebral column. This discrepancy increases as one proceeds from cervical to lumbar regions. The spinous process of the third thoracic vertebra, for example, lies over the fourth and fifth thoracic segments of the cord; the spinous processes of the 11th and 12th thoracic and the 1st lumbar vertebrae overlie all the lumbar and sacral segments of the cord. The dorsal and ventral nerve roots of the two lower cord segments must travel for some distance before they reach the intervertebral foramen through which they exit. This bundle of long nerve roots, called the *cauda equina,* occupies that part of the vertebral canal formed by the last four lumbar vertebrae and the entire sacrum.

The spinal cord has an average length of 45 cm in the adult male and is about 43 cm in length in the female.

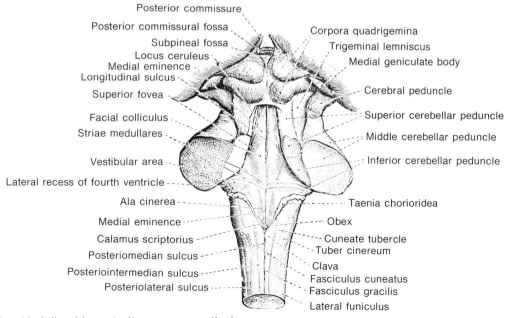

FIG. 6-12C. *Medulla oblongata (transverse section)*

142 ■ BASIC ANATOMY FOR THE ALLIED HEALTH PROFESSIONS

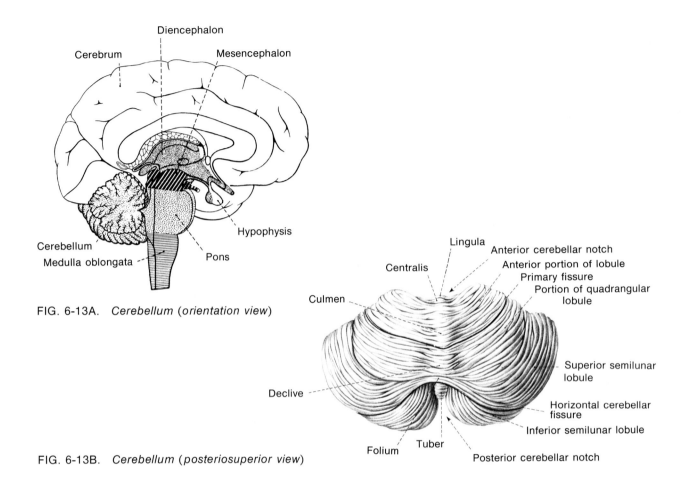

FIG. 6-13A. *Cerebellum (orientation view)*

FIG. 6-13B. *Cerebellum (posteriosuperior view)*

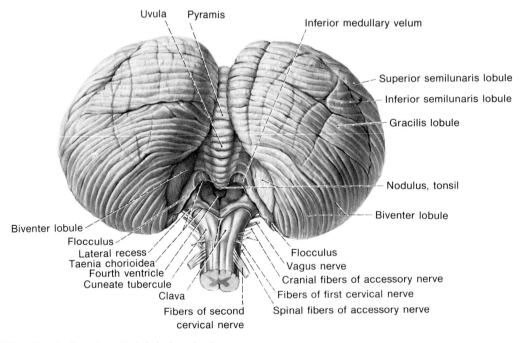

FIG. 6-13C. *Cerebellum (posteroinferior view)*

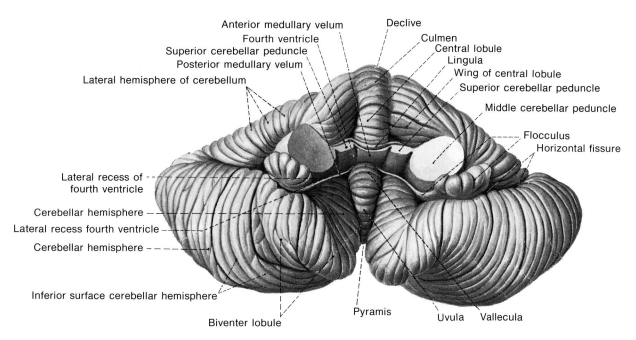

FIG. 6-13D. *Cerebellum [anterior (peduncular) view]*

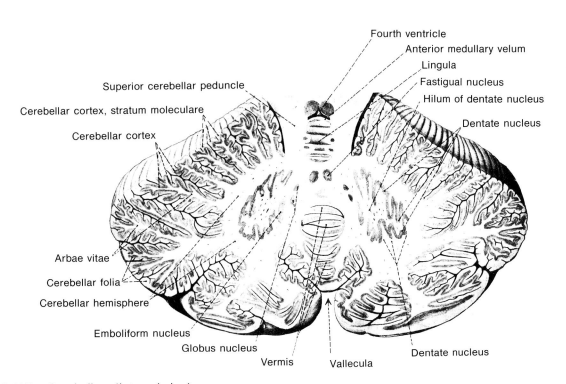

FIG. 6-13E. *Cerebellum (internal view)*

144 ■ BASIC ANATOMY FOR THE ALLIED HEALTH PROFESSIONS

The lower cervical and the lumbar portions of the cord are broadened into the *cervical* and *lumbar enlargements*. The spinal nerves attached to these portions of the cord, serving the skin and musculature of the upper and the lower limbs, are larger than the remaining spinal nerves.

The cervical enlargement begins at the level of the third cervical vertebra and extends to the second thoracic vertebra. It is widest opposite the lower part of the fifth cervical vertebra, which corresponds to the attachment of the sixth pair of cervical nerves.

The *lumbar enlargement* begins at the level of the

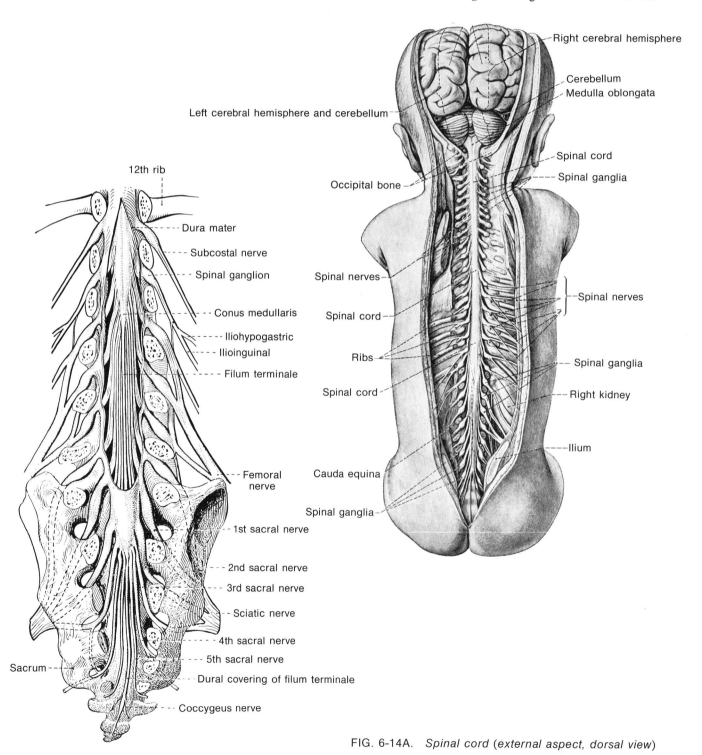

FIG. 6-14A. *Spinal cord (external aspect, dorsal view)*

9th or 10th thoracic vertebra and attains its greatest transverse diameter at the level of the 12th thoracic vertebra, corresponding to the attachment of the 4th lumbar nerve. Below this level it rapidly diminishes into the *conus medullaris.*

The cord is divided into symmetrical right and left halves by the *anterior median fissure* and the *posterior median sulcus.* The anterior median fissure is deep and relatively wide. It contains blood vessels and pia mater. The posterior median sulcus is shallow, forming along the lower two-thirds of the cord little more than a line that marks the extension to the surface of the posterior median septum, which divides the two halves of the dorsal part of the cord internally. Each lateral half of the cord is subdivided into posterior, lateral, and anterior divisions called *funiculi,* by the posterolateral and anterolateral sulci. The *posterolateral sulcus* forms a small groove lateral to the posterior median sulcus. It delimits the posterior from the lateral funiculus. Along this sulcus the dorsal root fibers enter the cord as small bundles in a linear series. The anterolateral sulcus forms an irregular, linear area rather than a groove. This constitutes the surface boundary of the lateral funiculus from the anterior funiculus. It is the zone along which the ventral root fibers emerge from the cord to be assembled into the ventral nerve root. In the cervical and upper thoracic portions of the cord, the external surface of the posterior funiculus is subdivided into a medial and a lateral area by a small longitudinal groove, the *posterointermediate sulcus.* These areas represent the surfaces of the *fasciculus gracilis* and the *fasciculus cuneatus,* respectively.

The posterior, lateral, and anterior funiculi are composed of longitudinally directed fascicles of myelinated fibers. The funiculi constitute its white substance, which forms an external white collar for the cord.

Internal Structure of the Spinal Cord

The *gray substance* consists of a continuous column extending the entire length of the cord. The *white substance* surrounds the column of gray substance, but, because of the uneven surface of the gray column, as seen in the cross-section of the cord, the layer of white substance varies in thickness (fig. 6-15).

The volume of white and of gray substance varies, not only in total amount, but also relatively at different levels of the cord. The total amount, both of white and of gray substance, increases at the lumbar and cervical enlargements. In the cord as a whole the rela-

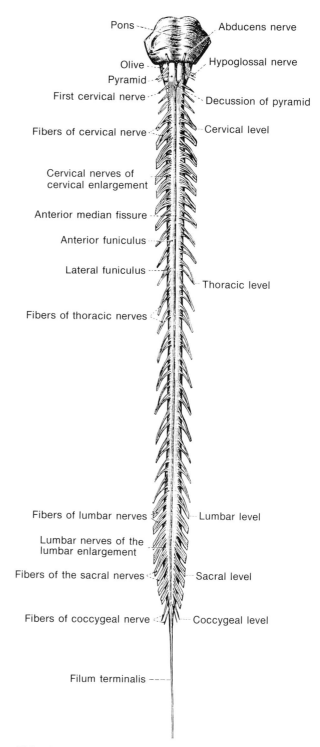

FIG. 6-14B. *Spinal cord (external aspect, ventral view)*

tive amount of white substance gradually increases from the conus medullaris to the medulla oblongata.

The *central canal* in the adult cord extends throughout its length and upward into the lower portion of the medulla oblongata, opening into the fourth ventricle. The immediate wall of the central canal is formed of tall, ependymal cells. Outside the layer of ependyma the canal is surrounded by gray substance.

The column of gray substance is made up of two symmetric masses, one in each lateral half of the cord, joined across the midplane by a narrow band of gray substance, in the midst of which the central canal is situated. Each lateral mass, as seen in transverse section, has a concentric outline. The column of the crescent extending dorsally and somewhat laterally toward the surface of the cord is known as the *dorsal* or *posterior gray column;* that extending ventrally is the *ventral* or *anterior gray column.* The transverse band across the midplane is known as the *gray commissure.* The outline of the gray substance, as a whole, in transverse section roughly resembles the letter H. The central canal divides the gray commissure into an *anterior gray commissure,* lying ventral to the canal, and a *posterior gray commissure,* lying dorsal to it. The substance of the gray commissure around the central canal is gelatinous in appearance and is known as the *central gelatinous substance.*

The white substance of the spinal cord consists of fibers referred to as the fasciculi, which are subdivisions of a funiculus. Just anterior to the anterior gray commissure is a bundle of transverse fibers, the *anterior white commissure,* composed of crossing fibers from various nerve cells.

Spinal Nerves

From the sides of the cord 31 pairs of nerves are attached: *8 cervical, 12 thoracic, 5 lumbar, 5 sacral,* and *1 coccygeal* (fig. 6-16). The cord is accordingly subdivided into cervical, thoracic, lumbar, and sacral regions. This division is based on the relationship of parts of the cord to the issuing spinal nerves, and not on their correspondence to vertebrae. Each spinal root is attached to the cord by a ventral and a dorsal root.

The Meninges and Cerebrospinal Fluid

The brain and spinal cord are protected by three membranous coverings known as the *meninges* (fig. 6-17). The most external layer is known as the *dura mater.* A thin, cobweblike middle layer constitutes the *arachnoid.* The most internal layer of connective tissue fibers is the *pia mater.* The pia mater and arachnoid are often regarded as a single structure, the *leptomeninx* or *leptomeninges.*

The spinal dura mater is separated from the periosteum by the narrow *epidural space,* in which are found anastomosing venous channels lying in loose connective tissue. Between the dura and the arachnoid is the capillary *subdural space,* which has no direct communication with the *subarachnoid space.* Spinal dura is attached to the outer surface of the arachnoid by threadlike trabeculae.

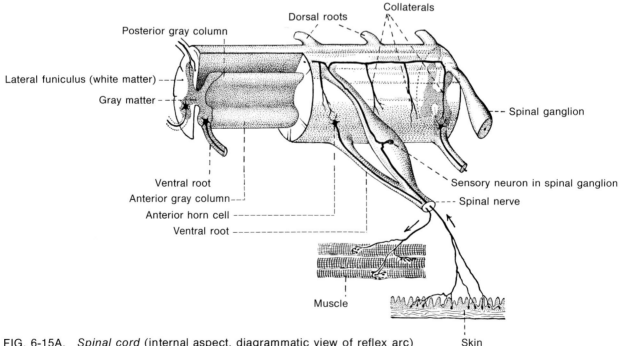

FIG. 6-15A. *Spinal cord* (internal aspect, diagrammatic view of reflex arc)

The spinal dura extends as a closed, tough sac from the margins of the foramen magnum above to the level of the second sacral vertebra below. At the level of the second sacral vertebra, the dura forms an investing sheath around the filum terminale to form a thin fibrous cord, the *coccygeal ligament.* The spinal cord terminates at the upper border of the second lumbar vertebra, and the lower portion of the dura is occupied by the filum terminale and cauda equina.

The cerebral dura contains both an investing sheath for the brain and periosteum for the inner layer of the cranial vault. It consists of two layers, an inner layer and an outer layer, which form the periosteum. Between the two layers are found the large

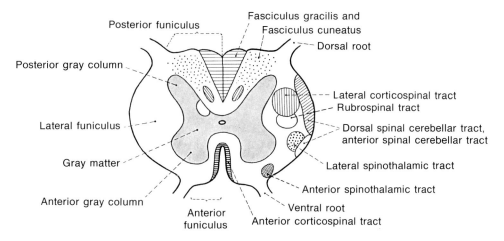

FIG. 6-15B. *Spinal cord (internal aspect, diagrammatic view of sensory and motor pathways)*

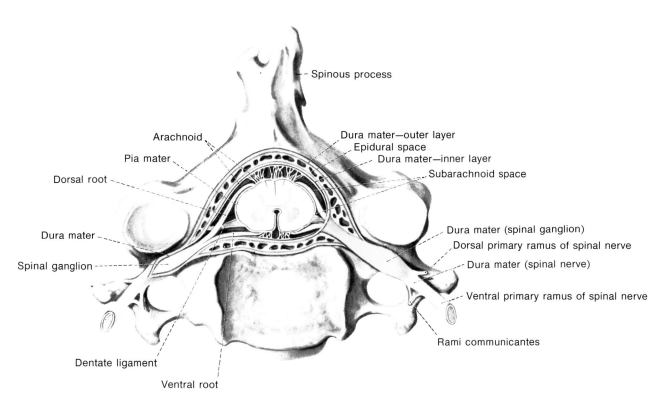

FIG. 6-15C. *Spinal cord (transverse section of vertebral column, spinal cord and its covering)*

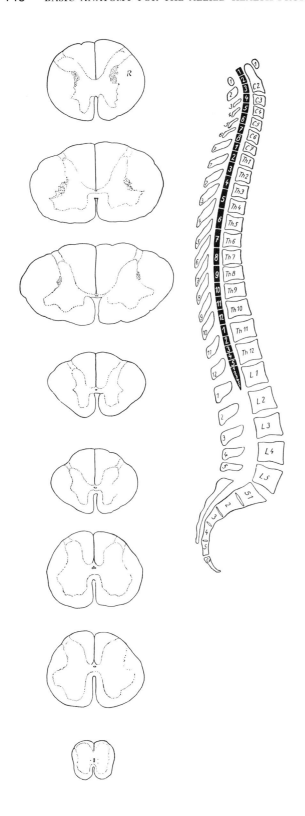

FIG. 6-15D. *Spinal cord (transverse sections at various levels of spinal cord)*

venous sinuses of the brain. Cerebral dura forms septa that divide the cranial cavity into incomplete compartments. The *falx cerebri* is a sickle-shaped septum extending from the crista gala to the internal occipital protuberance, which separates the two cerebral hemispheres. The *tentorium cerebelli* is a transverse, arched septum separating the occipital lobes from the cerebellum. The free border of the tentorium cerebelli forms the *tentorial incisure,* through which the brain stem passes.

The arachnoid is partly separated from the pia by fluid spaces and by trabeculae that pass from pia to arachnoid, which constitute the *subarachnoid space.* The space is filled with cerebrospinal fluid and is in direct communication with the fourth ventricle of the brain by three apertures. A median aperture known as the *foramen of Magendie* is located in the caudal part of the thin ventricular roof. Two lateral apertures, known as the *foraminae of Luschka,* open into the pontine subarachnoid cisterns.

In its transition to the spinal cord at the base of the brain, the arachnoid becomes widely separated from the pia in certain places, giving rise to the *subarachnoid cisterns.* The posterior aspect of the medulla oblongata is surrounded by a large subarachnoid space commonly referred to as the cisterna magna.

The inner layer of the dura is sometimes penetrated by cerebral pia-arachnoid prolongations that protrude into a venous sinus. Such granulations are commonly known as *arachnoid granulations* (Pacchionian bodies). These granulations are the major sites of fluid transfer from the subarachnoid space to the venous system.

The *pia mater* adheres to the underlying nervous tissue and sends fibrous septa into the spinal cord. At the point where blood vessels enter or leave the central nervous system, the pia mater is invaginated to form the outer wall of a *perivascular space.* The spinal cord is attached to the dura mater by two lateral series of flattened bands of pia mater called the *denticulate ligaments.* Each ligament is attached medially to the lateral aspects of the cord, midway between the dorsal and ventral roots along the entire length of the spinal cord from the medulla to the conus medullaris. There may be as many as 24 of the dentate ligaments anchoring the spinal cord to the dura. The pia continues caudally as the *filum terminale.*

The cerebrospinal fluid is a clear, colorless fluid that provides a fluid pathway for chemical substances to reach the intercellular spaces of the brain, and for neural metabolites to be returned to the venous system. Cerebrospinal fluid is drained from the lateral and third ventricles by the cerebral aqueduct, which

THE NERVOUS SYSTEM ▪ 149

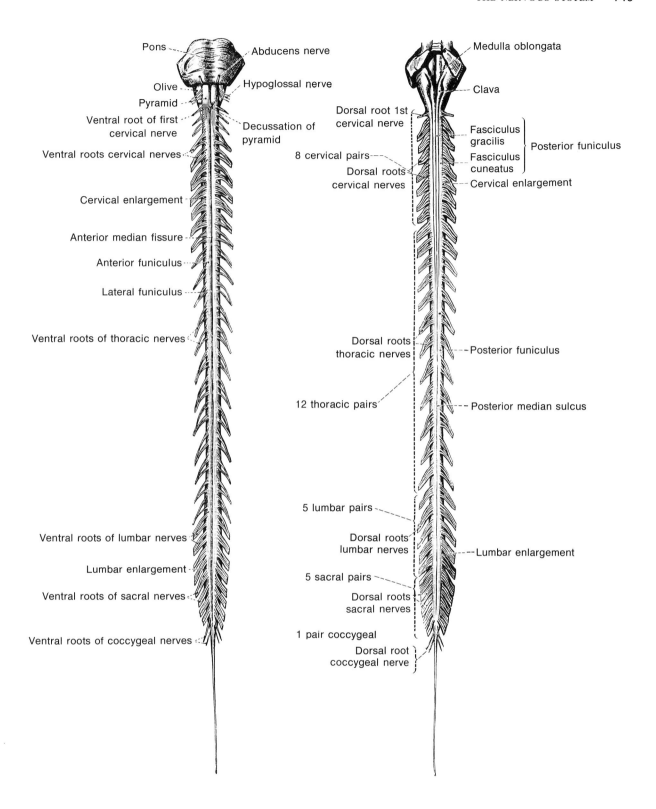

FIG. 6-16A. *Spinal nerves (spinal cord segments)*

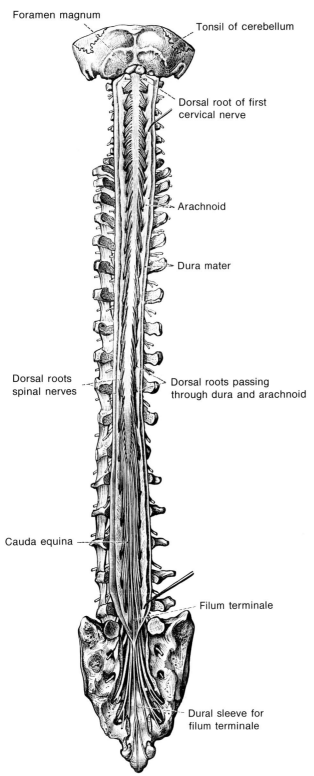

FIG. 6-16B. *Spinal nerves (dorsal view of vertebral canal, spinal cord and spinal nerves)*

opens into the fourth ventricle. Through the medial and lateral apertures of the fourth ventricle the fluid circulates into the cisterns and subarachnoid spaces surrounding the brain and spinal cord. The bulk of the cerebrospinal fluid is then returned to the venous system through the arachnoid granulations.

Blood Supply of the Central Nervous System

BLOOD SUPPLY OF THE SPINAL CORD The *blood supply of the spinal cord* (fig. 6-18) is derived from two principal sources: (1) the *vertebral arteries* and (2) the *radicular arteries,* derived from segmental vessels (deep cervical, intercostal, lumbar, and sacral arteries). Most of the cervical spinal cord is supplied by small branches of the two vertebral arteries. Radicular arteries course along the spinal nerves, pass through the intervertebral foramina and divide into small *anterior* and large *posterior radicular arteries.*

The vertebral arteries give off two arteries as they ascend along the anterolateral surface of the medulla. The first, or most inferior branch, the *posterior spinal artery,* turns dorsally and descends as a discrete vessel on the posterior surface of the spinal cord. As the two posterior spinal arteries descend, they receive a variable number of contributions from the posterior radicular arteries. The second, or more superior branch, is the anterior spinal descending *anterior spinal artery.*

The anterior spinal artery gives off a number of *sulcal* branches, which enter the anterior median fissure of the cord and pass alternately to the right and left. The anterior sulcal arteries are most numerous in the lumbar region and least numerous in the thoracic region, where the segmental blood supply is poorest.

Through its sulcal branches, the anterior spinal artery supplies the anterior gray column, lateral gray columns, the central gray column, and the base of the posterior gray columns. It also supplies the anterior and lateral funiculi, including the lateral pyramidal tract. The posterior spinal arteries supply the posterior gray column and the posterior funiculus.

The *venous distribution* is similar to that of the arteries. The posterior radicular veins form a more or less distinct posterior median spinal vein or trunk along the entire extent of the cord. Smaller paired posterolateral trunks are also formed from the anterior radicular veins.

The posterior radicular veins drain the posterior funiculus, the posterior gray columns (including their base), and a narrow strip of lateral funiculus immediately adjacent to the posterior gray columns. The an-

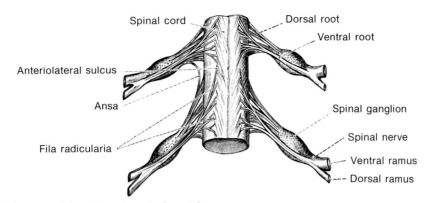

FIG. 6-16C. *Spinal nerves (dorsal and ventral roots)*

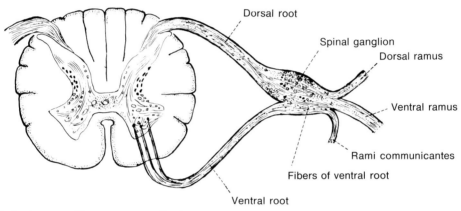

FIG. 6-16D. *Spinal nerves (transverse section)*

terior spinal vein, through the sulcal vessels, drains the sulcal marginal white matter and the medial portion of the anterior gray column.

ARTERIAL BLOOD SUPPLY OF THE BRAIN Four main large arterial trunks carry the bulk of the blood to the brain (fig. 6-19). They are the two *internal carotid arteries* and the two *vertebral arteries*.

Each *carotid artery* divides into a larger branch, the *middle cerebral artery,* which may be traced into the deep cleft of the lateral fissure; a smaller branch, the *anterior cerebral artery,* which runs medially and then is lost from sight in the ventral extension of the dorsal longitudinal fissure; and a short trunk, the *posterior communicating artery,* which is directed backward along the inner border of the cerebral peduncle to join the posterior cerebral artery (a branch of the basilar artery).

The *vertebral arteries,* as they enter the inner cranial cavity through the foramen magnum, unite at the superior border of the medulla, forming the *basilar artery.* The *basilar artery* continues forward in the basilar groove and terminates at the upper border of the pons by dividing into two large terminal branches, the *posterior cerebral arteries*. Each posterior cerebral artery, a short distance from its origin, receives an anastomotic branch from the internal carotid, the posterior communicating artery. The anterior communicating artery joins the two anterior cerebral arteries in the depth of the anterior extension of the longitudinal fissure. This circular anastomosis at the base of the brain between the branches of the vertebral arteries and those of the internal carotid is known as the *circulus arteriosus* (the circle of Willis).

Anteriorly, the *anterior communicating arteries* bring into continuity two *anterior cerebral arteries.* The vessels that enter the formation of this circle give rise to a large series of branches that are distributed to deep and superficial structures of the brain. On the basis of this distribution the arterial branches are generally considered as constituting two systems of vessels: the *cortical* or *superficial* and the *central* or *deep.* The cortical vessels ramify in the cerebral cortex

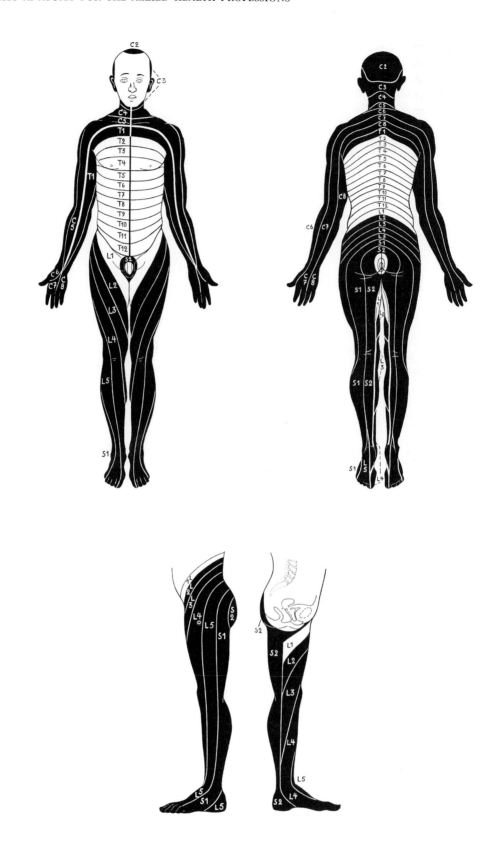

FIG. 6-16E. *Spinal nerves (dermatome chart for cutaneous innervation)*

and adjacent subcortex at various depths. The central or slender twigs, which generally come off in crowded groups, pierce the substance of the brain stem to gain the deep-seated structures.

Vertebral Artery and Its Branches The *vertebral artery* is a branch of the subclavian. It pierces the dura mater and arachnoid of the spinal cord in the upper part of the vertebral column, proceeds upward, and enters the cranial cavity by way of the foramen magnum. The vertebral anterior arteries join at the posterior border of the pons to form the basilar artery. The *branches* of the vertebral artery are: (1) the *posterior spinal;* (2) the *posterior inferior cerebellar;* and (3) the *anterior spinal.*

Basilar Artery From its origin at the lower border of the pons, where it is formed by a union of two vertebral arteries, the *basilar artery* continues upward in the median (basilar) groove of the pons and, on reaching the anterior part of the pons, splits into two large trunks, the *posterior cerebral arteries.* The basilar artery gives rise to several branches, which spring from its side and extend laterally somewhat obliquely. They are: (1) the *pontine;* (2) the *labyrinthine;* (3) the *anterior inferior cerebellar;* (4) the *superior cerebellar;* and (5) the *posterior cerebellar.*

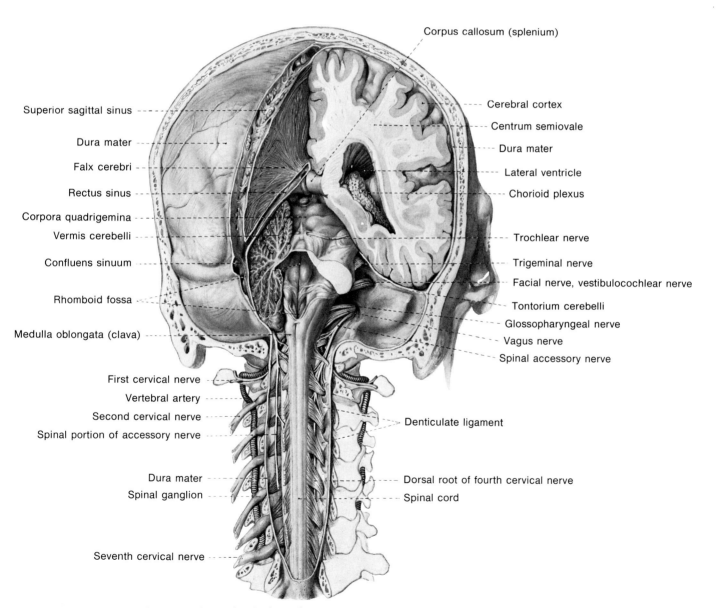

FIG. 6-17A. *Meninges (brain and spinal cord)*

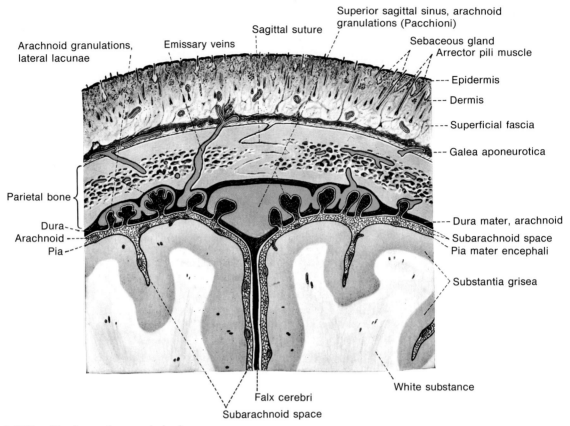

FIG. 6-17B. *Meninges (coronal view)*

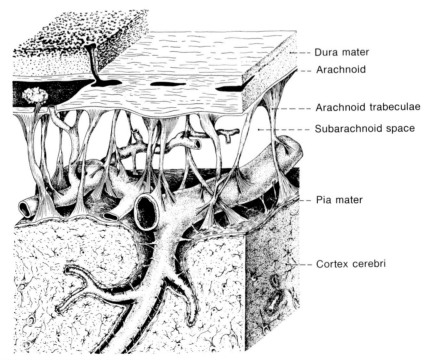

FIG. 6-17C. *Meninges (diagrammatic view)*

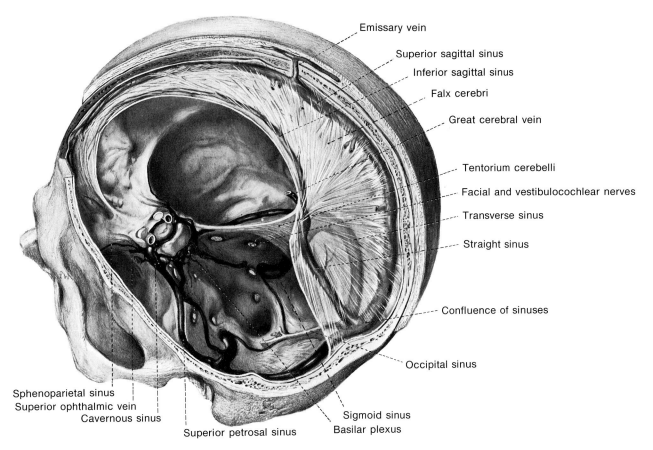

FIG. 6-17D. *Meninges (dural fold)*

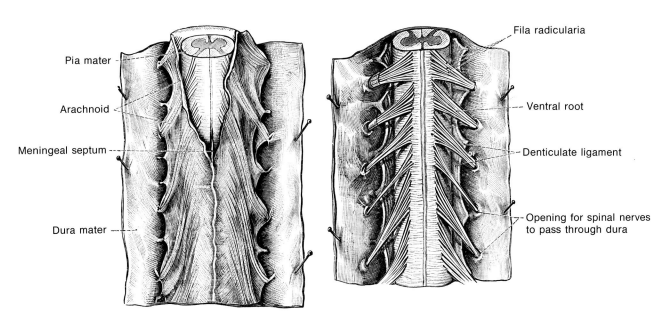

FIG. 6-17E. *Meninges (Spinal cord)*

156 ■ BASIC ANATOMY FOR THE ALLIED HEALTH PROFESSIONS

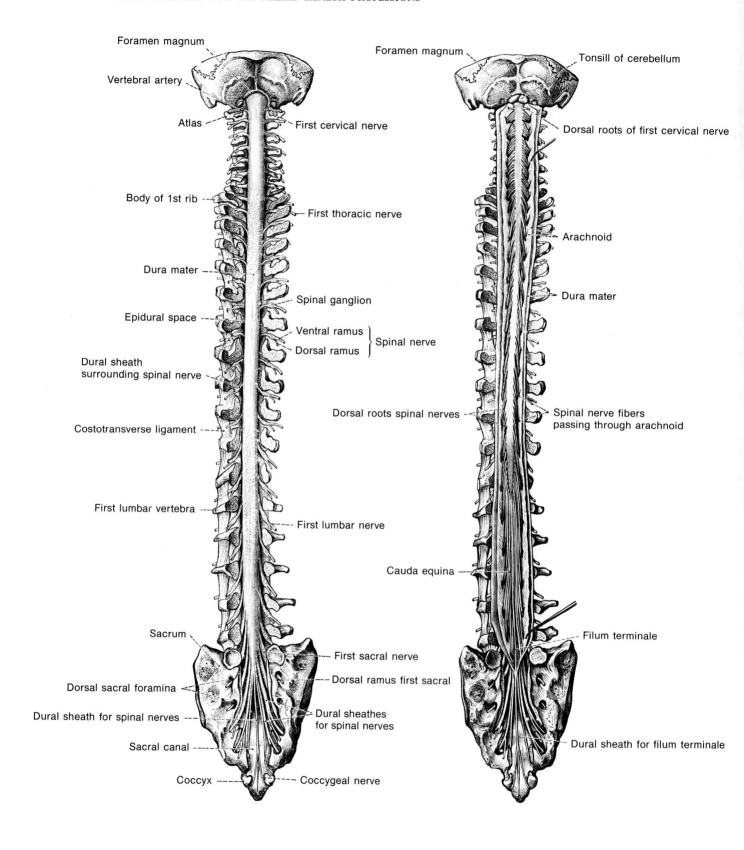

FIG. 6-17F. *Meninges* (*dissected*)

THE NERVOUS SYSTEM ■ 157

Anterior Cerebral Artery As it arises from the internal carotid artery the *anterior cerebral artery* takes an oblique course directly in front of the optic chiasma toward the median plane. Then it bends sharply, turning upward in the depth of the anterior extension of the longitudinal fissure directly in front of the lamina terminalis, and continues on the medial aspect of the hemisphere. It curves around the genu of the corpus callosum, reaches the upper limits of the corpus callosum, and then turns backward on the medial surface of the hemisphere following the curve of the upper surface of the corpus callosum as far as the parietal occipital fissure. In the neighborhood of the lamina terminalis it is joined to the corresponding artery of the opposite side by means of a short trunk, the *anterior communicating artery.* In the longitudinal fissure, the two vessels lie close to one another. Each vessel gives rise to two groups of branches, the cortical and central branches.

The central branches of the anterior cerebral artery comprise a small set of arterial twigs that enter the anterior perforated substance to supply the deep structures in the region (rostrum of the corpus callosum, lamina terminalis, head of the caudate, the column of the fornix, the anterior part of the septum pellucidum, and the anterior commissure).

Medial Cerebral Artery This vessel is the largest branch of the internal carotid. Directly after its origin, it passes laterally along the stem of the lateral fissure. It follows this fissure, gradually gaining greater depth, and upon reaching the surface of the insula breaks up into several terminal branches. In the posterior part of the insula, the middle cerebral artery divides into its terminal branches, the parietotemporal and the temporal. Here, too, the branches of the main trunk form two groups, the cortical and the central branches.

The central group of the middle cerebral artery

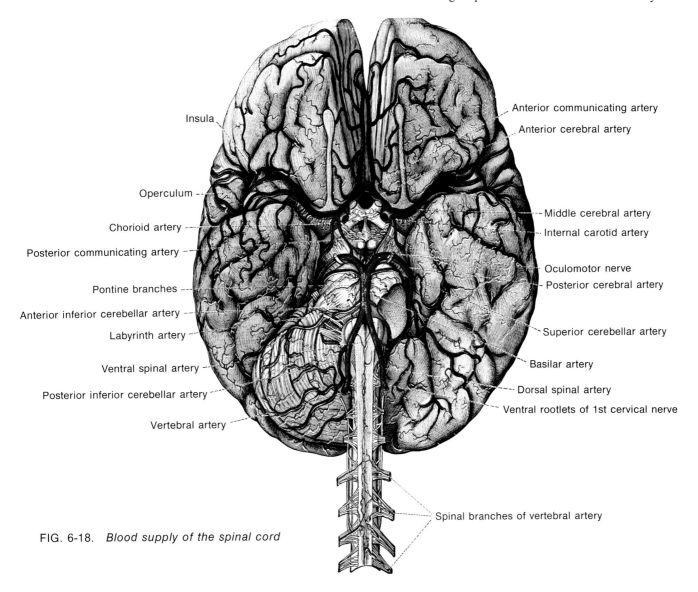

FIG. 6-18. *Blood supply of the spinal cord*

arise as numerous small twigs in the neighborhood of the anterior perforated space and group themselves into two sets, the medial and the lateral striate arteries.

The *medial striate arteries* pass through the anterior perforated space upward into the globus pallidus and the internal capsule, and end in the corpus caudatum, supplying the anterior part of these structures.

The *lateral striate arteries* pass through the anterior perforated space more laterally, enter the putamen, and then divide into the anterior and posterior twigs. The former terminate in the caudate nucleus, the latter in the thalamus.

Posterior Communicating Artery This is a relatively short arterial trunk, which springs from the posterior aspect of the internal carotid artery nearest termination. It runs backward below the optic tract, passes above the oculomotor nerve, and joins the posterior cerebral artery. In its course it distributes branches to the optic chiasma, optic tract, cerebral peduncle, interpeduncular structures, internal capsule, and anterior portion of the thalamus.

Chorioidal Branch of the Internal Carotid Artery This is a small branch that arises near the termination of the internal carotid. It passes backward between the cerebral peduncle and the uncus to the lower part of the chorioidal fissure, enters it, and terminates in the chorioidal plexus of the inferior horn of the lateral ventricle.

VENOUS CHANNELS OF THE BRAIN The *cerebral veins* (fig. 6-20) have some features that distinguish them from veins elsewhere. Their walls are exceedingly thin; they are almost completely devoid of muscular fibers, and have no valves. They do not parallel the course of arteries. The cerebral veins may be considered as forming two systems, grouped as *superficial* and *deep*. The superficial cerebral veins are the most conspicuous and most numerous and are spread in the pia over the entire free surface of the brain. Unlike the arteries, whose main trunk originates at the base of the brain, the veins are formed mainly on the dorsolateral surface by the confluence of smaller venous channels, which originate in the subjacent cortex and subcortex. The deep veins of the brain receive the return blood from the deep-seated structures in the brain stem. The superficial and deep cerebral veins converge ultimately to enter one of the cerebral dural sinuses, which in turn open into the internal jugular vein.

The *superficial* system of *veins* includes: (1) the *superior cerebral veins;* (2) the *middle cerebral veins;* (3) the *inferior cerebral veins;* (4) the *deep middle cerebral vein;* and (5) the *basal vein.*

DURAL SINUSES AND OTHER VENOUS CHANNELS

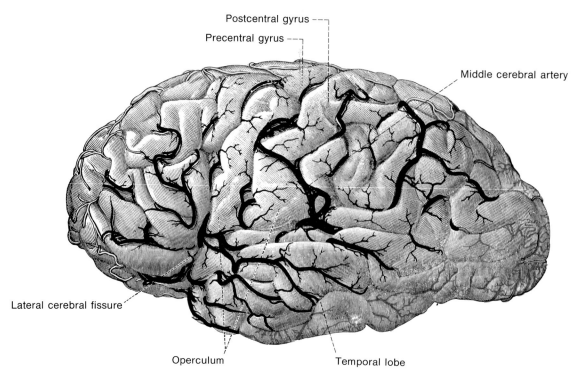

FIG. 6-19A. *Blood supply of the brain (lateral view)*

Between the layers of the dura, channels are formed that convey the venous blood returning from the brain and constitute the venous sinuses (fig. 6-21).

The *superior sagittal sinus* extends from the foramen cecum to the external occipital protuberance, lying along the attached border of the falx cerebri, and constantly increases in caliber as it proceeds caudally. In its middle portion it gives off a number of lateral diverticula, the *venous lacunae,* to which the arachnoid villi protrude. The narrow and short *inferior sagittal sinus* extends caudally along the free border of the falx. On reaching the anterior border of the tentorium it is joined by the *great cerebral vein (of Galen)*, which drains the deep structures of the brain; the two veins form the *sinus rectus*. The latter runs backward and downward along the line of attachment of the falx and tentorium and joins the superior sagittal sinus near the internal occipital protuberance. From this place the two *transverse sinuses* arise, each of which passes laterally and forward in the transverse groove of the occipital bone. On reaching the occipitopetrosal junction, each sinus curves sharply caudally and, as the *sigmoid sinus,* leaves the skull through the jugular foramen. The point of union of the superior sagittal, straight, and transverse sinuses is known as the *sinus confluence*. It also receives the *small unpaired occipital sinus* coming from the region of the foramen magnum and ascending in the falx cerebelli.

The important *cavernous sinus* is a large irregular space located on the side of the sphenoid bone, lateral to the sella turcica. It is a network of intercommunicating cavernous channels enclosing the internal carotid artery, the oculomotor, trochlear, and abducens nerves, and the ophthalmic division of the trigeminal nerve. The cavernous sinus of each side is connected with the other by venous channels that pass anterior and posterior to the hypophysis and by the *basilar venous plexus*. The lateral venous plexus extends along the basilar portion of the occipital bone as far caudally as the foramen magnum where it communicates with the venous plexus of the vertebral canal. The venous ring, surrounding the hypophysis and composed of the two cavernous sinuses and their

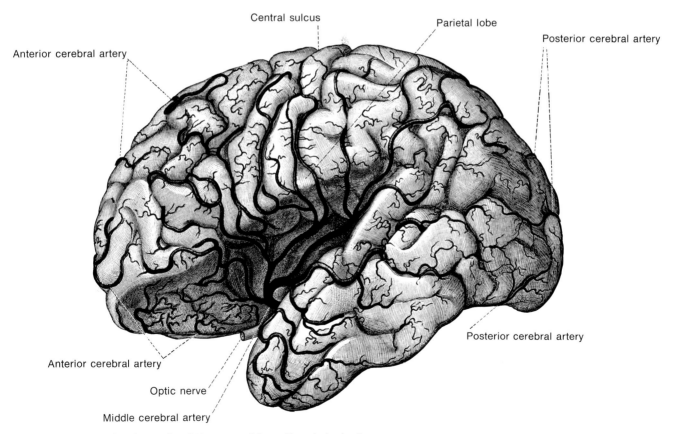

FIG. 6-19B. *Blood supply of the brain [deep (insular) view]*

connecting channels, often is designated as the *circular sinus.* Rostrally it receives the two ophthalmic veins through the orbital fissure and the small *sphenoparietal sinus,* which runs along the undersurface of the lesser wings of the sphenoid. Posteriorly it empties into the superior and inferior petrosal sinus, through which it is connected, respectively, with the transverse sinus and the bulb of the internal jugular vein.

The dural sinuses communicate with the extracranial veins by a number of emissaries. Thus the superior sagittal sinus is connected with the frontal and nasal veins through the frontal diploic veins and the emissaries of the *foramen cecum.* It also sends a *parietal emissary* to the superficial temporal vein. The cavernous sinus, besides receiving the ophthalmic veins, is connected with the internal jugular vein and with the pterygoid and pharyngeal plexuses by a fine venous net that passes through the various foramina of the skull.

PERIPHERAL NERVOUS SYSTEM

The *peripheral nervous system* (fig. 6-22) is customarily divided into *cranial nerves, spinal nerves,* and the *ganglia of the autonomic nervous system.*

The outstanding characteristics of a spinal nerve include a dorsal sensory root with a ganglion, a ventral motor root, and a segmental arrangement.

Nerve fibers conduct impulses, afferents in one direction and efferents in the other. They may have a sheath of myelin (myelinated or medullated fibers) or may not (unmyelinated or nonmedullated); intermediate conditions also exist.

Cranial Nerves

The *cranial nerves* (fig. 6-23A) are frequently grouped into three categories according to their functions as follows.

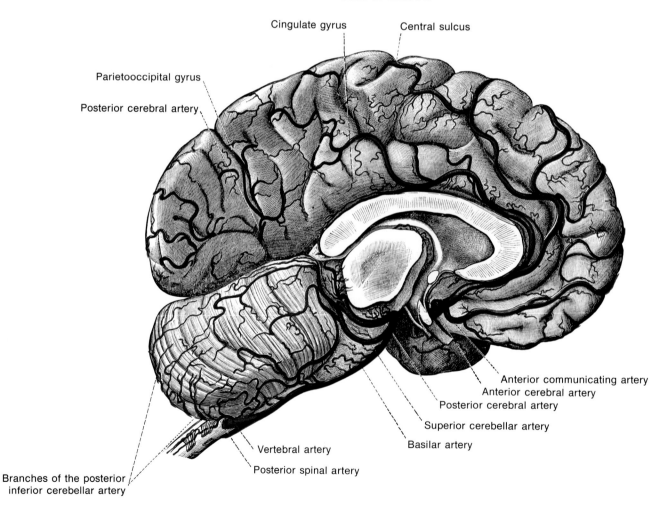

FIG. 6-19C. *Blood supply of the brain* (*sagittal view*)

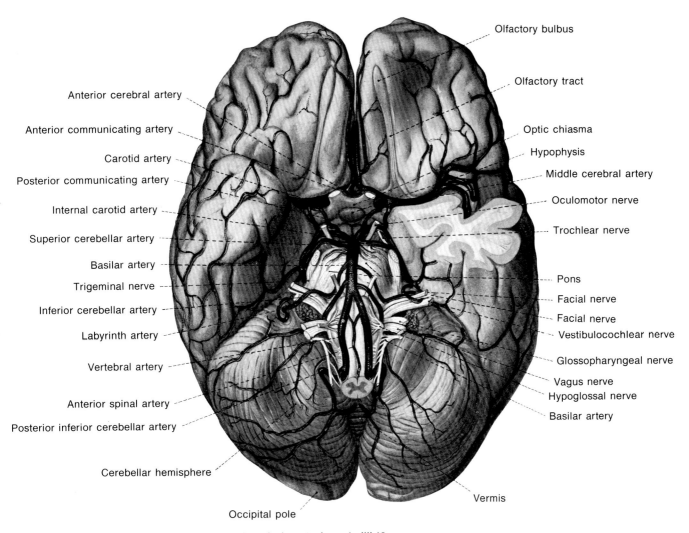

FIG. 6-19D. *Blood supply of the brain [circle arterious (willis)]*

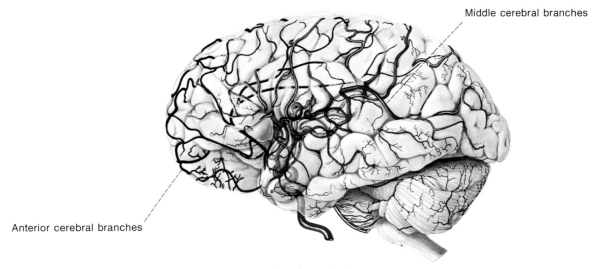

FIG. 6-19E. *Blood supply of the brain (Anastomosing branches)*

Special Sensory
    I. Olfactory
    II. Optic
    VIII. Vestibulocochlear

Somatic Motor
    III. Oculomotor
    IV. Trochlear
    VI. Abducens
    XII. Hypoglossal

Branchiomeric
    V. Trigeminal
    VII. Facial
    IX. Glossopharyngeal
    X. Vagus
    XI. Accessory

More detail regarding each sort of fiber carried by the cranial nerves follows.

Somatic motor: Oculomotor, trochlear, abducens, and hypoglossal nerves carry these fibers.

Branchiomeric motor (special visceral motor): Trigeminal, facial, glossopharyngeal, vagus, and spinal accessory nerves carry these fibers.

Parasympathetic motor (general visceral motor): Oculomotor, facial, glossopharyngeal, and vagus nerves carry these fibers. Because these are autonomic fibers there are two sorts of neurons concerned: preganglionic in the central nervous system, and postganglionic in the parasympathetic ganglia of the head and trunk.

Special sensory: These fibers must be divided into five categories: (1) smell, carried by the olfactory nerve; (2) sight, carried by the optic nerve; (3) taste, carried by the glossopharyngeal, facial, and vagus nerves; (4) hearing, carried by the cochlear part of the VIIIth cranial nerve; and (5) equilibrium, carried by the vestibular part of the VIIIth cranial nerve. The cell bodies of the olfactory nerve are in the olfactory epithelium; of the optic, in the ganglionic layer of the retina; and of the other special sensory components, in definite peripheral ganglia of the respective nerves involved. It should be mentioned that the vestibular sense of equilibrium usually is not classified as a special sense, but as proprioceptive.

General somatic sensory: Trigeminal, facial, glossopharyngeal, and vagus nerves, all with ganglia, carry these fibers. The trigeminal is the nerve chiefly concerned with the sensibility of the head. In addition, there is an area around the external auditory meatus and in the middle ear that has somatic sensory innervation furnished by the facial, glossopharyngeal, and vagus nerves.

General visceral sensory: Glossopharyngeal and vagus nerves, both with ganglia, carry these fibers. The former supplies sensibility to the pharynx, from the pillars of the fauces to the epiglottis, and the latter to the larynx and superior part of the esophagus.

Proprioceptive sensory: The end organs of proprioceptive fibers are the muscle spindles. These occur in all somatic muscles, and sparingly in the branchiomeric muscles. The center for the cranial proprioceptive neurons, as far as is known, is exceptional in being not a ganglion, but the mesencephalic nucleus of the trigeminal nerve in the brain stem. The muscles innervated by the trigeminal nerve are the only ones with proprioceptive fibers that have been traced to this nucleus, although fibers probably reach this destination from the other cranial muscles as well.

GANGLIA The *ganglia* (fig. 6-23*B*) in the head are of two sorts: *parasympathetic* and *sensory*. The *parasympathetic* comprise *ciliary, pterygopalatine, submandibular,* and *otic ganglia*.

The *cranial sensory ganglia* contain neurons of three categories of fibers: *special sensory, somatic sensory,* and *visceral sensory;* but not proprioceptive as mentioned above. There are no ganglia for the senses of smell or of sight. The ganglia of hearing (cochlear ganglion of cranial nerve VIII) and of equilibrium (vestibular ganglion of cranial nerve VIII) are simple, having but one sort of cell.

The *vagus* and *glossopharyngeal* each have two *ganglia,* a proximal *root ganglion* and slightly more distally a *trunk ganglion*. The root ganglion of the vagus is the *jugular* and its trunk, the *nodose*. Similarly, in the glossopharyngeal the root ganglion is the *superior* and the trunk ganglion the *petrosal,* but in this nerve the two are often fused or partly so. In both nerves the trunk ganglia (nodose and petrosal) contain neurons of visceral sensibility and taste, while the root ganglia (jugular and superior) contain somatic sensory cell bodies. The ganglion of the facial nerve (*geniculate*) is single, containing mainly cell bodies of taste fibers, and a small proportion of somatic sensory neurons. Similarly, the ganglion of the trigeminal nerve (*semilunar*) is single, and composed only of somatic sensory cell bodies.

CRANIAL NERVE I—OLFACTORY The *olfactory nerve* (fig. 6-24) carries the special sensory smell fibers. It is not a typical cranial nerve, for it develops by a central ingrowth to the olfactory bulb of processes from the olfactory cells in the olfactory epithelium of the nose. Its cells are situated in the mucosa over a small area in the superior part of the nasal septum and the superior concha, with extremely short processes reaching the surface of the mucosa.

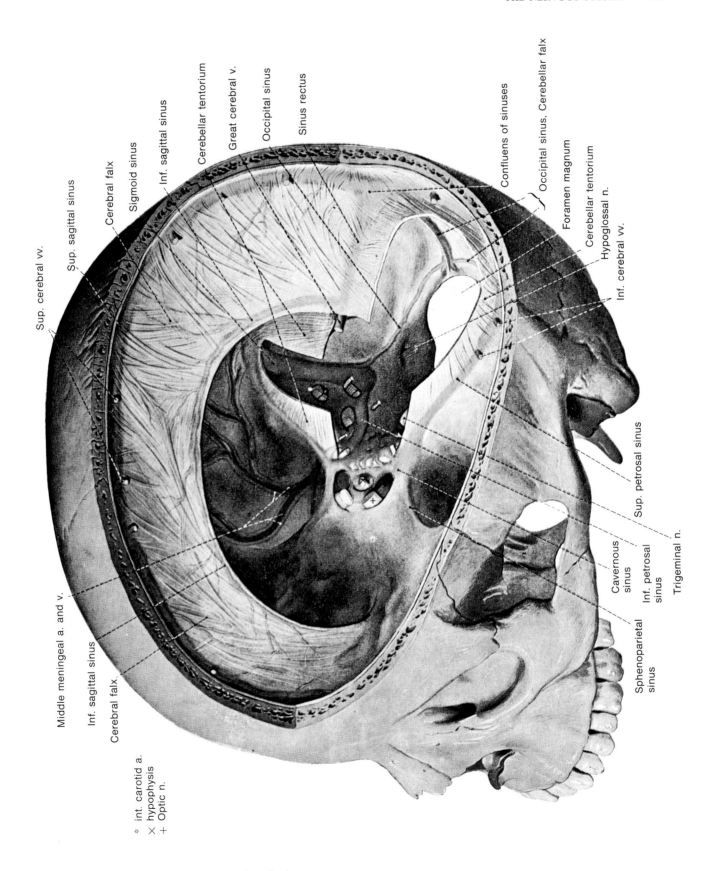

FIG. 6-20A. *Cerebral veins (superior view)*

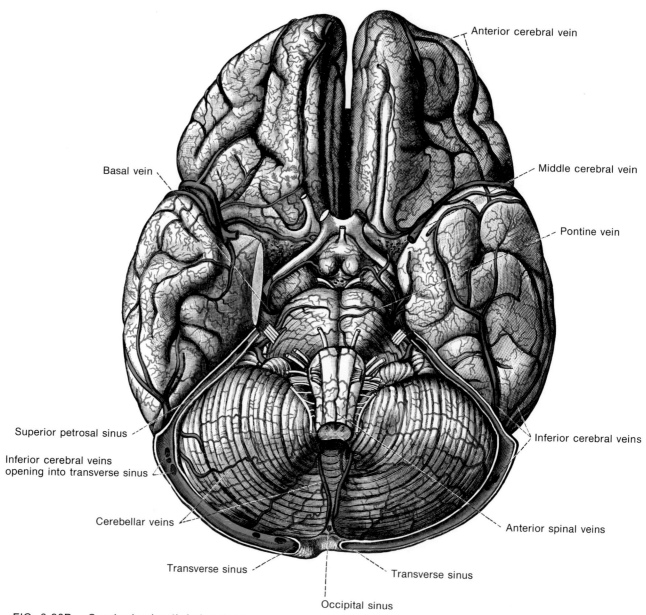

FIG. 6-20B. *Cerebral veins (inferior view)*

Their axons are gathered into about 20 small filaments, which pass through the perforations of the cribriform plate of the ethmoid bone to the olfactory bulb.

CRANIAL NERVE II—OPTIC The *optic nerve* (fig. 6-24) carries special sensory sight fibers. It also is not a true cranial nerve, but rather the projected stalk of the optic vesicle. Its neurons constitute the ganglionic cells of the retina, from which the optic nerve courses posteriorly in the orbit to the optic foramen, through which it passes to the optic chiasma in the middle cranial fossa, where there is a partial decussation (a chiasm or crossing) of fibers. The optic tract extends from chiasma to the midbrain and thalamus. Each tract carries fibers from both eyes, as well as commissural (crossing) fibers. The optic nerve is ensheathed by prolongations of the cranial dura and arachnoid, and these are attached to the sclera of the eyeball.

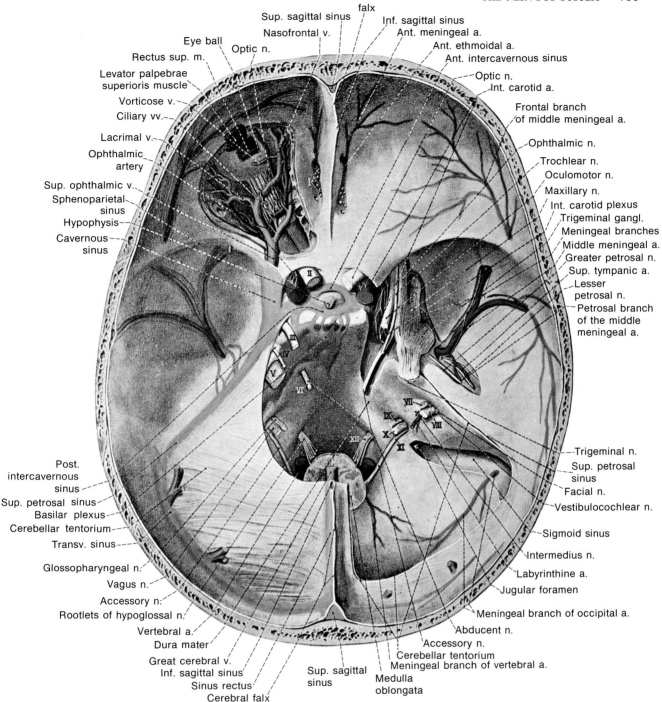

FIG. 6-21. *Dural sinuses and other venous channels*

CRANIAL NERVE III—OCULOMOTOR The *oculomotor nerve* (fig. 6-25) carries somatic motor fibers to the levator palpebrae superioris muscle and all extrinsic muscles of the eye except the lateral rectus and superior oblique, and parasympathetic motor fibers to the ciliary ganglion. This nerve emerges from the midbrain by several rootlets enclosed for a short distance by a sheath of arachnoid. It pierces the dura, anterolaterally to the posterior clinoid process, in the lateral wall of the cavernous sinus. It passes through the superior orbital fissure and then through the tendinous ring from which arise the ocular muscles. It then divides into two branches.

The smaller *superior ramus* extends superior to the optic nerve and supplies the superior rectus and levator palpebrae superioris muscles. The *inferior ramus* extends inferior to the level of the optic nerve and lateral to it. It innervates the medial and inferior recti muscles and the inferior oblique. It sends a short root to the ciliary ganglion, by which parasympathetic fibers are supplied.

CRANIAL NERVE IV—TROCHLEAR The *trochlear nerve* (fig. 6-26) carries somatic motor fibers to the superior oblique muscle. It arises at the junction of the midbrain and hindbrain and, as the smallest of the cranial nerves, runs anteriorly to pierce the free border of the tentorium near the posterior clinoid process. It then passes through the external wall of the cavernous sinus and superior orbital fissure. In the orbit it lies superiorly, and to the medial side, of the frontal nerve. The word *trochlea* means pulley, referring to the tendinous pulley of the superior oblique muscle. Hence this is the nerve of the pulley muscle.

CRANIAL NERVE V—TRIGEMINAL The *trigeminal nerve* (fig. 6-27A) is the nerve of the first or mandibular arch. It is a compound nerve, consisting of a motor root and a much larger sensory root, both of which are from the hindbrain and enter the dura of the posterior cranial fossa inferior to the tentorium. Upon the anteromedial part of the petrous temporal bone they occupy a cavity in the dura, and here is situated the *semilunar* or *Gasserian ganglion* of the sensory root (fig. 6-27B). This ganglion is composed exclusively of somatic sensory cell bodies, and from it diverge the three divisions of the trigeminal nerve (fig. 6-27C and D).

The *motor root* is branchiomeric motor, innervating the muscles of mastication (temporalis, masseter, and medial and lateral pterygoids), tensor palatini, tensor tympani, anterior belly of the digastric and mylohyoid. It is situated medial to the sensory root and passes from the cranial cavity through the foramen ovale. The *sensory root* carries only somatic and proprioceptive sensory fibers. From the semilunar ganglion it breaks up into three trunks as follows.

The *ophthalmic division of the trigeminal nerve* supplies sensibility to the eye, lacrimal glands, conjunctiva, part of the mucosa of the nasal cavity, eye-

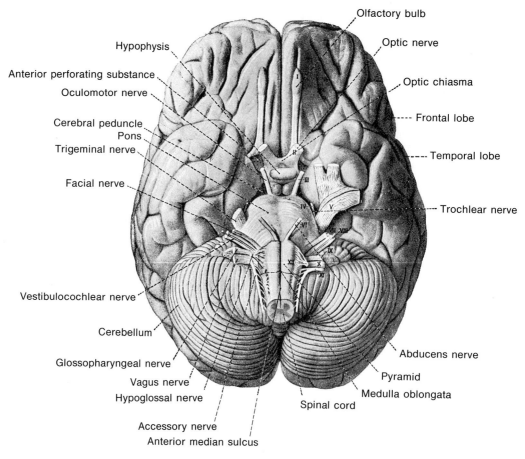

FIG. 6-22A. *Peripheral nervous system (cranial nerves)*

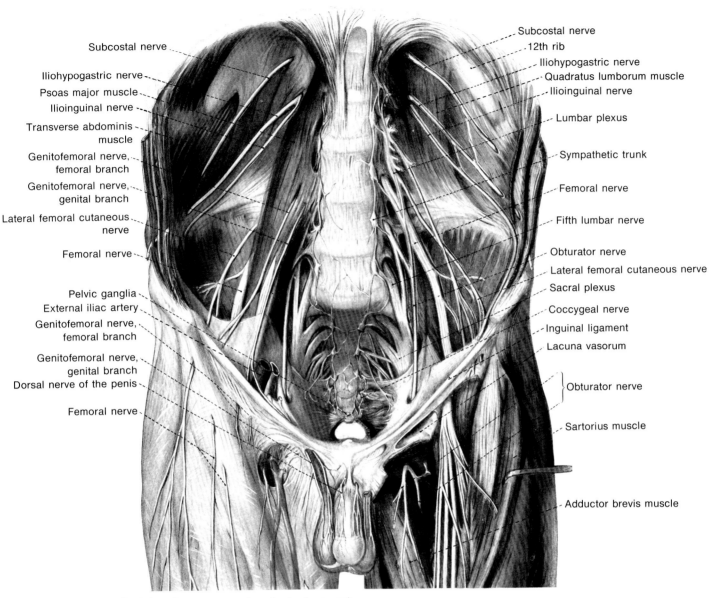

FIG. 6-22B. *Peripheral nervous system (spinal nerves)*

lids, nose, and forehead. It enters the dura in the middle cranial fossa and passes through the lateral wall of the cavernous sinus to the superior orbital fissure, where it divides into three branches: frontal, lacrimal, and nasociliary.

(1) The *frontal nerve* passes superior to the orbital muscles to the anterior part of the orbit, where it becomes the *supraorbital nerve,* passing through the supraorbital foramen and ending in medial and lateral branches to the forehead, scalp, and upper lids. In the orbit it gives off the *supratrochlear nerve,* which reaches the face adjacent to the medial canthus of the eye and innervates the surrounding skin.

(2) The *lacrimal nerve* is small. It also passes anteriorly in the orbit, superior to the muscles, but in its lateral part. It supplies sensibility to the lacrimal gland, conjunctiva, and lateral part of upper lid.

(3) The *nasociliary nerve* pierces the lateral rectus muscle, passes between the two divisions of the oculomotor nerve and runs inferior to the superior rectus and superior oblique muscles. Then it courses in the medial orbit to the anterior ethmoid foramen. Here as the *ethmoidal anterior* it passes into the cranial cavity, through the nasal slit lateral to the cribriform plate, where it terminates in the nasal mucosa. In the orbit it gives off first a *short ciliary* filament to the ciliary ganglion and *long ciliary* twigs to the eyeball. Both carry sensibility from the eye, and the latter also sends sympathetic fibers to the dilator of the pupil. Finally, the *infratrochlear nerve* is distributed to the

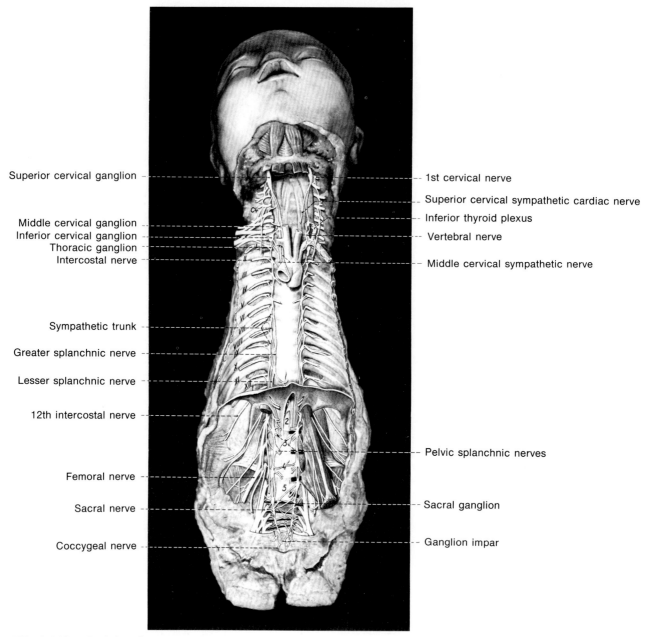

FIG. 6-22C. *Peripheral nervous system (autonomic)*

skin below the medial canthus of the eye. *Posterior ethmoidal branches* may be present in the orbit.

The *maxillary division of the trigeminal nerve* supplies sensibility to the upper lip, lower lid, lateral nose, anterior cheek, anterior temple, upper teeth, pharynx, palate, part of the nasal cavity, and part of the periosteum of the orbit. In the middle cranial fossa it gives off a *meningeal branch* to the meninges. The main nerve passes through the foramen rotundum and crosses the pterygopalatine fossa. It then passes through the inferior orbital fissure to the orbit, where it becomes the *infraorbital nerve,* which leaves the orbit by passing through the infraorbital foramen and is distributed by *superior labial, inferior palpebral,* and *external nasal branches* to the upper lip, lower lid, and lateral nose. In its course it gives off branches in three main groups: (1) in the pterygopalatine fossa, two short pterygopalatine branches connect to the ganglion of the same name, from which sensory branches are distributed to the pharynx (*pharyngeal branches*) and palate (*greater* and *lesser palatine nerves*). The nasal cavity receives branches through the sphenopalatine foramen by way of the nasopalatine, lateral, and medial nasal nerves. (2) The *zygomatic nerve* in the orbit sends a communicating branch to the lacrimal nerve by which the lacrimal gland is supplied with parasympathetic fibers and the periosteum with sensory fibers. The nerve then divides into the *zygomaticofacial nerves,* piercing the zygomatic foramen to the skin over the cheek bone, and *zygomaticotemporal nerves,* which also pierce the zygomatic bone to terminate over the anterior temple. (3) Several *superior alveolar branches* are sent to the roots of the upper teeth.

The sensory part of the *mandibular division of the trigeminal nerve* supplies sensibility to the skin of the temporal region and about the ear, on the lower lip, chin, and middle cheek, and to the mucosa of the cheek, lower gums, and anterior two-thirds of the tongue. It is joined, however, by the parasympathetic fibers from the facial (via the chorda tympani) to submandibular ganglion, and by taste fibers on their way to the facial nerve.

As it leaves the semilunar ganglion this division

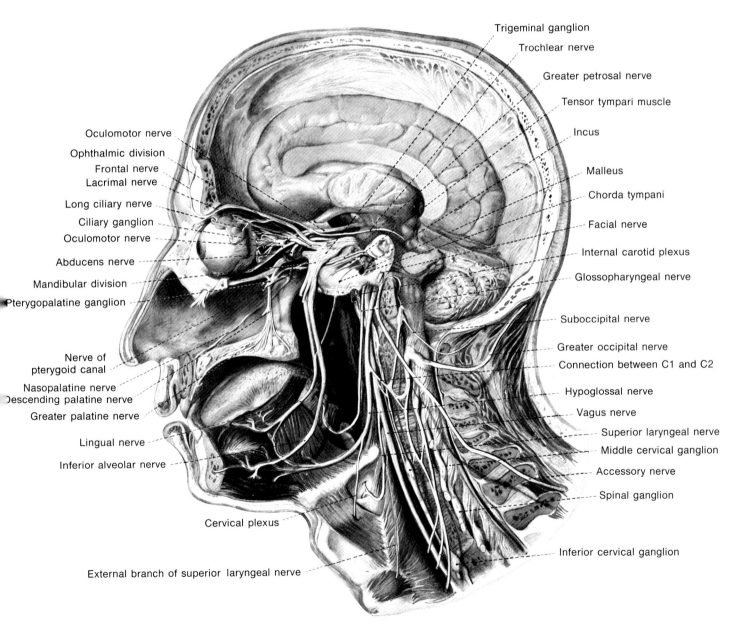

FIG. 6-23A. *Cranial nerves*

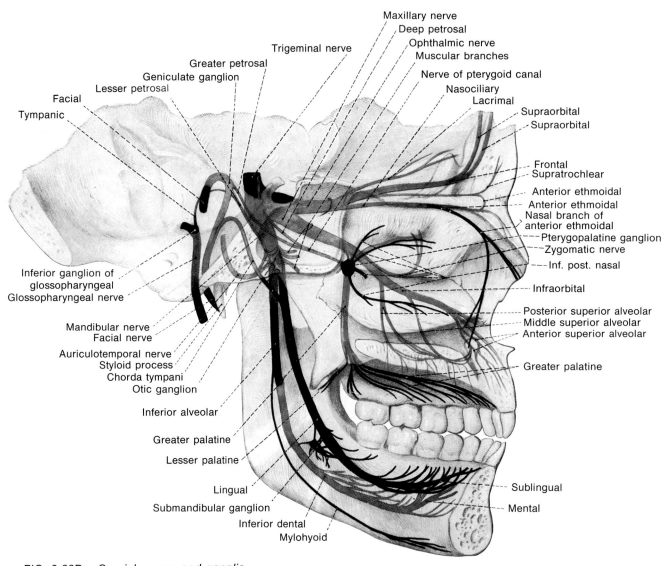

FIG. 6-23B. *Cranial nerves and ganglia*

receives on its deep surface the motor fibers for the muscles of mastication. Here it gives off a recurrent meningeal branch, which reenters the cranium through the foramen spinosum, and a motor twig to the medial pterygoid muscle that is connected with the otic ganglion. It is convenient to consider the remainder of the division as comprising four main branches.

(1) An anterior trunk coursing anterolaterally just lateral to the lateral pterygoid; a motor twig to the masseter, which reaches the muscle by passing over the mandibular notch between the articular and coronoid processes; two or three ascending motor twigs to the temporalis that pass over the angle of bone between the temporal and infratemporal fossae; and the terminal sensory *buccal branch,* which pierces the buccinator to supply sensibility to the cheek, partly upon the external but mostly the internal or buccal surface.

(2) The *lingual* branch is basically sensory, but is joined by parasympathetic and taste fibers (in the chorda tympani nerve) of the facial nerve. It is the smaller and more anterior of the two contiguous terminal divisions of the mandibular division of the trigeminal. It courses inferiorly at first between the two heads of the lateral pterygoid, where it receives the

chorda tympani at an acute angle. It then is situated between the medial pterygoid and the mandibular ramus, here giving off a filament to the tonsil and surrounding mucosa. It then courses anteriorly, passing between the mylohyoid and hyoglossus muscles, where it gives and receives branches to and from the submandibular ganglion, and transmits sensibility from the anterior two-thirds of the tongue, acting as a vehicle also for autonomic and taste fibers.

(3) The *inferior alveolar nerve* (inferior dental nerve) and the lingual nerve constitute the two terminal divisions of the mandibular division of the trigeminal nerve. The former extends inferiorly just posterior to the lingual and in the same situation. It emerges from beneath the lateral pterygoid muscle and gives off the *mylohyoid nerve* before entering the mandibular foramen. In the mandibular canal it gives off *dental branches*. In the anterior part of the canal it is known as the *mental nerve,* and emerges from the mental foramen to supply sensibility to the skin and lower lip. The *mylohyoid nerve* is motor. It pierces the sphenomandibular ligament and contin-

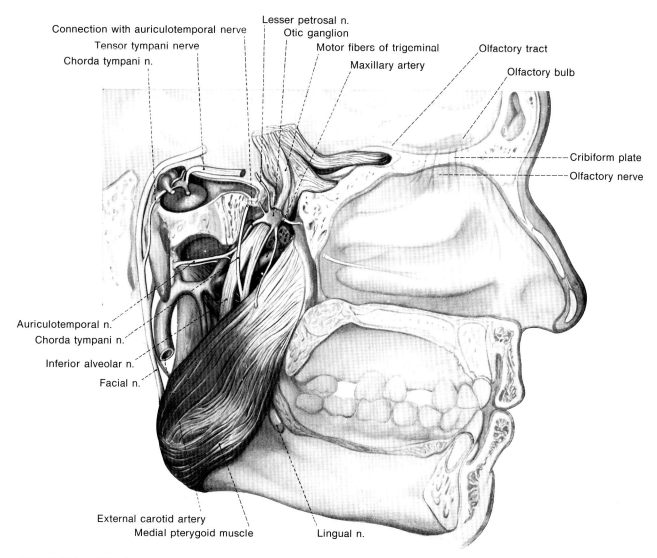

FIG. 6-24A. *Olfactory nerve*

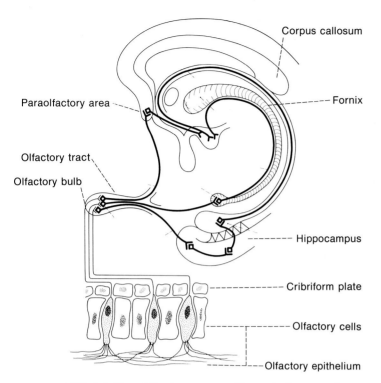

FIG. 6-24B. *Olfactory nerve with central connections*

ues inferiorly to innervate the anterior belly of the digastric and mylohyoid muscles.

(4) The *auriculotemporal nerve* carries fibers of sensibility, and also autonomic fibers for a short distance. It is a small but important branch formed of two roots, diverging posteriorly from the mandibular nerve. It extends posteriorly deep to the lateral pterygoid muscle, then between the sphenomandibular ligament and the neck of the mandible. Then the nerve courses superiorly and emerges from beneath, or even pierces, the parotid gland, extending superiorly just anterior to the auricular cartilage. Distribution is to the skin over the posterior temporal region. Near origin it receives autonomic fibers from the tympanic plexus via the otic ganglion, destined for the parotid gland, and it gives off sensory twigs to the mandibular joint, parotid gland, tympanic membrane and adjacent skin, and the ear.

Communications of the mandibular nerve, except with autonomic ganglia, comprise connection of the lingual nerve with chorda tympani of the facial nerve, and with the hypoglossal nerve just anterior to the submandibular gland; and of the auriculotemporal nerve with the facial nerve in the substance of the

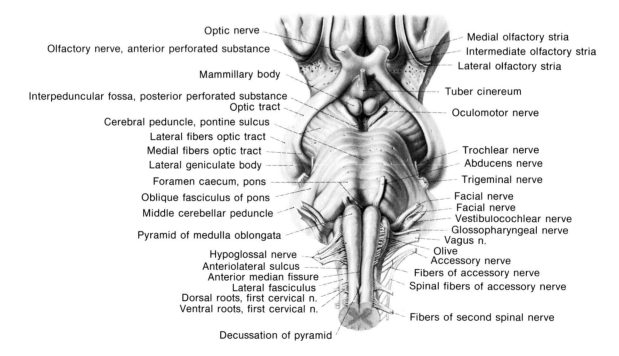

FIG. 6-25. *Optic and oculomotor nerves*

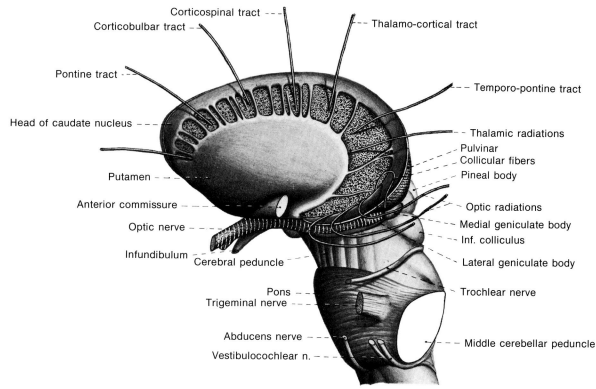

FIG. 6-26. *Trochlear nerve*

parotid gland. Although fibers of the trigeminal nerve are associated with ciliary, pterygopalatine, submandibular, and otic ganglia, these are autonomic ganglia and are described with that system.

CRANIAL NERVE VI—ABDUCENS The *abducens nerve* (fig. 6-28) is somatic motor and is the nerve to the abductor of the eyeball, the lateral rectus muscle. It arises from the hindbrain and extends lateral to the dorsum sellae with the oculomotor and trochlear nerves but is separated from them by dura. It passes through the cavernous sinus and the superior orbital fissure to the orbit, in the lateral part of which it innervates its muscle.

CRANIAL NERVE VII—FACIAL The *facial nerve* (fig. 6-29) is a compound nerve consisting of a *motor root* and a mixed root, the *intermedius nerve* with taste and parasympathetic fibers. The motor root supplies the facial muscles of expression, including the platysma, stapedius, posterior belly of the digastric, and stylohyoid muscles. The mixed root transmits: (1) taste fibers usually considered as from the anterior two-thirds of the tongue, passing via the lingual nerve and the chorda tympani; (2) somatic sensibility, which is carried from the external ear and tympanic membrane; and (3) parasympathetic fibers to the pterygopalatine ganglion (via the greater petrosal nerve) and to the submandibular ganglion (via the chorda tympani and lingual nerves). The facial nerve enters the internal auditory meatus, there lying superior to the vestibulocochlear nerve. It then passes largely laterally, abruptly turns posteriorly in the facial canal, the geniculate ganglion being situated at the angle, and then turns inferiorly to its emergence from the skull through the stylomastoid foramen. The *geniculate ganglion* is formed of taste and somatic sensory cell bodies. From the region of the ganglion is given off the greater petrosal nerve, carrying preganglionic parasympathetic fibers to the pterygopalatine ganglion. In the facial canal a minute anastomotic branch is also given off to the lesser petrosal nerve of the glossopharyngeal nerve; a branch to the stapedius muscle; and the *chorda tympani,* carrying taste and parasympathetic fibers, which leaves the nerve at an angle, traverses the middle ear between malleus and incus, and then passes through a small canal to the infratemporal fossa, where it joins the lingual nerve, deep to the lateral pterygoid. It thereby transmits parasympathetic fibers to submandibular and sublingual glands, and receives taste fibers from the anterior two-thirds of the tongue.

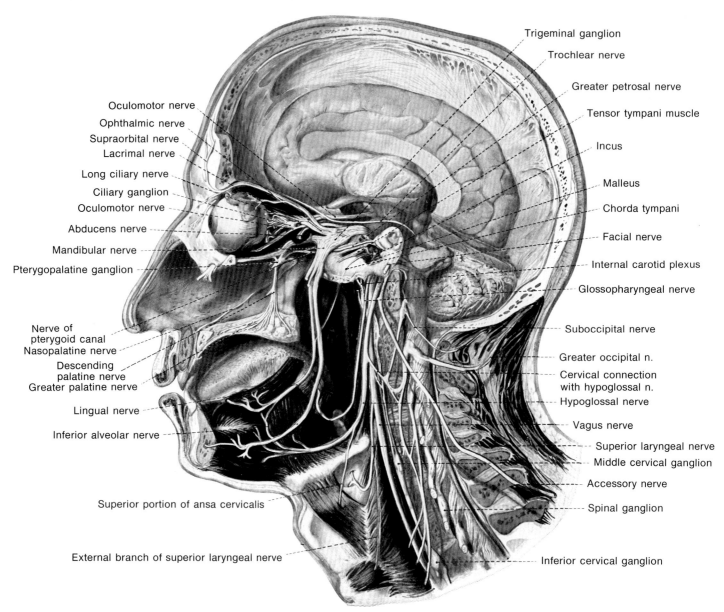

FIG. 6-27A. *Trigeminal nerve and ganglion (lateral view)*

After emergence from the stylomastoid foramen, the facial nerve gives off a small motor branch to the posterior auricular, the posterior belly of the digastric, and the stylohyoid muscles.

The facial nerve then courses anteriorly through the substance of the parotid gland and then emerges in several branches. These branches comprise: (1) an inferior one at the angle of the jaw with a *cervical branch* to platysma, a *mandibular* and a *buccal branch* to the muscles of those regions; and (2) a superior group, consisting of *temporal* and *zygomatic branches*, the latter to the muscles of upper lip and nose.

CRANIAL NERVE VIII—VESTIBULOCOCHLEAR The *vestibulocochlear nerve* (fig. 6-30) is compound, consisting of cochlear and vestibular divisions. Both are from the hindbrain and enter the internal auditory meatus inferior to the facial nerve.

The cochlear division carries the special sensory fibers of hearing. It separates from the vestibular nerve in the internal auditory meatus, it being the more inferior of the two, and passes to the central

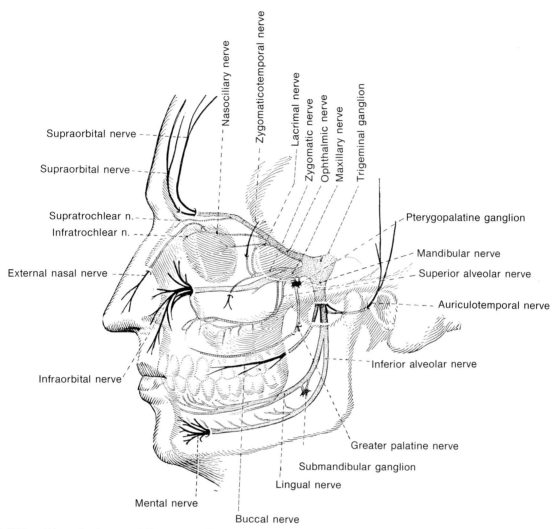

FIG. 6-27B. *Trigeminal nerve (diagrammatic view)*

pillar of the cochlea. The spiral ganglion is located here in the modiolus where filaments are distributed to the *organ of Corti*.

The vestibular division has its *vestibular ganglion* in the internal auditory meatus. Here it separates from the cochlear division into two branches, and often sends an anastomosing branch to the facial nerve. The two branches pass to the sacculus, utriculus, and the ampullae of the semicircular canals.

CRANIAL NERVE IX—GLOSSOPHARYNGEAL The *glossopharyngeal nerve* (fig. 6-31A) is motor to the stylopharyngeus muscle; parasympathetic motor to the parotid gland; taste from the posterior one-third of the tongue; somatic sensibility from the middle ear and medial surface of the tympanic membrane; and visceral sensibility from the posterior one-third of the tongue and mucosa of the pharynx. The glossopharyngeal shares common nuclei with the vagus nerve, and arises from the medulla by several filaments in series with those of the vagus. The nerve passes from the skull through the jugular foramen and is considered there to have two ganglia, a *superior* and *inferior ganglion*, although the superior ganglion, always small, is frequently absent or fused with its neighbor. The superior contains cells of somatic sensibility, from the external ear, cells of visceral sensibility and taste, both from the posterior one-third of the tongue.

From the inferior ganglion is given off the important *tympanic branch*, which passes through the tympanic canaliculus and in the middle ear contributes to the *tympanic plexus*. Here it supplies the mucosa, and that of the auditory tube, and then continues as the *lesser petrosal nerve*, carrying parasympathetic fibers to the otic ganglion. Distal to the ganglia the glossopharyngeal extends inferior and anteriorly between the external and internal carotid arteries and curves around the posterior border of the stylopharyngeus, which it innervates, and then passes deep into the hyoglossus muscle, there receiving twigs from the vagus and sympathetic, and sending branches for receiving visceral sensibility and taste from the posterior one-third of the tongue, and sensibility from the tonsil, soft palate, and pillar of the fauces (fig. 6-31*B*).

CRANIAL NERVE X—VAGUS The *vagus* (fig. 6-32) is in series with the glossopharyngeal nerve, carrying motor fibers to muscles of the pharynx and larynx; parasympathetic motor to all but the most inferior of the viscera; taste from thyroepiglottis; somatic sensibility from the external auditory meatus and lateral tympanic membrane; visceral sensibility from the larynx and upper esophagus; and proprioceptive sensory from the muscles of the pharynx and larynx.

The vagus arises in two parts, both by a number of filaments from the medulla. The minor part is that portion arising in series with the spinal accessory but within the foramen magnum and diverging from it to join the ganglia of the vagus. Its fibers diverge later to become the recurrent laryngeal nerve. The pharyngolaryngeal fibers of this part, as with the vagus, arise from the nucleus ambiguus, thus supplying axons to the vagal field of striated musculature.

Both parts pass through the jugular foramen, where the *jugular ganglion* is situated. The ganglion contains somatic sensory neurons. Slightly more distal is the larger trunk ganglion, the *nodose ganglion,* containing cells of taste and visceral sensibility.

From the nodose ganglion the vagus descends in the neck in the carotid sheath, between the common carotid artery and the internal jugular vein, and lateral to the cervical sympathetic trunk. The nerve enters the mediastinum, on the right side ventral to the subclavian artery and between that and the brachiocephalic vein, at the root of the lung forming the posterior pulmonary plexus. Upon the left side it passes between the common carotid and the subclavian arteries, but ventral to both, dorsal to the left brachioce-

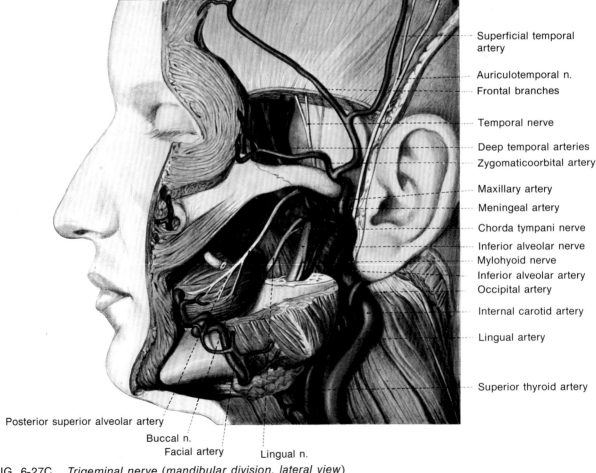

FIG. 6-27C. *Trigeminal nerve (mandibular division, lateral view)*

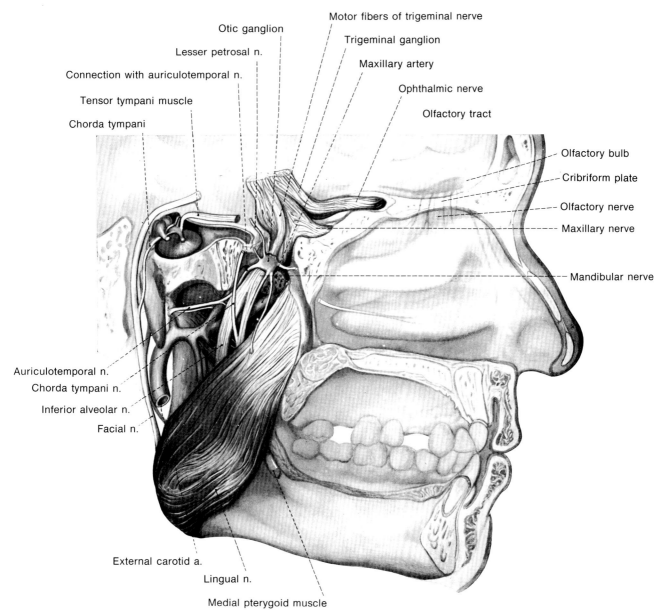

FIG. 6-27D. *Trigeminal nerve (mandibular division, medial view)*

phalic vein. It then passes ventral to the aorta and to the posterior pulmonary plexus at the root of the left lung. From this situation and dorsal to the bronchi, two nerves on each side constitute the continuation of the vagus. They converge through the esophageal hiatus of the diaphragm to the inferior esophagus, both vagi contributing to form the esophageal plexus. Inferior to the stomach the vagal fibers continue in plexiform manner to furnish parasympathetic fibers to all the viscera superior to the level of the transverse colon.

Branches given off the vagus comprise the following.

Recurrent meningeal branch: from the jugular ganglion.

Auricular branch: from the jugular ganglion passing through the mastoid canaliculus in the jugular fossa, carrying somatic sensory fibers from the inferior part of the external auditory meatus and ear.

Pharyngeal branch: from the nodose ganglion or just distal to it, which passes between the external and internal carotid arteries to the pharynx, where it forms a plexus in combination with glossopharyngeal and sympathetic fibers, and innervates the muscles of the pharynx.

178 ■ BASIC ANATOMY FOR THE ALLIED HEALTH PROFESSIONS

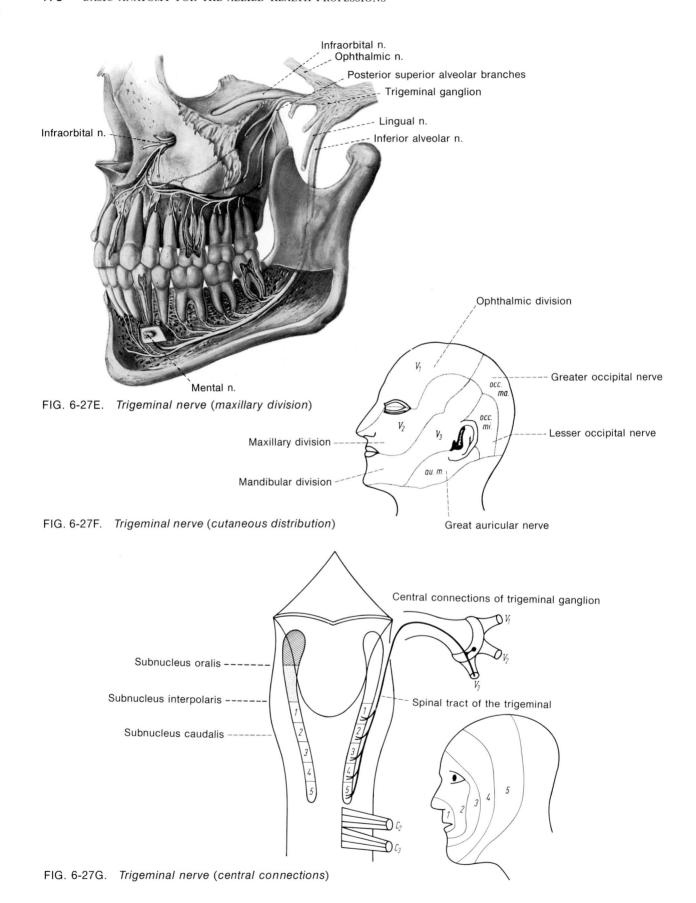

FIG. 6-27E. *Trigeminal nerve (maxillary division)*

FIG. 6-27F. *Trigeminal nerve (cutaneous distribution)*

FIG. 6-27G. *Trigeminal nerve (central connections)*

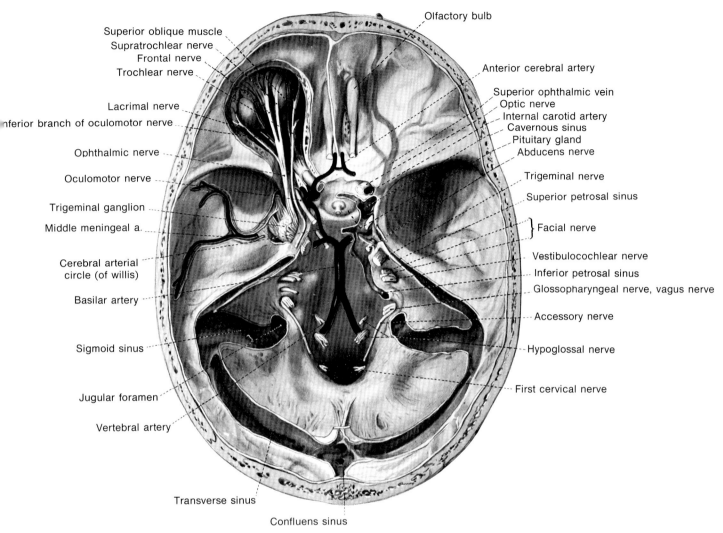

FIG. 6-28. *Abducens nerve*

Superior laryngeal branch: from the nodose ganglion is distributed as an *external* and *internal* part to the mucosa of the laryngeal region, inferior constrictor muscle of larynx, and the cricothyroideus muscle.

Cardiac branches in two groups: The *superior branches* (two or three) diverge from the vagus in the upper and lower neck. On the right, both pass into the mediastinum dorsal to the subclavian artery and to the deep cardiac plexus. On the left, the more superior branch follows this course, but the more inferior passes ventral to the aorta and to the superficial cardiac plexus. The *inferior cardiac branches* on the right arise in the superior mediastinum from the vagus and its recurrent laryngeal branch; those on the left from the recurrent nerve alone. Both end in the deep cardiac plexus.

Recurrent laryngeal branch: On the right this branch leaves the vagus at the level of the subclavian artery, passes beneath this artery, and then proceeds cranially dorsal to the common carotid artery in the sulcus between the trachea and esophagus. On the left it diverges at the level of the aorta, curves below and dorsally of it, and courses similarly to the right branch. They supply fibers to the muscles of the larynx except the cricothyroid and parasympathetic fibers to the smooth musculature of the trachea and esophagus. The recurrent laryngeal nerves provide inferior laryngeal branches at the level of the cricoid cartilage.

CRANIAL NERVE XI—SPINAL ACCESSORY The *spinal accessory nerve* (fig. 6-33) is considered as comprising only the spinal part of the complex, the bulbar

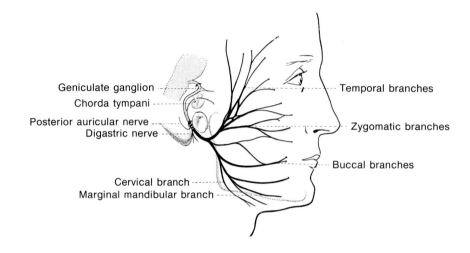

FIG. 6-29. *Facial nerve*

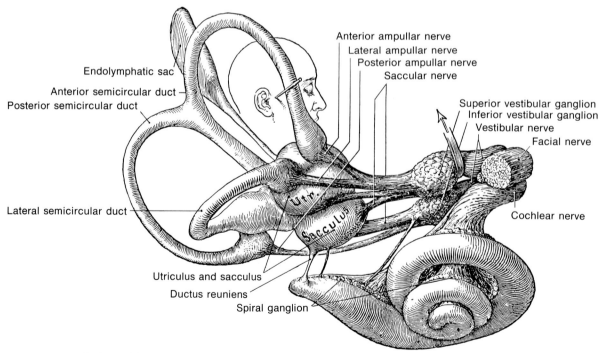

FIG. 6-30. *Vestibulocochlear nerve*

THE NERVOUS SYSTEM ■ 181

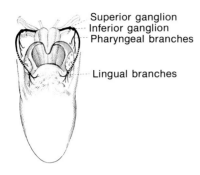

FIG. 6-31A. *Glossopharyngeal nerve*

portion belonging with the vagus. Its name is derived from the fact that it is accessory to the vagus. It is motor in composition, supplying fibers to the trapezius and sternocleidomastoideus muscles.

From the jugular foramen the nerve passes into the neck between the internal carotid artery and the internal jugular vein. It lies on the deep belly of the sternocleidomastoid muscle or frequently pierces it, sending a branch to innervate this muscle. It then crosses to the border of the trapezius muscle, beneath which it passes and continues inferiorly on its deep surface, with branches to the muscle substance.

CRANIAL NERVE XII—HYPOGLOSSAL NERVE The *hypoglossal nerve* (fig. 6-34) is somatic motor to the tongue musculature. It arises by a number of fila-

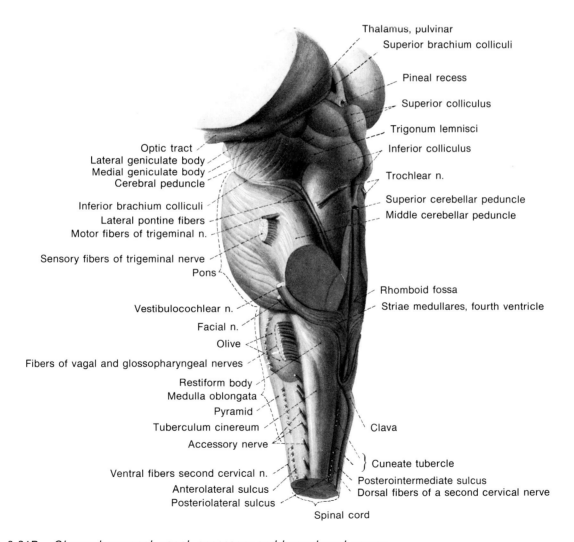

FIG. 6-31B. *Glossopharyngeal, vagal, accessory and hypoglossal nerves*

ments from the medulla and pierces the hypoglossal foramen. Extracranially, after giving off a minute recurrent meningeal branch, it lies at first between the internal carotid and internal jugular vein. It courses inferiorly and then anteriorly between the posterior belly of the digastric and hyoglossus muscles, and then between the mylohyoid and genioglossus muscles. Motor branches also extend to the hyoglossus, styloglossus, genioglossus, and the intrinsic lingual musculature.

Spinal Nerves

The *spinal nerves* (fig. 6-35A) include 8 pairs of cervicals, 12 pairs of thoracics, 5 pairs of lumbars, 5 pairs of sacrals, and 1 pair of coccygeal. The nerves are numbered to correspond with the vertebra above except in the cervical series, which take the number of the vertebra below, so that there are eight cervical nerves although only seven vertebrae. In the spinal canal the roots of the nerves at the cranial end of the

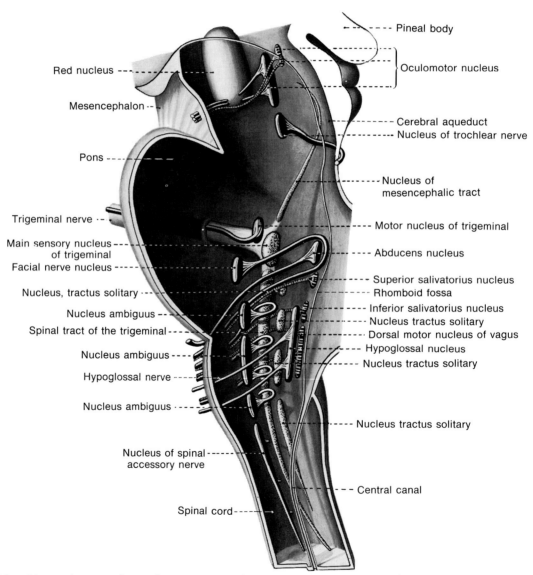

FIG. 6-32. *Glossopharyngeal, vagal, accessory and hypoglossal nerves and nuclei*

series are directed laterally, with a gradual change in direction until the inferior lumbar and sacral nerves course caudally, to form the cauda equina.

A typical spinal nerve has a *dorsal sensory root* derived by the convergence of filaments connected with the dorsal column of gray matter of the cord, and a smaller *ventral motor root,* derived by the convergence of filaments from the ventral column of gray matter. The dorsal roots include ganglia situated in the intervertebral foramina (except those of the sacral and coccygeal series), within the canal, and the first and second cervical ganglia, which lie on the vertebral arches.

The two roots join before leaving the intervertebral canal. Central to this point each root is enclosed in separate sheaths of the meninges. Peripheral to this point, where sensory and motor fibers intermingle, there is a single sheath of dura and pia, which blends with the connective tissue covering of the nerve. Just before leaving the intervertebral foramen the single nerve trunk gives off a minute *recurrent meningeal filament,* which receives a fine contribution from the sympathetic system and is distributed to the arteries of the vertebral canal. As the main trunk leaves the foramen it splits into two primary rami, a *dorsal ramus,* and a *ventral ramus.* Each ramus carries both sensory and motor fibers. The dorsal supplies the dorsal deep back musculature and overlying skin, and the ventral, the lateroventral musculature of the trunk and all the musculature of the upper and lower limbs, and corresponding sensory areas (fig. 6-35*B*).

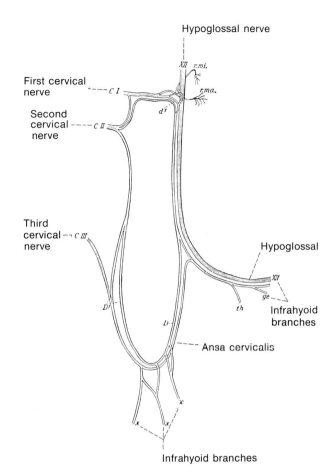

FIG. 6-34. *Hypoglossal nerve*

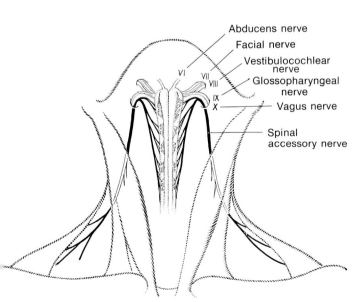

FIG. 6-33. *Spinal accessory nerve*

DORSAL RAMI OF SPINAL NERVES The posterior rami (fig. 6-35*C*) are directed dorsally and except for those of the first cervical, lower sacral, and coccygeal nerves, they divide into medial and lateral branches.

Cervical Region Although the dorsal root of the first cervical nerve is small or even absent, the dorsal ramus is larger than the ventral, and it does not divide into lateral and medial branches. It enters the suboccipital triangle, sends a communicating branch to innervate the posterior rectus capitis major and superior oblique muscles, and contributes to the innervation of the semispinalis capitis, inferior oblique, and posterior rectus capitis minor muscles.

The dorsal ramus of the second cervical nerve is the largest of all dorsal rami, and larger than the ventral. It emerges inferior to the inferior oblique muscle which it helps innervate, and then divides into a small lateral and a large medial branch. The medial branch constitutes the *greater occipital nerve.* It ascends and pierces the semispinalis capitis and trapezius at a var-

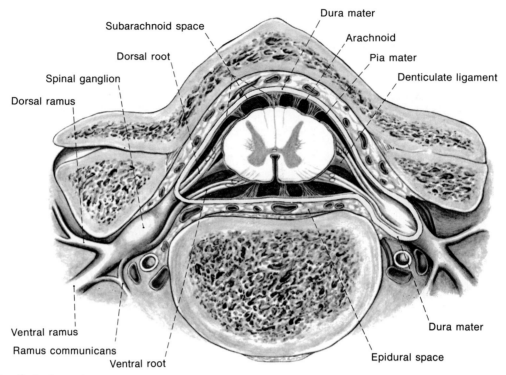

FIG. 6-35A. *Spinal nerves*

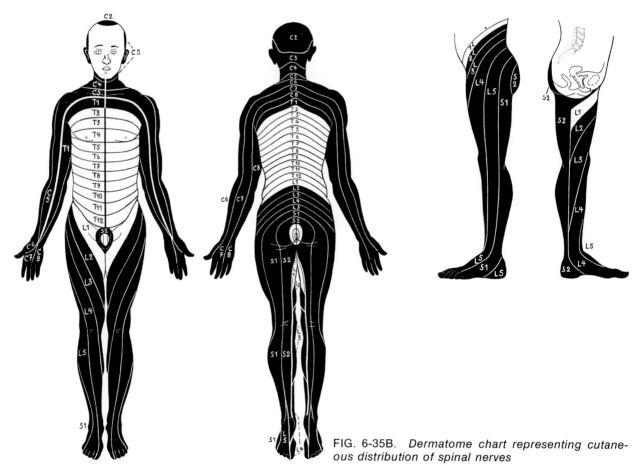

FIG. 6-35B. *Dermatome chart representing cutaneous distribution of spinal nerves*

iable point near the midline, and is distributed over the scalp, anastomosing terminally with the lesser occipital, greater auricular, posterior auricular, and third cervical branches.

The dorsal ramus of the third cervical nerve is small. The medial branch innervates adjacent back muscles and then extends superiorly, piercing the trapezius usually just medial and inferior to the greater occipital nerve; it helps innervate the skin of the occipital region.

The dorsal rami of the fourth to eighth cervical nerves divide into *lateral branches,* carrying motor fibers to the adjacent back musculature, and *medial branches* to the skin near the midline, although those of the inferior two or three nerves seldom reach the surface.

Thoracic Region The dorsal rami of the thoracic nerves divide into *lateral* and *medial branches.* The medial branches of the superior six carry some motor fibers to the dorsal musculature, or sacrospinae muscles, but are chiefly sensory reaching the skin near the midline, the more superior of the series sending long, and the more inferior sending shorter twigs laterally over the back. In the lower six lumbars, however, the medial branches are chiefly motor and seldom send distinguishable sensory filaments to the skin. The lateral branches of the superior six nerves are chiefly muscular, and the inferior six chiefly sensory.

Lumbar Region The dorsal rami of the lumbar nerves divide into *lateral* and *medial branches.* The latter are muscular to the multifidus division of the

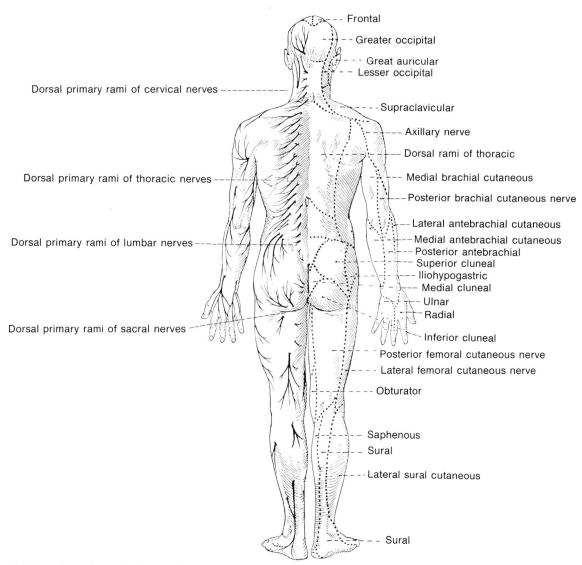

FIG. 6-35C. *Dorsal rami of the spinal nerves*

erector spinae. Lateral branches are partly muscular, to the erector spinae, but the superior three emerge subcutaneously just cranial to the iliac crest as long sensory branches known as *superior cluneal*, and descend over the lateral part of the buttock. The lateral branches of the last two lumbar nerves are usually exclusively muscular.

Sacral and Coccygeal Regions The dorsal rami of the sacral and coccygeal nerves are small and short. Those of the first three divide into medial muscular branches, and lateral branches that supply the skin of the sacral region. Those of the last two sacral and the coccygeal nerves do not divide, but intercommunicate with branches to the skin of the coccygeal region.

VENTRAL RAMI OF SPINAL NERVES The *ventral ramus* of a typical spinal nerve (fig. 6-35D) continues from the intervertebral foramen and at once gives off a short communicating branch carrying fibers from the sympathetic system to the spinal nerve, the gray ramus communicans, present in every spinal nerve. In addition, in the case of the thoracic and first three lumbar nerves only, there is associated with the gray ramus a white ramus communicans, carrying fibers from the spinal cord to the sympathetic trunk. The rami communicantes are not entirely uniform; a gray

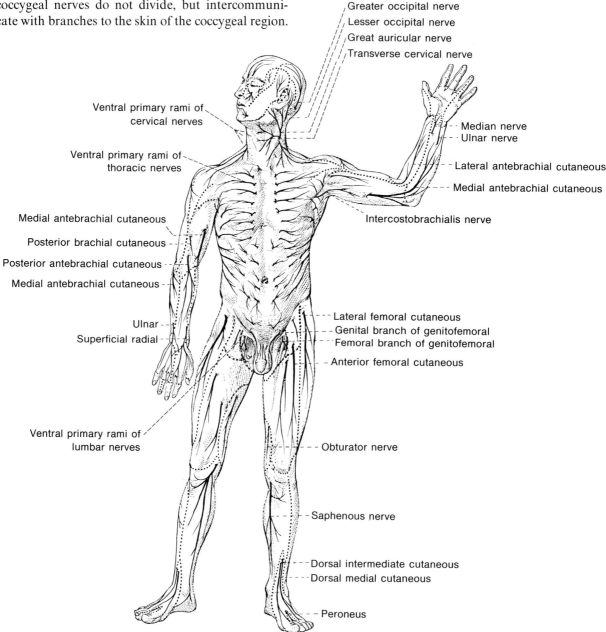

FIG. 6-35D. *Ventral rami of the spinal nerves*

may be partly fused with a white ramus, or one may even be absent, its fibers being carried by a neighboring ramus.

The ventral ramus in the thorax follows the curve around the trunk, giving off muscular and cutaneous branches in its course. Muscular branches, to lateral and ventral musculature, are given off at irregular intervals. At about the middle of its course, however, a lateral cutaneous branch is given off that enters the subcutaneous tissue on the side of the body and is distributed by a dorsal and a ventral twig. The ventral ramus terminates in a ventral cutaneous branch, distributed to the skin of the ventral aspect of the trunk by medially and laterally directed twigs.

In the cervical, lumbar, and sacral regions, however, the ventral rami are not of the above typical pattern, but communicate to form plexuses. The reason for plexus formation is not known, but probably has to do with diversification of action of individual muscles used in movements in which numerous other muscles also take part. The plexuses innervating the limbs are derived from the ventral rami of the spinal nerves concerned.

The *ventral rami of the cervical nerves* are conveniently divided into a superior series, comprising C1 to C3 and part of C4, which enter into the formation of the *cervical plexus;* and an inferior series, from C4 (partly) to T1, divided into nerves to the trunk and nerves to the limb (*brachial plexus*).

Cervical Plexus Cutaneous branches of the cervical plexus (fig. 6-36) are derived from C2 to C4 and consist of a superior series of fibers from C2 and C3, and an inferior series similarly derived from C3 and C4.

The superior series consists of branches with three names: (1) *lesser occipital nerve,* coursing largely in line with the posterior border of the sternocleidomastoid muscle in a dorsocranial direction to supply sensibility to the lateral occipital region; (2) *great auricular nerve* (C2–C3) obliquely crosses the sternocleidomastoid muscle and sends twigs of sensibility dorsal to the ear and to the angle of the jaw; (3) *transverse cervical* (C2–C3) passes ventrally deep to the external jugular vein and breaks up into twigs supplying sensibility to the ventral part of the neck. All three nerves course deep to, and in intimate contact with, the platysma muscle, piercing that muscle toward their terminations.

The inferior or descending series consists of three branches of the *supraclavicular nerve* (C3–C4), termed the *medial, intermediate,* and *lateral branches.* All course upon the deep surface of the platysma muscle until near final distribution, which is the cutaneous innervation of the base of the neck, superior pectoral region, and the shoulder, both in ventral and dorsal aspects.

The *ansa cervicalis* is the inferior ramus from C2 and C3 and a superior ramus from C1. The superior ramus seems to descend from the hypoglossal nerve to innervate the thyrohyoid muscle. The ansa cervicalis innervates the thyrohyoid, omohyoid, sternohyoid, and sternothyroid muscles.

The *phrenic nerve* is from C3, C4, and C5. It passes ventral to the scalene anterior and enters the thorax between the subclavian artery and vein. In the mediastinum it continues ventral to the root of the long, between the pericardium and pleura. The right nerve is shorter and deeper than the left. It supplies mostly somatic motor, but also sensory (to the central part) and vasomotor (sympathetic, from the middle and inferior cervical ganglia) fibers to the diaphragm.

Brachial Plexus Although C3 and T2 both supply cutaneous fibers to the root of the arm, the *brachial plexus* (fig. 6-37A) is considered to be formed by branches of C5 to T1. C4 usually contributes a small branch (to C5) as well.

The brachial plexus extends caudolaterally, emerging between the anterior and middle scalene muscles and extending dorsal to the clavicle, and farther distally, dorsal to the subclavian vessels as well.

The brachial plexus is constructed on a basic plan that may be explained as follows. The *roots* (first zone) of C5 joins C6, and C8 joins T1, resulting in an arrangement that may be likened to the device V I V. From these extend three *cords* (second zone), the upper with fibers from C5–C6, the middle with fibers from C7 only, and the lower from C8–T1. In a third zone, comprising *divisions,* each of the cords breaks up so that all posterior fibers separate from all the anterior fibers, and these fibers pass into a fourth zone of *trunks.* Thus all three cords send posterior fibers through the division zone converging to the posterior trunk; the upper and middle cords send their anterior fibers through the divisions to the lateral trunk; and the lower cord sends its ventral fibers through the division zone to the medial trunk.

The fourth zone comprises *lateral* and *medial* (both anterior) and *posterior trunks.*

Anterior branches of the brachial plexus (fig. 6-37B) supply muscles belonging to the primitive flexor or anterior group, and ventral areas of the skin of the appendage. Units include: the subclavius nerve, medial and lateral pectoral nerves, suprascapular nerve, musculocutaneous nerve, median nerve, ulnar nerve, medial brachial cutaneous nerve, and medial antebrachial cutaneous nerve.

(1) The *subclavius* (C5–C6) is a small nerve arising from the ventral nerve of the upper cord or just distal to it. It passes dorsal to the clavicle and innervates the subclavius muscle.

(2) The *medial and lateral pectoral nerves* arise from a loop that is attached at one end to the lateral trunk and at the other to the medial trunk. Branches are given off the loop and these innervate the pectoralis major (C5, C6–C8) and pectoralis minor (C7–C8) muscles. Branches reach the major division by piercing both the clavipectoral fascia and the pectoralis minor.

(3) The *suprascapular nerve* (C5–C6) is given off the upper cord, and is the first or most superior of the nerves coursing dorsolaterally from the plexus. It extends to the cranial border of the scapula, passes through the suprascapular notch beneath the transverse scapular ligament to the dorsal surface of the bone, innervates the supraspinatus and the shoulder joint, and then traverses the scapular notch to end in the infraspinatus muscle.

(4) The *musculocutaneous nerve* (C5–C7) is given off the lateral trunk of the plexus. It innervates the coracobrachialis muscle. The musculocutaneous nerve then pierces this muscle obliquely to innervate the biceps brachii and brachialis muscles. The muscular part ends at these muscles, but a sensory part passes between them and emerges, lateral to the biceps, in the lower brachium as the *lateral antebrachial cutaneous,* which supplies the skin toward the radial side of the palmar aspect of the forearm distally as far as the wrist.

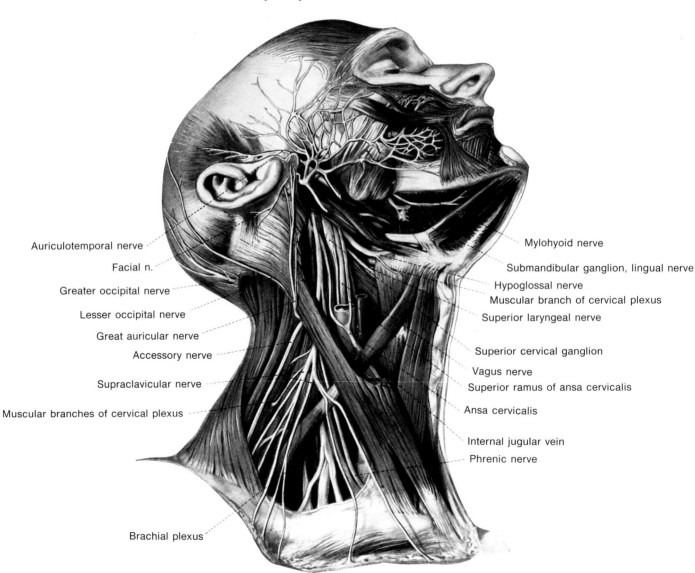

FIG. 6-36. *Cervical plexus*

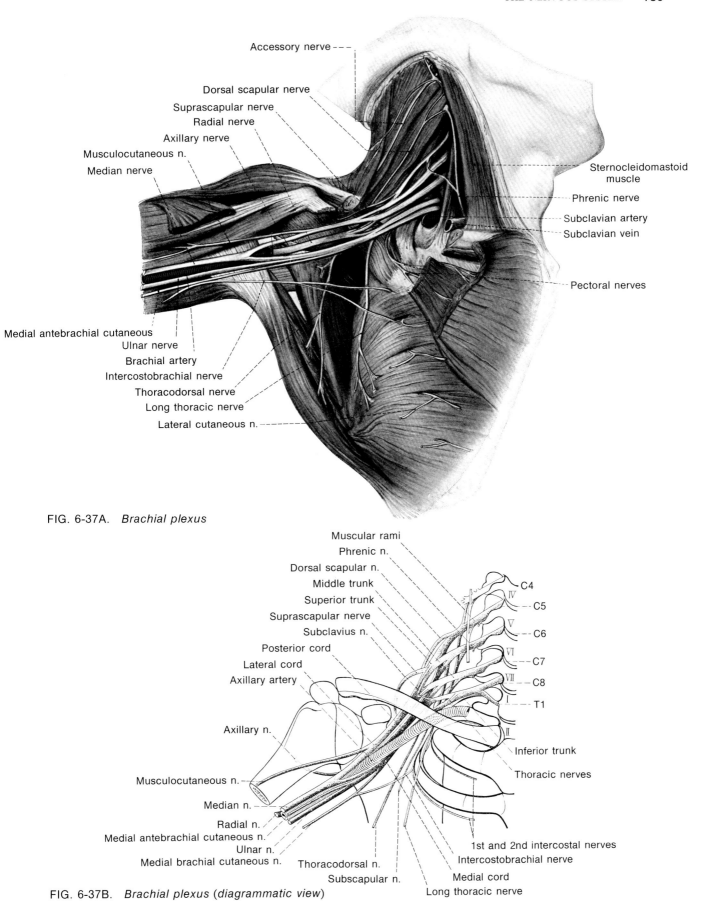

FIG. 6-37A. *Brachial plexus*

FIG. 6-37B. *Brachial plexus (diagrammatic view)*

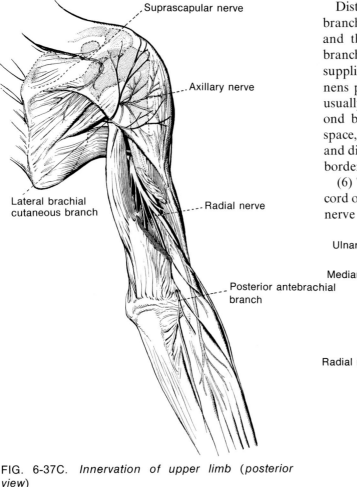

FIG. 6-37C. *Innervation of upper limb (posterior view)*

Distal to the wrist the median nerve splits into two branches that are dorsal to the palmar aponeurosis and the superficial palmar arterial arch. The first branch courses toward the first interdigital space, supplies twigs to the abductor pollicis brevis, opponens pollicis, and the flexor pollicis brevis. The last usually supplies the first lumbrical muscle. The second branch extends toward the second interdigital space, supplies a twig to the second lumbrical muscle, and divides into cutaneous branches for the adjoining borders of the second and third digits.

(6) The *ulnar nerve* (C8–T1) arises from the lower cord of the plexus. The terminal branches of the ulnar nerve pierce the transverse carpal ligament, going

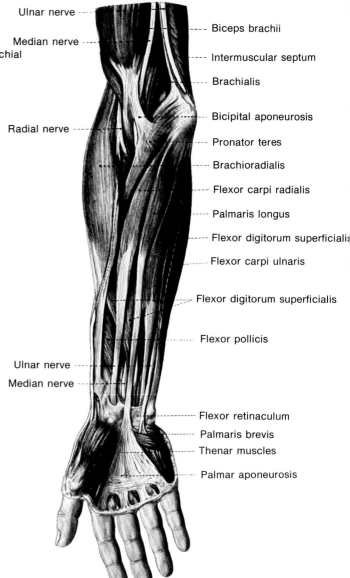

(5) The *median nerve* (C5–T1) arises by a root from the lateral trunk, and another from the medial trunk, between which passes the brachial artery. These join in some cases sooner than in others. In its course along the medial brachium it passes from the anterior to the posterior side of the brachial artery, then passes to the forearm medial to the biceps and deep to the biceps aponeurosis.

As it enters the forearm it gives off an articular twig to the elbow joint, passes between the two heads of the pronator teres muscle, and extends between the superficial and deeper flexor muscles of the forearm to the wrist. Here it courses deep to the transverse carpal ligament and superficial to the flexor tendons to the digits.

The median nerve gives off branches to the pronator teres, flexor carpi radialis, palmaris longus, flexor digitorum superficialis, flexor pollicis longus, flexor digitorum profundus, and pronator quadratus muscles.

FIG. 6-37D. *Innervation of upper limb (anterior view)*

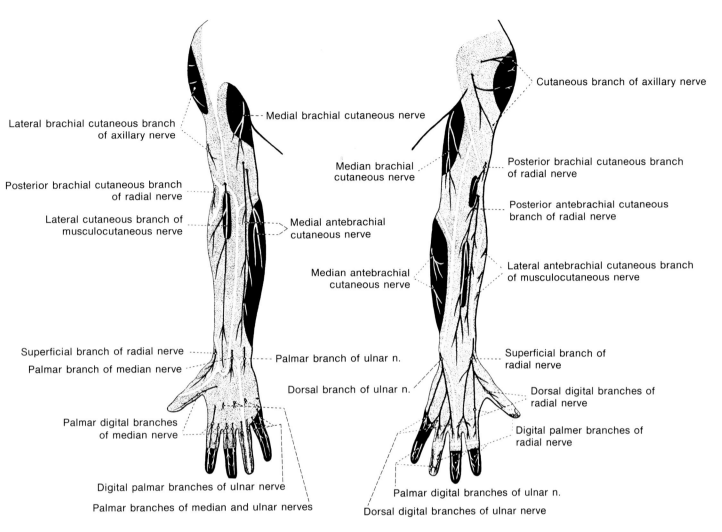

FIG. 6-37E. *Cutaneous innervation of upper limb*

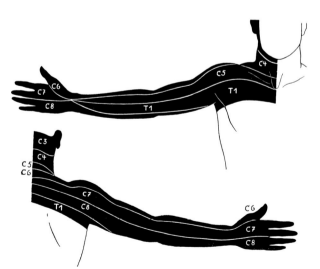

FIG. 6-37F. *Dermatome chart representing cutaneous innervation of upper limb*

deep between the origin of the short abductor and flexor of the little finger. Dorsal to the long flexor tendons it then arches radialward, giving off in its course twigs to the abductor, flexor, and opponens digiti minimi, interossei, fourth and third lumbrical, and adductor pollicis muscles.

(7) The *antebrachial cutaneous nerve* (C8–T1) arises from the medial trunk of the brachial plexus. It accompanies the other brachial nerve to the medial side of the upper arm and emerges subcutaneously in the middle of the arm at the point of disappearance of the basilic vein. Then it is distributed to the skin over the ulnar aspect of the forearm.

Posterior branches of the brachial plexus supply muscles belonging to the primitive extensor or posterior groups. As with the ventral branches, even the most proximal twigs should be considered as belonging with the dorsal nerve element of the limb.

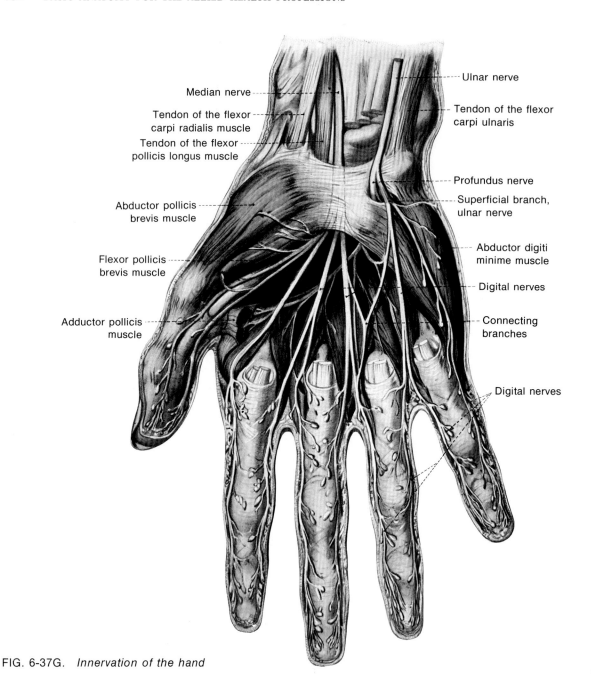

FIG. 6-37G. *Innervation of the hand*

(1) The three roots of the *long thoracic nerve* (C5–C7) fuse as they pass inferiorly in close contact with the thorax, and the nerve thus formed gives off twigs to innervate the digitations of the serratus anterior muscle.

(2) The *dorsal scapular nerve* (C5) passes deep to the levator scapulae and is distributed to the deep surfaces of the rhomboidei major, minor, and levator scapulae muscles.

(3) The *subscapular nerves* (C5–C7) are somewhat variable in arrangement and number. There is usually an *upper, middle,* and a *lower nerve,* from the posterior division of the plexus. They extend to the deep surface of the subscapularis: the upper nerve innervates the teres major.

(4) The *thoracodorsal nerve* (C6–C8) arises from the posterior cord of the plexus, often in company with the lower subscapular nerve, and passes inferiorly upon the deep surface of the latissimus dorsi muscle, which it innervates.

(5) The *axillary nerve* (C5–C6) diverges from the radial nerve and leaves the axilla by passing through

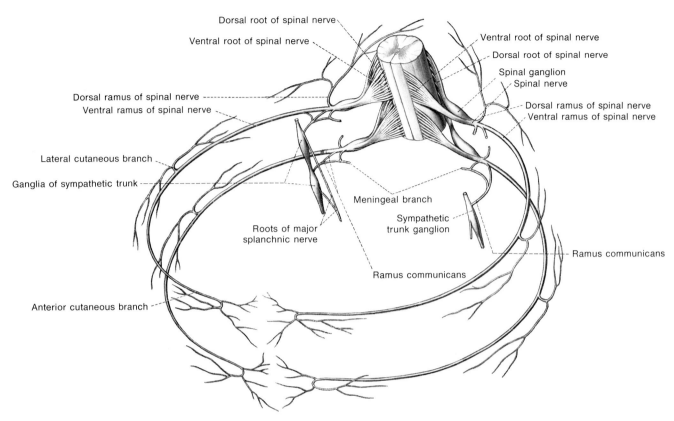

FIG. 6-38. *Typical thoracic spinal nerves*

the space bounded by the teres minor, long head of the triceps, and teres major muscles, and the humerus, and innervates the teres minor and deltoid muscles. Cutaneous fibers also curve around the dorsal border of the deltoid muscle to become the lateral brachial cutaneous nerve.

(6) The *radial nerve* (C5–T1) is the direct continuation of the posterior trunk of the plexus. In the upper brachium it leaves the other nerves of the arm and obliquely curves posterolaterally and then anteriorly around the humerus, deep to the origin of the lateral head of the triceps, and reaches the level of the elbow joint between the brachioradialis and brachialis. In its course in the upper arm it gives off muscular and cutaneous branches. The posterior brachial cutaneous nerve curves caudolaterally to innervate the skin over the superior part of the triceps. The antebrachial cutaneous is given off deep to the lateral head of the triceps and becomes subcutaneous laterally and somewhat distal to the deltoid. It is distributed to the skin of the elbow and forearm distal to this point.

The muscular branches of the radial nerve supply all divisions of the triceps, and branches are given off to the anconeus, brachioradialis, extensor carpi radialis longus, and brachialis.

In the forearm the radial nerve splits into two parts at the elbow: (1) The *superficial cutaneous branch* extends distally deep to the brachioradialis to the wrist on the radial border of the forearm and becomes subcutaneous. It is distributed to the dorsum and radial border of the thumb, both sides and dorsum of the basal part of the index finger, and adjoining border of the third finger, with some variation in the number of digits supplied. (2) The *deep radial nerve* is entirely muscular. It pierces the supinator and traverses the forearm between the layers of superficial and deep flexor muscles. In its course in the forearm it innervates the following muscles: extensor carpi radialis longus, supinator, extensor carpi radialis brevis, extensor digitorum, extensor digiti minimi, extensor carpi ulnaris, abductor pollicis longus, extensor pollicis longus, extensor pollicis brevis, and extensor indicis.

Cutaneous innervation of the upper limb is shown in figure 6-37C. The superior part of the shoulder is supplied by branches of the supraclavicular nerve (C3–C4) and the axillary region by the intercostobrachial nerve [(T1) T2 (T3)]. The skin of the arm is supplied by three groups of nerves, two posterior and one anterior: medial brachial cutaneous (anterior), posterior

194 ■ BASIC ANATOMY FOR THE ALLIED HEALTH PROFESSIONS

brachial cutaneous (posteior), and lateral brachial cutaneous (posterior). The first and last of these encroach upon the anterior aspect. The elbow and forearm likewise are supplied by three sets of similar cutaneous nerves, but two of them are anterior and one posterior: medial antebrachial cutaneous, lateral antebrachial cutaneous, and posterior antebrachial cutaneous.

In the hand, the ulnar two-fifths of the palm is supplied by the ulnar nerve, and the medial three-fifths, as well as the dorsum of the second to fourth digits, by the median nerve. Upon the dorsal surface, the ulnar two-fifths is supplied by the ulnar, and the remainder by the radial nerve.

Application of the above statements may be demonstrated with the arm extended laterally, palm forward (fig. 6-37D). The cutaneous innervation of the neck gradually merges with that of the preaxial border of the upper arm, lower arm, and hand; then with postaxial border of the hand, lower arm, upper arm, axilla, and trunk. Thus C3 innervates the lower neck; C4, the shoulder; C5, the preaxial brachium ob-

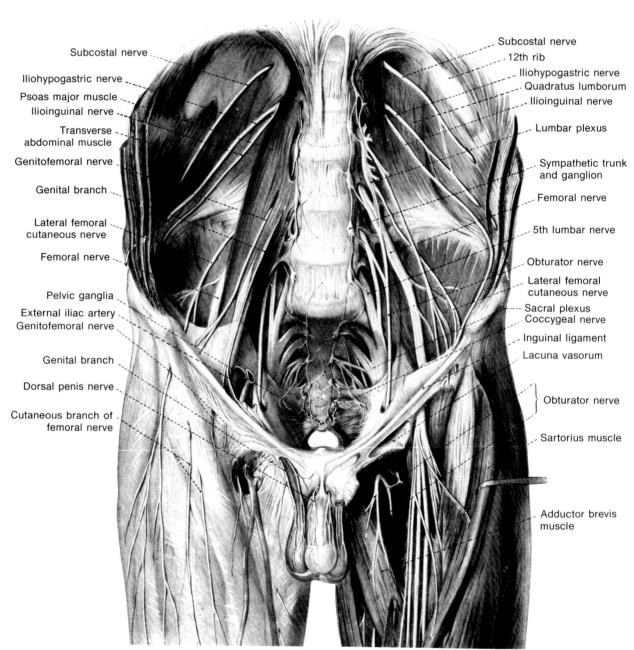

FIG. 6-39A. *Lumbosacral plexus*

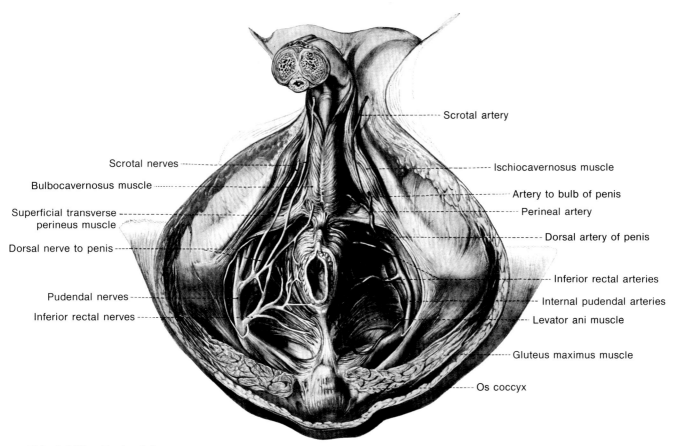

FIG. 6-39B. *Pudendal nerve*

liquely; C6, the preaxial lower brachium and forearm; C7, the radial border of the hand; C8, the ulnar border of the hand; T1, the postaxial (ulnar) forearm; T2, the postaxial brachium; and T3, the postaxillary region.

Thoracic Nerves There are 12 pairs of thoracic nerves (fig. 6-38), each pair below the corresponding vertebra. Their *ventral rami* conform to the pattern already described as typical, except for those of the 1st, 2nd, 3rd, and 12th vertebrae, which are exceptional in the respects mentioned. In the case of the remainder, the ventral ramus follows the intercostal space, at first between the pleura and intercostal fascia, but shortly thereafter between the external and internal intercostal muscles. Toward the ventral surface of the body the nerves extend obliquely through the substance of the internal intercostal muscle until, near the medial termination of the costal cartilage, the nerve is entirely deep to that muscle. In the case of the more inferior thoracic nerves, the ventral rami extend obliquely to supply the entire abdominal wall, passing to the interchondral ligaments and between the internal oblique abdominal muscle and the transversus abdominis to the sheath of the rectus, which they perforate.

In their course, the ventral rami of the thoracic nerves give off muscular branches to all of the lateroventral musculature of the trunk. As they approach the lateral aspect they also give off cutaneous branches, which pierce the intervening muscles and appear subcutaneously along a line roughly from the axilla to the anterior spines of the ilium. These are the *lateral cutaneous branches* of the ventral rami. Each divides into a *posterior* and *anterior branch*. Superiorly, the former are long and the latter short; inferiorly, their condition is reversed. Finally, the ventral rami end in *anterior cutaneous branches*, which become subcutaneous superiorly in the niches between costal cartilages and sternum, and inferiorly in irregular situations in the ventral abdomen. They, too, divide into *lateral* and *medial branches*, although the latter are often too small to be readily distinguished. The lateral branches are longer superiorly and shorter inferiorly, to correspond with neighboring branches of the lateral divisions.

The ventral ramus of the first thoracic nerve differs from the above plan in that it sends a large branch obliquely across and ventral to the neck of the first rib to join the brachial plexus. The remaining intercostal portion rarely has a lateral or a ventral branch.

The ventral ramus of the second, and often the third, thoracic nerve differs from the typical plan in that its lateral division receives a communication from the medial brachial cutaneous and sends cutaneous branches—*intercostobrachial nerves*—to the axillary region.

The ventral ramus of the 12th thoracic, or subcostal nerve, differs from the typical thoracic plan in receiving a proximal contribution from T11. Its lateral division supplies the skin over and inferior to the iliac crest, and its ventral division supplies the inferior abdomen near the pubis.

VENTRAL RAMI OF LUMBOSACRAL NERVES The ventral rami of the lumbosacral nerves (fig. 6-39A) may be divided into three groups, comprising (1) lumbar nerves to the structures of the inferior trunk, (2) nerves to the structures of the lower limb (lumbosacral plexus), and (3) pudendal and coccygeal plexuses.

Lumbar Nerves to the Trunk The trunk muscles in the lumbar region comprise the quadratus lumborum, innervated by short deep branches from T12 to L3. In addition, the first lumbar nerve behaves largely as a thoracic nerve, but receives a proximal contribution from T12. It has one branch, termed the *iliohypogastric nerve*, that conforms to the pattern of a tho-

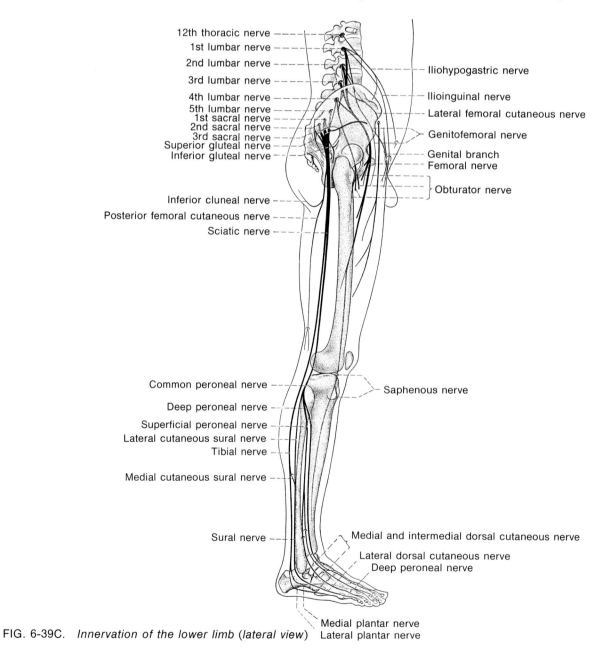

FIG. 6-39C. *Innervation of the lower limb (lateral view)*

racic nerve, with a lateral cutaneous division coursing inferiorly over the crest of the ilium and a ventral division that courses at first between the internal oblique abdominal and transversus abdominis muscle, then between the oblique muscles, to the skin over the superficial inguinal ring. L1 has another proximal branch, the *ilioinguinal nerve,* which passes between the abdominal muscles giving off branches to the muscles and to the peritoneum, passes through the superficial inguinal ring, and supplies sensory twigs to the neighboring medial thigh as well as anterior scrotal branches to the superior part of the scrotum (labium majus).

The *genitofemoral nerve* (L1, L2) comes off the root of L2, but receives a contribution from L1. It does not follow the periphery of the abdomen, as do the more superior nerves, but courses inferiorly, perforating the psoas major muscle and passing dorsal to the ureter. In the abdominal cavity it separates into two branches: (1) the *femoral branch,* which passes

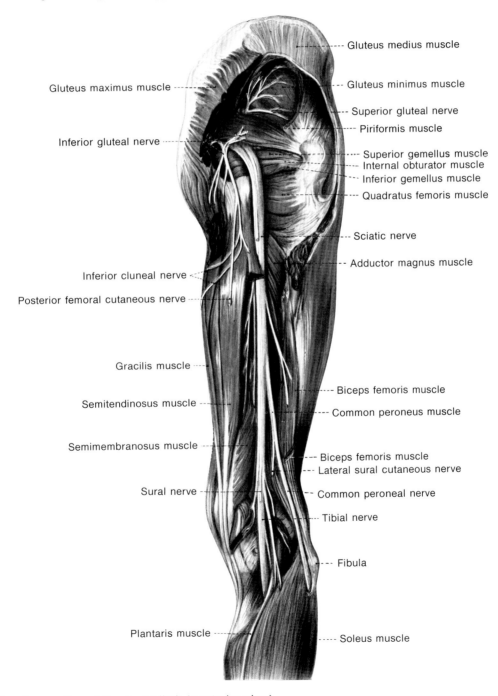

FIG. 6-39D. *Innervation of the lower limb (posterior view)*

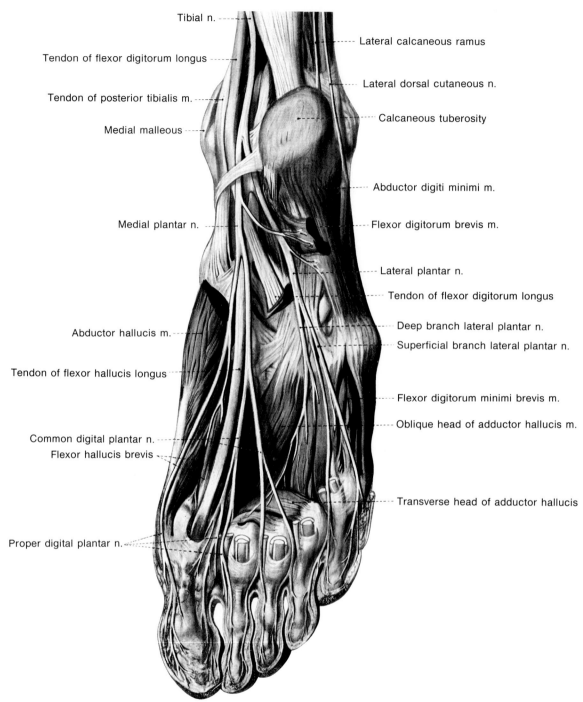

FIG. 6-39E. *Innervation of the foot*

through the lacuna vasorum and innervates the skin around the fossa ovalis; and (2) the *genital branch*, which passes through the deep inguinal ring and sends twigs to the cremaster muscle, the tunica dartos, and the skin of the scrotum.

The nerves of the lower limb, like the muscles, are divisible into four groups, and four only, which are known by the names of their chief nerves. The *pre-pelvic group of nerves* constitutes the *lumbar plexus*, the dorsal element being the femoral group of nerves, and the ventral element, the obturator group of nerves. It so occurs that neither femoral nor obturator nerves innervate any muscle distal to the knee.

The *postpelvic group of nerves* constitutes the *sacral plexus*, the dorsal element being the peroneal group of nerves, and the ventral element, the tibial group.

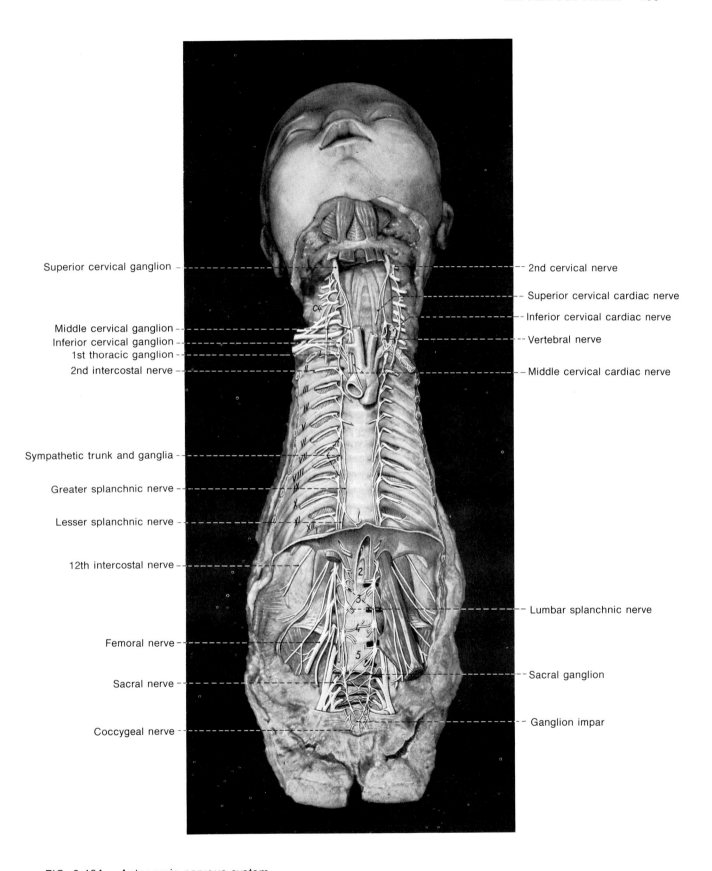

FIG. 6-40A. *Autonomic nervous system*

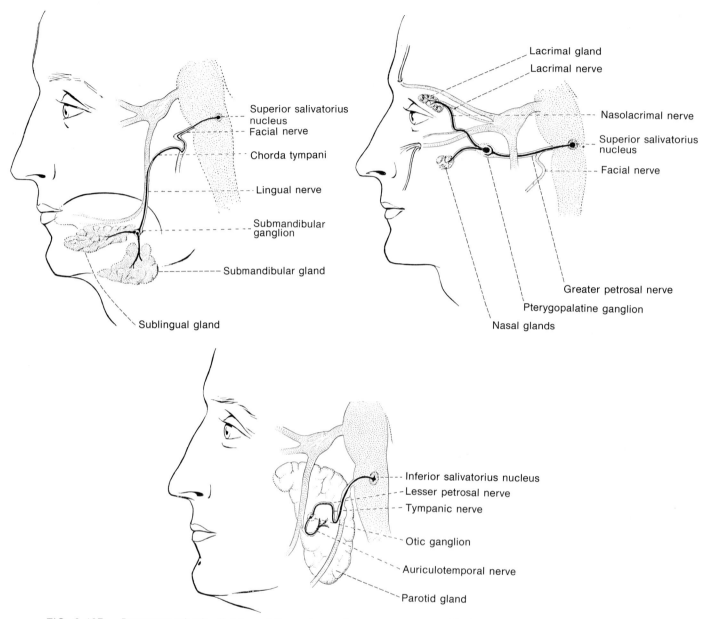

FIG. 6-40B. *Parasympathetic division of the autonomic nervous system (facial and glossopharyngeal nerves)*

All branches of the sacral plexus belong to one or the other of these groups. For some distance, peroneal and tibial nerves are enclosed in a common sheath, and this is known as the sciatic nerve. Tibial and peroneal nerves innervate muscles of the hip and thigh, and the leg and foot.

Lumbar Plexus The *lumbar plexus* (fig. 6-39B) of prepelvic nerves is divisible into dorsal (femoral), and ventral (obturator) elements. It is formed of contributions from the ventral rami of L2, L3, and L4. The shortest branches of the *femoral nerves* are those to the psoas major and minor, from L2, L3, and usually L4. They penetrate the muscle directly. The nerve to the iliacus muscle is also a high branch of the femoral nerve, which enters the muscle directly.

The *lateral femoral cutaneous nerve* (L2–L3) is a cutaneous branch of the femoral nerve. It passes around the dorsal margin of the psoas major, follows the medial surface of the iliacus, and courses beneath the inguinal ligament. Then it emerges subcutaneously upon the thigh, in two or more branches, and supplies the skin of the lateral part of the anterior aspect of the thigh distally as far as the knee.

The proximal part of the *femoral nerve* contains obturator elements that follow within the same sheath for a short distance. These are described with

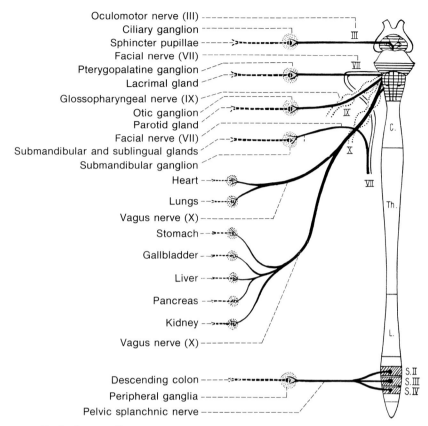

FIG. 6-40C. *Parasympathetic innervations*

the obturator group. The femoral nerve emerges into the pelvis between the psoas major and iliacus and then passes from beneath the inguinal ligament through the lacuna musculorum in the same relationship with these muscles, and lateral to the femoral sheath and vessels. At this point (called the femoral triangle) it breaks up into a number of branches, one of which innervates the hip joint; two sets of branches, short and long, go to the sartorius; numerous branches innervate the quadriceps femoris and articular branches from these go to the knee; and there are additional anterior femoral cutaneous branches. The latter occur in a high intermediate group, from lateral to the fossa ovalis distally to the knee; and a medial group, in the more medial and distal part of the anterior thigh.

The *obturator nerve* (L2–L4) pierces the psoas major muscle near its medial border, and then enters the small pelvis, passing over the medial surface of the obturator internus, below the pelvic brim in company with the obturator vessels, and by the obturator canal through the obturator foramen. At this point it divides into two branches:

(1) The *anterior ramus* gives off an articular branch to the hip, and passes anterior to the obturator externus and posterior to the pectineus and adductor longus muscles. It innervates the latter muscle and, often, the former. It then passes anterior to the adductor brevis and gracilis, sending twigs to both these muscles. The branch ends in cutaneous twigs that reach the medial thigh anterior to the gracilis.

(2) The *posterior ramus* often perforates the obturator externus, which it innervates; passes medial to the adductor brevis; supplies branches to the adductor brevis and a part of the adductor magnus; and ends in an articular branch to the knee.

The *saphenous* (L2, L3, and L4) is the main cutaneous nerve of the obturator group, a high branch that is not included in the part of the obturator that becomes enclosed by the obturator foramen, but instead follows in the sheath with the femoral nerve. It follows the femoral vessels down the medial thigh, becomes subcutaneous just behind the medial epicondyle, and continues down the leg with the great saphenous vein to the ankle and the medial side of the foot. In its course it gives off branches to the medial side of the knee, leg, and ankle.

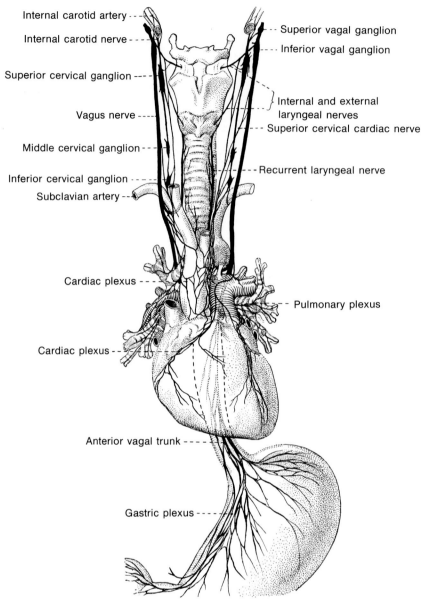

FIG. 6-40D. *Parasympathetic innervation (vagus nerve)*

Sacral Plexus The *sacral plexus* of postpelvic nerves is divisible into dorsal (peroneal) and ventral (tibial) elements, each of which innervates muscles in the hip and thigh, and leg and foot. It is formed of contributions from the ventral rami of L4 (thus overlapping the lumbar plexus), L5, S1, S2, and S3. The nerves join to form a broad plate soon after they emerge into the pelvis, and this narrows to leave the pelvis over the greater sciatic notch, so that in the hip all of them enter the gluteal region deep to the piriformis. The main nerves (peroneal and tibial), in a common sheath, course down the posterolateral thigh between the dorsal and ventral musculature of this segment.

The muscles of *peroneal nerve* division that may be classed as hip muscles are those of the gluteus groups. Accordingly, their nerves are classed together under three headings. They are all high. The remaining divisions are those of the common peroneal nerve proper.

(1) *Piriform nerves* (S1, S2) are the highest and are usually two in number. They enter the piriformis.

(2) The *superior gluteal nerve* (L4, L5, S1, S2) diverges superiorly from the mass of the plexus beneath the piriformis muscle and then branches between the glutei medius and minimus. It innervates the last two muscles and a twig continues to supply tensor fasciae latae upon its deep surface.

(3) The *gluteal nerve* (L5, S1, S2) is given off the mass of the plexus inferior to the beginning of the sciatic cord, deep to the piriformis, from the inferior border of which it emerges, and at once innervates the gluteus maximus.

As one of the two components of the sciatic cord, the common peroneal nerve (L4, L5, S1, S2) passes over the sciatic notch from the pelvis, emerges from the inferior border of the piriformis, and courses distally, at first beneath the gluteus maximus, between the greater trochanter and ischial tuberosity, and then deep to the long head of the biceps femoris. In the superior popliteal space the common peroneal nerve leaves the sciatic sheath, and just above the knee is deflected to the lateral side, passes over the origin of the lateral head of the gastrocnemius muscle and behind the head of the fibula, where it divides into two branches, superficialis and profundus, both of which pass through the origin of the peroneus longus muscle.

Well above the knee the common peroneal nerve gives off a muscular branch to the biceps femoris and, in the distal part of the thigh, an articular branch to the knee. The lateral cutaneous sural nerve is given off upon the medial side of the nerve in the popliteal space and innervates the skin upon the lateral side of the posterior aspect of the lower leg. Toward the ankle, at the lateral side of the common flexor tendon, it anastomoses with the medial sural cutaneous to form the sural nerve.

In the leg, the more anterior branch of the common peroneal nerve, the superficial peroneus, at first courses distally between the fibula and peroneus longus, innervating that muscle and the peroneus brevis. In the middle of the crus it divides into two terminal branches: (1) The dorsal cutaneous pedis intermedius nerve becomes subcutaneous anterior and proximal to the lateral malleolus, extends in a line with the fourth metatarsal, and supplies sensibility to the dorsum of the third to fifth digits. (b) The dorsal medial cutaneous pedis nerve becomes subcutaneous more proximally, upon the lateral side of the anterior crus, and extends toward the great toe, dividing into branches that supply sensibility to the medial side of the foot and great toe, and the adjoining margins of the second and third digits.

In the leg the more posterior branch of the common peroneal nerve, the *peroneus profundus,* gives off several branches adjacent to the neck of the fibula before passing deep to the origins of the peronei and extensor digitorum longus muscles. It courses distally between the tibialis anterior and the long extensors, then deep to the transverse ligaments of the ankle and the extensor hallucis brevis, and emerges subcutaneously in the first interdigital space, to supply the lateral border of the first digit and the medial border of the second digit. In the leg the deep peroneal nerve supplies an articular twig to the knee, and innervates the extensor digitorum longus tibialis anterior, extensor hallucis longus, and extensor digitorum brevis muscles.

The nerves of the tibial group to the hip comprise the following.

The *nerve to the quadratus femoris and gemellus inferior* (L4, L5, S1), which passes from the pelvis over the greater sciatic notch and supplies twigs to the deep surfaces of the above muscles, as well as to the hip joint.

The *nerve to the superior gemellus and obturator internus* (L5, S1, S2), which passes over the greater sciatic notch and, lateral to this, innervates the superior gemellus, but then enters the pelvis again over the lesser sciatic notch and innervates the obturator internus.

The nerves of the tibial group to the thigh are two; one cutaneous, the other chiefly muscular.

The *posterior femoral cutaneous* (S1, S2, S3) is located in the hip dorsomedial to the sciatic nerve and courses distally, deep to the gluteus maximus. In this situation it gives off *perineal rami* to the skin of the thigh near the perineum and scrotum (labium), and *inferior lateral cluneal rami* to the skin at the inferior margin of the buttock. The posterior femoral cutaneous nerve then emerges from the inferior border of the gluteus maximus and traverses the middle of the posterior aspect of the thigh, sending cutaneous branches to the neighboring skin as far distally as the proximal part of the crus.

The *nerve to the hamstring muscles* (L4, L5, S1, S2, S3) is the high branch of the tibial nerve to this group of muscles. It is located medial to the sciatic nerve and sends twigs distally to innervate the semitendinosus, semimembranosus, biceps femoris, and (in part) the adductor magnus muscles.

The nerves of the tibial group to the leg and foot are carried by the tibial nerve.

The tibial nerve (L4, L5, S1, S2, S3) is one of the two elements composing the sciatic, extending from the inferior border of the piriformis, beneath the gluteus maximus, and inferiorly upon the posterior thigh, covered by the biceps femoris. In the inferior part of the thigh it leaves the peroneus communis and continues through the popliteal space and between the two heads of the gastrocnemius muscle. In the popliteal space it gives off articular twigs to the knee joint to innervate the interosseus membrane of the

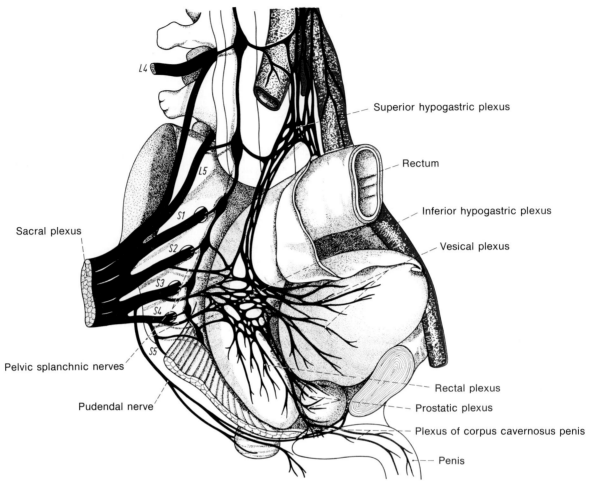

FIG. 6-40E. *Parasympathetic innervation (pelvic splanchnic nerves)*

crus and associated structures; rami musculares to the gastrocnemius, soleus, plantaris, and popliteus muscles; and the *medial sural cutaneous nerve.* Above the level of the ankle it receives an anastomosis from the lateral sural cutaneous and is thereafter known as the sural. This passes behind the lateral malleolus, innervates the lateral side of the heel, and continues over the lateral part of the foot as the *lateral dorsal cutaneous pedis nerve* to end upon the dorsum of the fifth digit.

In the crus the tibial nerve descends deep to the soleus, diverges to the medial side of the tendo calcaneus, and curves around the back of the medial malleolus, there dividing into lateral and medial plantar nerves. In the crus it gives off muscular branches to the flexor hallucis longus, flexor digitorum longus, and tibialis posterior, articular twigs to the ankle joint, and sensory twigs to the skin of the sole of the posterior part of the foot.

The *lateral plantar nerve* continues medial to the calcaneus onto the sole of the foot and diverges toward the lateral side between the quadratus plantae and flexor digitorum brevis muscles. Here it divides into a superficial and a deep branch. Before this point it innervates the abductor digiti minimi and quadratus plantae, and sends an articular branch to the tarsus. The *superficial ramus* sends twigs to the lateral side of the fifth digit, and the margins of the toes bordering the fourth interosseous space, as well as to the dorsum of the terminal phalanges of these toes. In its course it gives off twigs to the opponens and flexor digiti minimi muscles and the interossei of the fourth interdigital space. The *profundus ramus* courses dorsal to the quadratus plantae and next to the interossei. It innervates the adductor hallucis, the two or three lateral muscles and the interossei of the first to third interosseous space.

The *medial plantaris* (fig. 6-39*C,D*) continues distally on the medial side of the sole, lying upon the plantar aspect of the tendon of the flexor digitorum longus muscle and then of the flexor hallucis longus. In the middle of the foot it divides into branches, one to the medial side of the great toe, and the others to

There is a wedge-shaped area over the side of the hip supplied by the rami dorsales of T12 to L3, and an area lateral to the genitalia innervated by the genitofemoral nerve. Otherwise, the cutaneous innervation of the thigh is, approximately: anteriorly by the anterior femoral cutaneous; laterally by the lateral femoral cutaneous; posteriorly by the posterior femoral cutaneous; and medially by the obturator nerve. In the knee and lower leg there are only three (rather than four) such areas: medioanteriorly by the saphenous; laterally by the peroneal; and posteromedially by the tibial nerve. Thus in the thigh the cutaneous innervation is mostly dorsal, and in the leg, mostly ventral. Basically, the dorsum of the foot has peroneal, and the heel and sole, tibial innervation.

Pudendal Plexus The *pudendal plexus* (6-39A) overlaps the sacral plexus, and may be considered as a part of that, but concerned with the perineum and associated structures. Roots of S1, S2, S3, and S4 contribute to its formation. Its nerves leave the pelvis through the greater sciatic foramen and lie deep to the inferior border of the piriformis, but then pass over the sacrospinous and deep to the sacrotuberous ligaments, through the lesser sciatic foramen, and into the ischiorectal fossa. Here the nerves pass through the *pudendal canal* in the obturator fascia. It may be considered to have the following parts:

Deep visceral branches, carrying sensibility to the bladder, inferior rectum, and associated structures.

Deep muscular branches, deep in the ischiorectal fossa, innervating the levator ani and coccygeus muscles.

The medial inferior cluneal (S2, S3) is a dorsal cutaneous nerve, diverging from the base of the plexus, coursing with the pudendal nerves, and innervating the medial part of the buttock superior to the anus.

Inferior rectal branches, which are sensory to the vicinity of the anus and muscular to the sphincter ani externus.

Scrotal (labial) and perineal branches, passing superficial to the urogenital trigone and distributed to the perineum and scrotum, or labia in the female.

Deep muscular and dorsal penis branches, which are associated. They pierce the trigone, some of the twigs supplying neighboring muscles, with the dorsal penis nerve coursing for a short distance between the fascial layers of the trigone before emerging at the root of the penis (clitoris) and being distributed upon its dorsum. In its course, the transversus perinei (including the sphincter urethrae membranaceae), bulbocavernosus, and ischiocavernosus muscles are supplied.

Coccygeal Plexus The *coccygeal plexus* (fig. 6-39A) may in turn be considered as the terminal part

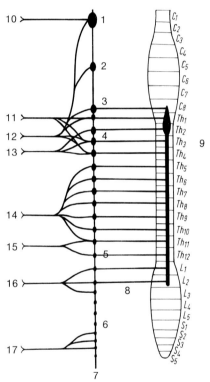

1—Superior cervical sympathetic ganglion
2—Middle cervical sympathetic ganglion
3—Inferior cervical sympathetic ganglion
4—Thoracic sympathetic ganglion
5—Lumbar sympathetic ganglion
6—Sacral sympathetic ganglion
7—Sympathetic trunk and ganglion
8—Rami communicantes
9—Spinal segments of spinal cord
10—Head and facial regions
11—Neck
12—Heart
13—Lungs
14—Liver, gallbladder, pancreas
15—Kidneys
16—Large and small intestines
17—Pelvic organs

FIG. 6-40F. *Sympathetic division of the autonomic nervous system*

the first, second, and third interdigital spaces and the adjoining digital margins. The branch to the last interdigital space receives an anastomosis from the lateral plantaris. The dorsal aspect of the terminal phalanges of these digits is also supplied. In its course the medial plantar nerve innervates the flexor digitorum brevis, flexor hallucis brevis, and abductor hallucis muscles.

It is impossible in a brief resume to give more than the general boundaries of the *cutaneous innervation of the inferior limbs,* and these overlap to a considerable extent.

206 ■ BASIC ANATOMY FOR THE ALLIED HEALTH PROFESSIONS

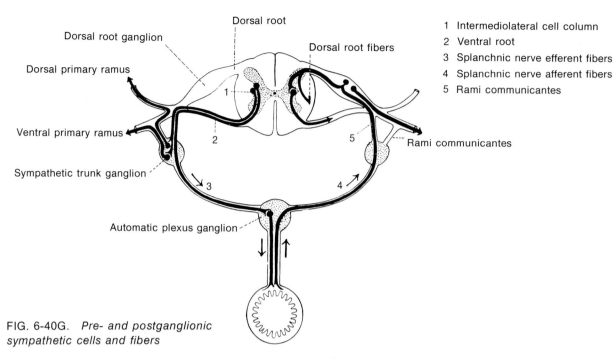

FIG. 6-40G. *Pre- and postganglionic sympathetic cells and fibers*

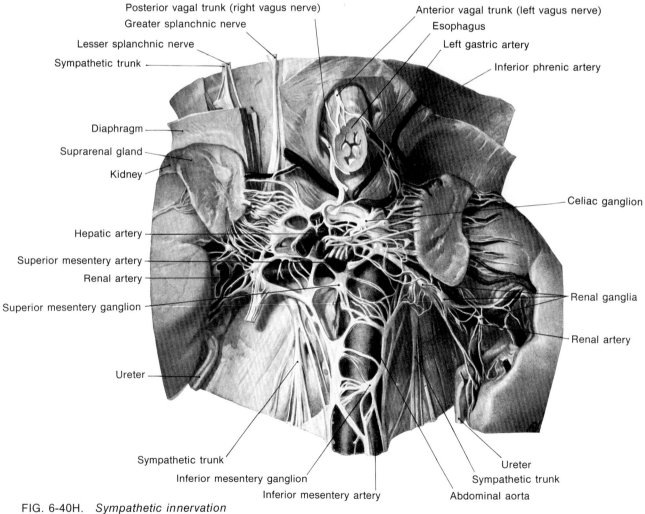

FIG. 6-40H. *Sympathetic innervation*

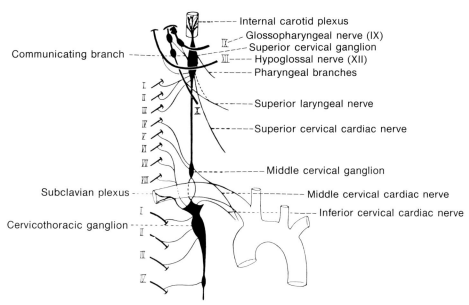

FIG. 6-40I. *Cervical sympathetic trunk and ganglia*

of the pudendal plexus, from S4, S5, and Coc. 1. It lies ventral to the origin of the coccygeus, innervates the coccygeus, and sends twigs to the viscera, sympathetic system, and the skin of the coccygeal region.

AUTONOMIC NERVOUS SYSTEM

The *autonomic nervous system* (fig. 6-40A) is the term applied to that part of the nervous system concerned with the smooth musculature, cardiac muscle, and glandular tissue.

The autonomic nervous system may be separated anatomically from the somatic nervous system only in the regions where the latter does not occur, as in most parts of the thoracic and abdominal cavities, and in small twigs of nerves to structures not supplied by the somatic system, such as filaments from nerves to arteries. Almost all nerves carry autonomic fibers, but not all nerves carry somatic fibers.

There are two motor neurons in the autonomic arc, a *preganglionic neuron* with cell body always in the central nervous system, and a *postganglionic neuron* with cell body outside the central system.

The autonomic system is divisible into *parasympathetic* or craniosacral and *sympathetic* or thoracolumbar portions.

Parasympathetic System

The *parasympathetic system* (fig. 6-40B) is composed of parasympathetic fibers carried by four cranial nerves (oculomotor, facial, glossopharyngeal, and vagus) and by three sacral nerves (S2, S3, and S4).

The four parasympathetic cranial ganglia are the *ciliary, pterygopalatine, submandibular,* and *otic.* They are formed of postganglionic cell bodies. However, every structure with parasympathetic innervation also receives sympathetic fibers, and accordingly, associated with each cranial parasympathetic ganglion are postganglionic sympathetic fibers, whose cell bodies are located in the superior cervical ganglion and whose fibers reach their destinations by following the carotid plexus and its continuations.

Parasympathetic fibers of the oculomotor (preganglionic cell bodies in the Edinger-Wesphal nucleus) follow the third nerve and then pass to the *ciliary ganglion* by its short motor root. The ganglion is situated toward the back of the orbit between the optic nerve and the lateral rectus muscle. Through it pass sensory fibers received from the nasociliary nerve by the long sensory root, and postganglionic sympathetic fibers from the cavernous plexus by the sympathetic root. Sensory, sympathetic, and postganglionic parasympathetic fibers are distributed by the short ciliary nerves to the ciliary muscle of accommodation and sphincter of the iris (both parasympathetic), and the dilator of the pupil (sympathetic).

Parasympathetic fibers of the facial (preganglionic cell bodies in the superior salivatory nucleus) (fig. 6-40C) follow two pathways:

(1) Fibers leave the geniculate ganglion and follow the greater petrosal nerve through the facial canal

hiatus in the petrous temporal, cross the foramen lacerum and enter the pterygoid (Vidian) canal, are joined there by the deep petrosal nerve carrying sympathetic fibers from the carotid plexus. The junction of the two forms the Vidian nerve (nerve of the pterygoid canal), which enters the pterygopalatine fossa, and joins the *pterygopalatine ganglion,* situated therein at the level of the base of the middle nasal concha. Through the ganglion also course sensory fibers received from the maxillary by way of the short pterygopalatine nerve. Numerous nasal and palatine nerves are distributed from the ganglion, anteriorly and inferiorly, to the mucous glands and arteries of the paranasal sinuses, nose, palate, uvula, and tonsils, and fine ascending orbital branches carry secretory fibers to the lacrimal gland and sensory fibers to the surrounding periosteum. Lateral nasal nerves course anteriorly to the conchae; medial nasal nerves, to the nasal septum; and palatine nerves, inferiorly to the palatine foramina.

(2) The second pathway taken by the parasympathetic fibers of the facial nerve is by the chorda tympani, through the middle ear to its junction with the lingual nerve, then following the latter to the *submandibular ganglion,* lying deep to the mandible and anterior to the medial pterygoid. The ganglion does not lie in the lingual nerve, but is connected with it by communicating branches. Postganglionic parasympathetic fibers pass from the ganglion (and sympathetic fibers through it) to the submandibular gland, while similar fibers destined for the sublingual gland reach their destination by joining the peripheral part of the

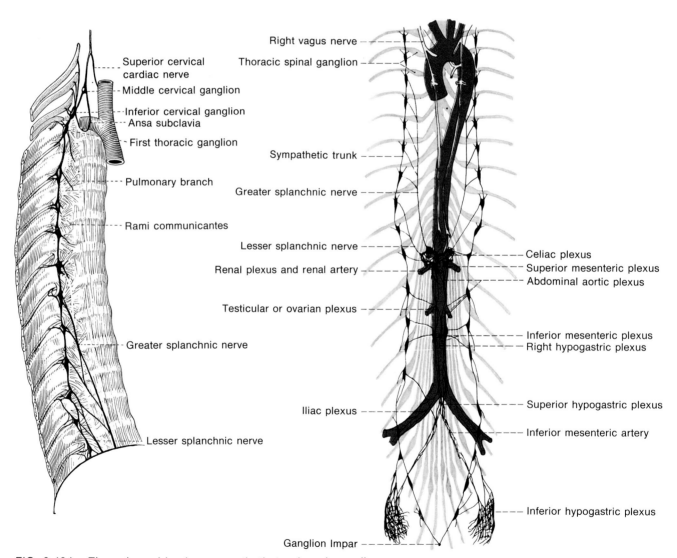

FIG. 6-40J. *Thoracic and lumbar sympathetic trunk and ganglia*

lingual nerve. Postganglionic sympathetic fibers associated with this ganglion course from the superior cervical ganglion via the carotid plexus and maxillary artery; and somatic sensory fibers via the lingual nerve.

Parasympathetic fibers of the glossopharyngeal (preganglionic cell bodies in the inferior salivatory nucleus) (fig. 6-40C) pass from the ninth nerve through the canaliculus tympanicus and through the middle ear as the tympanic nerve, there receiving sympathetic branches from the carotid plexus; this complex is known as the tympanic plexus. It leaves the ear as the lesser petrosal nerve, and passes through the foramen lacerum, or foramen ovale. As it passes close to the geniculate ganglion it receives a filament from the facial nerve, which probably, however, adds no parasympathetic fibers. The lesser petrosal nerve then continues to the *otic ganglion,* which is situated upon the medial side of the root of the lingual nerve just inferior to the foramen ovale. Passing through the ganglion are sympathetic fibers received from the plexus surrounding the middle meningeal artery, and sensory fibers from the mandibular nerve. Communicating branches of the otic ganglion are quite numerous, but most of them probably represent merely fusion of nerve sheaths with the ganglion. Secretory fibers are distributed from the ganglion to the parotid gland via the auriculotemporal nerve.

Parasympathetic fibers of the vagus (preganglionic cell bodies in the dorsal motor nucleus of the vagus) are shown in figure 6-40D. There is no collateral ganglion of the vagus corresponding to the ganglia of the other parasympathetic cranial nerves; its postganglionic cell bodies are instead located on the organs innervated.

Parasympathetic fibers of the sacral nerves (pelvic splanchnic nerve) (fig. 6-40E) are carried by S2, S3, and S4. The preganglionic fibers course to the hypogastric and its continuing pelvic plexuses, where most of the postganglionic neurons are located; however, some of the postganglionic neurons, supplying the bladder, are located upon the walls of that viscus. This division supplies the distal one-fourth of the transverse colon, descending colon, rectum, and urogenital organs except the kidneys.

Sympathetic System

The *sympathetic outflow* (fig. 6-40F) or thoracolumbar part of the autonomic nervous system comprises autonomic nerve fibers associated with the thoracic and first three lumbar nerves. The grossly detectable parts consist of an interconnected chain of ganglia—the *sympathetic trunk*—on either side of the vertebral column; a series of interconnected paraxial ganglia in association with the visceral organs; and nerves between these ganglia; and between them and the spinal nerves on the one hand, and the visceral organs on the other. Visceral afferent fibers pass to their cell bodies in the dorsal root ganglia, those from the viscera coursing through the white rami communicantes.

The *thoracolumbar sympathetic trunk* (fig. 6-40G) consists of interconnected segmentally arranged ganglia on either side of the vertebral column. It occupies the region from the first thoracic to the third lumbar segment. In one direction it extends throughout the neck, and in the other, to the coccyx.

The sympathetic afferent cell bodies are located in the dorsal root ganglia of the spinal nerves. Their afferent processes reach them via the splanchnic nerves and white rami communicantes if from the viscera, and via the spinal nerves if from the periphery. The *preganglionic sympathetic motor cell bodies* are located in the visceral motor column of the spinal cord. The *postganglionic cell bodies* are located in ganglia (or exceptionally scattered on a viscus) outside the central nervous system, either in the sympathetic trunk or the paraxial ganglia of the abdominal cavity.

The sympathetic trunk ganglia are connected with the roots of the spinal nerves by short rami communicantes. There is a *white ramus communicans* (myelinated) to the trunk ganglion of each thoracic and the first three lumbar (but not cervical or sacral) segments. These carry visceral afferent fibers (from the viscera) and preganglionic visceral efferent fibers from the central nervous system to the sympathetic ganglia. There is also a *gray ramus communicans* (unmyelinated) to each trunk (including cervical and sacral) ganglion, carrying postganglionic (motor) fibers from the trunk ganglia to the spinal nerves and thus to the periphery (vasomotor, pilomotor, and skin glands) and afferent fibers as well.

The typical *sympathetic reflex arc* then comprises an afferent fiber from the viscus (passing through a trunk ganglion and a white ramus) or from the periphery (via a spinal nerve) to its cell body in a root ganglion. Its dendrite synapses in the spinal cord presumably with a collateral neuron, which synapses with a preganglionic neuron also in the cord. The axon of the latter then passes through a white ramus to a trunk ganglion. Depending upon the destination, any one of several things may happen. If the effector is in the periphery, there will be a synapse in the trunk ganglion with a postganglionic neuron, whose axon will pass through a gray ramus en route to its destination. If the effector is in a viscus, it will pass through the trunk ganglion, either direct or up or down the trunk on the way to a collateral ganglion, where it synapses with a postganglionic neuron. Axons of the superior part of the sympathetic system are prone to pass up the trunk, and, conversely, those of the inferior part tend to pass down it.

It seems possible to state two rules concerning the position of postganglionic sympathetic neurons: if the effector is in the periphery or superior to the abdominal cavity the postganglionic cell body is situated in one of the ganglia of the sympathetic trunk; if the effector is in the abdominal cavity, the postganglionic cell body is in a paraxial collateral ganglion (splanchnic, celiac or one of its subdivisions, or inferior mesenteric). There are exceptions to this rule, however, as in the adrenal medulla, which is directly innervated by preganglionic neurons.

The cervical part of the sympathetic system (fig. 6-40*H*) is located in the neck between the carotid sheath and the longus colli muscle, and just medial to the vagus nerve. It is characterized by the absence of white rami communicantes and a modification of the segmental arrangement of the ganglia, resulting in the concentration of the cell bodies concerned into three ganglia. All preganglionic fibers reaching these cell bodies pass through the more superior of the thoracic white rami, and some of them may come from as far inferiorly as the ninth thoracic nerve.

The *superior cervical ganglion* is at the level of the second and third cervical vertebrae and located between the sheath of the carotid and longus colli. It represents the fused first four segmental ganglia, and accordingly has four gray rami, to the first four cervical nerves, carrying postganglionic fibers to the periphery. Other postganglionic fibers are distributed as follows. To the glossopharyngeal, vagus, and hypoglossal nerves, near their emergence from the skull; superiorly to the internal carotid artery and anteriorly to the external carotid artery, and by the plexuses surrounding these arteries to all parts of the head; anteriorly to form, with the glossopharyngeal and vagus nerves, the pharyngeal plexus; and inferiorly by the superior cardiac nerves to the cardiac plexuses. The cardiac plexuses descend the neck and usually pass ventral to the inferior thyroid artery. On the right side the nerves pass either ventral or dorsal to the subclavian artery and to the dorsal aspect of the aorta, where they join the deep cardiac plexus. On the left they course ventral to the common carotid and the aorta, toward the superficial cardiac plexus.

The *middle cervical ganglion* regularly represents the fusion of the fifth and sixth ganglia, but occasionally one of these, or even both, are fused with the inferior cervical ganglion. It is located in the sympathetic trunk at the level of the sixth cervical vertebra and close to the inferior thyroid artery. It sends two gray rami to the fifth and sixth cervical nerves. It gives off a loop, *ansa subclavia,* around the subclavian artery to the inferior cervical ganglion, with twigs to the artery. The *middle cardiac nerve* arises from the ganglion or below it, and courses to the deep cardiac plexus. Filaments also are given off the middle cervical ganglion to the thyroid gland.

The *inferior cervical ganglion* regularly represents the fused seventh and eighth ganglia, but at times includes the middle cervical ganglion, and may be fused with the first thoracic ganglion. It is situated just superior to the first rib. It is connected with the middle ganglion not only by the trunk, but by the ansa subclavia as already mentioned. It gives off gray rami communicantes to the seventh and eighth cervical nerves, filaments to surrounding arteries, and the *inferior cardiac nerve,* carrying fibers from the first thoracic ganglion as well. On the left side this frequently anastomoses with the middle cardiac nerve, to the deep cardiac plexus. Frequently there is communication with the phrenic nerve.

The thoracolumbar part of the sympathetic trunk, from T1 to L3, contains ganglia segmentally arranged, each with a gray and a white ramus to the corresponding spinal nerve. It lies between pleura in the thorax and the psoas major in the abdomen, and the heads of the ribs superiorly, or vertebral body inferiorly. Peripheral branches from these to the viscera carry fibers with complex arrangement. Each contributes filaments to the aorta. The first thoracic ganglion contributes fibers to the inferior cardiac nerve. The remaining branches are as follows.

Pulmonary branches are given off by the second, third, and fourth ganglia to the posterior pulmonary plexus, with postganglionic cell bodies in the sympathetic trunk ganglia.

The *greater splanchnic nerve* is formed by contributions from the fifth to ninth thoracic ganglia, which join to form a stout nerve that courses obliquely to pierce the diaphragm and end in the celiac plexus; but fibers come to this from as far superiorly (in the sympathetic trunk) as the first or second thoracic ganglion. In the greater splanchnic nerve just superior to the diaphragm is the *splanchnic ganglion,* which supplies twigs to the esophagus and aorta.

The *lesser splanchnic nerve,* from the 9th and 10th thoracic ganglia, follows the greater through the diaphragm, and ends in the inferior part of the celiac plexus—the aorticorenal plexus, with postganglionic cell bodies in the latter situation.

The *least splanchnic nerve* is not infrequently fused with the lesser. Its preganglionic fibers pass through the 11th and 12th thoracic ganglia. The nerve pierces or passes dorsal to the diaphragm and ends in the renal plexus, where its postganglionic cell bodies are situated.

The *lumbar splanchnic nerves* are irregular in form, as are their ganglia. They comprise preganglionic fibers to the inferior mesenteric plexus, where the postganglionic cell bodies are located, supplying the descending colon, rectum, and urogenital system except the kidneys.

The pelvic part of the sympathetic system is an inferior extension of the thoracolumbar portion. It has gray rami but not white rami, and its fibers supply the peripheral structures, rather than those within the abdominal cavity.

MAJOR PLEXUSES OF THE SYMPATHETIC SYSTEM Upon and near the visceral organs the autonomic fibers form an extremely intricate system of interconnected plexuses that are located largely at the bases of the organs. These consist of both sympathetic and parasympathetic fibers, with postganglionic cell bodies of one or the other according to the situation. The following major plexuses are briefly described.

The *pharyngeal* and the communicating *laryngeal plexuses* are only partly autonomic in character, for they contain also branchiomeric fibers (cranial nerves IX and X) fibers. They are located on the pharynx and larynx, and communicate inferiorly with the esophageal plexus.

Cardiac plexuses are two in number. The *superficial cardiac plexus* is located on the ventral aspect of the aorta and on its concave curvature, and is small. It receives the cardiac postganglionic nerves of the left superior cervical ganglion and the inferior cervical cardiac branch (preganglionic) of the left vagus. It communicates with the deep cardiac plexus. The *deep cardiac plexus* is much larger and consists of bilateral (connected) halves placed between the aorta and trachea. It is formed of these cardiac branches of the sympathetic and vagus that do not contribute to the superficial cardiac plexus; or, in other words, the cardiac branches of the right superior, middle, and inferior, and of the left middle and inferior cervical ganglia; the right superior and inferior cervical and thoracic cardiac branches of the vagus; and the left superior cervical and thoracic branches of the vagus. All cell bodies in the cardiac ganglia are parasympathetic (vagal).

The *dorsal* and *ventral pulmonary plexuses* are really continuous with the deep cardiac plexus. They are formed by the pulmonary (preganglionic) branches of the vagus, and postganglionic fibers from the second, third, and fourth thoracic sympathetic ganglia in which the postganglionic sympathetic cell bodies are located.

The *celiac plexus* is the largest and most important of the visceral plexuses and is situated about the celiac branch of the aorta at the level of the first lumbar vertebra. It receives preganglionic fibers from both vagi. In addition, the greater splanchnic nerve (preganglionic and postganglionic fibers) enters the upper part, and the lesser splanchnic nerve enters the lower, termed the *aorticorenal plexus*. Minor parts, or extensions, of the celiac plexus have been termed gastric, phrenic, suprarenal, renal, hepatic, splenic, spermatic (or ovarian), superior mesenteric, and aortic plexuses. The celiac plexus is continued inferiorly along the aorta by the intermesenteric nerves.

The *inferior mesenteric plexus* is at the base of the inferior mesenteric artery. It receives its parasympathetic fibers from the sacral outflow, with postganglionic cell bodies on the viscus, and its sympathetic fibers from the lumbar nerves, with postganglionic cell bodies in the inferior mesenteric plexus.

The *Hypogastric plexus* is situated at the level of the lumbosacral junction and consists of bilateral parts. Its inferior continuation forms the *pelvic plexuses*, subdivided into *middle rectal vesicular, prostatic,* and *cavernous* in the male and *uterine* and *vaginal* in the female. It supplies the urogenital organs and rectum. It receives postganglionic sympathetic fibers from the inferior mesenteric plexus and its cell bodies seem to be exclusively postganglionic parasympathetic, from the sacral nerves.

It should be mentioned that all the larger, more central arteries are invested by a sympathetic network of nerves, which is continued along their peripheral branches. In addition, however, the peripheral arteries receive sympathetic twigs from the spinal nerves.

Located within the walls of the stomach and intestines are autonomic nervous mechanisms that function, in peristalsis and perhaps in other ways, to some extent independently of the central system, for they react to the mechanical extension of an extirpated part of the gut. These are the *myenteric plexus (of Auerbach)* and the *submucosal plexus (of Meissner)*. They are supplied by postganglionic sympathetic fibers from the celiac plexus, and preganglionic parasympathetic fibers from the vagus, containing within themselves the postganglionic parasympathetic cell bodies.

CLINICAL APPLICATIONS

Cerebral palsy is a disorder of the motor system commonly manifested by spastic weakness of the upper and lower limbs. Most patients afflicted with cerebral palsy exhibit some degree of mental retardation and exaggerated tendon reflexes.

Multiple sclerosis involves demyelination in various regions of the brain and spinal cord. The patient with multiple sclerosis experiences progressive weakness, incoordination, jerking movements, and disturbances of speech and vision.

Parkinsonism is a chronic disorder marked by tremor, muscular rigidity, and slowness of movement.

Epilepsy is characterized by recurrent attacks of convulsions whereby the patient has altered states of consciousness.

Cerebrovascular accidents result from hemorrhage, thrombosis, embolism, or other vascular insufficiencies.

Tumors of the brain are associated with pressure. Approximately 50% of all primary tumors of the brain are *gliomas,* or growths involving the neuroglia.

Inflammation reactions include *meningitis* (inflammation of the meninges) and *encephalitis* (inflammation of the brain).

Infections of the nervous system include *bacterial* (*tetanus*) and *viral* (*poliomyelitis, rabies*).

REVIEW QUESTIONS:

1. The nervous system is divided into _____ and _____ portions.
2. The deeper furrows of the cerebral hemispheres are called _____ and those less pronounced are called _____.
3. The _____ fissure separates the cerebrum into right and left hemispheres.
4. The _____ fissure separates the frontal and temporal lobes of the cerebral hemisphere.
5. The inferior frontal gyrus may be subdivided into _____, _____, and _____ parts.
6. The _____ is a somewhat triangular, convex prominence that is concealed when the lateral cerebral fissure is closed.
7. The _____ gyrus is located between the inferior temporal sulcus and the collateral fissure.
8. Posteriorly, the corpus callosum is folded on itself to form the _____.
9. The _____ is the wedge-shaped area located between the parietooccipital and calcarine fissures.
10. The _____ and parahippocampal gyri connected by an isthmus beneath the splenium of the corpus callosum form the limbic lobe.
11. The _____ is a thin membrane attached to the overlying corpus callosum, genu, and the underlying fornix.
12. The lentiform nucleus includes a lateral mass known as the _____ and a lighter colored medial mass known as the _____.
13. The _____ is a thin plate of gray substance embedded in the white matter separating the insula from the putamen.
14. The caudate and lentiform nucleus are separated by the _____.
15. The third ventricle is continuous with the fourth ventricle by means of the _____.
16. The diencephalon includes the _____, _____, _____ and _____.
17. The third ventricle communicates with the lateral ventricles of the cerebral hemispheres by the _____.
18. The _____ is an unpaired, cone-shaped structure that projects over the midbrain and lies in a groove between the superior colliculi.
19. The _____ is an oval mass of gray substance forming the largest subdivision of the diencephalon.
20. That portion of the midbrain dorsal to the cerebral aqueduct is known as the _____.

21. The _____ nerves leave the brain stem from the interpeduncular fossa on the medial side of the cerebral peduncle.
22. The pons consists of two distinct parts, known as the _____ and _____ portions.
23. The _____, _____ and _____ nerves exit from the brain stem just dorsal to the olive.
24. The middle cerebellar peduncle connects the _____ with the cerebellum.
25. The cerebellum shows a large lateral hemisphere on each side and a narrow median portion, the _____.
26. The distal end of the spinal cord is known as the _____.
27. The dorsal root fibers enter the spinal cord along the _____ surface.
28. The _____ occupies that part of the vertebral canal formed by the last four lumbar vertebrae and the entire sacrum.
29. The pia mater and arachnoid are often regarded as the _____.
30. The spinal cord is attached to the dura mater by two lateral series of flattened bands of pia mater called the _____ ligaments.
31. The pia mater continues caudally as the _____.
32. The blood supply of the spinal cord is derived from the _____ and _____ arteries.
33. The _____ is a large irregular space located on the side of the sphenoid bone, lateral to the sella turcica.
34. The _____ nerve innervates the superior oblique muscle.
35. The lingual nerve is a branch of the _____ division of the trigeminal nerve.
36. The _____ nerve innervates the lateral rectus muscle.
37. The _____ nerve innervates the stylopharyngeus muscle.
38. The _____ nerve innervates the tongue musculature.
39. Cutaneous branches of the cervical plexus are derived from _____ to _____.
40. The _____ nervous system is concerned with innervation of smooth musculature, cardiac muscle and glandular tissue.
41. List four parasympathetic cranial ganglia.
    a.
    b.
    c.
    d.
42. The _____ system is lined with ependymal cells and contains cerebrospinal fluid.
43. The floor and medial wall of the inferior horns of the lateral ventricles are formed by the _____ and _____.
44. The lateral ventricular wall above the hypothalamic sulcus is formed by the _____.
45. Immediately below the inferior colliculi, the _____ nerve exits from the midbrain.
46. The _____ capsule is a thin lamina of white matter separating the claustrum from the putamen.
47. The rostrum of the corpus callosum extends downward as the _____.
48. The pituitary floor is connected with the floor of the hypothalamus by way of the _____.
49. The _____ is the thickened anterior end of the hippocampal gyrus, which bends medially around the hippocampal fissure.
50. The _____ is the thalamic termination of the auditory pathway.

7. The Special Senses

STUDENT OBJECTIVES
After you have read this chapter, you should be able to:
1. Locate the receptors for olfaction.
2. Describe the neural pathway for smell.
3. Identify the gustatory receptors.
4. Describe the neural pathway for taste.
5. List the structural divisions of the eyeball.
6. Describe the visual pathways.
7. Define the anatomical subdivisions of the ear.
8. Identify the receptor organs for equilibrium.

The senses comprise mechanisms by which messages are received from the environment. They may be divided into two categories: apparatuses for general sensibility, consisting of receptors of touch, pressure, pain, heat, cold, and proprioception; and the *organs of the special senses* of *smell, taste, sight, hearing,* and *equilibrium,* although the latter may be regarded as a specialized part of the proprioceptive apparatus.

SMELL (OLFACTORY TRACT)

The peripheral olfactory receptors are found in specialized areas of the nasal mucosa known as the *olfactory epithelium* (fig. 7-1A). The receptor cells have elongated processes that extend through the cribriform plate as the *olfactory nerve* (fig. 7-1B). The olfactory nerves then pierce the surface of the *olfactory bulb* to synapse with neurons in the rostral area. The olfactory bulbs contain *mitral, tufted,* and *granule* cells, which give rise to fibers that form the *olfactory tracts* at the caudal ends of the bulbs. The tracts then bifurcate into *medial* and *lateral olfactory striae* (fig. 7-1C).

SIGHT (EYEBALL)

The organ of the special sense of sight is the *eyeball* (fig. 7-2A and B). Associated with it are other structures so it is reasonable to consider the contents of the orbit, eyelids, and lacrimal duct as parts of the organ of sight.

The optic nerve is not a cranial nerve, strictly speaking, but with the retina, its end organ, is a cup-shaped outgrowth of the brain.

The *eyeball* is a subspheroid body with three concentric layers. It has a smaller anterior *corneal segment* and a larger posterior *scleral segment*. There is recognized an *anterior pole*, at the central point of the cornea, a *posterior pole* at the central point of the sclera, and an *equator,* comprising an imaginary line surrounding the eyeball midway between the poles. The optic nerve, within the optic sheath, does not enter the eyeball at the posterior pole, but medial to it. The sclera, but not the cornea, is surrounded by a fascial sheath within which is a lymph space, the *interfascial space.*

The *eyeball* (fig. 7-2B) consists of three concentric layers, which enclose three refracting media. The outermost layer is a fibrous coat that forms the sclera and

THE SPECIAL SENSES ■ 215

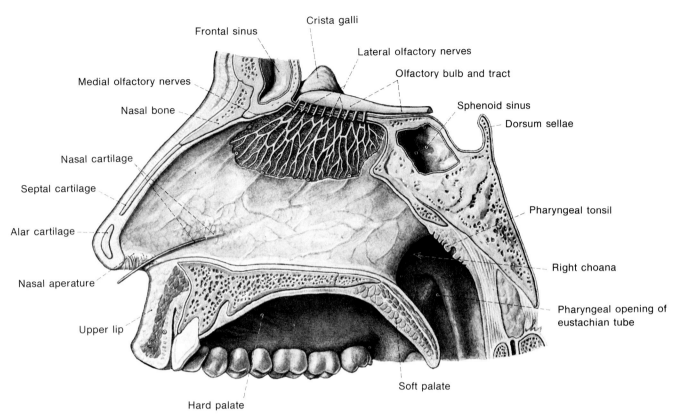

FIG. 7-1A. *Olfactory nerves; septum of the nasal cavity*

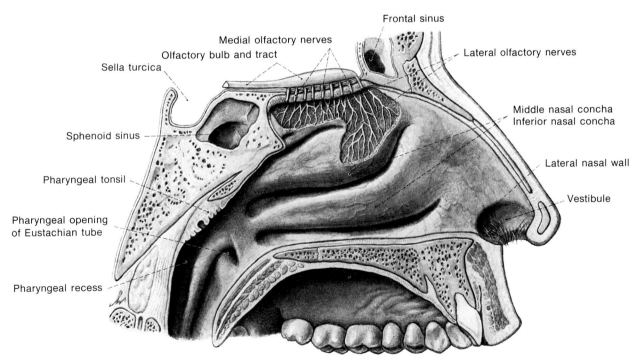

Fig. 7-1B. *Olfactory nerves; lateral wall of the nasal septum*

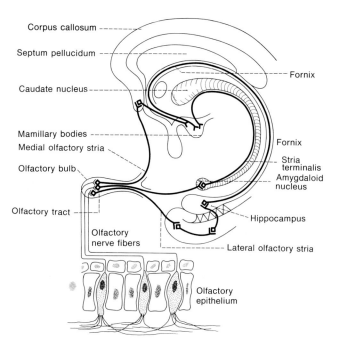

FIG. 7-1C. *Olfactory pathways*

the cornea; the middle layer is vascular and includes the choroid, ciliary body, and iris; the innermost layer is termed the retina.

The *sclera* covers the posterior five-sixths of the eyeball. Anteriorly it continues with the cornea at the corneoscleral junction, and distal to the margin of the cornea the sclera may be observed through the conjunctiva as "the white of the eye." The sclera is pierced by the blood vessels that ramify in the choroid coat. Inferior and medial to the posterior pole of the eyeball the sclera is pierced by the optic nerve and, surrounding the point of entry of the nerve, by the posterior ciliary arteries. The blood distributed by the ciliary arteries is usually returned by four *vorticose veins.* The anterior ciliary arteries enter the sclera near the corneoscleral junction.

The *cornea* consists of transparent fibrous tissue that is continuous with the sclera. The superficial aspect of the cornea is covered by a layer of stratified epithelium that is continuous with the conjunctiva. Posteriorly, it is covered by a *posterior elastic membrane,* which in turn is covered by a layer of mesothelium. At its peripheral margin the fibers of the posterior elastic membrane separate into three layers. The innermost fibers run medially into the iris; the middle layer of fibers give origin to the ciliary muscle; and the outermost layer of fibers blend with fibers of the sclera. The cornea derives its nourishment from the circulating lymph. Note that pain is the only sensation that can be aroused from the cornea. Its nerve supply is derived from the ophthalmic division of the trigeminal through the short ciliary nerves, and the fibers form both a subepithelial and an intraepithelial plexus, which send free nerve endings between the cells of the epithelium covering its free surface.

At the corneoscleral junction the circular *sinus venosus sclerae* is situated in the deeper part of the coat. It communicates with the scleral veins and with the aqueous humor.

The *choroid layer* consists of an outer pigmented layer in contact with the sclera, and an inner vascular layer, in which the short posterior ciliary arteries ramify. Anteriorly the choroid continues as the *ciliary body,* which lies behind the peripheral part of the iris. It consists of a number of *ciliary processes* separated from one another by radial furrows that have free,

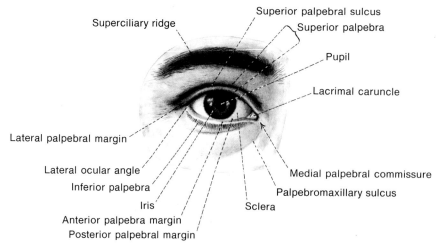

FIG. 7-2A. *Eyeball* (anterior view)

rounded, central extremities. The *ciliary muscle* (fig. 7-2C) is located in the ciliary body. Its fibers arise at the corneoscleral junction and radiate backward and inward to the ciliary processes. There are also circular fibers placed medial to the rest of the muscle.

The *iris* (fig. 7-2D), which also forms part of the intermediate layer of the eyeball, is a contractile diaphragm with a central aperture. Peripherally it is continuous with the ciliary body, while its central free margin bounds the pupil. The anterior aspect of the iris is covered by a layer of mesothelium that is continuous with the mesothelium on the posterior aspect of the cornea. Its posterior aspect is covered by a double layer of pigment containing columnar epithelium.

The iris contains a circular sphincter muscle and a radial dilator muscle that lies near its posterior surface.

The *retina* (fig. 7-2E) contains an outer pigmented layer and an inner nervous layer that is in contact with the vitreous body. The nervous layer lines the posterior three-fourths of the eyeball, and its anterior limit forms a wavy ring, termed the *ora serrata*. It continues forward, as a single layer of columnar epithelium, over the ciliary processes and the posterior aspect of the iris in company with the outer pigmented layer.

Inferomedial to the posterior pole of the eyeball, the optic nerve reaches the retina and its fibers spread out over its surface. The point of entrance is termed the *optic disk,* which constitutes the *blind spot.* Superior and lateral to the optic disk, at the posterior pole of the eye, there is a yellowish area devoid of blood vessels. This is the *macula lutea* and over its depressed central portion, the *fovea centralis,* the rods are absent and only cones are present. It is the part of the retina that is most sensitive to light stimuli. The rods predominate in the peripheral retina. The *central artery* (fig. 7-2F) appears at the optic disk, and its retinal branches may easily be observed in the living eye by ophthalmoscopic examination.

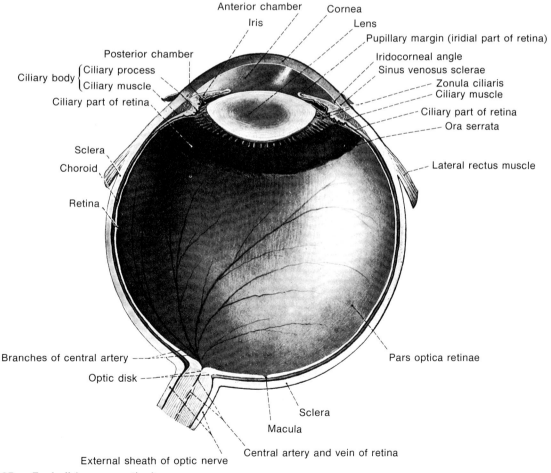

FIG. 7-2B. *Eyeball (cross-section)*

The three *refracting media* that are enclosed by the layers of the eyeball include (1) the aqueous humor, (2) the lens, and (3) the vitreous body.

The *aqueous humor* is a clear fluid that occupies the space between the cornea anteriorly, and the lens posteriorly. This space includes both the *anterior and posterior chambers of the eye.* The former is limited posteriorly by the anterior aspect of the iris and the central part of the lens; the posterior chamber lies between the posterior aspect of the iris and the lens. The two chambers therefore communicate freely with each other through the aperture of the iris. The aqueous humor is secreted into the posterior chamber by the ciliary body and circulates into the anterior chamber through the aperture of the pupil. From the anterior chamber, it drains away into the sinus venosus sclerae.

The *vitreous body* fills the area of the eyeball posterior to the lens. It is a transparent, jellylike substance, enclosed by the *hyaloid membrane.* The hyaloid membrane becomes somewhat thickened anteriorly and presents a number of radial furrows and elevations. The furrows receive the ciliary processes, and the elevations fit into the corresponding furrows in the ciliary body. Near the margin of the lens this part of the hyaloid membrane divides into a posterior layer, which lines the hollowed-out anterior aspect of the vitreous body, and an anterior layer, which is attached to the anterior aspect of the lens, constituting its *suspensory ligament.*

The *lens* (fig. 7-2G) is a transparent, biconvex body, which lies between the aqueous humor anteriorly and the vitreous body posteriorly. Circular in outline, it consists of concentric laminae of elongated cells known as lens fibers. The entire lens is enclosed within an elastic capsule. When the eye is at rest the anterior surface of the lens is flattened by the tension of the suspensory ligament. Contraction of the ciliary muscle relaxes the ligament by drawing forward the hyaloid membrane, and the anterior surface of the lens becomes more convex, because of the elasticity of the lens substance.

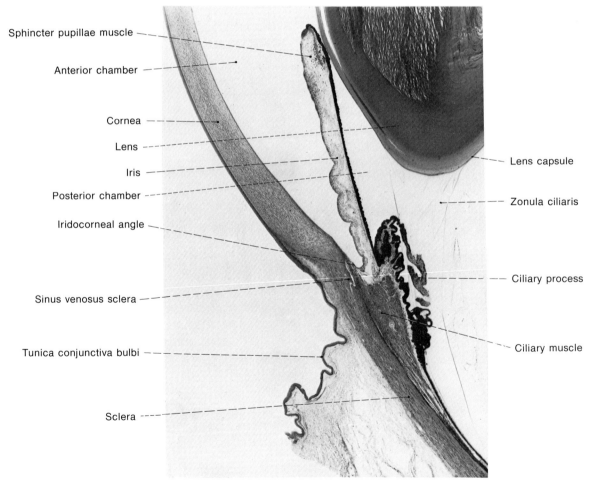

FIG. 7-2C. *Ciliary muscle*

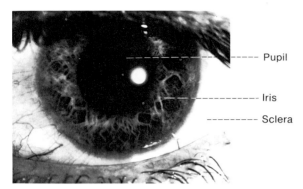

FIG. 7-2D. *Iris*

Ciliary Muscle

Both the sphincter and dilator muscles of the iris and the ciliary muscle are supplied by the *short ciliary nerves* from the ciliary ganglion. The fibers that supply the sphincter of the iris and the ciliary muscle are parasympathetic, and travel in the oculomotor nerve from their origin in the Edinger-Westphal nucleus, whereas those that supply the dilator muscle of the iris are sympathetic in origin.

The *ciliary muscle* (fig. 7-2C) has *circular* and *meridional fibers* stretching from the inner surface of the sclera that act upon the *zonula ciliaris* at the margin of the lens to change its convexity and thus bring about accommodation. The muscle fibers of the iris include a *sphincter pupillae* and a *dilator pupillae*. The latter is innervated by sympathetic fibers and the former by the parasympathetic from the oculomotor nerve.

Bony Orbit

The *bony orbit* (fig. 7-2H), lined with periosteum, contains the eyeball and the soft parts concerned with it, chiefly muscles, nerves, vessels, and a gland, as well as soft fat and connective tissue. The four *recti* muscles are attached to the eyeball anterior to its equator upon the superior, inferior, lateral, and medial aspects, while the two *oblique muscles* are attached posterior to the equator upon the superior and lateral aspects. All except the inferior oblique arise from the apex of the orbit, as does the *levator palpebrae superioris,* the attachment being to a tendinous ring, the *common annular tendon* surrounding the optic nerve and the optic foramen; this tendon is continuous with the periosteum. The recti adjoin the walls of the orbit, except for the superior rectus, which is covered by the levator palebrae.

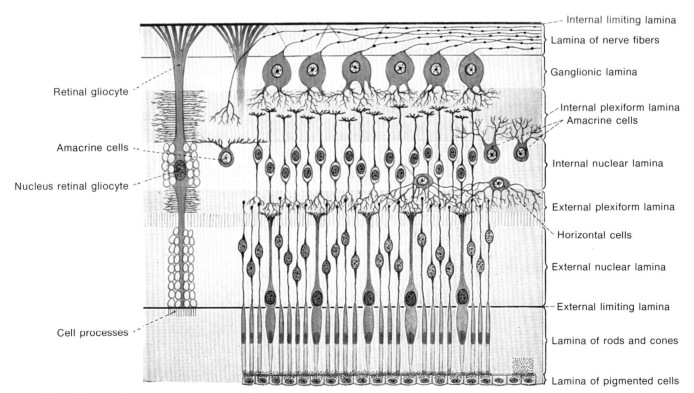

FIG. 7-2E. *Retina (diagrammatic view)*

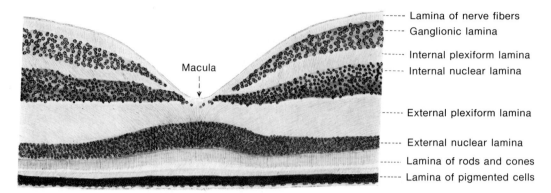

FIG. 7-2F. *Retina (transverse section)*

Ophthalmic Artery

The *ophthalmic artery* (fig. 7-2*I*) enters the orbit through the optic foramen inferolateral to the optic nerve. It passes over the optic nerve to the medial and anterior part of the orbit, where it anastomoses with the angular and ethmoidal arteries. The veins largely correspond, but twigs unite to form the *superior* and *inferior ophthalmic veins.* The inferior vein communicates with the *pterygoid plexus* and both drain into the *cavernous sinus* through the *superior orbital fissure.* It also communicates with the *angular vein,* and with the cerebral veins through the foramen cecum.

Lacrimal Gland

The *lacrimal gland* (fig. 7-2*J*), the shape and size of an almond, is located in the anterior part of the orbit laterosuperior to the eyeball, lying in a shallow fossa of the frontal bone. It tends to be divided into a larger superior part and a smaller palpebral part associated with the upper lid. There are several secretory ducts, opening upon the superior fornix conjunctivae.

The contents of the orbit (fig. 7-2*I*) receive special sensory, somatic sensory, proprioceptive, somatic motor, parasympathetic, and sympathetic innervation. Special sensory (sight) innervation is carried by the optic nerve, passing through the optic foramen and entering the eyeball. Somatic sensory fibers are carried by the ophthalmic division of the trigeminal nerve, which enters the orbit through the superior orbital fissure in three divisions: frontal, lacrimal, and nasociliary. The first of these passes superomedial to the muscles, the second traverses the medial part of the orbit, and the third passes between the two heads of the lateral rectus muscle and between the two branches of the oculomotor nerve, then crosses to

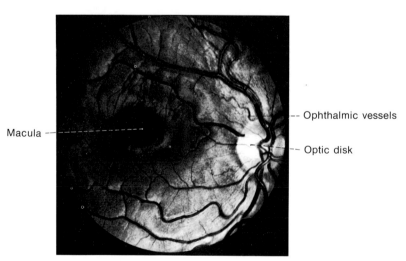

FIG. 7-2G. *Central artery*

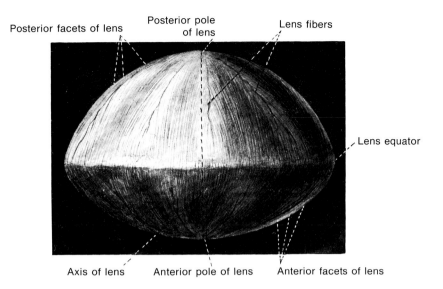

FIG. 7-2H. *Lens*

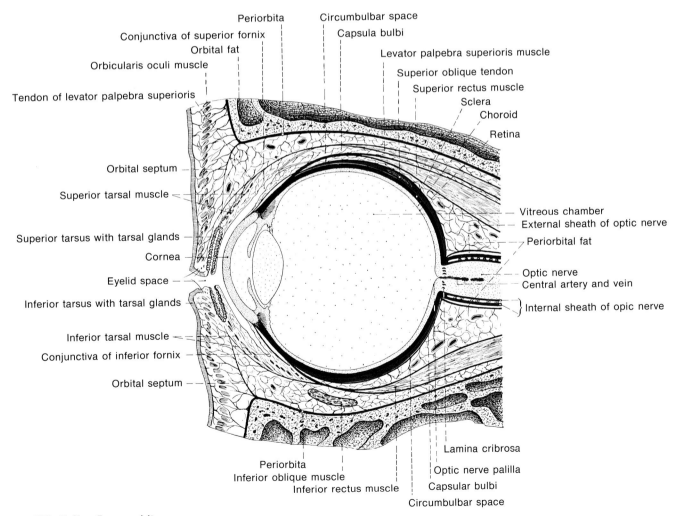

FIG. 7-2I. *Bony orbit*

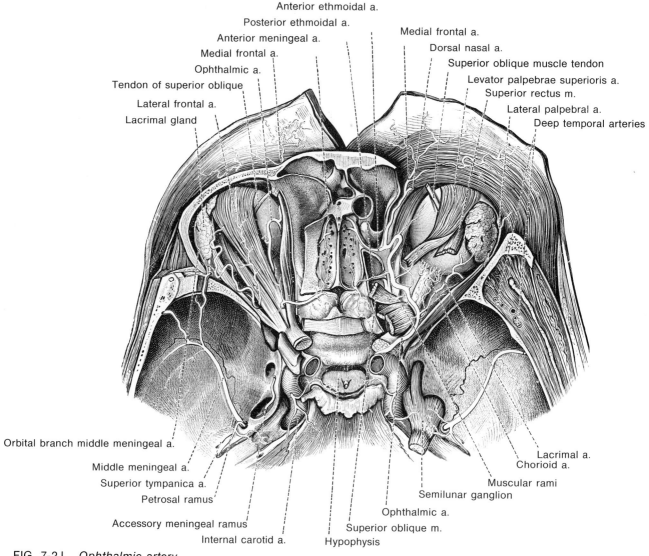

FIG. 7-2J. *Ophthalmic artery*

the medial wall of the orbit between the optic nerve and the superior rectus muscle. The zygomatic branch of the maxillary division of the trigeminal nerve also enters the orbit through the *inferior orbital fissure* and anastomoses with the lacrimal nerve before leaving the orbit. Somatic motor fibers are carried from and to the extrinsic muscles of the eye by the *oculomotor, trochlear,* and *abducens* nerves, all entering the orbit through the *superior orbital fissure*. The oculomotor nerve reaches the orbit in a superior and an inferior division; both pass between the two heads of the lateral rectus muscle. The superior branch passes between the optic nerve and the superior rectus muscle, which it innervates, as well as the levator palpebrae superioris. The inferior branch courses inferior to the optic nerve and supplies the inferior and medial recti and the inferior oblique muscles. The trochlear nerve passes superior to the lateral rectus and then anteromedially, superior to all muscles, to supply the superior oblique muscle. The abducens nerve passes between the two heads of the lateral rectus muscle and distally in the orbit to innervate that muscle. Parasympathetic fibers are carried by the oculomotor nerve to the ciliary ganglion, between the optic nerve and the lateral rectus muscle, and then to the smooth muscles of the eyeball by short ciliary branches. Parasympathetic fibers are also carried from the facial nerve through the ptery-

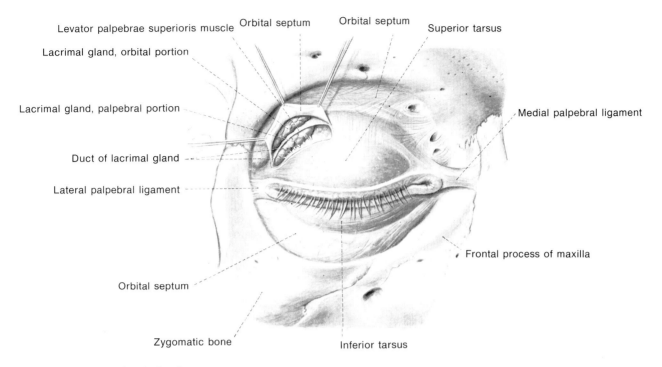

FIG. 7-2K. *Lacrimal gland*

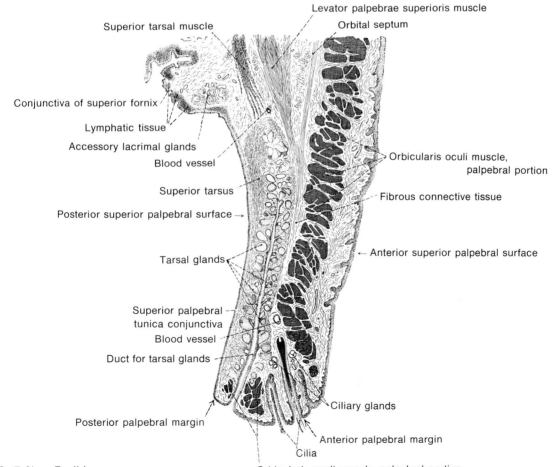

FIG. 7-2L. *Eyelids*

224 ■ BASIC ANATOMY FOR THE ALLIED HEALTH PROFESSIONS

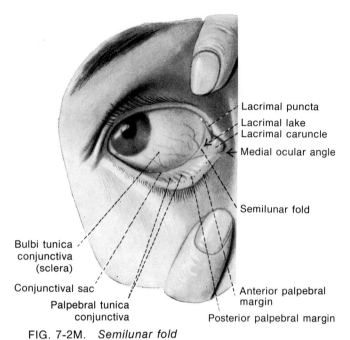

FIG. 7-2M. *Semilunar fold*

gopalatine ganglion to the orbital periosteum and lacrimal gland. Sympathetic fibers reach the orbit from the internal carotid plexus partly by the ophthalmic artery and partly by fibers that join the nerves before their entry into the orbit.

The eyebrows are merely areas of thickened skin upon the superciliary ridges that have retained a covering of hair. They are of some use in shading the eyes.

Eyelids

The *eyelids* (fig. 7-2K) are two sliding flaps of skin that can cover, protect, moisten, and clean the cornea. Closure of the lids is effected by contraction of the orbicularis oculi, pulling upon either angle of the eye, while opening the lids is accomplished by contraction of the levator palpebrae superioris muscle, inserting into the *superior tarsus*. The upper lid is the larger and more movable. In each there is a *tarsal part* adjoining the free border and an *orbital part* next to the orbital margin. In the upper lid the junction of the

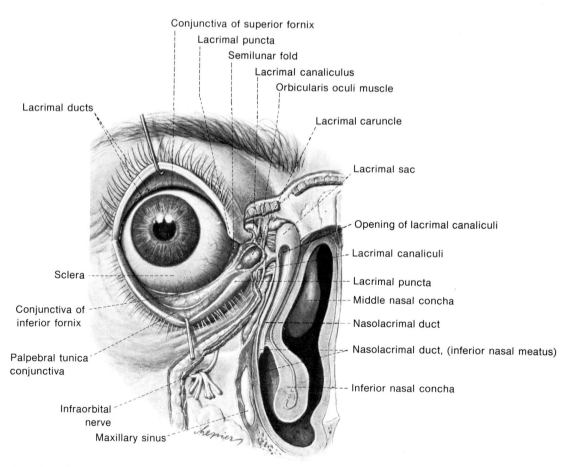

FIG. 7-2N. *Lacrimal caruncle*

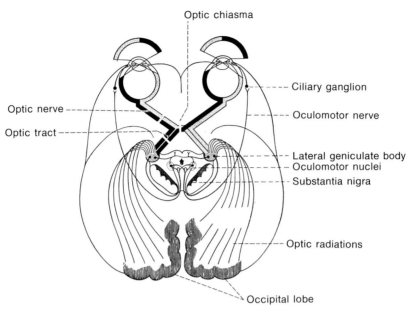

FIG. 7-2O. *Visual pathways*

two parts is marked by the *superior palpebral fold,* but this line is not noticeable upon the lower lid. In both lids there are smooth muscle fibers.

Each lid is covered by skin and lined with a mucous membrane termed the *palpebral conjunctiva.* It is reflected at the *conjunctival fornix* onto the sclera, and then over the cornea, where it is transparent, without blood vessels, and closely adherent. Medially this reflection of the conjunctiva is seen as the *semilunar fold* (fig. 7-2L). In the tarsal part of the lid is the tough fibrous tissue termed the *tarsus,* as well as the *tarsal glands* opening upon the margin of the lid. At the margin is also a row of bristles or *cilia,* the eyelashes, behind the roots of which are the *ciliary glands.* The superior and inferior tarsi are joined at either end by the *lateral raphe* and the *medial palpebral ligament,* which are both anchored to adjacent parts of the bone.

The space between the upper and lower lids is the *palpebral aperture,* and the junctions of the two lids are termed the *angles* or *canthi* of the eye—the *medial* or nasal, and the *lateral* or temporal.

Lacrimal Apparatus

Within the medial canthus is the slight prominence of the *lacrimal caruncle* (fig. 7-2M). This is bordered above and below, upon the margins of the adjoining part of the lids, by the *lacrimal papillae,* upon each of which opens a *lacrimal punctum.* From either punctum extends a *lacrimal duct,* by which fluid is drained from the eye into the nasal cavity. The ducts extend almost vertically, then with a slight ampulla as they bend medially, converge and often join at the *lacrimal sac.* The sac, several millimeters in diameter, lies in the *lacrimal fossa,* in the medial wall of the orbit. Its superior extremity or *fundus* is blind, above the junction of the lacrimal ducts, while the inferior extremity continues into the *nasolacrimal duct,* which in turn opens into the inferior meatus of the nasal cavity beneath the inferior concha.

The visual pathways are simplified for basic understanding in figure 7-2N. Note that stimuli to the retinae are transmitted to the optic nerves, optic chiasma, optic tracts, lateral geniculate bodies, optic radiations, and occipital cortex. Note the lesions at points *1, 2,* and *3* in the figure and determine the visual loss.

TASTE (GUSTATORY ORGANS)

The *gustatory organs* (fig. 7-3) are the taste buds. In man, taste is transmitted sparingly from the *anterior two-thirds of the tongue* in a single chain comprising the lingual nerve, *chorda tympani,* and the *facial nerve,* with cells in the *geniculate ganglion;* from the *posterior two-thirds of the tongue* by the *glossopharyngeal nerve* with nerve cells in the glossopharyngeal ganglion; from the dorsal surface of the epiglottis; and scantily from the oral surface of the soft palate by the vagus nerve, with nerve cells in the nodose ganglion.

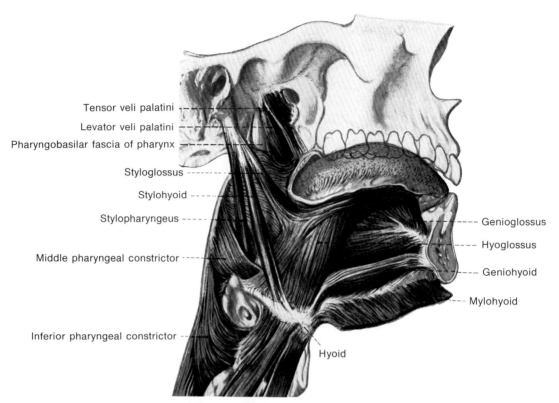

FIG. 7-3A. *Gustatory organs: taste buds (the tongue)*

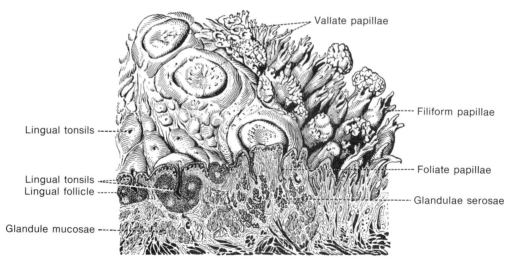

FIG. 7-3B. *Gustatory organs: taste buds (the surface of the tongue)*

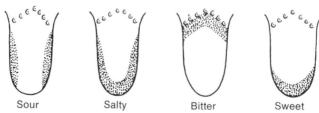

FIG. 7-3C. *Gustatory organs: taste buds (localization of taste bud sensation on the tongue)*

HEARING AND EQUILIBRIUM

Hearing is concerned with the *cochlear* portion of the vestibulocochlear nerve; equilibrium is concerned with the *vestibular* part of the same cranial nerve. The organ of hearing is complicated and consists of receiving, transmitting, and recording portions. The first has to do with the external, the second with the middle, and the third with the internal ear.

External Ear

The *auricle* (fig. 7-4A) is an irregular flap of yellow fibrocartilage equipped with several vestigial muscles of the facial field and covered by skin that partly surrounds the external acoustic meatus. It has a free border, the *helix* being the superior part and the *lobule* the inferior. The enclosed depression is bounded mostly by an inverted rim which is usually marked by the *tragus*, anterior to the meatus, and the *antitragus*, inferior to the meatus. The deepest part of the central depression, partly continuous with the meatus, is the *concha*, divided into an inferior and a superior part by the *crus of the helix*. The concha is bounded posteriorly by the *antihelix*, a ridge that divides, as it passes superiorly, into an *upper* and a *lower crus*, between which is the *triangular fossa*. Just within the helix is the *scaphoid fossa*.

The *external acoustic meatus* (fig. 7-4B) is the passage extending from the concha to the tympanic membrane. It is lined with skin, beneath which is cartilage in the more lateral part and bone in the more medial part, the elements of the latter being the tympanic ring, together with a part of the petrous temporal bone. The cartilage is connected with that of the auricle by a narrow *isthmus*. The skin covering it is supplied with numerous *ceruminous glands,* and the lateral portion bears bristly hairs in men as they advance in age (fig. 7-4C). At the extreme medial end of the meatus is the *tympanic membrane.* At birth this is almost horizontal but it later becomes more oblique, sloping inferomedially.

The *tympanic membrane* (fig. 7-4D) is obliquely orientated, so that its outer surface looks downward and forward as well as laterally. Therefore, the floor and anterior wall of the meatus are longer than the roof and posterior wall. The membrane is concave laterally, and the point of deepest concavity, the *umbo*, corresponds to the tip of the handle of the malleus. If the tympanic membrane is examined by reflected light, the anteroinferior quadrant is always strongly illuminated, and is known as "the cone of light." The apex of the cone lies at the umbo, and from this point the shadow thrown by the handle of the malleus may be traced upward and forward to a small whitish area, which corresponds to the lateral process of the malleus.

The tympanic membrane consists of: (1) an outer, cuticular layer; (2) an inner, mucous layer; and (3) an intermediate fibrous layer containing a majority of radiating fibers, which pass from the umbo to the periphery. Except for its upper part, the circumference of the membrane is thickened and fits into a groove in the bone. The gap in the upper part of the groove is termed the *tympanic notch*. The greater portion of the tympanic membrane, *pars tensa*, is held taut by the tonus of the tensor tympani muscle. The upper portion, which is attached to the tympanic notch, is referred to as the *pars flaccida.*

Middle Ear

The *middle ear* (fig. 7-4E) comprises the tympanic cavity, the tympanic antrum, and the auditory tube. The *tympanic cavity* is an irregular cavity in the temporal bone separated from the external ear by the tympanic membrane. It is lined with an entodermal mucous membrane, as the abdominal cavity is lined

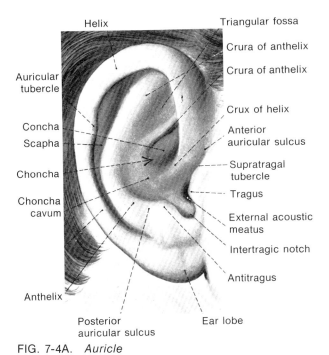

FIG. 7-4A. *Auricle*

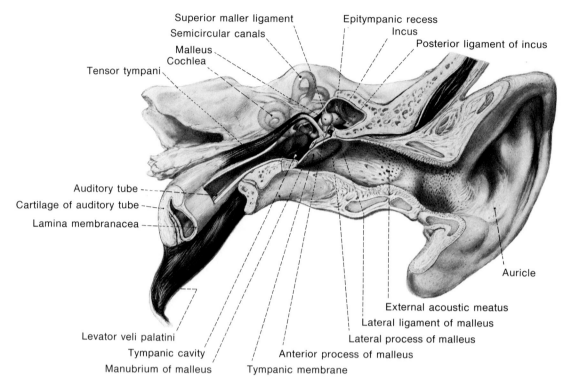

FIG. 7-4B. *External acoustic meatus*

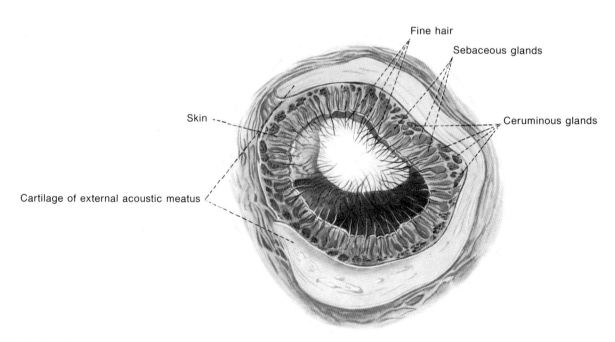

FIG. 7-4C. *External acoustic meatus (cross-section)*

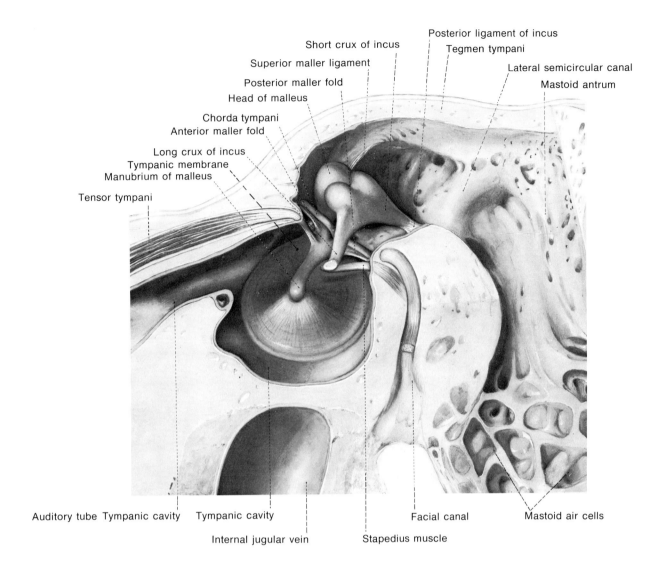

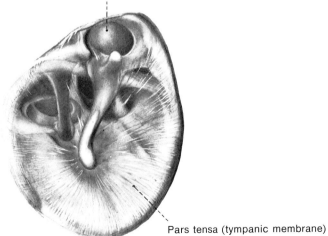

FIG. 7-4D. *Tympanic membrane*

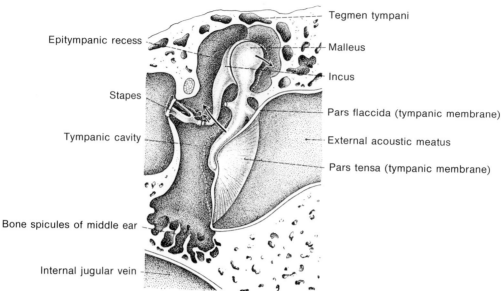

FIG. 7-4E. *Middle ear*

with peritoneum, and between this membrane and the walls of the cavity are the ossicles and two nerves. The cavity continues for a slight distance superior to the level of the membrane as the *epitympanic recess* and extends to the oropharynx medioinferiorly as the auditory tube. The cavity has a *roof,* the *tegmen tympani,* a *floor,* adjoining the jugular fossa, and a *medial wall,* characterized by a convexity, the *promontory,* caused by the underlying cochlea, and three depressions. The most superior of these is the *fenestra vestibuli* (oval window), into which fits the footplate of the stapes. Superior to this is the prominence of the canal of the facial nerve, and just inferior to the fenestra is the slight depression of the *tympanic sinus.* The third and most inferior depression is the *fenestra cochleae* (round window), covered by the secondary tympanic membrane. In the cleaned skull the fenestra vestibuli opens into the *vestibular labyrinth,* and the fenestra cochlea into the *scala tympani* of the cochlea. Superoanterior to the fenestra vestibuli is a pulley, the *cochleariform process,* for the tendon of the tensor tympani muscle, giving access to the canal for the belly of this small muscle. Between the fenestra vestibuli and the tympanic sinus is a small process, the *pyramidal eminence,* for the stapedius muscle.

The ossicles of the middle ear consist of the malleus, articulating with the tympanic membrane; the stapes, articulating with the fenestra vestibuli; and the incus, the intermediate bone in the chain, articulating with both the others. The *malleus* (hammer), the largest of the ossicles, has a *handle* and a *lateral process* articulating with the tympanic membrane. The handle has a slight projection for the insertion of the tensor tympani muscle. The malleus also has a rounded *head,* articulating with the incus, subtended by a restricted *neck,* and a long, slender *anterior process.* The bone is anchored by three small ligaments. The *incus* (anvil) has a *body* with a saddle-shaped process for articulation with the head of the malleus, a *short crus* for the attachment of an anchoring ligament, and a *long crus* with a knoblike end for articulation with the stapes. An anchoring ligament is also connected with the body. The *stapes* (stirrup) is of a shape indicated by its name, with a *capitulum,* articulating with the incus, two *crura,* and a *base* or footplate anchored into the fenestra vestibuli. The *stapedius* muscle inserts into the capitulum.

Of the two muscles of the middle ear, the slender *tensor tympani muscle* arises from the cartilage of the tuba, is held by the greater wing of the sphenoid and the walls of the bony canal, parallels the auditory tube, and inserts into the handle of the malleus. It is innervated by the trigeminal nerve. The *stapedius muscle* arises from within the pyramidal eminence, from which its tendon emerges to insert into the neck of the capitulum of the stapes. It is derived from the musculature of the second gill arch and is innervated by the facial nerve.

Between the mucous membrane and walls of the middle ear are also two nerves with other destinations. The glossopharyngeal nerve sends an important branch, the *tympanic nerve,* into the middle ear, this forming the *tympanic plexus* over the promontory. It gives off twigs to the neighboring mucous membrane and a communication with the carotid plexus, and then continues as the *lesser petrosal nerve*

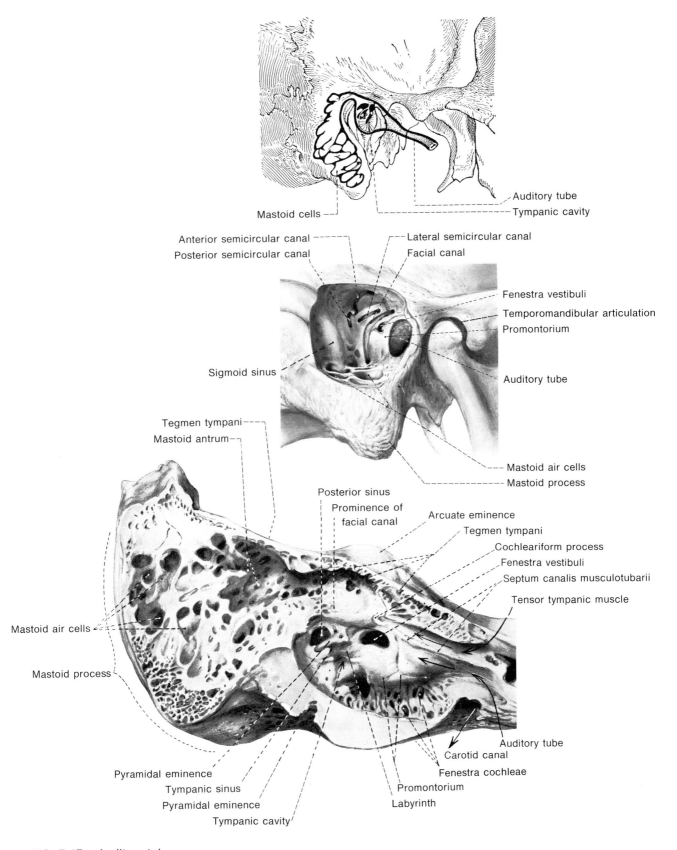

FIG. 7-4F. *Auditory tube*

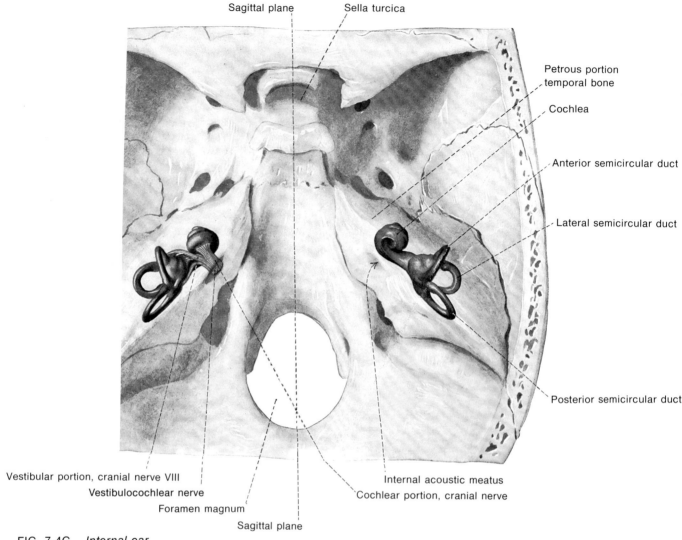

FIG. 7-4G. *Internal ear*

to the otic ganglion. In addition, the *chorda tympani*, from the facial nerve, enters the middle ear superoposterior to the tympanic membrane, passes between the handle of the malleus and the long crus of the incus, and leaves the tympanic cavity through the petrotympanic fissure, anterosuperior to the membrane.

The *tympanic antrum* is a small chamber with associated mastoid cells, posterior, and partly superior and lateral, to the tympanic cavity, that communicates with the epitympanic recess through the *aditus*.

The *auditory tube* (fig. 7-4F) is the passage from the middle ear to the nasopharynx. It is lined with mucous membrane and the more lateral part is bony, while the more medial is cartilaginous. The latter is not tubular but is folded to form a groove, roofed with fibrous tissue.

Internal Ear

The *inner ear* (fig. 7-4G), within the petrous part of the temporal bone, comprises the more essential parts of the organs of hearing and of equilibrium. It consists of a membranous labyrinth and associated soft parts within a bony labyrinth. The bony labyrinth is made up of a vestibule and of cochlear and semicircular canals.

The *vestibule* (fig. 7-4H) is the central, enlarged part of the labyrinth, communicating anteriorly with the cochlea and posteriorly with the semicircular canals. Its lateral wall is pierced by the fenestra vestibuli, into which fits the footplate of the stapes, and its medial wall by the cribriform openings for the vestibular nerve.

The *cochlea* (fig. 7-4I) has the form of a spiral snail

shell with two-and-one-half turns around a central pillar, the *modiolus*. The base of the coils is at the fundus of the *internal acoustic meatus* and its apex or *cupula* is directed anteriorly. The osseous *spiral lamina* is a shelf winding around the modiolus like the threads on a screw and helping, with the spiral ligament, to divide the cochlear canal into two ducts, a more inferior *scala tympani* and a more superior *scala vestibuli*. The scala tympani begins at the fenestra cochlea, within the orifice of which opens the *cochlear aqueduct,* which extends to the jugular fossa for the transmission of the *perilymphatic duct,* while the *vestibular aqueduct* for the *endolymphatic duct,* extends from the vestibule to the posterior surface of the petrous temporal bone.

The three *semicircular canals* (fig. 7-4*J*) each describes about two-thirds of a circle, opens into the vestibule at both ends, and has a dilatation, the *ampulla,* at one end. Each is at right angles to the others. The *superior,* the highest, which usually causes a slight elevation upon the endocranial surface of the petrous temporal, is directed vertically in a plane passing anterolaterally at 45° to the cranial axis. The *posterior* semicircular canal is also vertical, but is directed posterolaterally at an angle of 90° to the superior canal, and at a level slightly inferior to it. The *lateral* semicircular canal is essentially horizontal, directed posterolaterally in the angle between, and at right angles to, the other two.

With the inner ear should be mentioned the *internal acoustic meatus.* It is a short canal coursing laterally in the petrous temporal and ending in a *fundus* divided by a *transverse crista* into a superior and an inferior area. The anterior part of the former receives the facial nerve, and the posterior, a part of the vestibular nerve through several small openings. The anterior part of the inferior area receives the cochlear nerve, and the posterior, a part of the vestibular nerve, both through cribriform openings.

The *membranous labyrinth* is a system of membranous tubes within and largely conforming to the shape of the bony labyrinth, containing the terminal parts of the vestibulocochlear nerve. It also contains the *endolymphatic fluid* and is surrounded by *perilymphatic fluid.*

The membranous labyrinth forms two parts, utriculus and sacculus, united by a slender tube. The *utriculus* is a vesicle within the vestibulum and into it

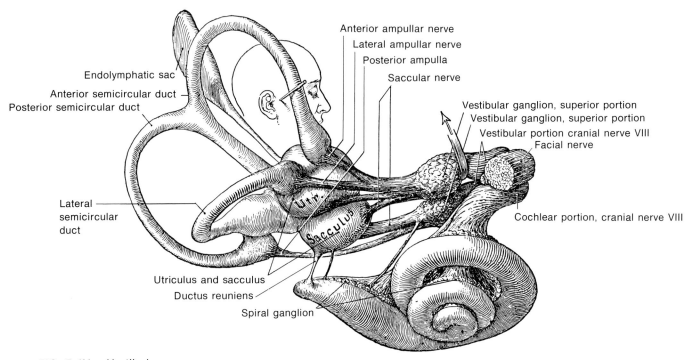

FIG. 7-4H. *Vestibule*

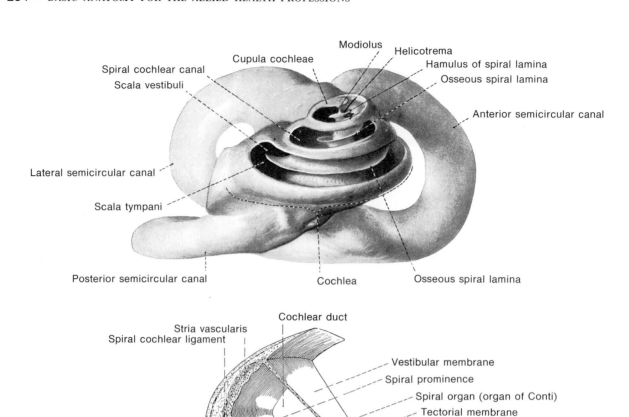
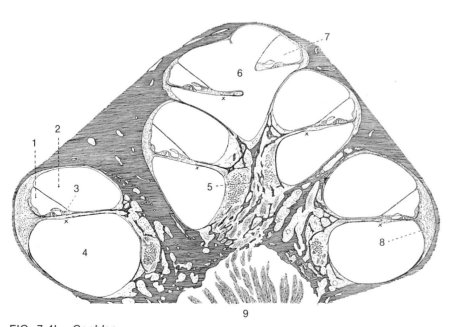

FIG. 7-41. *Cochlea*

1—Cochlear duct
2—Scala vestibuli
3—Limbus laminae spiralis
4—Scala tympani
5—Modiolus
6—Helicotrema
7—Cupula cecum
8—Spiral cochlear ligament
9—Spiral ganglion

open the three semicircular ducts, each with a dilated *ampulla* about one-third the diameter of their bony canals. Connection to the bone is by connective tissue, and the ducts lie upon the concave walls of the canals.

The *sacculus* is a smaller, flattened vesicle in the vestibule, communicating by a restricted *ductus reuniens* with the *cochlear duct,* which begins in the *cochlear recess* of the vestibule and follows the spirals of the cochlea to its apical part. It is largely triangular and occupies the more peripheral part of the scala vestibuli. The whole has attachment to the bony walls but is largely surrounded by perilymph.

The utriculus and sacculus are connected by short tributaries of the *endolymphatic duct,* which passes through the canal of the same name and ends in the blind, dilated *endolymphatic sac,* between the posterior aspect of the petrous temporal bone and the dura mater.

The membranous labyrinth is a closed space with no variation in its pressure. The perilymphatic space, on the other hand, communicates with the subarachnoid space by the *perilymphatic duct,* the minute pore of which is lined by modified dura and arachnoid of extreme fineness. The pressure of the perilymph is the same as that of the cerebrospinal fluid.

The cochlear duct (fig. 7-4*K*) is the membranous part of the bony cochlea and is continuous with the remainder of the membranous tubules through the canalis reuniens connecting to the sacculus. The cochlear duct appears triangular in cross-section with a *basilar membrane* forming one boundary, the *vestibular membrane* forming the second, and the *thickened periosteum* of the bony cochlea forming the third boundary. The *spiral organ* (organ of Corti) sits on the basilar membrane and stimulation of the hair cells of the spiral organ results in auditory perception.

The arterial supply of the ear is by a multiplicity of small twigs, mostly of the external carotid artery, but also by the internal carotid artery and the basilar artery. External carotid branches are: inferior tympanic branches of the ascending pharyngeal artery to the tympanic cavity and tube; auricular branches of the occipital artery and the auricular artery to the auri-

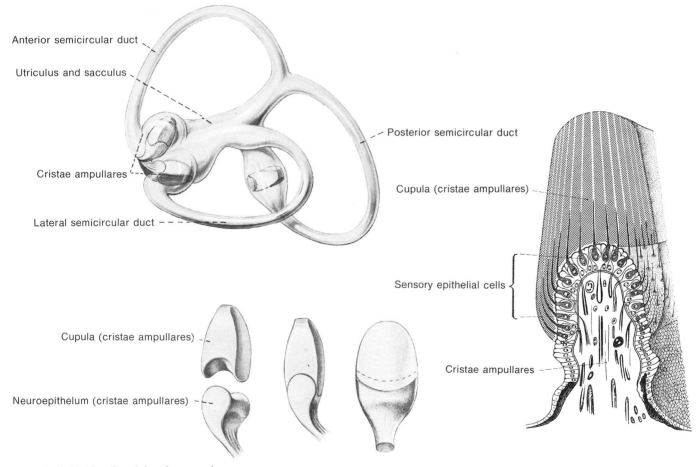

FIG. 7-4J. *Semicircular canals*

cle; several stylomastoid (to the meatus and middle and inner ear) and the auricular (to the auricle) branches of the posterior auricular artery; and deep auricular (to the meatus), anterior tympanic and middle meningeal (to the tympanic cavity), and palatine and pterygoid canal (to the tube) branches of the maxillary artery. The internal carotid supplies a tympanic branch to the tympanic cavity; and the basilar artery supplies an internal auditory branch to the middle ear. Venous drainage of the external ear is by anterior auricular branches of the facial, and by posterior auricular veins, as well as by an emissary vein into the transverse sinus; drainage of the tympanic cavity and tube is into the petrosal sinus, posterior facial vein, and pterygoid plexus; and drainage of the inner ear is by the internal auditory vein to the sigmoid or inferior petrosal sinus, by the internal acoustic meatus.

Somatic innervation of the ear is by the great auricular and lesser occipital branches of the cervical plexus. In addition to the above mentioned nerves the auriculotemporal branch of the trigeminal nerve provides innervation to the meatus. The vagus also innervates the external surface of the auditory membrane by its auricular branch; from the medial tympanic cavity and auditory tube by the glossopharyngeal nerve, through its tympanic plexus. Special sensory, or hearing, is received through the organ of Corti in the cochlea through the cochlear nerve. Special proprioception, or equilibrium, is received through the semicircular canals through the vestibular portion of the vestibulocochlear nerve. Motor is supplied to the facial musculature of the auricle by the facial nerve and to the muscles of the middle ear by the trigeminal and facial nerves. Sympathetic fibers are supplied to the mucous membrane by way of communications between the internal carotid plexus and the tympanic plexus.

The *intrapetrous part of the facial nerve* comprises the following. The facial nerve and its sensory root accompany the vestibulocochlear nerve into the internal acoustic meatus. At the bottom of the meatus it enters the facial canal, which runs laterally above the vestibule of the labyrinth to the medial wall of the middle ear. It then bends at right angles and runs backward along the medial wall above the promontory and the fenestra vestibuli. At the medial wall of the aditus to the tympanic antrum the canal curves downward, and eventually opens on the inferior surface of the temporal bone at the stylomastoid foramen.

In the first part of its course the facial nerve and its sensory root run side by side. At the right-angled bend, however, the *geniculate ganglion of the facial nerve* is situated, and in it the sensory root ends. As it runs backward, the facial nerve is separated from the middle ear and the aditus to the tympanic antrum only by a *thin plate of bone,* and in the latter situation it is related superiorly to the ampulla of the lateral semicircular canal. In the descending part of its

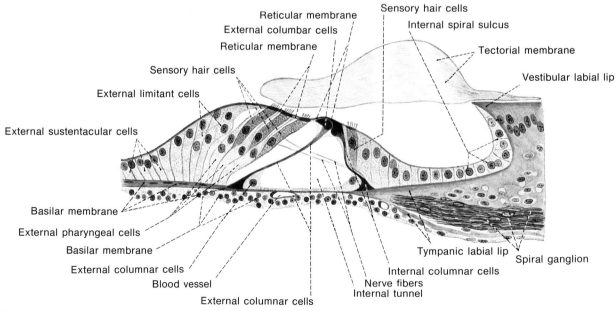

FIG. 7-4K. *Spiral organ (organ of Corti)*

course the facial nerve gives off a small branch to the stapedius muscle, and also gives origin to the *chorda tympani nerve.*

The *chorda tympani nerve* passes forward through the bone and emerges on the posterior wall of the middle ear. It then crosses the upper part of the tympanic membrane and the handle of the malleus, covered on its medial side by the mucous membrane of the middle ear. At the anterior border of the tympanic membrane the nerve enters a minute canal, which conducts it to the exterior of the skull at the petrotympanic fissure. It unites with and is distributed by the lingual nerve. Traced centrally, the afferent fibers of the chorda tympani end in cells of the facial ganglion, and the centrally running processes of those cells constitute the sensory root of the facial nerve.

CLINICAL APPLICATIONS

Following inflammation of the iris, adhesions may form between the iris and the front of the lens, obstructing the outflow of the aqueous humor from the posterior to the anterior chamber. When this happens, secretion of the aqueous humor continues and the intraocular pressure soon rises to a dangerous degree—the condition of *acute glaucoma.* Only immediate operative treatment to restore the drainage by the removal of a part of the iris can then save the sight of the eye.

A rise in cerebrospinal fluid pressure in the subarachnoid space compresses the walls of the retinal vein resulting in congestion of the retinal vein, edema of the retina, and *papilledema.*

Blindness in one-half of each visual field is known as *hemianopsia.*

In complete oculomotor paralysis, the eye cannot be moved upward, downward, or inward. The eye looks laterally because of activity of the lateral rectus and superior oblique muscles. The patient sees double (*diplopia*). Drooping of the upper eyelid (*ptosis*) occurs because of the paralysis to the palpebrae superioris. The pupil is dilated because of the paralysis of the sphincter pupillae and the unopposed action of the dilator pupillae.

A trochlear nerve paralysis is most obvious when the patient's eye turns medially and downward. An abducens nerve paralysis prevents the patient from turning the eye laterally. The unopposed medial rectus causes *internal strabismus.*

Infections of the tympanic cavity (*otitis media*) produce bulging and redness of the tympanic membrane. Further spread of the infection may lead to *acute mastoiditis, meningitis, facial nerve palsy,* and *vertigo.*

REVIEW QUESTIONS:

1. The olfactory bulbs contain _____, _____ and _____ cells.
2. The blood distributed by the ciliary arteries is usually returned by four _____ veins.
3. The _____ fills the area of the eyeball posterior to the lens.
4. The ciliary muscle has _____ and _____ fibers.
5. The sphincter pupillae is innervated by the _____ nerve.
6. The ophthalmic artery enters the orbit through the _____ foramen.
7. The ophthalmic veins drain into the _____ plexus.
8. The eyelids are covered by skin and lined with a mucous membrane termed the _____.
9. The space between the upper and lower lids is the _____.
10. The _____ organs are the taste buds.
11. Taste to the anterior two-thirds of the tongue is provided by the _____ nerve.
12. Taste to the posterior one-third of the tongue is provided by the _____ nerve.
13. Hearing is concerned with the _____ portion of the vestibulocochlear nerve.
14. The concha of the ear is bounded posteriorly by the _____.
15. The _____ is the passage extending from the concha to the tympanic membrane.
16. The greater portion of the tympanic membrane is held taut by the tonus of the _____ muscle.

17. The ossicles of the middle ear consist of the _____, _____, and _____.
18. The stapedius muscle is innervated by the _____ nerve.
19. The chorda tympani leaves the tympanic cavity through the _____ fissure.
20. The essential parts of the internal ear consist of a _____ and a _____ labyrinth.
21. The _____ labyrinth contains the endolymphatic fluid and is surrounded by _____ fluid.
22. The _____ and _____ membranes form boundaries of the cochlear duct.
23. The tympanic plexus is formed by the _____ nerve.
24. The _____ is the passage from the middle ear to the nasopharynx.
25. The eyeball consists of _____ concentric layers.

8. The Digestive System

STUDENT OBJECTIVES
After you have read this chapter, you should be able to:
1. Identify the organs of the gastrointestinal tract and the accessory organs of digestion.
2. Define the mesentery, mesocolon, falciform ligament, lesser omentum, and greater omentum.
3. Define deciduous and permanent dentition.
4. Identify the parts of a typical tooth.
5. Discuss the morphology and location of the stomach.
6. Describe the location and structure of the small intestine.
7. Discuss the location of the large intestine.
8. Describe the location and morphology of the pancreas.
9. Identify the various lobes of the liver.
10. Discuss the morphology of the liver.
11. Describe the location and structure of the gallbladder.
12. Discuss the esophagus.
13. Describe the location and structure of the pharynx.
14. Discuss the oral cavity.

The *digestive system* (fig. 8-1) begins at the lips and ends at the anus. It comprises the *mouth* and its contents, the *pharynx* (shared with the respiratory system), *esophagus, stomach, small intestine,* and *large intestine,* together with the associated glands, including the *salivary glands, liver,* and *pancreas.*

MOUTH OR ORAL CAVITY

The *mouth* (fig. 8-2) is formed by the cheeks, hard and soft palates, and tongue, and extends from the lips to the pharynx. The lips are covered externally by skin and internally by mucous membrane, with the orbicularis oris and labial glands located between the two. The reddish external part of the lips, which is of variable extent, is of modified mucous membrane. The junction of upper lip with cheeks is marked externally on either side by the *nasolabial sulcus,* the groove between the border of the nose and the angle of the mouth; and between lower lip and chin by the *mentolabial sulcus.* In the middle of the external upper lip is the *philtrum,* a slight groove bounded by two low ridges. Below the groove the *labial tubercle* of the upper lip is often apparent.

The mouth is divisible into vestibule and oral cavity proper.

The *vestibule* is bounded externally by the lips and cheeks and internally by the teeth and gums. The oral cavity proper is separated from the vestibule by the intrusion of the teeth and gums. In the midline, between lips and gums, are two folds of membrane, the *superior* and *inferior labial frenula.* Upon the oral surface of lips and cheeks are the ducts of glands, minute for the most part except for the large duct of the parotid gland, which opens on a small papilla at the level of the second upper molar.

Teeth

The *teeth* of mammals are of four recognized sorts according to shape: incisors, canines, premolars, and molars. As a rule they occur in this class in two successive sets: milk or deciduous, and permanent teeth. A tooth has a *crown,* projecting beyond the gum, a *neck* between the gum and the border of the bony dental arch, and one, two, or three *roots,* set in a bony socket or *alveolus* and surrounded by periosteum. A tooth is formed of dentine, cement, and

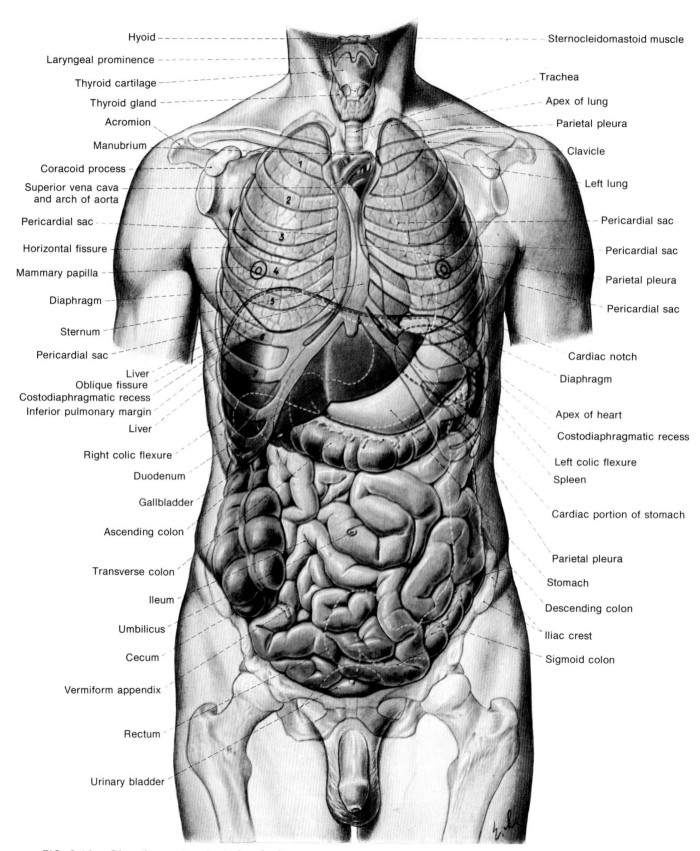

FIG. 8-1A. *Digestive system (anterior view)*

THE DIGESTIVE SYSTEM ■ 241

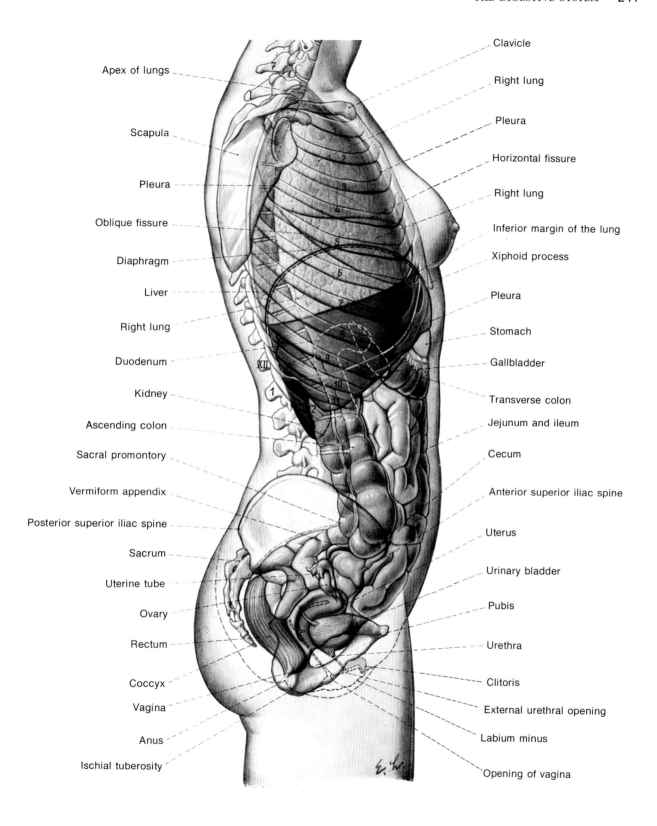

FIG. 8-1B. *Digestive system (right lateral view)*

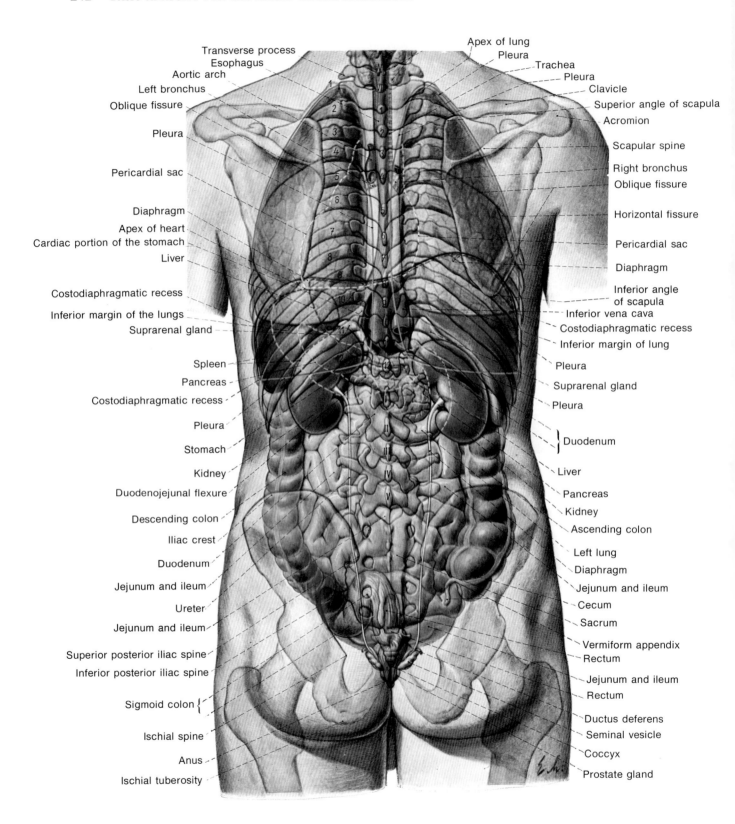

FIG. 8-1C. *Digestive system (posterior view)*

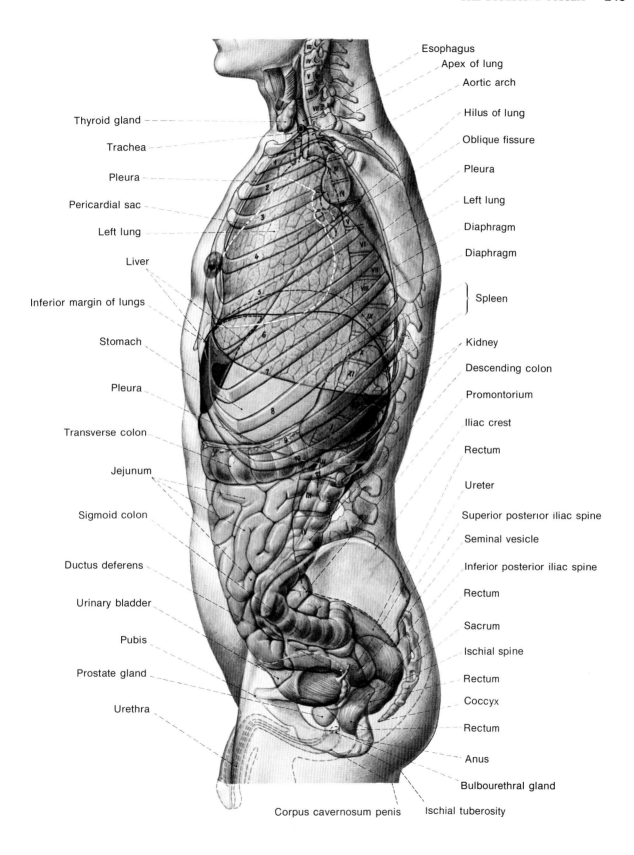

FIG. 8-1D. *Digestive system (left lateral view)*

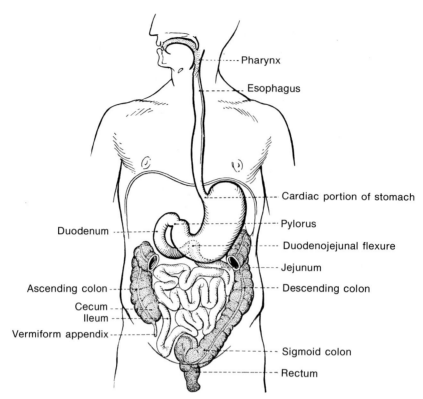

FIG. 8-1E. *Digestive system (diagrammatic view)*

enamel. The *dentine* constitutes the main portion of the tooth. It is derived from the *dental pulp*, which occupies the *central pulp cavity*, an expanded portion of the *root canal*, which carries the nerve of the tooth from the *root foramen* at the tip of the root. The nerves are branches of the alveolar branches of the trigeminal nerve. The *cementum* surrounds the neck and root of the tooth, while the *enamel*, formed by the enamel organ and of ectodermal derivation, covers the crown.

The 20 *deciduous teeth* (fig. 8-3A) comprise 2 incisors, 1 canine, and 2 molars, on either side, above and below. They are smaller than the corresponding permanent teeth, with smaller necks, more slender roots, and thicker enamel, while the crown pattern of the molars is different and the first molar is smaller than the second. The upper molars have three roots, the lower molars two, and the remaining teeth one. The time of eruption is somewhat variable but the following average may be given: medial incisors, at age 7 months; lateral incisors, 9 months; first molars, 14 months; canines, 18 months; and second molars, 22 months. The deciduous teeth are lost in the order of appearance of the permanent teeth.

The 32 *permanent teeth* (fig. 8-3B–E) appear, as a rule, in the following order, at the approximate ages indicated: *first molars*, 6 years; *medial incisors*, 7 years; *lateral incisors*, 8 years; *first premolars*, 10 years; *canines*, 11 years; *second premolars*, 11 years; *second molars*, 12 years; and *third molars*, 12–25 years or even later, occasionally failing to appear at all.

The *oral cavity proper* extends from the dental arches to the isthmus of the fauces, the opening between the oral cavity and the pharynx. The palate constitutes its roof, and its floor is formed anteriorly by the mucous membrane inferior to the apex of the tongue, and posteriorly by the base of the tongue.

Palate

The *palate* (fig. 8-4) is the partition between the nasal and oral cavities. The roof of the mouth is formed by the *hard palate*, consisting of the palatine processes of the maxillae and of the palatine bones, covered by periosteum and mucous membrane, and by the anterior part of the *soft palate*. The oral surface of the hard palate is concave in both directions, and has a *median palatine raphe,* or ridge, terminating anteriorly in a small *incisive papilla,* as well as several *transverse palatine plicae* upon its anterior part. Its free border or *velum* is arched and at its center is the *uvula.*

Within the soft palate are five muscles (four paired and one single) that can shorten, tense, elevate, or depress the palate and constrict either nasopharynx or oropharynx. The levator veli palatini elevates it and constricts the nasopharynx and its companion muscle, the tensor veli palatini, by virtue of the fact that it passes at a right angle around the hamular process and tautens the velum. The uvulae muscle shortens the soft palate as well as the uvula. The depressors and the constrictors of the oropharynx are the palatopharyngeus and the palatoglossus muscles. A fold containing the palatoglossal muscle extends from the side of the soft palate to the root of the tongue and is known as the *palatoglossal arch* (one of the paired pillars of the fauces) which bounds the isthmus of the fauces. The fauces is the passageway between the mouth and pharynx. Posterior to this point is the pharynx. Posterior to this arch, the palatopharyngeus forms a similar *palatopharyngeal arch.* Contraction of the palatoglossal muscle causes elevation of the tongue and constriction of the isthmus; of the palatopharyngeal, constriction of its arch and elevation of the pharynx.

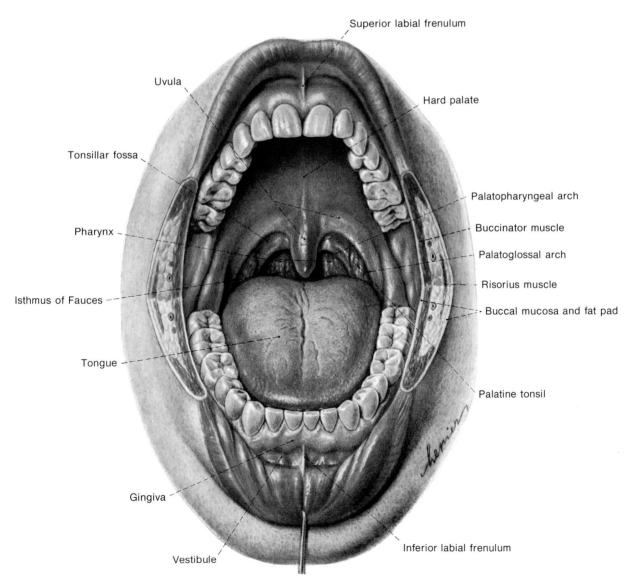

FIG. 8-2. *Oral cavity*

Tongue

The *tongue* (fig. 8-5) is a structure with a free apical and an anchored basal part, covered with mucous membrane, devoid of a skeletal framework, and with interlacing muscles so arranged that the free part can be moved in all directions, shortened, and even lengthened by the contraction of the fibers in the coronal plane. The apical part is used mostly in phonation and in the control of food while chewing; the basal part is used mostly in swallowing. At the posterior end of the free part it is anchored in the midline

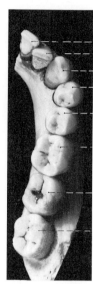

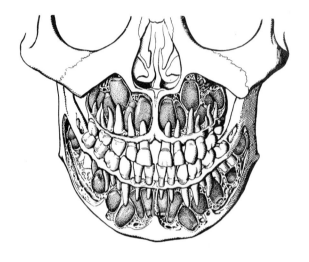

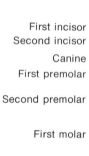

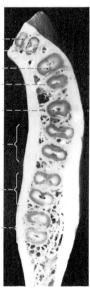

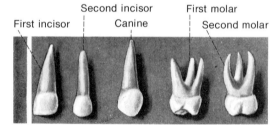

FIG. 8-3A. *Teeth (deciduous teeth)*

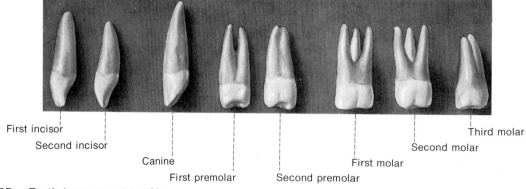

FIG. 8-3B. *Teeth (permanent teeth)*

THE DIGESTIVE SYSTEM ■ 247

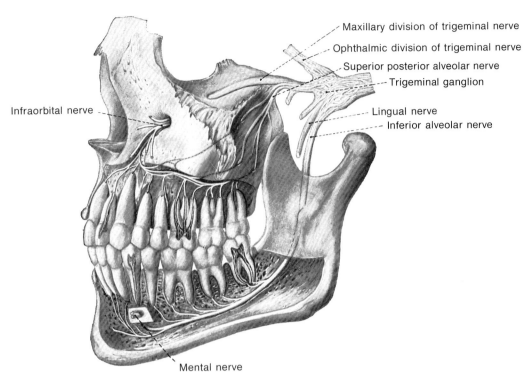

FIG. 8-3C. *Teeth (permanent teeth with innervation)*

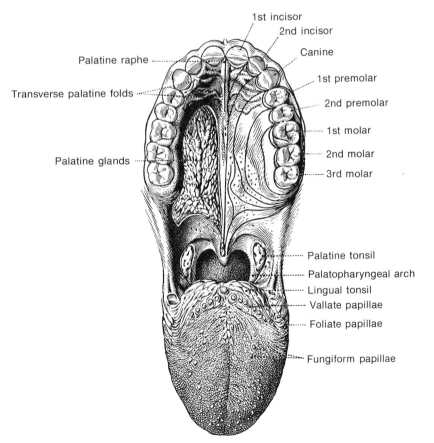

FIG. 8-3D. *Teeth (tooth eruption)*

to the floor of the mouth by the root, first by an anteriorly directed fold of membrane, the *frenulum*, and then by the broad basal part. Near the ventral termination of the frenulum is a small *sublingual papilla* upon which open the ducts of the submandibular gland. Just lateral to the frenulum, upon the undersurface of the tongue, there may be seen a slight ridge marking the deep lingual vein, and laterally to this, the *fimbriated fold*. The dorsum of the anterior tongue is sometimes smooth, while in some individuals it has well-defined shallow fissures. It is covered with minute *filiform* and larger *fungiform papillae*, the latter most numerous toward the tip and margins and associated with taste buds of the facial nerve field. The larger *circumvallate papillae* are arranged in a broad, shallow V just anterior to the foramen cecum. Around these are clustered many taste buds with glossopharyngeal innervation. These mark the boundary between the anterior two thirds and posterior one third of the tongue. Just posterior to them in the midline is found the *foramen cecum*, the remains of the thyroglossal duct, and diverging from this anterolaterally on either side, the short grooves of the *sulcus terminalis*. Posterior to this sulcus the tongue is more rugose. It is continuous laterally with the walls of the pharynx and inferiorly is connected with the

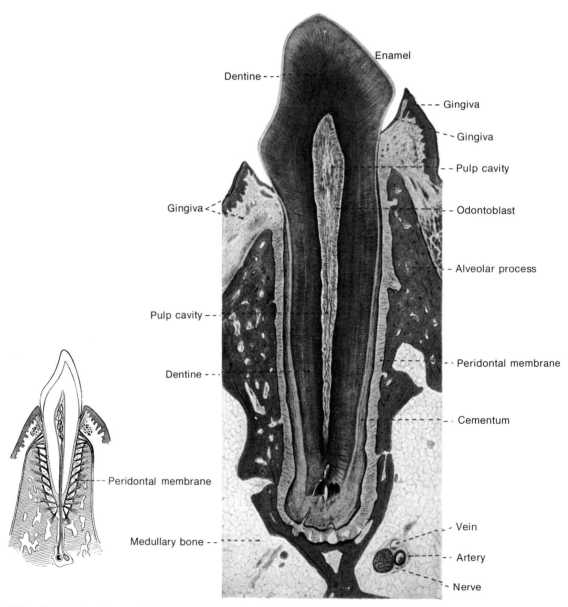

FIG. 8-3E. *Teeth* (*tooth morphology*)

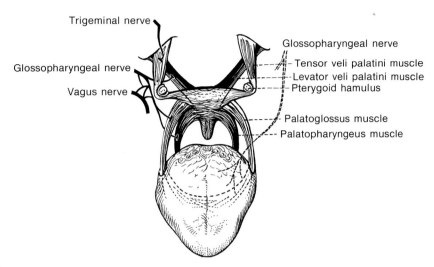

FIG. 8-4. *Palate*

larynx. The base of the tongue is innervated by the glossopharyngeal nerve. The vagus innervates the extreme posterior part of the tongue base, adjoining the epiglottis. There are mucous and serous glands embedded between the muscle fibers of the tongue. These serous glands occur in relation to the taste buds; the function of the two is interrelated. The internal part of the tongue is composed mostly of a number of somatic muscles of the hypoglossal field that have invaded the area from the anterior neck.

Upon the floor of the mouth the sublingual folds lie lateral to the frenulum. On either side of the frenulum open the ducts of the sublingual glands.

The blood supply of the cheeks, palate, tongue, and lips is by buccal, labial, and alveolar branches of the facial and maxillary arteries. The venous drainage is by corresponding veins.

Innervation of the mouth is complicated. The hypoglossal nerve supplies somatic motor fibers to the muscles of the tongue, and the facial nerve provides motor fibers to the muscles of the cheeks and lips; the facial nerve supplies parasympathetic fibers, via the pterygopalatine ganglion, to the mucous glands of the palate, and via the submandibular ganglion to the mucous glands of the floor of the mouth; the trigeminal nerve supplies somatic sensibility for the lips, cheeks, and the anterior two-thirds of the tongue and walls of the posterior mouth, while the teeth are innervated by the superior and inferior alveolar branches. Visceral sensibility is supplied to the posterior one-third of the tongue by the glossopharyngeal nerve. Taste is supplied to the anterior two-thirds of the tongue by the facial nerve, and to the region around the circumvallate papillae and the posterior one-third of the tongue by the glossopharyngeal nerve.

Salivary Glands

The *parotid gland* is located subcutaneously anterior and inferior to the external ear, extending anteriorly over the masseter and invading the upper neck adjacent to the submandibular gland. It is the largest of the *salivary glands* (fig. 8-6). It is usually pierced by the external carotid artery, the retromandibular vein, and by branches of the facial nerve. It is covered by the parotid fascia attached superiorly to the zygomatic arch. The parotid duct extends anteriorly across and superficial to the masseter muscle, pierces the buccinator, passes between the buccinator muscle and the mucous membrane of the cheek, and opens into the mouth by a small aperture at the level of the second upper molar. The fluid from the parotid gland contains no mucous but is rich in starch-splitting enzymes. Parasympathetic fibers are received from the glossopharyngeal nerve by its tympanic branch, and through the lesser petrosal nerve. The postganglionic neurons are located in the otic ganglion. The sympathetic postganglionic fibers are received from the superior cervical ganglion by way of the carotid plexus.

The *submandibular gland* is located medial to the inferior margin of the mandible anterior to its angle. When the salivary glands are large, this one may be separated from the adjacent part of the parotid gland only by fascia. There is a larger, superficial part called the body and a small, deeper part curving around the posterior border of the mylohyoid muscle. The submandibular duct extends from the deep surface of the body, along the deeper process of the gland medial to the mylohyoid muscle, medial to the sublingual gland, and to the *sublingual papilla* beside the frenulum of the tongue.

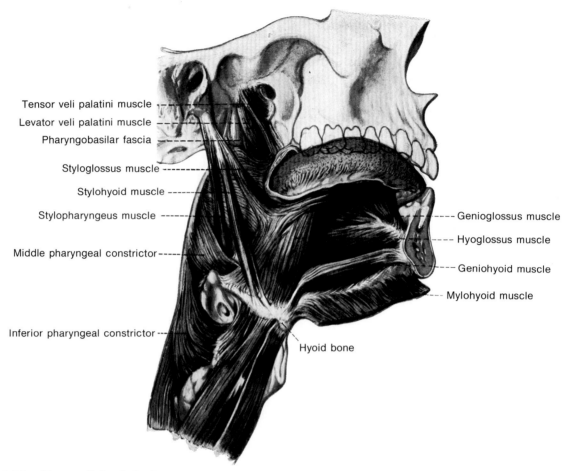

FIG. 8-5A. Tongue (*lateral view*)

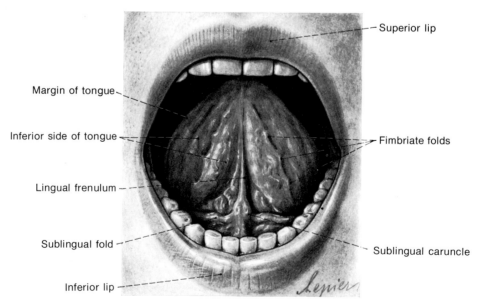

FIG. 8-5B. Tongue (*ventral surface*)

The *sublingual gland,* the smallest of the three salivary glands, is situated beneath the mucosa of the floor of the mouth, between the mandible laterally and the genioglossus medially. The gland discharges into the mouth by a number of small ducts with apertures upon the sublingual folds. The submandibular and sublingual glands produce mucous and a watery secretion. The innervation of both glands is the same. Parasympathetic fibers are received from the facial nerve via the chorda tympani, which travels with the lingual nerve. The postganglionic neurons are located in the submandibular ganglion. Sympathetic postganglionic fibers are received from the superior cervical ganglion by way of the carotid plexus.

PHARYNX

The *pharynx* (fig. 8-7) constitutes the part of the digestive system between the mouth and the esophagus, extending from the isthmus of the fauces to the level of the superior border of the cricoid cartilage, or the level of the sixth cervical vertebra. The pharynx is encroached upon superiorly by the soft palate, which divides this portion into nasopharynx and oropharynx.

The *nasal portion of the pharynx* extends from the *basal* or posterior part of the skull to the level of the soft palate. Upon its lateral wall is the curved *pharyngeal aperture* of the *Eustachian tube.* Posterior to the tubal aperture is the fold of the *pharyngeal recess,* posterior to which is an aggregation of lymphoid tissue termed the *pharyngeal tonsil.*

The *oral portion of the pharynx* extends from the soft palate to the *aryepiglottic fold.* Continuing inferiorly from the lateral termination of the soft palate is the *palatopharyngeal arch,* and between this and the palatoglossal arch is the *tonsillar sinus,* occupied by the *palatine tonsil.* Both of these arches are formed by the muscles of the same name. The tongue occupies the anterior wall of the oral portion of the pharynx, and inferiorly it is connected to the epiglottis by an unpaired *glossoepiglottic fold,* on either side of which is a depression, the *vallecula.* The *laryngeal portion of the pharynx* extends inferiorly from the aryepiglottic fold to the level of the cricoid cartilage.

The blood supply of the pharynx is by the ascending pharyngeal branch of the external carotid and the ascending palatine branch of the maxillary artery. The venous drainage is mostly by the pharyngeal and laryngeal plexuses, communicating inferiorly with the internal jugular vein and superiorly with the pterygoid plexus. Innervation of the pharynx includes sympathetic fibers from the superior cervical ganglion, via the carotid plexus, to the mucous glands; parasympathetic fibers from the facial nerve via the pterygopalatine ganglion to the glandular structures; motor fibers from the vagus to most of the muscles of the region; and from the glossopharyngeal nerve to the stylopharyngeus muscle.

ESOPHAGUS

The *esophagus* (fig. 8-8) is the part of the digestive system extending from the level of the cricoid cartilage to the cardiac end of the stomach. It is a flattened

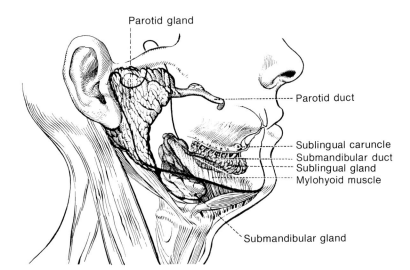

FIG. 8-6. *Salivary glands*

tube composed of an investing coat of fibroareolar tissue, within which is a layer of longitudinal and circular smooth muscle fibers, a submucous, and a mucous coat. Striated (vagal) musculature also extends from the pharynx in the outer layer of the esophagus for a variable distance. The esophagus has three constrictions: at the cricoid cartilage; where it passes dorsal to the left bronchus; and where it pierces the diaphragm. There is an appreciable dilatation of the esophagus superior to the diaphragm.

In the superior half of its course it is situated between the trachea and the center of the vertebrae. In the groove between it and the trachea passes the recurrent laryngeal branch of the vagus, and bordering it laterally in the neck is the carotid sheath. It passes through the esophageal opening of the diaphragm, forms a groove on the liver, and ends at the cardia of the stomach.

The blood supply of the esophagus is by twigs of the inferior thyroid artery, and by the inferior phrenic and left gastric arteries. Venous drainage is to the thyroid, azygos, and left gastric veins. Motor innervation is entirely autonomic, by the recurrent laryngeal and esophageal branches of the vagus, and by cervical and superior thoracic sympathetic branches. Visceral sensibility from the superior part is carried by the vagus. The remainder of the esophagus is devoid of pain sensation except in response to stretching.

ABDOMINAL CAVITY

The *abdominal cavity* (fig. 8-9A) is that part of the trunk cavities inferior to the diaphragm. It may be considered to include the pelvic cavity. It is bounded superiorly by the diaphragm; inferiorly by the pelvic

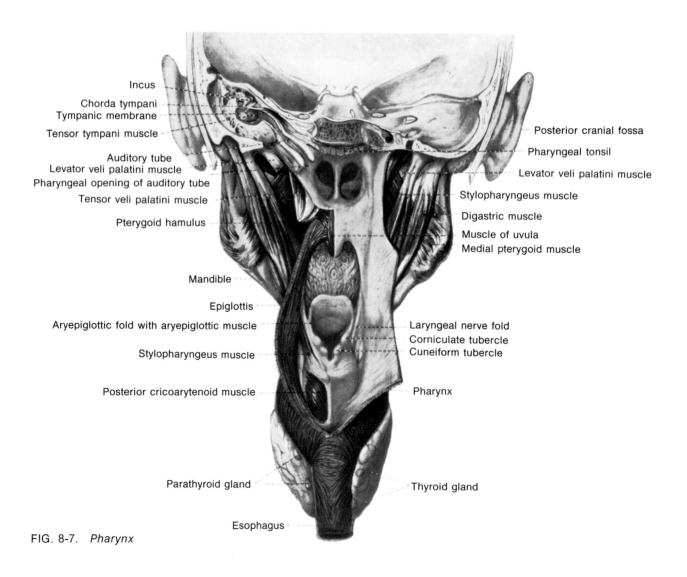

FIG. 8-7. *Pharynx*

diaphragm; dorsally by the vertebral column, part of the diaphragm, psoas and quadratus lumborum, and dorsal pelvic structures; and ventrally and laterally by the muscular walls of the abdomen and the pelvic structures. The abdominal cavity is lined by the parietal layer of the *peritoneum,* and contains most of the digestive system, which includes the distal esophagus to the inferior rectum, with accompanying diverticula, glands, vessels, nerves, and lymphatics.

Upon the inner surface of the ventral abdominal wall there extends superiorly from the umbilicus to the liver a fold of peritoneum, the *falciform ligament,* containing the *ligamentum teres hepatis,* the remnant of the umbilical vein. Radiating inferiorly from the umbilicus are three obliterated vessels: in the midline to the bladder, the *median umbilical ligament (urachus)* or remnant of the allantoic stalk; and upon either side the paired *lateral umbilical ligaments* (obliterated umbilical arteries). Lateral to these, but coursing in the same general direction, are the inferior epigastric vessels.

When the undisturbed contents of the abdominal cavity are exposed, the following structures may be seen. The *liver* protrudes from beneath the right costal margin, and between its right and left lobes the *gallbladder* may sometimes be seen. Inferior to and to the left of the liver is the *stomach,* from the greater curvature of which hangs the *greater omentum.* Dorsal to this and adjoining the stomach is the *transverse colon,* which is superior to the coils of the *small intes-*

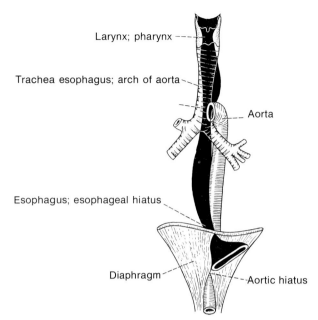

FIG. 8-8B. *Esophagus (esophagus and related structures)*

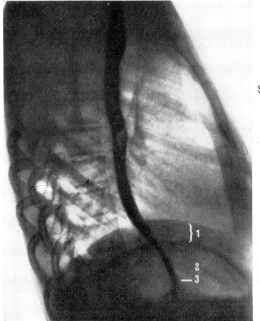

FIG. 8-8A. *Esophagus (radiogram)*

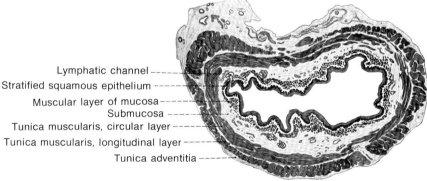

FIG. 8-8C. *Esophagus (cross section)*

1 Diaphragm
2 Fundus of stomach
3 Cardiac opening of stomach

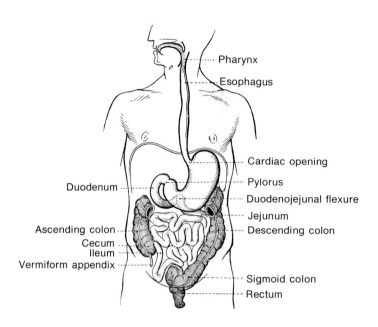

FIG. 8-8D. *Esophagus (diagrammatic view)*

tine. In most cases a part of the *ascending colon* is to be seen upon the right, and at times the *descending colon* upon the left.

PERITONEUM

The *peritoneum* (fig. 8-9B–D) is a closed sac without openings, except for the apertures of the fallopian tubes in the female, and it contains nothing but a trace of serous fluid. Various structures bulge into the abdominal cavity from the dorsal aspect, carrying the peritoneum before them. If they bulge for a sufficient distance from the dorsal abdominal wall to be completely covered by peritoneum and thus to have a mesentery, they are said to be *intraperitoneal;* if they bulge for a distance so short that they are covered by peritoneum only on one surface and thus have no mesentery, they are said to be *retroperitoneal.*

Mesenteries are double folds of peritoneum between which lie particular visceral structures, with their vessels and nerves, together with some connective tissue.

The *greater omentum* hangs down over the small intestine in a double fold. Each fold is composed of two layers of peritoneum, so the double fold is composed of four layers fused together, but thinned in spots, and often loaded with fat. In reality it passes from the stomach to the dorsal wall of the abdominal cavity, but the part lying upon the transverse colon and mesocolon has fused with these structures. Hence in the cadaver the greater omentum "appears" to extend loosely from the stomach to the colon, and is sometimes termed the *gastrocolic ligament.* It forms the inferior wall of the *omental bursa.* It continues on the right into the first part of the duodenum, and upon the left it is continuous with the gastrolienal ligament.

The *gastrolienal ligament* is the superior continuation upon the left of the greater omentum, between the greater curvature of the stomach and the spleen.

The *phrenicolienal ligament* is the short fold between the diaphragm, spleen and the left kidney, also forming a very short section of the wall of the omental bursa.

The *gastrophrenic ligament* is the short fold of peritoneum extending from the left border of the esophageal orifice of the diaphragm to the cardia of the stomach.

The *phrenicocolic ligament* is the part of the left border of the greater omentum that joins the left flexure of the colon to the diaphragm.

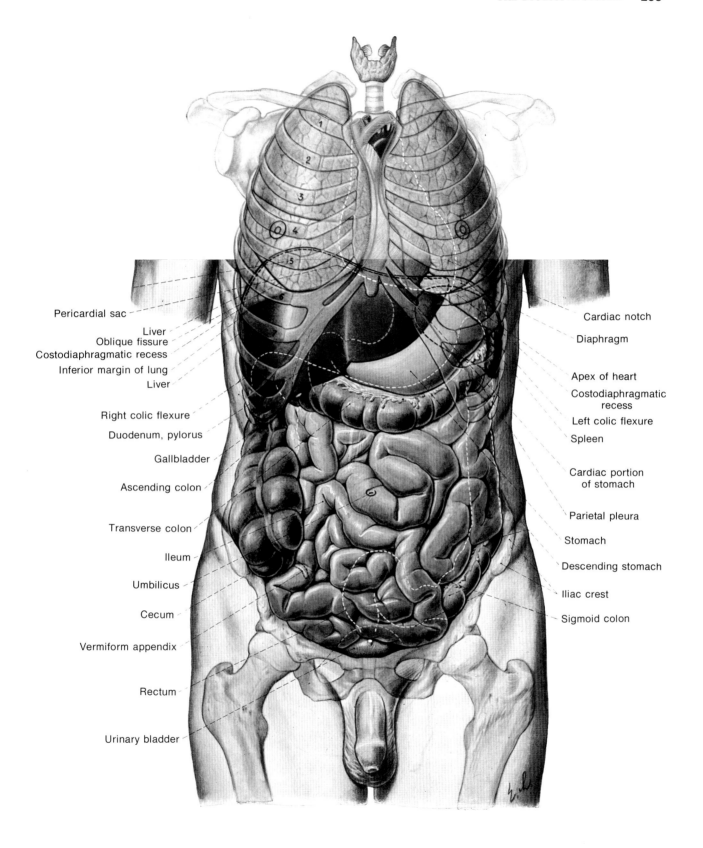

FIG. 8-9A. *Abdominal cavity (internal organs of abdominal cavity)*

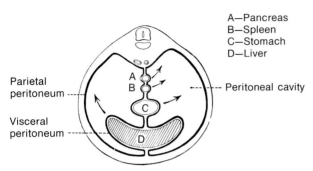

FIG. 8-9B. Abdominal cavity: peritoneum of embryo (*cross section*)

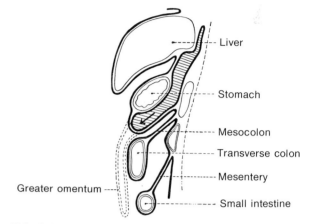

FIG. 8-9E. Abdominal cavity (*formation of peritoneal folds*)

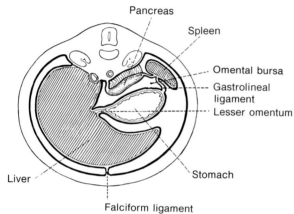

FIG. 8-9C. Abdominal cavity: peritoneum of adult (*cross section*)

The *mesentery* refers to the fold of peritoneum concerned with the jejunum and ileum. Its root extends from the duodenum obliquely and inferiorly to the right iliac region. From the root it fans out to contain the jejunum and ileum, and it contains the branches of the superior mesenteric vessels, lymphatics, fat, and connective tissue. Often the mesentery is continuous between the distal ileum and the root of the appendix as the *mesoappendix*.

Usually the *ascending colon* is retroperitoneal. The *transverse mesocolon* has a root crossing the duodenum and a part of the pancreas. It is fused with the overlying part of the greater omentum and contains

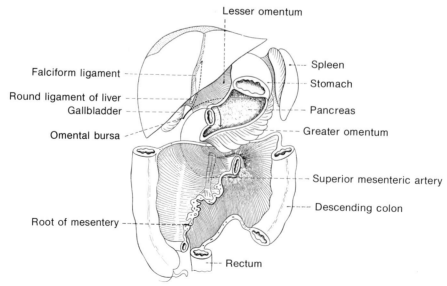

FIG. 8-9D. Abdominal cavity (*peritoneal fold*)

1—Esophagus
1a—Cardiac portion of stomach
2—Fundus of stomach
3a—Lesser curvature of stomach
3b—Greater curvature of stomach
4—Duodenal bulb
5a—Descending portion of duodenum
5b—Horizontal portion of duodenum
5c—Ascending portion of duodenum
5d—Duodenojejunal flexure
6—Jejunum
Between 7a and 7b—Peristaltic contraction XII—12th thoracic vertebra

FIG. 8-10D. *Stomach*
(*tall individual with long stomach*)

the middle colic vessels. The *descending colon* is also retroperitoneal. The *sigmoid mesocolon* is regularly present but is variable in arrangement.

The *lesser omentum* constitutes the *hepatogastric ligament*, between the lesser curvature of the stomach and the liver, and the *hepatoduodenal ligament*. The hepatoduodenal ligament contains the bile duct, hepatic artery, and portal vein. The lesser omentum forms a part of the wall of the omental bursa.

Peritoneal recesses occur customarily in three general areas. The *duodenal recesses* occur about, most frequently to the left of, the duodenojejunal flexure as this part of the small intestine changes from retro- to intraperitoneal. *Ileocolic recesses* occur in the mesentery above both the ileum and appendix. An *intersigmoid recess* frequently occurs upon the left side of the sigmoid mesocolon.

The peritoneal cavity is partially divisible into two portions of unequal size. The *lesser sac* or omental bursa is the part of the cavity enclosed by the stomach, lesser omentum, left lobe of the liver, and greater omentum. It opens into the remainder of the peritoneal cavity through the *epiploic foramen* (or *foramen of Winslow*), located just inferior to the porta of the liver, between the bile duct, hepatic artery, and portal vein ventrally, and the inferior vena cava dorsally. Just within the foramen is the *vestibule*. From this there is a *superior recess* dorsal to the caudate lobe of

258 ■ BASIC ANATOMY FOR THE ALLIED HEALTH PROFESSIONS

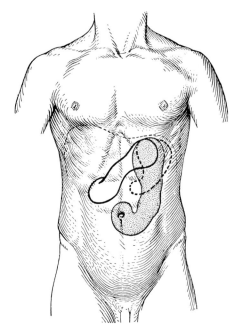

FIG. 8-10B. *Stomach (positions of the stomach)*

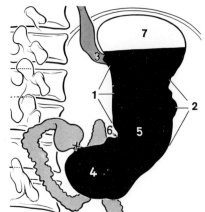

FIG. 8-10C. *Stomach (portions of the stomach)*

1—Lesser curvature
2—Greater curvature
3—Cardiac portion
4—Pyloric portion
5—Body of stomach
6—Angular incisure
7—Fundus of stomach
X—Pylorus

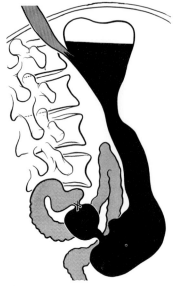

FIG. 8-10D. *Stomach (tall individual with long stomach)*

FIG. 8-10E. *Stomach (short individual with short stomach)*

FIG. 8-10F. *Stomach (radiogram of long stomach)*

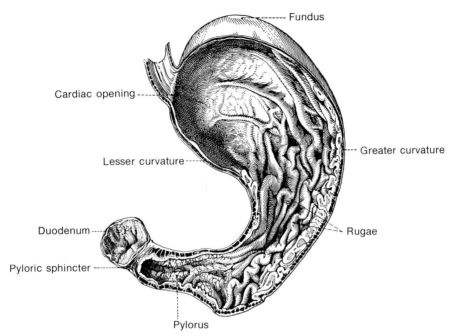

FIG. 8-10G. Stomach (*internal aspect*)

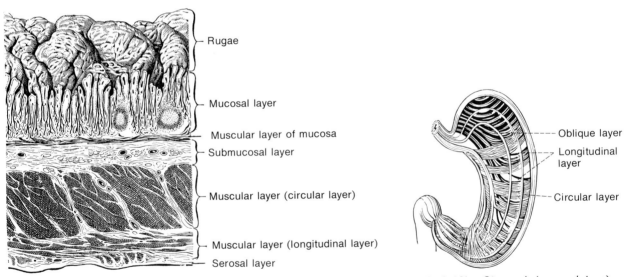

FIG. 8-10H. Stomach (*histology*)

FIG. 8-10I. Stomach (*musculature*)

the liver, while the *inferior recess* is below and to the left of the foramen.

The *greater sac* constitutes the part of the peritoneal cavity other than the omental bursa.

STOMACH

The *stomach* (fig. 8-10) is a dilated portion of the foregut between the esophagus and the duodenum for the reception and accumulation of food, which within the stomach is reduced to a semiliquid mass of chyme, in preparation for its discharge into the duodenum through the pyloric valve. It is located in the left hypochondriac region, inferior to the left side of the diaphragm and the left lobe of the liver, and usually has the shape of a comma.

The stomach communicates with the esophagus by the cardia, and with the duodenum by the pylorus. The fundus is the enlarged portion bulging to the left of and superior to the cardia, and the constricted infe-

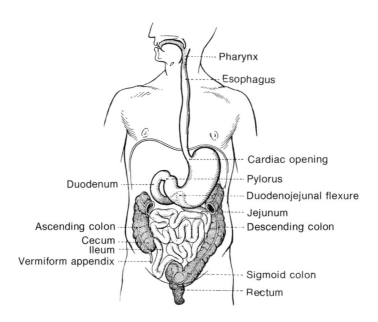

FIG. 8-11A. *Small intestine (diagrammatic view)*

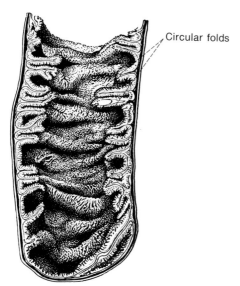

FIG. 8-11B. *Small intestine (longitudinal view)*

rior part is the pyloric end. The generally accepted divisions of the stomach are not grossly separable, but are based on function. The *cardia* is the part whose mucous membrane contains cardiac glands and is the part in a narrow zone (with relatively smooth internal surface) about the cardiac orifice. The *pylorus* of the stomach is the narrower, distal part, containing pyloric glands. The *fundus,* with wrinkled inner surface, is the part with fundic glands and is the part between the cardia and pylorus. The cardiac and pyloric glands produce mucous with no enzymes. The fundic glands produce mucous, hydrochloric acid, pepsin, and other enzymes. The stomach has a convex *greater curvature,* directed inferiorly and marked by the attachment of the greater omentum, and a concave *lesser curvature,* directed superiorly and marked by the attachment of the lesser omentum.

The stomach presents a smooth external surface. Internally the mucosa is thrown into coarse folds called *rugae.* These folds may be obliterated during distention of the organ. At the pylorus the inner surface is thickened to form a sphincter, the *pyloric valve.*

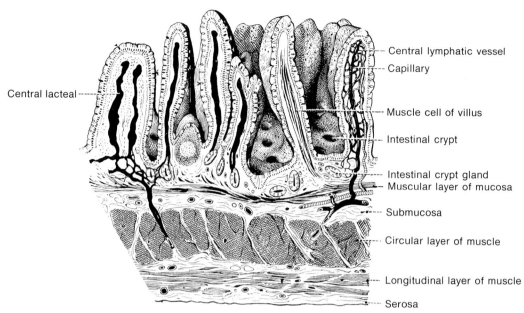

FIG. 8-11C. *Small intestine (microscopic view of villae)*

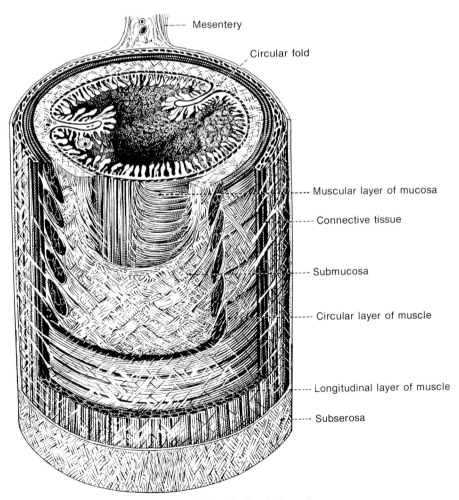

FIG. 8-11D. *Small intestine (diagrammatic view of transverse and longitudinal planes)*

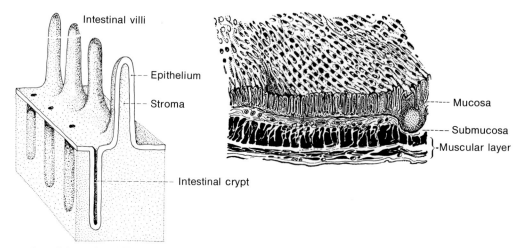

FIG. 8-11E. *Small intestine (diagrammatic view of villae)*

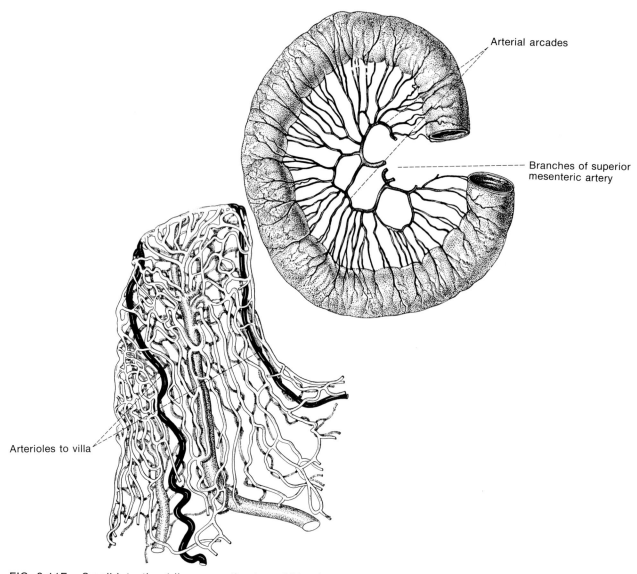

FIG. 8-11F. *Small intestine (diagrammatic view of blood supply to small intestine and villae)*

The blood supply of the stomach is by the celiac trunk: the left gastric artery and the right gastric branch of the hepatic artery supply the lesser curvature and vicinity; and the left gastroepiploic and short gastric branches of the splenic artery and the right gastroepiploic branch of the hepatic artery supply the greater curvature. The blood drains into the portal system, the coronary (of the lesser curvature) and the pyloric veins into the portal vein direct, the left gastroepiploic into the splenic vein, and the right gastroepiploic into the superior mesenteric vein. Innervation is by the sympathetic system via the splanchnic nerves and celiac ganglion, and by the parasympathetic system via the anterior and posterior gastric nerves, from both vagi.

SMALL INTESTINE

The *small intestine* (fig. 8-11) extends from the stomach to the cecum, and comprises the duodenum, jejunum, and ileum. Beneath the serous layer the intestine is composed of a longitudinal and a circular coat of smooth muscle, a submucous and a mucous coat. Upon the internal surface are minute *villi*, giving a plushlike appearance to the surface. In addition, there are circular folds of the mucosa, the *circular folds* of which some are permanent but most are obliterated with the stretching of the intestinal walls. These folds, as well as the almost constant motility, increase the efficiency of the absorptive and secretory processes by expanding the exposed surface. Small lymph nodes occur in the mucosa, either solitarily or in aggregations known as *Peyer's patches*. These patches, from one-half to three or more inches in length, but narrower, number as many as 35 or more in the distal part of the ileum of young subjects.

The small intestine secretes mucous and a great variety of enzymes of importance in carbohydrate, fat, and protein digestion. These digestive functions supplement those of the pancreatic juice and bile, poured into the lumen of the intestine. The small intestine is also the site of the absorption of the products of digestion.

The *duodenum* is the part of the small intestine between the pylorus of the stomach and the jejunum. It is C-shaped and is customarily divided for convenience into four parts. The first part, adjoining the stomach and at the inferior margin of the epiploic foramen, forms the *superior flexure* and is intraperitoneal. The second or descending portion passes inferiorly at the level of the mesocolon, usually in contact with the right kidney. The third portion or *inferior flexure* crosses the aorta toward the left, just inferior to the level of the mesocolon and the body of the pancreas. The fourth portion ends at the duodenal-jejunal junction as the small intestine becomes intraperitoneal. The second and third parts are retroperitoneal.

The internal surface of the first part of the duodenum is relatively smooth, but the second and third parts become increasingly ridged. Upon the left wall of the descending portion is a *longitudinal fold*, at the

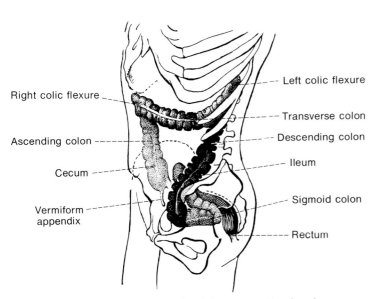

FIG. 8-12A. *Large intestine (diagrammatic views)*

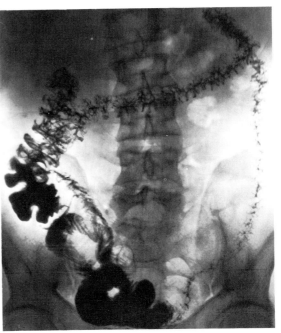

FIG. 8-12B. *Large intestine (radiogram)*

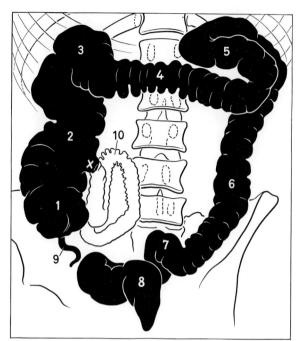

FIG. 8-12C. *Large intestine (radiogram diagrammatic view)*

superior termination of which is the *major duodenal papilla* where the common bile duct and pancreatic duct open either together or separately. Just above this point and slightly ventral to it is the *minor duodenal papilla,* where the accessory pancreatic duct opens.

The *jejunum* and *ileum* are located between the duodenum and the cecum, and constitute the intraperitoneal part of the small intestine, supported entirely by the mesentery. It is about 20 feet long when freshly extended, but is much shorter in the embalmed cadaver, and is arranged in a complicated and variable pattern of coils between the transverse colon and the pelvis. In man the point of juncture of jejunum with ileum is grossly indistinguishable. There is merely a gradual change from the proximal to the distal end of this part of the small intestine. As compared with the distal ileum, the proximal jejunum has thicker walls, more circular folds, no Peyer's patches, and less fat in the mesentery. Occasionally (2%) there occurs *Meckel's diverticulum,* the remains of the yolk stalk, upon the free border of the ileum located approximately 30 inches from the cecum. It is slightly narrower than the ileum and its most frequent length appears to be about 2 inches.

The blood supply of the duodenum is by the pancreaticoduodenal loops between the gastroduodenal and superior mesenteric arteries; the jejunum and ileum are supplied by the intestinal branches of the superior mesenteric artery. The venous drainage is by the corresponding veins to the portal system. The innervation is parasympathetic by vagi, and sympathetic by way of the greater and lesser splanchnic nerves, through the celiac and superior mesenteric plexuses.

LARGE INTESTINE

The *large intestine* (fig. 8-12) is the part of the digestive system between the ileum and the anus, and is composed of the cecum and appendix; the ascending, transverse, descending, and sigmoid colons; and the rectum. It is usually of considerably larger diameter than the small intestine and is larger at its proximal than its distal part (except for the ampulla of the rectum), but when empty it is frequently constricted to a small size, which is particularly the case in the transverse and descending portions. It presents a somewhat different picture from the small intestine. Suspended from the colon (but not the rectum) are small pouches of peritoneum, usually filled with fat—the *appendices epiploicae*. Throughout the colon the longitudinal muscle fibers occur in three narrow, glistening bands—the *teniae coli.* These tend to shorten the colon so that between them are sacculations or bulges

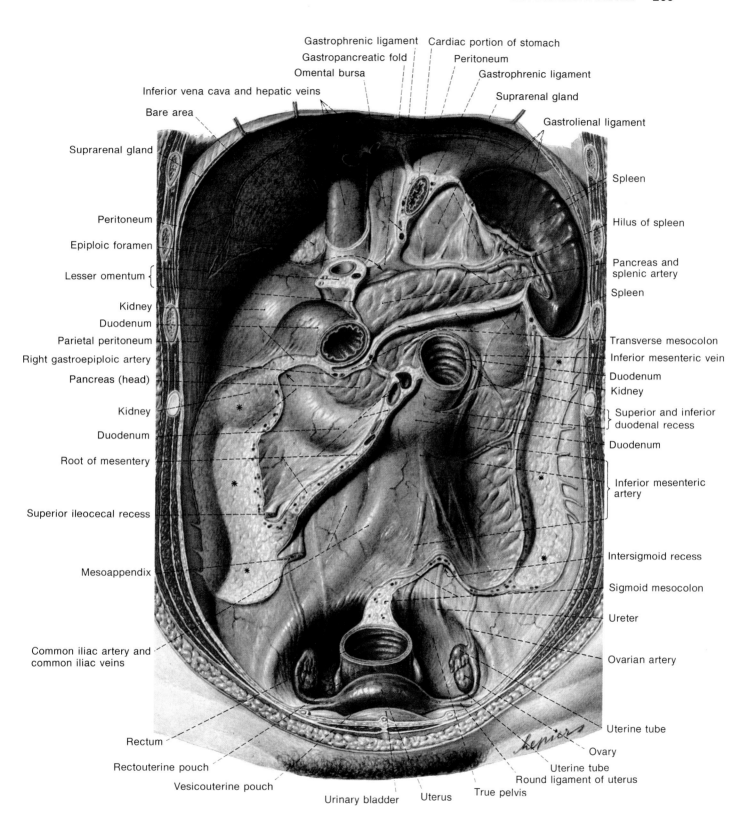

FIG. 8-12D. *Large intestine (attachments of large intestine to posterior abdominal wall)*

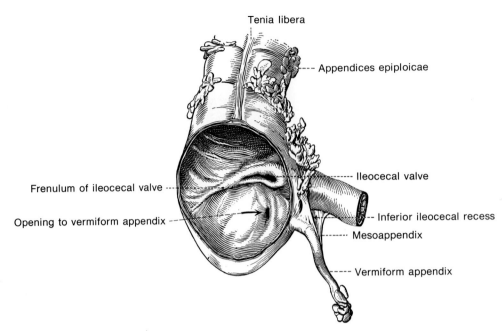

FIG. 8-12E. *Large intestine (cecum)*

of the walls—the *haustra.* One band, the tenia mesocolica, at the attachment of the mesocolon, is dorsal on the transverse colon and dorsomedial on the ascending and descending colons, while the other two teniae are evenly spaced. The large intestine functions chiefly in absorption, but also in secretion.

The termination of the cecum often tends to be free, with a fold of peritoneum from it to the iliac fossa. The ascending and descending colons and rectum are customarily retroperitoneal; the transverse and sigmoid colons, intraperitoneal.

The *cecum* and *appendix* together constitute a diverticulum of the large intestine inferior to its junction with the ileum. The appendix is the terminal portion, but during development it fails to follow the rate of growth of the remainder of the cecum. The cecum normally lies in the right iliac fossa but may become somewhat displaced. The variability of its form may be considered to follow three main patterns: a fetal or conical type, with cecum passing gradually into the appendix; an infantile type, fairly symmetrical, in which the appendix comes off the expanded cecum like the stem off a goblet; and an adult type, in which the appendix comes off the ileal side of the cecum and seems to bear an asymmetrical relation to it because of a disproportionate bulging of the lateral cecal wall. In all types, the three teniae coli continue onto the appendix, upon which the longitudinal muscle coat continues uninterrupted.

The *vermiform process* of the cecum or *appendix* is a blind tube diverging from the cecum. It is of small diameter and variable length. Its precise position is highly variable and it may be disposed in any direction.

The *ascending colon* extends from the cecum superiorly to the right or *hepatic flexure* of the colon. Usually it occupies the extreme dorsolateral part of the abdominal cavity and is retroperitoneal, but occasionally it is located somewhat more ventrally and has a distinct mesocolon. The *hepatic flexure,* where the ascending colon bends left to become the trans-

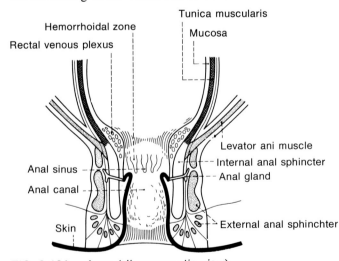

FIG. 8-13A. *Anus (diagrammatic view)*

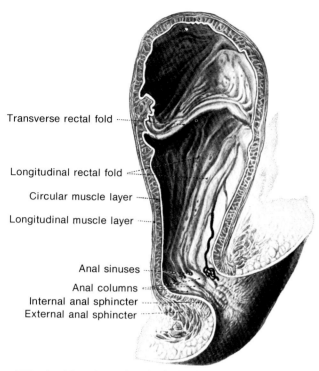

FIG. 8-13B. Anus (sagittal view)

verse colon, is situated at the inferior surface of the right lobe of the liver, lateral to the second part of the duodenum, and ventral to the right kidney.

The *transverse colon* extends from the right flexure to the left or *splenic flexure*. It is suspended by the mesocolon. For the most part the transverse colon lies inferior to the liver and stomach.

The *descending colon* extends from the splenic flexure almost to the pelvis, where it becomes the sigmoid colon. It usually is retroperitoneal, of smaller diameter than the ascending colon, and is suspended at the splenic flexure by the phrenicocolic ligament previously described. It is situated upon the left side of the abdominal cavity. Inferiorly it crosses the psoas major to join the sigmoid colon.

The *sigmoid colon* begins at the medial border of the psoas major and ends at the rectum. It is disposed in a loop of variable shape and lies within the pelvis.

The *rectum* is the part of the intestine, without a mesentery, between the sigmoid colon and the anus, situated ventral to the sacrum and the coccyx. It is characterized by the absence of appendices epiploicae, haustra, and teniae coli; the latter are absent because the longitudinal muscle fibers, confined to the teniae over the remainder of the colon, spread out to surround the rectum, but are thicker on the dorsal and ventral surfaces than on the sides. The rectum is divisible by a horizontal fold or slight constriction into a superior part, and a dilatable inferior part, the *ampulla*. Upon the mucous surface of the inferior part of the ampulla are vertical *rectal columns*, between which are the *anal sinuses*. The inferior termination of the columns marks the division between rectum and anus.

ANUS

The *anus* (fig. 8-13) is the constricted inferior termination of the digestive system. The anal orifice is customarily considered as surrounded by three muscle coats: the internal anal sphincter, the levator ani, and the external anal sphincter.

The blood supply of the appendix, cecum, and inferior part of the ascending colon is by the ileocolic branch. The middle part of the ascending colon is by the right colic branch; the superior end of the ascending and transverse colon is by the middle colic branch, all from the superior mesenteric artery. The descending colon is supplied by the left colic branch (anastomosing with the middle colic); the sigmoid colon by the sigmoid branches; and the superior rectum by the superior rectal branch, all of the inferior mesenteric artery. The superior rectal anastomoses with the middle rectal branch of the internal iliac, and this with the inferior rectal branches of the internal pudendal artery. Twigs of the medial sacral artery also reach the rectum. The superior and inferior mesenteric veins provide the venous drainage. Around the rectum the rectal veins form an external and an internal anal plexus, the latter in the submucous coat, largely in the rectal columns.

Innervation is by sympathetic and parasympathetic nerves. The former supply the ascending and most of the transverse colon through the celiac plexus (from inferior thoracic segments), and the left flexure, descending and sigmoid colons, and superior rectum through the inferior mesenteric plexus (from superior lumbar segments), while the rest of the rectum is supplied by the hypogastric plexus. The parasympathetic innervation of the ascending and most of the transverse colon is by the vagus; of the left flexure and the large intestine distal to it, by the sacral (S_2, S_3, and S_4) outflow. The anal canal and external sphincter receive somatic innervation by the pudendal nerve.

LIVER AND GALLBLADDER

The *liver* (fig. 8-14) is the largest gland in the body. It has the exocrine function of producing bile and

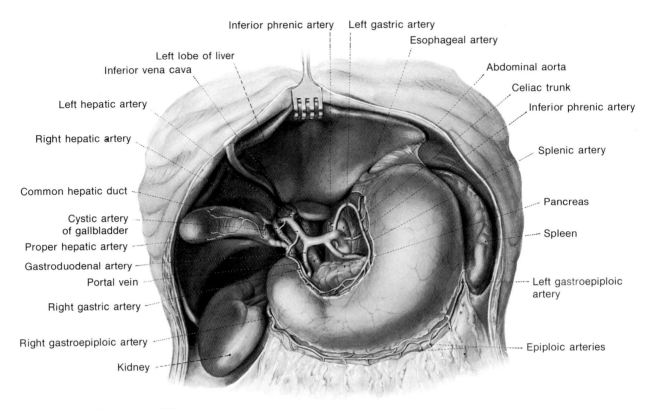

FIG. 8-14A. *Liver and gallbladder*

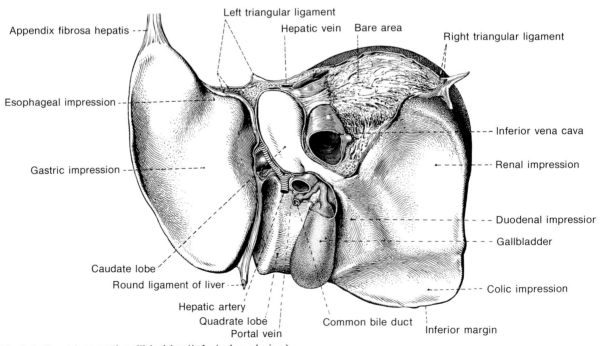

FIG. 8-14B. *Liver and gallbladder (inferiodorsal view)*

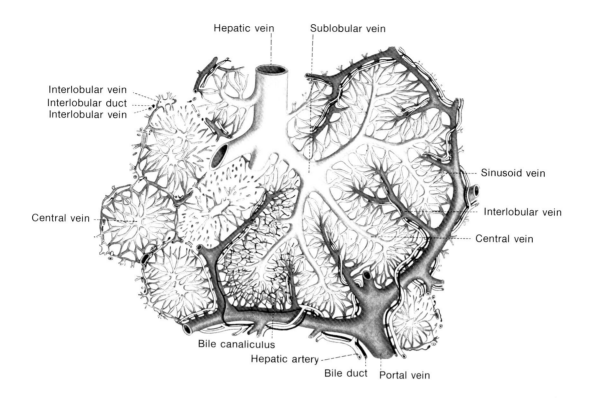
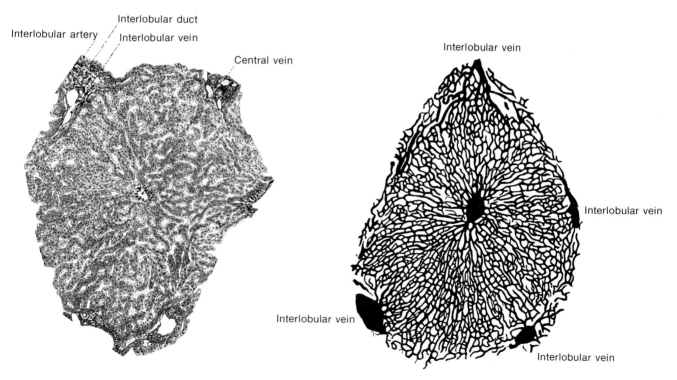

FIG. 8-14C. *Liver (diagrammatic microscopic view)*

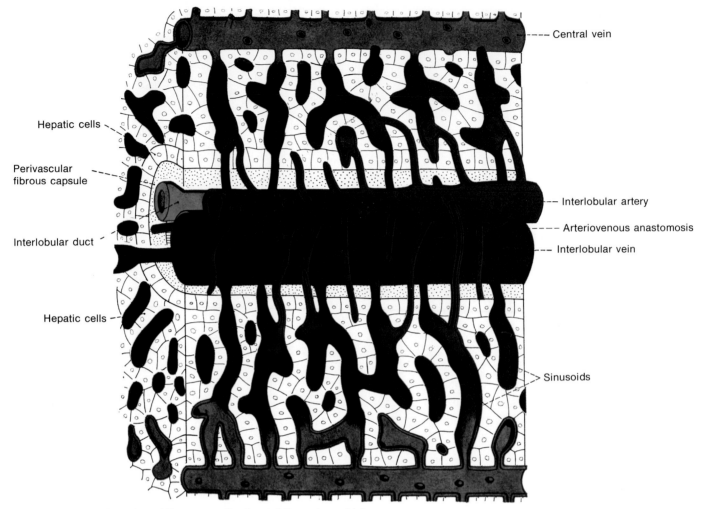

FIG. 8-14D. *Liver (diagrammatic view of liver sinusoids).*

endocrine functions in relation to the products glycogen, urea, heparin, and fibrogen.

The liver is connected to the inferior vena cava by the hepatic veins, to the aorta by the hepatic artery, to the duodenum by the common bile duct, to the veins of the intestines by the portal vein, and to the abdominal wall by the peritoneum.

The liver has four lobes; a large *right lobe,* a smaller *left lobe,* a small *quadrate lobe* upon the visceral surface to the left of the gallbladder, and a small *caudate lobe* to the left of the inferior vena cava, which here occupies a groove in the liver. The *porta hepatis,* by which vessels and nerves enter and leave the gland, is located upon the visceral surface between the caudate and quadrate lobes. The loose connective tissue about the vessels of the porta is continued inward as the fibrous *capsule of Glisson.*

The excretory duct of the liver or *common hepatic duct* leaves the porta and is joined by the *cystic duct* of the gallbladder. The *gallbladder,* which is a reservoir for the storage and concentration of bile, is located in the sulcus between the right and quadrate lobes upon the visceral surface of the liver. Its hepatic aspect lies against the liver, and its visceral aspect, covered by peritoneum, lies superior to the transverse colon and duodenum. The inner surface of the bladder is covered by fine honeycomb folds, the *plicae tunicae mucosae,* while that of the cystic duct has folds arranged as a *spiral valve.* After junction of the hepatic and cystic duct, the vessel is known as the *common bile duct* (*ductus choledochus*). This passes in the free border of the lesser omentum, ventral to the portal vein and to the right of the hepatic artery, then dorsal to the duodenum and head of the pancreas. It ends by piercing the medial wall of the second part of the duodenum in company with the pancreatic duct, and opening upon the major duodenal papilla.

The peritoneal covering of the liver is closely ad-

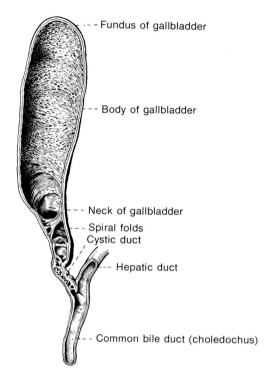

FIG. 8-14E. *Gallbladder (sagittal view)*

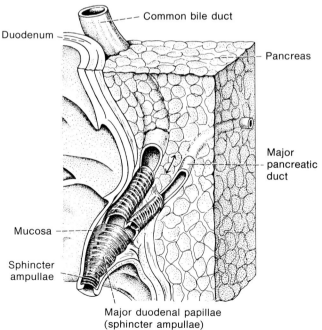

FIG. 8-14F. *Common bile duct*

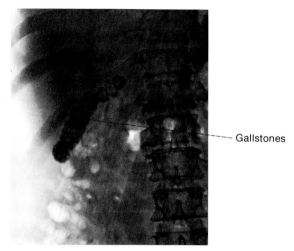

FIG. 8-14G. *Gallstones (radiogram)*

herent to the gland upon the ventral and visceral aspects of the right and left lobes, and upon the caudate and quadrate lobes. Ligaments of the liver, or folds where the peritoneum is reflected from the liver onto the abdominal wall, occur superiorly and dorsally, and at one point ventrally. The ventral fold occurs between the right and left lobes as the *falciform ligament*. The two layers of peritoneum forming this ligament diverge superiorly to comprise the superior right and left *coronary ligaments*. These follow a lateral course, and then each doubles back on itself medially to form a free border termed the right and left *triangular ligaments*. The more dorsal surface of these triangular ligaments thus comprises the inferior right and left coronary ligaments. Upon the left side the superior and inferior layers are close together, forming a narrow ligament; but upon the right there is a considerable space between the two that is not covered by peritoneum, and this is the *bare area* of the right lobe. This takes in the hepatic part of the inferior vena cava, and is continuous with a small bare area upon the superior part of the liver, between the right and left superior coronary ligaments.

The blood supply of the liver is by the hepatic artery and the portal vein. Both the artery and vein divide into right and left branches within the liver, and hepatic arterial twigs are accompanied by corresponding portal twigs. Venous drainage is by the right and left hepatic veins into the inferior vena cava, with accessory and minor connection between the left hepatic and paraumbilical veins through the falciform ligament. Innervation is autonomic; sympathetic from the celiac plexus via the network surrounding the hepatic artery, and parasympathetic from both vagi, through the (chiefly) anterior gastric nerve and its hepatic continuation.

PANCREAS

The *pancreas* (fig. 8-15) is a grayish gland, lobulated and soft, that is disposed transversely between the stomach and the great vessels along the course of the splenic vessels from duodenum to spleen. It has a hook-shaped *head* situated in the concavity of the duodenum. A slightly constricted *neck* separates the head from the *body* superiorly, and the body extends left as the *tail*. The pancreas is retroperitoneal. Between the surface are three borders, from the inferior of which is suspended the mesocolon. The *main pancreatic duct* (of Wirsung) extends within the substance of the gland from the tail of the pancreas to its head, where it may bend upon itself. Near its right termination it approaches and parallels the bile duct and, with the bile duct, it pierces the wall of the second part of the duodenum. It opens into the gut as the major duodenal papilla beside or in common with the bile duct. Just within the orifice there is often a slight dilatation, the *ampulla of Vater*. The *accessory pancreatic duct* (of Santorini) is smaller than the main duct and is variable in arrangement. It drains the head of the pancreas and also communicates with the main duct, but usually opens into the duodenum upon the *minor papilla*, a short distance superior to the *major papilla*.

The pancreas functions as an exocrine organ in secreting the pancreatic juice, which is poured into the duodenum to assist in carbohydrate, fat, and protein digestion, and as an endocrine organ producing insulin.

The blood supply of the pancreas is by the superior pancreaticoduodenal branch of the gastroduodenal artery; pancreatic branches of the splenic artery; and by the inferior pancreaticoduodenal branch of the superior mesenteric artery. Venous drainage is by the pancreatic branches of the splenic and the superior mesenteric veins. Innervation is autonomic, by sympathetic fibers from the celiac plexus accompanying the arteries to the gland and parasympathetic fibers from vagi.

CLINICAL APPLICATIONS

Oral Cavity

Dental caries, or tooth decay, involves the enamel, the dentine, and the pulp, in that order.

Trench mouth is an ulcerative infection of the mouth and throat.

Stomach

The stomach is the common site for *ulcers*, which occur most frequently along the lesser curvature and pylorus. *Hemorrhage* is commonly associated with stomach ulceration. Perforations caused by ulcers usually result in general *peritonitis*. Healing of ulcers near the pylorus may cause *stenosis*, resulting in distention.

Small Intestine

The small intestine may contain *diverticula*, which do not require treatment. A *Meckel's diverticulum* may be present in approximately 2% of individuals. It may be the site of ulcers or inflammation. It represents the remains of the vitellomesenteric duct, which extended from the umbilicus to the lower end of the small intestine.

Impaired absorption of nutrients from the small intestine may result in a complex array of symptoms that are referred to as *malabsorption syndromes*.

Large Intestine

A few of the more common clinical problems associated with the large intestine include the following. (1) *Diverticulosis* involves outpouchings in weakened areas of the intestinal wall. The sigmoid colon is vulnerable to this condition. Infection may develop inside these outpouchings, giving rise to *diverticulitis*. (2) *Hemorrhoids* are varicosities of the veins located beneath the mucosa of the anus and rectum. (3) *Appendicitis* or inflammation of the vermiform appendix usually calls for immediate surgery.

Cancer of the Digestive System

Cancer of the digestive system and associated organs and glands is a common threat to life. Other malignancies involving the digestive system include cancers of the oral cavity, throat, pancreas, liver, and gallbladder.

Liver

Cirrhosis refers to a fibrosis or scarring of the liver, which is a response to various types of injury. *Liver cancers* are predominantly secondary or metastatic.

THE DIGESTIVE SYSTEM ■ 273

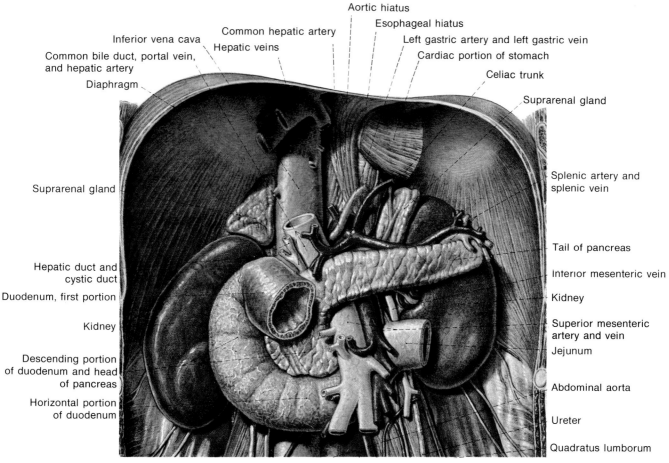

FIG. 8-15A. *Pancreas*

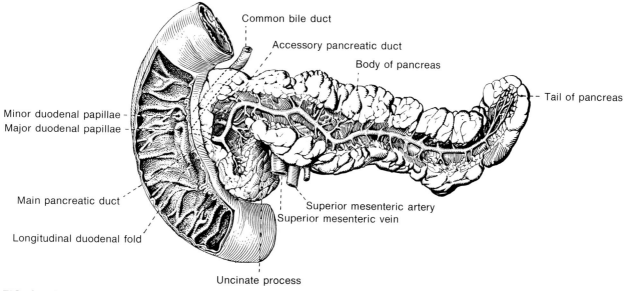

FIG. 8-15B. *Pancreatic ducts*

Pancreas

The pancreas is subject to inflammation, which may cause *necrosis* and *abscess*. It is also affected with *cysts* and *calculi*. *Cancer* of the pancreas usually affects the head of the gland.

REVIEW QUESTIONS:

1. The _____ is bounded externally by the lips and cheeks and internally by the teeth and gums.
2. The _____ is the partition between the nasal and oral cavities.
3. Near the ventral termination of the frenulum of the tongue is a small _____ papilla upon which open the ducts of the submandibular gland.
4. Just posterior to the circumvallate papillae in the midline is found the _____, the remains of the thyroglossal duct.
5. The _____, _____, and _____ comprise the salivary glands.
6. The _____ is the largest of the salivary glands.
7. The _____ portion of the pharynx extends from the soft palate to the aryepiglottic fold.
8. The esophagus extends from the level of the _____ cartilage to the cardiac end of the stomach.
9. The abdominal cavity is lined by the _____ layer of the peritoneum.
10. The _____ ligament contains the ligamentum teres hepatis.
11. The lesser omentum constitutes the _____ and _____ ligaments.
12. The hepatoduodenal ligament contains the _____, _____, and _____.
13. The greater peritoneal sac opens into the lesser peritoneal sac through the _____ foramen.
14. Internally, the mucosa of the stomach is thrown into coarse folds called _____.
15. The small intestine comprises the _____, _____, and _____.
16. The major duodenal papilla is the place where the _____ and _____ open into the second portion of the duodenum.
17. The minor duodenal papilla is where the _____ opens into the duodenum.
18. The _____ are suspended from the colon as small pouches of peritoneum, filled with fat.
19. The longitudinal muscle layer of the large intestine occurs in three narrow bands known as _____.
20. The _____ flexure is located where the ascending colon turns left to become the transverse colon.
21. The _____ is the constricted inferior termination of the digestive system.
22. The _____ is the largest gland in the body.
23. The _____ is located upon the visceral surface of the liver between the caudate and quadrate lobes.
24. The common hepatic duct leaves the porta hepatis and is joined by the _____ to form the common bile duct.
25. The _____ is located between the stomach and the great vessels along the course of the splenic vessels from duodenum to spleen.

9. The Cardiovascular System

> STUDENT OBJECTIVES
> After you have read this chapter, you should be able to:
> 1. Discuss the pericardial sac.
> 2. Describe the chambers of the heart.
> 3. Identify the coronary arteries.
> 4. Identify the cardiac veins.
> 5. Describe the valves of the heart.
> 6. Describe the initiation and conduction of a nerve impulse through the conduction system of the heart.
> 7. Identify the principal arteries and veins of systemic circulation.
> 8. Describe the route of blood in coronary circulation.
> 9. Describe the route of blood in pulmonary circulation.
> 10. Describe the route of blood in portal circulation.
> 11. Describe the route of blood in fetal circulation.
> 12. Describe the superficial veins of the upper and lower limbs.
> 13. Contrast the structure and function of arteries, arterioles, capillaries, venules and veins.
> 14. Describe the location of the heart.
> 15. Describe the composition of the heart wall.

The function of the cardiovascular system [heart, arteries, capillaries, and veins (fig. 9-1)] includes:

1. transport of nutrients
2. transport of oxygen
3. transport of carbon dioxide
4. transport of metabolic wastes
5. transport of hormones
6. regulation of body temperature
7. regulation of fluid and salt balance
8. protection of body against foreign particles that enter the body

SURFACE PROJECTIONS OF THE HEART

The *apex* of the heart (fig. 9-2) is left of the sternum in the fifth intercostal space. The *left border* is formed by the left ventricle, which extends along a curved line from the left second costochondral junction to the fifth costochondral junction. The *inferior border* or *diaphragmatic surface* is located at the level of the xiphisternal junction. The *base* of the heart is formed mostly by the left atrium opposite the fourth and fifth thoracic vertebrae. The *right border* of the heart is formed by the right atrium and extends from the xiphisternal junction to the right third costal cartilage.

The valves of the heart can also be identified by surface projections. The pulmonary semilunar valves lie approximately behind the inner end of the left third costal cartilage and the sternum. The aortic semilunar valve is located behind the left margin of the sternum, opposite the third intercostal space. The tricuspid valve lies behind the right margin of the sternum at the level of the fourth intercostal space. The bicuspid or mitral valve lies behind the right margin of the sternum opposite the fourth costal cartilage.

PERICARDIAL SAC (PERICARDIUM)

The *pericardial sac* (fig. 9-3) contains the heart. It is situated in the middle mediastinum extending between the levels of the sternal angle and the xiphisternal junction. It is usually separated from the sternum by the ventral folds of the pleurae. It also forms the ventral boundary of the posterior mediastinum, where it is in relationship with the aorta and esopha-

276 ■ BASIC ANATOMY FOR THE ALLIED HEALTH PROFESSIONS

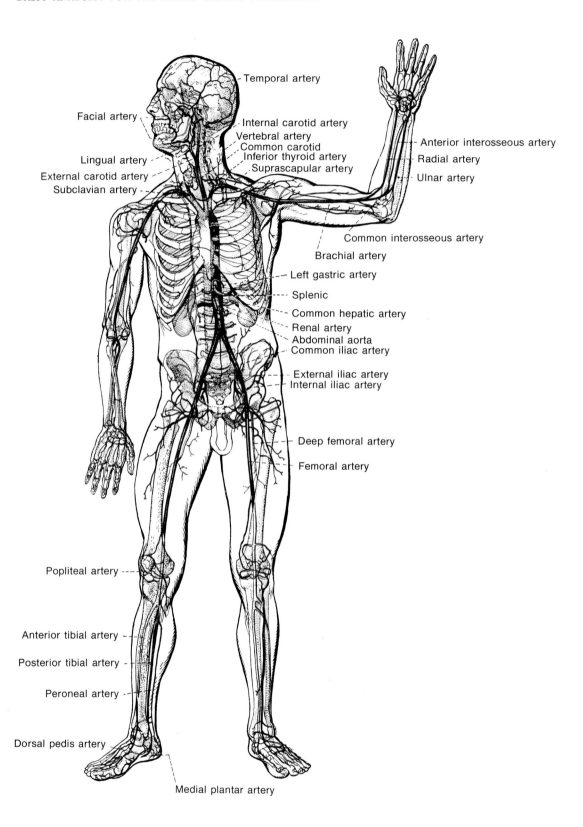

FIG. 9-1A. *Cardiovascular system (arteries)*

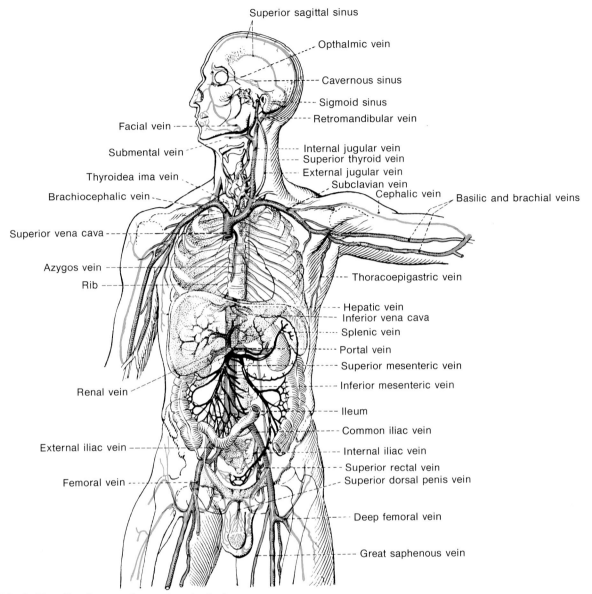

FIG. 9-1B. *Cardiovascular system (veins)*

gus. Laterally it is in contact with the pleurae, and inferiorly with the central tendon of the diaphragm, to which it is firmly anchored.

The pericardial sac consists of two layers referred to as the *fibrous* and *serous pericardium*. The fibrous pericardium constitutes the tough external layer. It is pierced by the great vessels of the heart and merges with their sheaths. The serous pericardium is a closed serous sac; the *inner* or *visceral layer* is the intimate investment of the heart, and the *outer* or *parietal layer* is closely adherent to the fibrous pericardium. The two are continuous and are reflected together at the roots of the great vessels. The *pericardial cavity* is the space between the visceral and parietal layers of the serous pericardium. This cavity usually contains *pericardial fluid*, which reduces friction between the heart and pericardial sac.

THE HEART

The *heart* (fig. 9-4) is a muscular saclike organ, located in the middle mediastinum, and surrounded by the pericardial sac. The *base* of the heart is located superiorly and is pierced by the great vessels. The *diaphragmatic surface* of the heart is formed by the right ventricle. The *apex* of the heart is the bluntly pointed end that extends somewhat to the left.

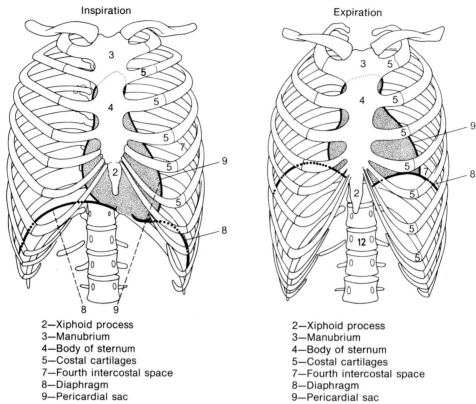

2—Xiphoid process
3—Manubrium
4—Body of sternum
5—Costal cartilages
7—Fourth intercostal space
8—Diaphragm
9—Pericardial sac

2—Xiphoid process
3—Manubrium
4—Body of sternum
5—Costal cartilages
7—Fourth intercostal space
8—Diaphragm
9—Pericardial sac

FIG. 9-2. *Surface projections of the heart (diagrammatic views)*

HEART WALLS

The *walls of the heart* (fig. 9-5) include three layers. The *epicardium* is the visceral layer of the pericardium. The *myocardium* consists of laminated layers of cardiac muscle fibers arranged in a complicated spiral manner to compress the heart with a wringing movement. It is thickest around the ventricles, thinnest around the atria. The wall of the left ventricle is thicker than that of the right ventricle because the blood in the left ventricle must be pumped over a greater distance. The *endocardium* is the lining of the heart and is continuous with and comparable to the inner coat of the blood vessels.

CHAMBERS OF THE HEART

The heart is divided into *four chambers* (fig. 9-6). The upper two are called the *right* and *left atria*. The atria are the receiving or venous chambers of the heart. "Atrium" and "auricle" are terms often, but erroneously, used interchangeably. The auricles are ear-shaped diverticula extending from the ventrosuperior aspect of each atrium. The internal surface of each atrium contains small ridges known as *pecti-*

nate muscle. The atria are separated by the *interatrial septum*. A depression known as the *fossa ovalis* is well marked on the right side of the interatrial septum. The *limbus fossae ovalis* is the thickened margin surrounding the fossa ovalis. The left atrium exhibits the remains of the foramen ovale upon the interatrial septum, which is called the *valvulae foramina ovalis.*

The right and left ventricles constitute the arterial portion of the heart. They are separated by the *interventricular septum*. Externally the *anterior* and *posterior interventricular sulci* separate the right and left ventricles. A groove known as the *coronary sulcus* separates the atria from the ventricles. Internally, ridges and folds of the ventricular walls are called *trabeculae carneae.*

PATH OF BLOOD FLOW THROUGH THE HEART

The right atrium receives blood from the upper portion of the body by way of the *superior vena cava* (fig. 9-7). The *inferior vena cava* returns blood from the lower portion of the body and the *coronary sinus* drains blood from the vessels supplying the heart walls. The right atrium contracts and squeezes blood

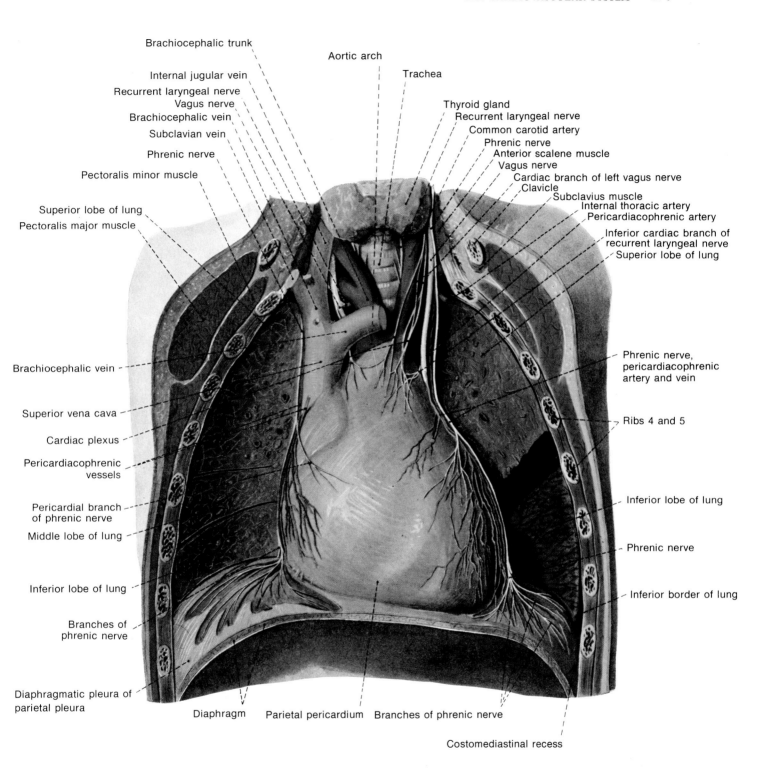

FIG. 9-3A. *Pericardial sac (anterior view)*

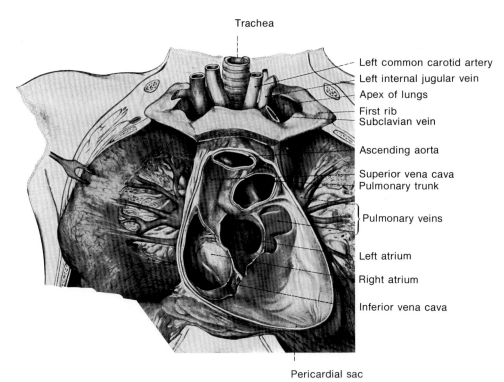

FIG. 9-3B. *Pericardial sac (heart removed)*

through the *right atrioventricular aperture* into the right ventricle, which pumps the blood into the *pulmonary trunk*. The pulmonary trunk divides into *right* and *left pulmonary arteries,* which carry blood to the lungs. The blood picks up oxygen and releases carbon dioxide in the lungs and returns to the left atrium by way of the *four pulmonary veins.* Blood is then squeezed through the *left atrioventricular aperture* into the left ventricle. The blood exits the left ventricle via the ascending aorta.

VALVES OF THE HEART

Heart valves (fig. 9-8) prevent the blood from flowing backwards. The right and left atrioventricular apertures are guarded by flaps of fibrous tissues that are covered with endocardium. The *right atrioventricular valve* is commonly called the *tricuspid valve* because it consists of three cusps. The *left atrioventricular valve* is called the *bicuspid* or *mitral valve.* From the atrial aspect these cusps are smooth, but on the ventricular surface they are roughened and are attached to papillary muscles by the *chordae tendineae. Pulmonary semilunar valves* are located in the opening where the pulmonary artery exits the right ventricle. The *aortic semilunar valve* is located at the opening between the left ventricle and the aorta.

CONDUCTION SYSTEM OF THE HEART

The heart has an intrinsic innervation termed the *conduction system* (fig. 9-9). The conduction system develops during embryological life from specialized myocardial fibers.

The *sinoatrial node,* commonly called the *SA node* or *pacemaker,* is located in the right atrium. The spread of excitation extends from the SA node to the *atrioventricular (AV) node* located beneath a leaflet of the tricuspid valve. The *atrioventricular bundle* connects the AV node and the interventricular septum. It continues down both sides of the septum as the *right* and *left branch bundles.*

ARTERIES

The *arterial system* (fig. 9-10) comprises efferent vessels carrying blood away from the heart to the capillary beds. With one exception (pulmonary arteries), arteries in the adult carry oxygenated blood. They are devoid of valves. The main arterial trunks are the *aorta* and *pulmonary* trunk. These give off branches that split into smaller branches until those of small caliber, termed *arterioles,* supply the capillary networks. Here blood plasma passes through the walls to bathe the surrounding cells, some of it reentering the capillaries at other points and the remainder passing into the lymphatic channels.

The arteries have three coats: an inner *tunica intima,* chiefly of endothelium; a middle *tunica media* of smooth muscle fibers, replaced partially on the largest vessels by elastic tissue and in the smaller ones gradually disappearing as the caliber decreases; and an outer *tunica adventitia* of fibrous and elastic tissue, at times with longitudinally disposed smooth muscle fibers. Inseparable from the tunica adventitia is a sheath of looser fibroareolar tissue carrying the *vasa vasorum,* or blood vessels, to and from the substance of the arterial walls and nerves.

Aorta and Its Branches

The blood of the systemic circulation is carried from the heart by the *aorta* (fig. 9-11*A* and *B*), which arises from the left ventricle. As it leaves the heart, it has a *semilunar valve* composed of three cusps, posterior, right, and left, similar to those of the pulmonary trunk, with three similar, shallow *aortic sinuses.* The *ascending aorta* is the short part adjoining the heart and is of greater diameter than the rest of the aorta, the size usually increasing with age, markedly so in some individuals. The ascending aorta gives rise to the *right* and *left coronary arteries.* The aorta then curves dorsally toward the left, forming the *arch of the aorta* in the superior mediastinum and crossing the trachea to the left side. To the concave curvature of the arch is attached the ligamentum arteriosum.

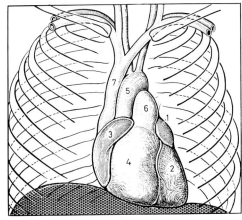

Anterior view

1—Left atrium
2—Left ventricle
3—Right atrium
4—Right ventricle
5—Arch of aorta
6—Pulmonary trunk
7—Superior vena cava

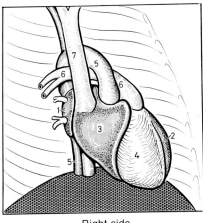

Right side

1—Left atrium
2—Left ventricle
3—Right atrium
4—Right ventricle
5—Arch of aorta
6—Pulmonary trunk
7—Superior vena cava

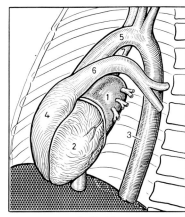

Left side

1—Left atrium
2—Left ventricle
3—Thoracic aorta
4—Right ventricle
5—Arch of aorta
6—Pulmonary trunk
7—Superior vena cava

FIG. 9-4A. *Heart (diagrammatic views)*

Where the arch turns inferiorly to become the descending aorta there is often a slight constriction, the *isthmus of the aorta,* followed by a slight dilatation, the *aortic spindle.* The arch of the aorta gives rise to the *brachiocephalic, left common carotid,* and the *left subclavian arteries.* The *descending aorta* lies first in the posterior mediastinum to the left of the bodies of the adjacent vertebrae, its situation becoming slightly more ventral as it passes through the aortic opening of the diaphragm. It is then known as the *abdominal aorta.* In the abdomen it lies ventral to the vertebral bodies, slightly to the left of the median line. It passes dorsal to the left renal vein to the level of the fourth lumbar vertebra, where it appears to bifurcate into the right and left common iliac arteries.

Right and Left Coronaries

The *right and left coronaries,* (fig. 9-11C) branches of the ascending aorta, arise from the right and left sinuses of the aorta. The right passes on the surface of the heart between the pulmonary trunk and the right atrium, and then courses in the right coronary sulcus to the dorsal aspect, where it anastomoses with the left coronary artery. In its course it gives off a number of branches, chief of which are the *marginal,* and dorsally, the *posterior interventricular branches.* The left coronary artery passes between the pulmonary trunk and the left atrium. It follows the left coronary sulcus dorsally and anastomoses with the right coronary artery. Its chief divisions comprise the *anterior interven-*

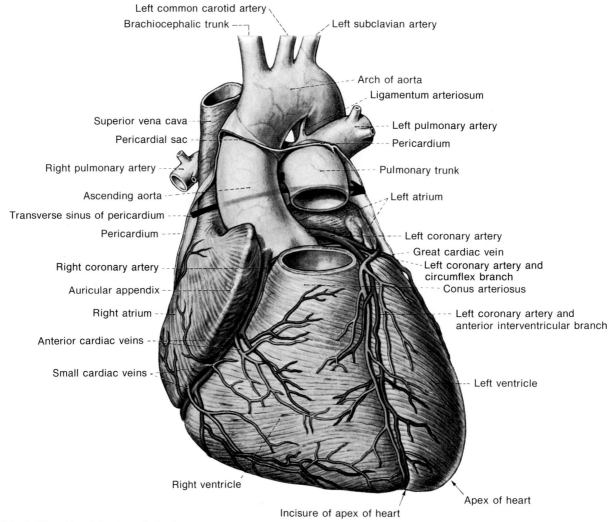

FIG. 9-4B. *Heart (external view)*

tricular branch toward the apex of the heart, and the *circumflex branch*, which follows the coronary sulcus, giving off a marginal branch before continuing to the dorsal surface and extending apically on the left ventricle.

Arch of the Aorta and Its Branches

The *brachiocephalic artery* arises toward the right part of the aortic arch (fig. 9-11*D*) and courses superiorly for a distance of 4 or 5 cm before it divides into *right subclavian* and *right common carotid arteries*. The *left common carotid artery* is the second branch of the arching aorta. The left subclavian artery is the third branch of the aortic arch.

COMMON CAROTIDS The *right common carotid* arises from the brachiocephalic artery dorsal to the sternoclavicular articulation, while the *left* arises in the mediastinum from the arch of the aorta between the brachiocephalic and left subclavian arteries (fig. 9-11*E*). In the lower neck each is situated deep to the infrahyoid and sternocleidomastoid musculature. The artery is enclosed with the internal jugular vein and vagus nerve in the carotid sheath. It ends at the level of the superior margin of the thyroid cartilage, where two branches are given off to form the *external and internal carotid arteries*. Here the common carotid is slightly enlarged, and this portion is termed the *carotid sinus*. It functions in aiding the regulation of blood pressure.

EXTERNAL CAROTID AND ITS BRANCHES The *external carotid* (fig. 9-12*A*) is the smaller of the two main branches of the common carotid. It is at first situated medial, but soon curves lateral, to the internal carotid. In the upper neck it passes deep to the

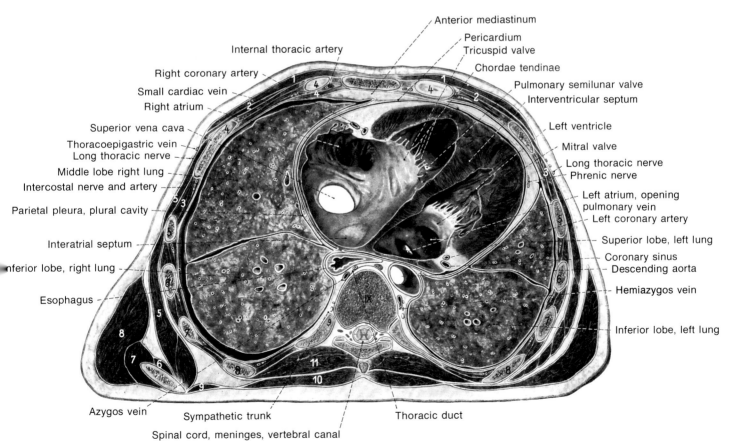

FIG. 9-5. *Heart (internal view)*

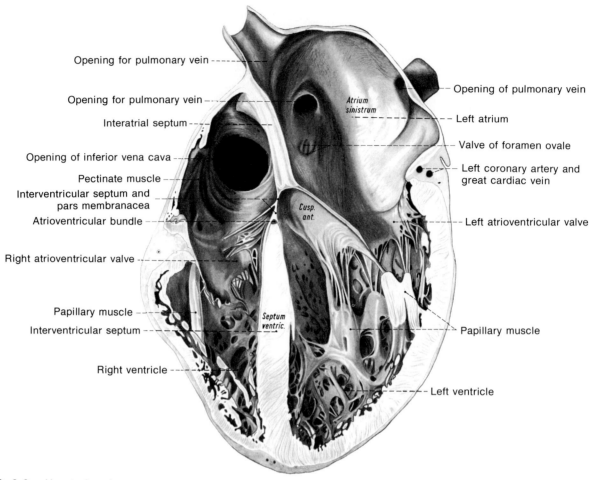

FIG. 9-6. *Heart chambers*

posterior belly of the digastric and stylohyoid muscles and thence into the substance of the parotid gland, where it breaks up into terminal branches. In its course it gives off the following branches, usually in the sequence listed: *ascending pharyngeal artery, superior thyroid artery, lingual artery, facial artery, occipital artery, posterior auricular artery, maxillary artery,* and *superficial temporal artery.*

The *ascending pharyngeal artery* is small and courses superiorly upon the pharynx to supply the longus colli and capitis, the walls of the pharynx, cervical lymph glands, tonsil, and Eustachian tube.

The *superior thyroid artery* is given off ventrally at the level of the hyoid bone and courses inferiorly to the thyroid gland, which it supplies by several terminal branches. It gives off a hyoid branch, deep to the thyrohyoid muscle.

The *lingual artery* is given off ventrally, at the level of the hyoid and extends to the tongue.

The *facial artery* is usually given off superior to the lingual artery. It passes obliquely superiorly in the superficial carotid triangle, covered only by the platysma and the anterior belly of the digastric muscles. It usually courses deep to the submandibular gland. It courses over the border of the mandible anterior to the masseter and winds tortuously toward the nose and orbit, where it ends in the *angular branch.* In its course it gives off the following branches that are worthy of note: the *ascending palatine branch* to the superior pharynx and its musculature, Eustachian tube, and tonsil; the *tonsillar branch* to the tonsil; the *submental branch* given off at the submandibular gland; the *inferior labial branch* to the lower lip and chin; the *superior labial branch,* at the angle of the

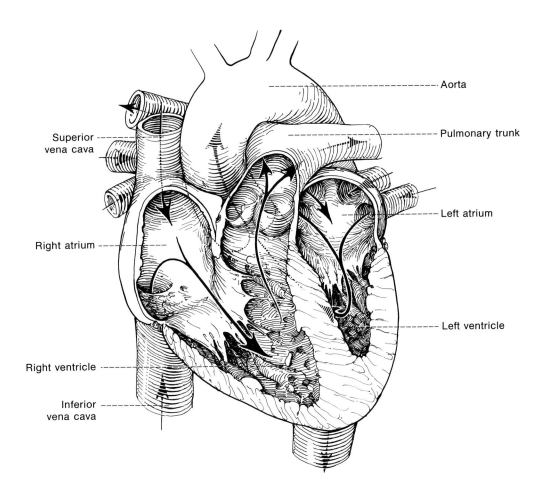

FIG. 9-7. *Path of blood flow*

mouth, to the upper lip; the *lateral nasal branch* to the side of the nose; and the *angular branch,* at the medial border of the orbit, entering the orbit and anastomosing with the dorsal nasal and palpebral branches of the ophthalmic artery, of much importance in collateral circulation of the orbit.

The *occipital artery* diverges dorsally and courses deep to the sternocleidomastoid, splenius capitis, and longissimus capitis, and ends in the occipital branch on the back of the head, usually located inferior to the posterior belly of the digastric muscle.

The *posterior auricular artery* diverges from the dorsal aspect of the external carotid deep to the posterior belly of the digastric. It pierces the parotid gland and passes just dorsal to the external auditory meatus. It may be observed coursing superior to the posterior belly of the digastric.

The *maxillary artery* (fig. 9-12B) arises near the posterior margin of the temporomandibular joint. It courses anteriorly medial to the neck of the mandible and usually between the lateral and medial pterygoid muscles and enters the pterygopalatine fossa where its terminal branches diverge. It usually gives off the following branches in the order stated: the *deep auricular,* to the tympanic membrane; the *anterior tympanic,* to the middle ear; the *middle meningeal,* the largest branch, coursing deep to the lateral pterygoid and entering the foramen spinosum, then being distributed over the endocranial surface and dura; the *inferior alveolar,* close to the sphenomandibular ligament, diverging inferiorly and dividing into two branches, one entering the mandibular foramen to supply the lower teeth and emerging through the mental foramen as the mental branch, and the other accompanying the lingual nerve to the mucous membrane of the mouth; the *masseteric,* near the neck of the mandible to the masseteric muscle; the *posterior deep temporal,* separate or a branch of the masseteric,

286 ■ BASIC ANATOMY FOR THE ALLIED HEALTH PROFESSIONS

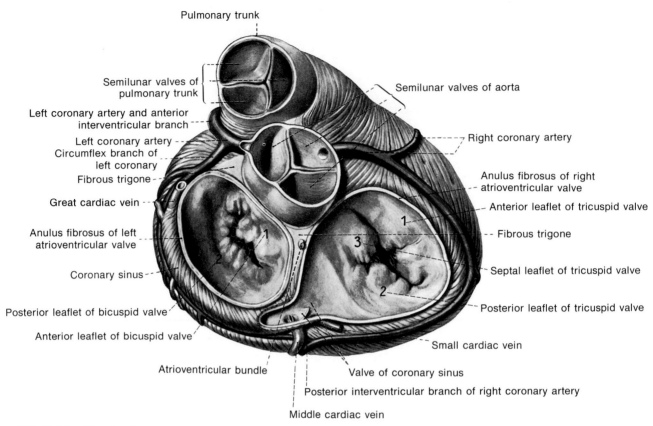

FIG. 9-8A. *Heart valves*

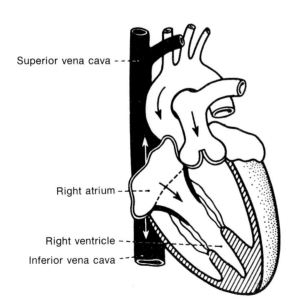

FIG. 9-8B. *Heart (atria emptying)*

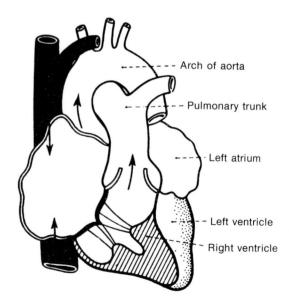

FIG. 9-8C. *Heart (atria filling)*

THE CARDIOVASCULAR SYSTEM ■ 287

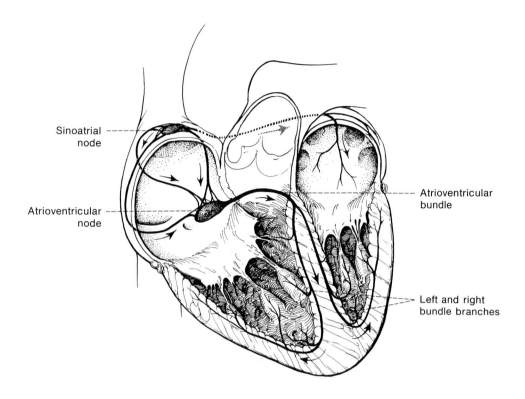

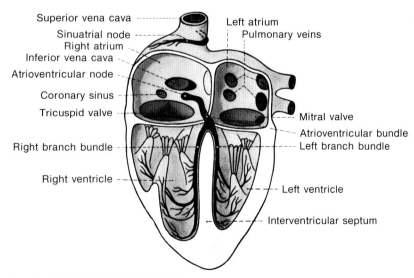

FIG. 9-9A. *Conduction system (diagrammatic view)*

to the temporal muscle; the *pterygoid,* to the pterygoid muscles; the *buccal,* accompanying the buccal nerve to the buccinator muscle; the *anterior deep temporal,* to the temporal muscle; the *superior alveolar branches,* to the upper teeth and surrounding structures; the *infraorbital,* through the inferior orbital fissure and the infraorbital foramen to the face; the *palatine branches,* accompanying the palatine nerves from the pterygopalatine ganglion to the palate; the *pterygoid canal branches,* through this canal to the roof of the pharynx, Eustachian tube, and middle ear; and the *sphenopalatine,* from the pterygopalatine fossa through the sphenopalatine foramen to the roof of the nasal cavity, nasal septum, and nasopharynx.

The *superficial temporal artery* is the terminal branch of the external carotid. It courses deep to the parotid gland and just anterior to the external auditory meatus, in company with the auriculotemporal nerve and superficial temporal vein. It passes deep to the branches of the facial nerve as it ramifies over the temporal and zygomatic regions. It gives off the *transverse facial* and *anterior and posterior branches.*

INTERNAL CAROTID ARTERY AND ITS BRANCHES
The *internal carotid artery* (fig. 9-13*A*) is the larger of the two main branches of the common carotid. From the level of the superior margin of the thyroid cartilage it courses superiorly between the external carotid and the transverse processes of the cervical vertebrae. It passes anterior to the styloid process to the carotid canal, in which it courses anteromedially to the foramen lacerum, then into the lateral wall of the cavernous sinus (fig. 9-13*B*), and then superiorly and medially beneath the anterior clinoid process, in its course describing a sigmoid curve. At the base of the brain it continues as the middle cerebral artery.

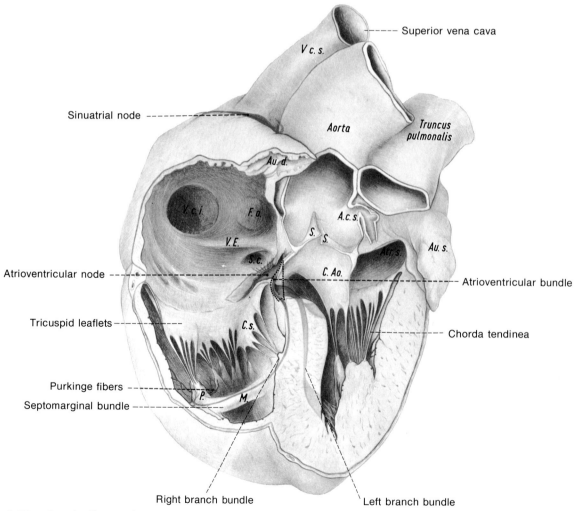

FIG. 9-9B. *Conduction system*

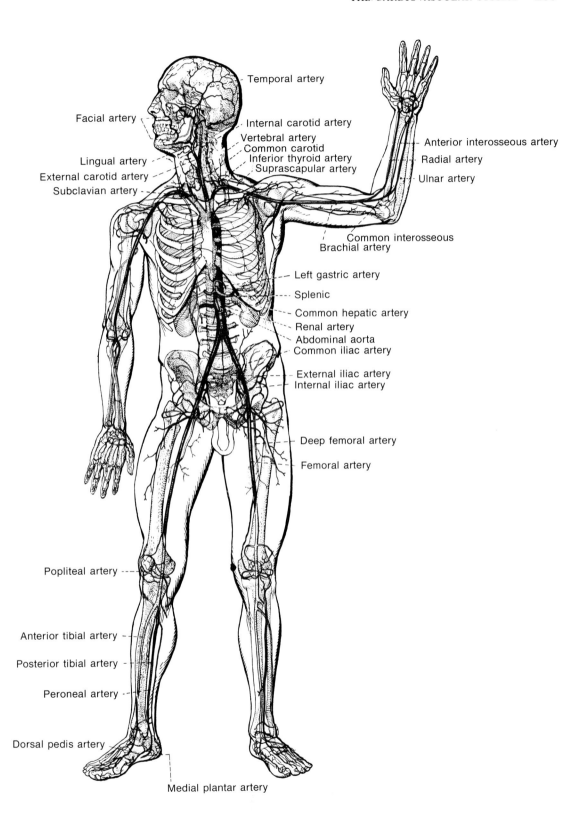

FIG. 9-10A. Arteries (*major arteries*)

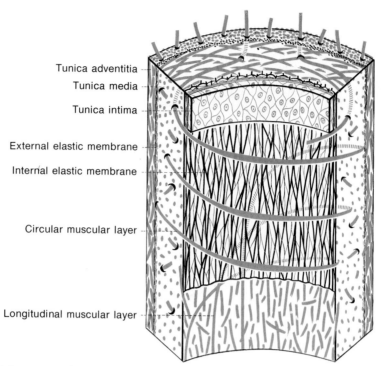

FIG. 9-10B. Arteries (*diagrammatic view*)

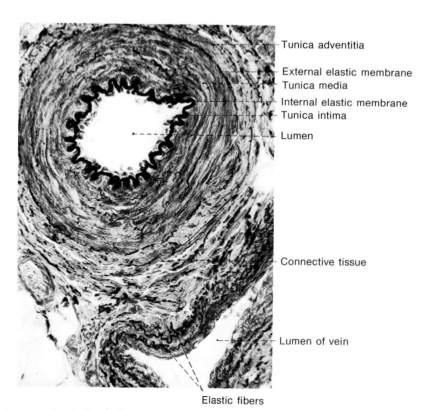

FIG. 9-10C. Arteries: small arteries (*microscopic cross-sectional view*)

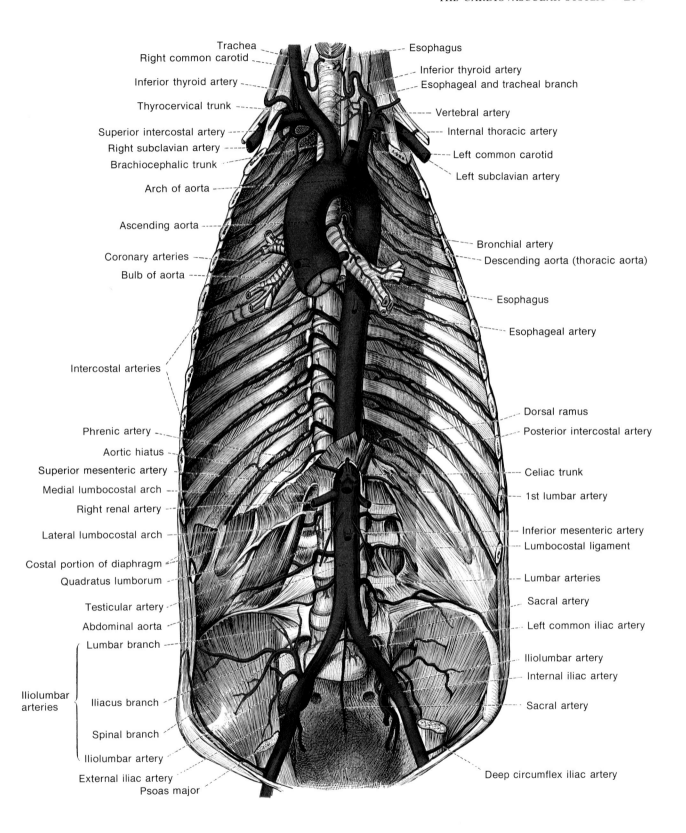

FIG. 9-11A. *Aorta (ascending arch, thoracic, and abdominal)*

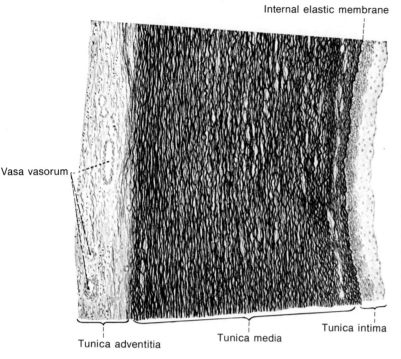

FIG. 9-11B. *Aorta (cross section)*

The *caroticotympanic artery* is a slender branch of the internal carotid artery (fig. 9-13C), diverging in the internal carotid canal. It passes through the carotid canal to the middle ear.

The *ophthalmic artery* arises just posteromedial to the anterior clinoid process and passes to the orbit through the optic foramen with, and inferolateral to, the optic nerve. *Branches* of the ophthalmic artery include the *central artery of the retina, lacrimal, supraorbital, supratrochlear, ciliary,* and *muscular arteries.*

CEREBRAL ARTERIES The *cerebral arteries* (fig. 9-14) join with the vertebral arteries at the base of the brain to form the *arterial circle of Willis.* The parts of the internal carotid concerned with this constitute the *anterior* and *middle cerebral* and the *anterior* and *posterior communicating arteries.* The middle cerebral is the direct continuation of the internal carotid. At the base of the brain it curves laterally and, by way of the Sylvian fissure, is distributed to the lateral part of the temporal, parietal, and frontal lobes of the brain. The anterior cerebral artery runs in the anterior longitudinal fissure of the cerebrum and supplies the structures of the brain more anteriorly situated. The posterior communicating artery is given off the internal carotid medial and inferior to the anterior cerebral artery. It courses posteriorly, with a slight medial slant, lateral to the infundibulum of the hypophysis, and joins the posterior cerebral artery, which is a continuation of the vertebral artery. Lateral to the posterior communicating artery is also given off the *choroid artery,* coursing posteriorly to end in the choroid plexus.

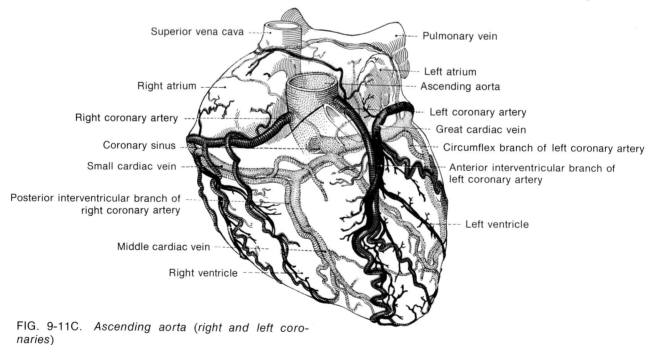

FIG. 9-11C. *Ascending aorta (right and left coronaries)*

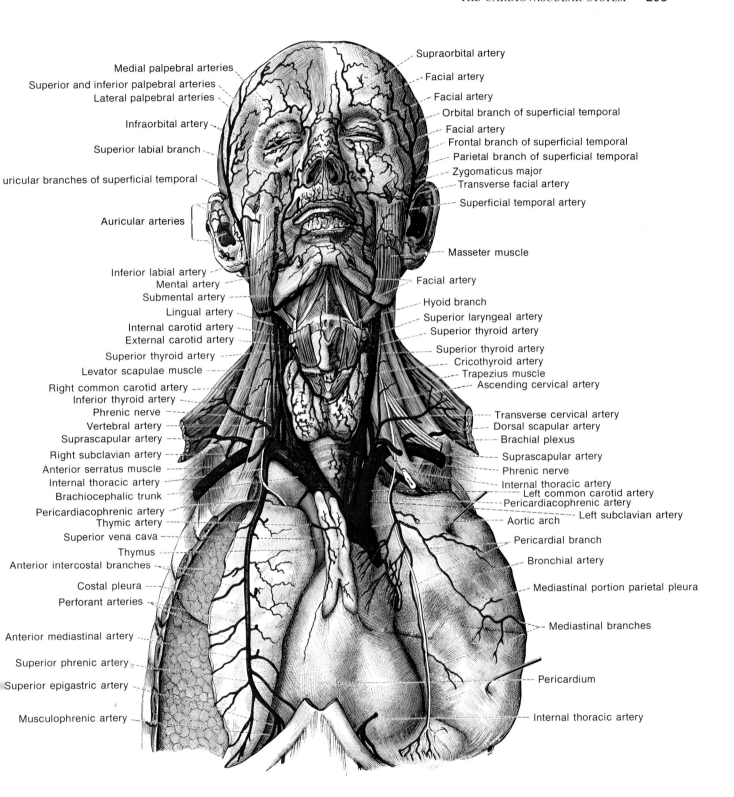

FIG. 9-11D. *Arch of the aorta and its branches*

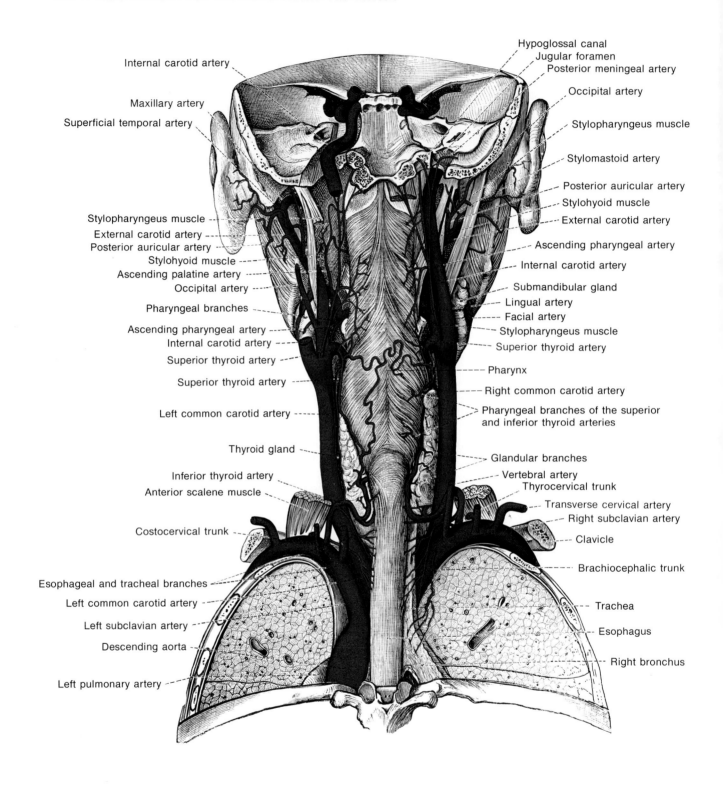

FIG. 9-11E. *Common carotids (posterior view)*

THE CARDIOVASCULAR SYSTEM ■ 295

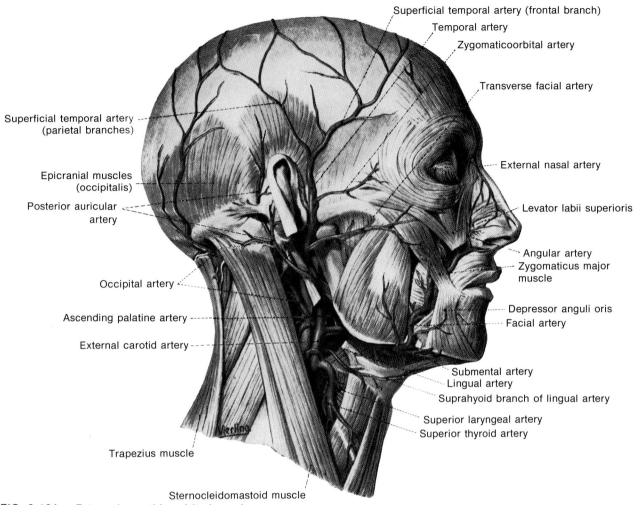

FIG. 9-12A. *External carotid and its branches*

Subclavian Artery

The *main arterial channel in the upper limb* (fig. 9-15A) is divisible for convenience into three zones, termed the *subclavian, axillary,* and *brachial arteries.*

The *subclavian portion* (fig. 9-15B) extends distally to the lateral border of the first rib. It is situated at the base of the neck and passes over the subclavian arterial groove on the superior surface of the first rib between the anterior and medial scalene muscles. The subclavian artery is crossed by the anterior scalene, and, for convenience in describing this muscle, is used to define three portions of the subclavian artery. The first portion of the subclavian artery lies medial to the anterior scalene. The second portion lies behind and the third portion lies lateral to the muscle. The *vertebral, thyrocervical,* and *internal thoracic trunks* arise from the first portion of the subclavian artery. The second portion gives rise to the *costocervical trunk,* and the *dorsal scapular artery* arises from either the second or third portion.

The *vertebral artery* is the most medial branch of the subclavian artery. It courses superiorly and through the foramina of the transverse processes of the first six cervical vertebrae, ventral to the spinal nerves. It then passes between the occipital condyles and atlas and pierces the dura mater, and bends forward and joins with its fellow of the opposite side to form the unpaired *basilar artery* between the pons and basioccipital bone. At the posterior margin of the dorsum sellae this divides into bilateral *posterior cerebral arteries.* It is joined with the internal carotid by the posterior communicating arteries, thus completing the circle of Willis.

The circle of Willis surrounds the infundibulum of the hypophysis, the optic chiasma, tuber cinereum, and mammillary bodies. It is formed on either side by the curve of the internal carotid artery as it becomes

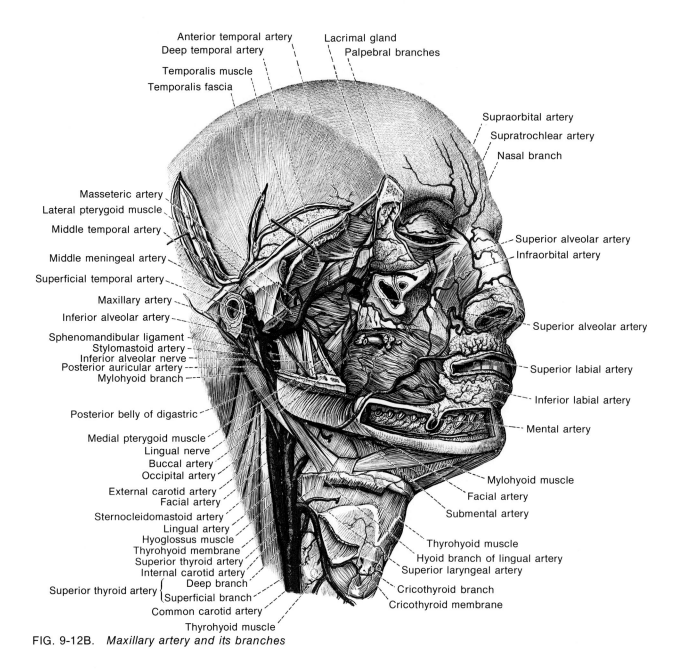

FIG. 9-12B. *Maxillary artery and its branches*

the middle cerebral by the anterior cerebral artery anteriorly, completed by the anterior communicating; and by the posterior communicating, which join the posterior cerebral, the terminal branch of the basilar, forming the posterior boundary of the circle.

The infrabasilar part of the vertebral artery gives off the *posterior and anterior spinal arteries*—the posterior curving over the medulla to the dorsal aspect of the spinal cord, and the anterior to the ventral aspect after uniting with its fellow of the opposite side—and the *posterior inferior cerebellar* artery to the cerebellum. The *basilar artery* gives off branches as follows: the *anterior inferior cerebellar* of the cerebellum; the *internal auditory,* accompanying the vestibulocochlear nerve to the inner ear; the *pontine,* to the pons; the *superior cerebellar,* immediately adjacent to the posterior cerebral branches, to the superior surface of the cerebellum and the choroid plexus of the third ventricle; and the *posterior cerebral artery,* to the occipital lobe, part of the temporal lobe, and adjacent parts of the brain.

The *thyrocervical trunk* arises from the superior

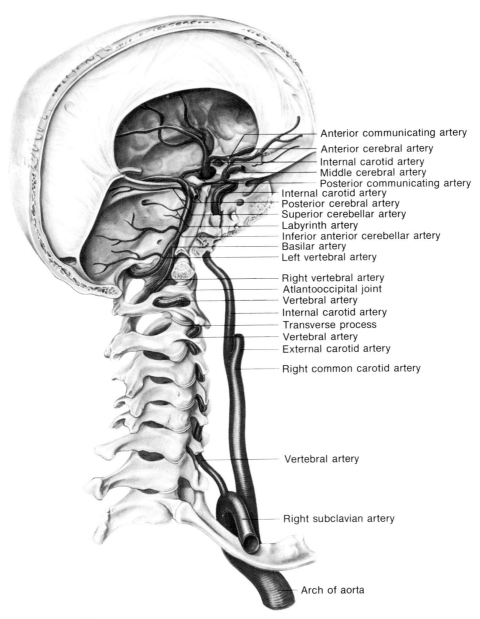

FIG. 9-13A. *Internal carotid*

part of the subclavian artery. It splits into its constituent parts as follows. The *ascending cervical* courses superiorly dorsal to the carotid sheath, and supplies the spinal canal and the deep muscles of the neck; the *inferior thyroid artery*, the largest branch, passes medially dorsal to the carotid sheath and to the inferior part of the thyroid gland; the *suprascapular artery* courses downward and laterally to the clavicle, then laterally to pass over the transverse ligament of the scapula to supply the supraspinatus and the infraspinatus muscles. The *transverse cervical artery* continues laterally over the scapula to supply part of the trapezius muscle.

The *internal thoracic artery* diverges inferiorly at about the level of the thyrocervical trunk. It passes dorsal to the clavicle and upon the inner aspect of the ventral wall of the thorax close to the sternum. Within the thorax it breaks up into several branches: twigs to the contents of the mediastinum, to the thymus gland, trachea, bronchi, diaphragm, pericardium, sternum, mammary twigs, and intercostal spaces. The internal thoracic artery at the level of the sixth intercostal space divides into the *musculophrenic and superior epigastric arteries*. The *inferior epigastric artery* anastomoses with the *superior epigastric artery* in the rectus sheath posterior to the rectus abdominis muscle.

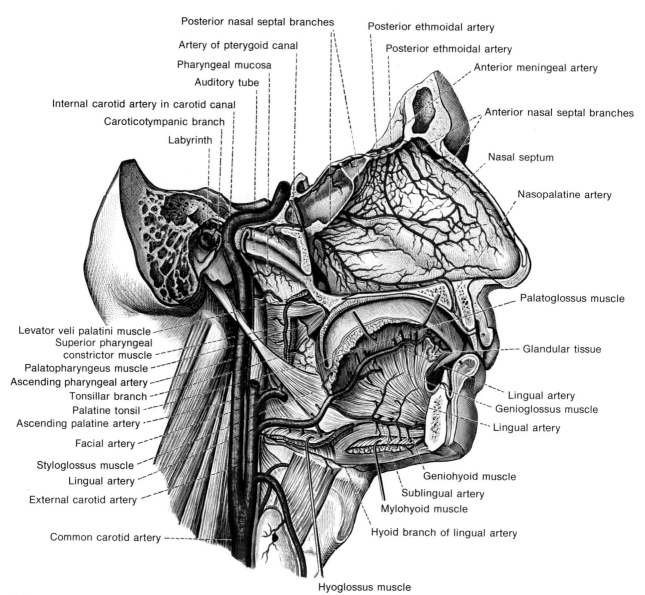

FIG. 9-13B. *Internal carotid in the carotid canal*

The *costocervical trunk* diverges from the dorsal aspect of the subclavian artery and splits into two branches, the *deep cervical,* with an ascending and a descending part in the erector spinae musculature, and the *highest intercostal,* passing around the ventral border of the first rib to the first, and sometimes to the second, intercostal space, with twigs to adjacent structures.

The *dorsal scapular artery* (fig. 9-15) descends along the medial border of the scapula, deep to the serratus anterior and rhomboid muscles. The dorsal scapular artery forms a major channel in collateral circulation of the scapula and shoulder area.

Axillary Artery

The *axillary artery* (fig. 9-15C) extends from the outer border of the first rib to the lower border of the teres major muscle where it becomes the *brachial artery*. The tendon of the pectoralis minor muscle crosses the axillary artery and therefore, for convenience in description, this is used to define the three parts of the artery.

The *first portion of the axillary artery* has one branch, the *highest thoracic artery,* which arises from the inferior aspect of the axillary artery at the lower border of the subclavian muscle.

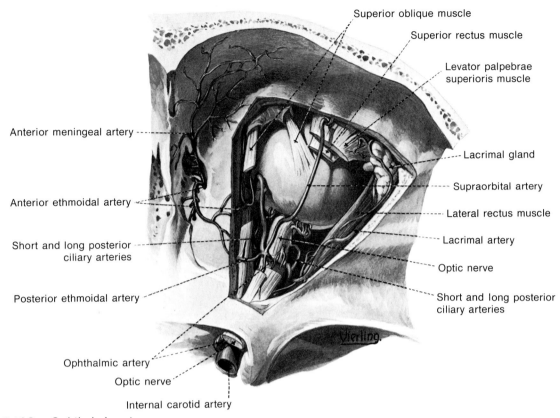

FIG. 9-13C. *Ophthalmic artery*

The *second portion of the axillary artery* lies behind the tendon of the pectoralis minor muscle and gives rise to the *thoracoacromial* and *lateral thoracic branches*. The *thoracoacromial* divides into the *deltoid*, the *acromial*, the *pectoral* and *clavicular branches*. The *lateral thoracic* diverges deep to the pectoralis minor and courses with the long thoracic nerve over the thorax to supply the serratus anterior muscle.

The *third portion of the axillary artery* extends from the lateral margin of the pectoralis minor to the lower border of the teres major. The *subscapular* and *anterior* and *posterior circumflex humeral arteries* arise from this portion of the axillary artery. The *subscapular* arises at the inferior border of the subscapularis and has two main branches: the *thoracodorsal*, coursing upon the deep aspect of the ventral border of the latissimus dorsi to supply chiefly that muscle and the teres major; and the *circumflex scapular*, which passes to the dorsal aspect of the shoulder. The *anterior circumflex humeral* diverges at the level of the surgical neck of the humerus and curves around the anterior aspect of the bone, supplying surrounding muscles and the shoulder joint. The *posterior circumflex humeral* is larger and curves around the posterior aspect of the surgical neck of the humerus and passes with the axillary nerve to the dorsal aspect of the shoulder through the quadrangular space between the teres minor and major and the long head of the triceps and humerus. It supplies these surrounding muscles and the shoulder joint.

Brachial Artery

The *brachial artery* (fig. 9-15D) extends from the inferior border of the teres major to just distal to the elbow, where it breaks up into its terminal branches. Its collateral ulnar branches supply blood to the ulnar side, and the profunda branch to the radial side of the upper arm. In the brachium it gives off the following branches, besides small muscular twigs.

The *deep brachial artery* arises at a level just proximal to the insertion of the deltoid muscle and follows a spiral course with the radial nerve between the origins of the lateral and medial heads of the triceps brachii. Above the lateral epicondyle it becomes superficial, as does the collateral radial artery, joining the arterial network around the elbow.

The *superior ulnar collateral artery* arises slightly

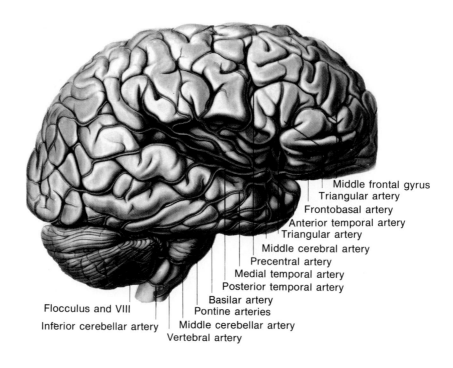
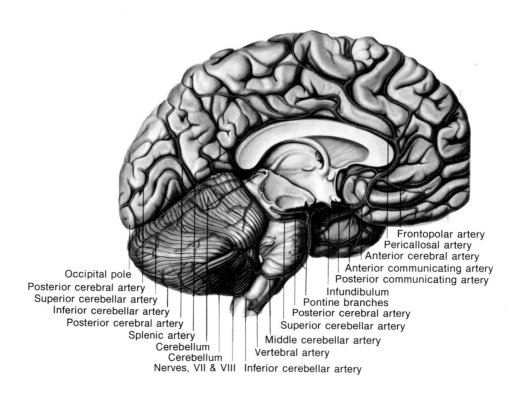

FIG. 9-14. *Cerebral arteries*

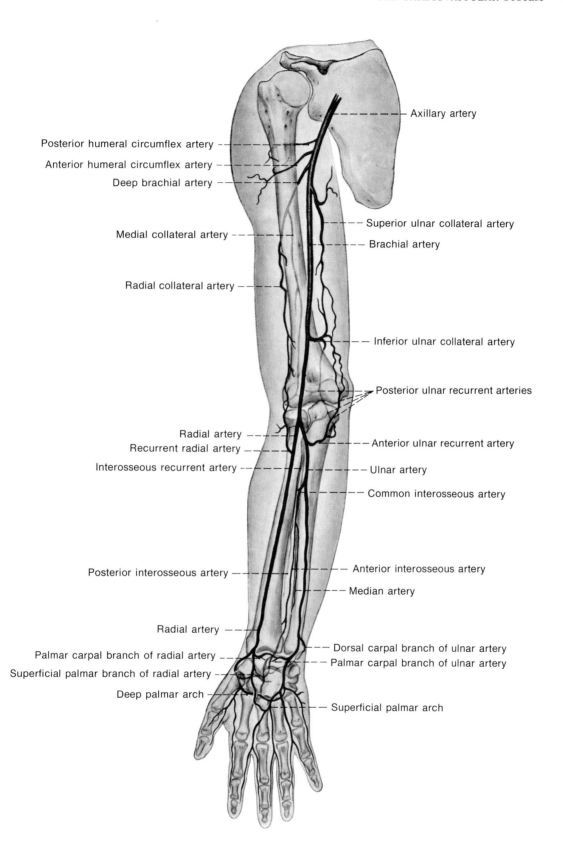

FIG. 9-15A. *Arteries of the upper limb*

302 ■ BASIC ANATOMY FOR THE ALLIED HEALTH PROFESSIONS

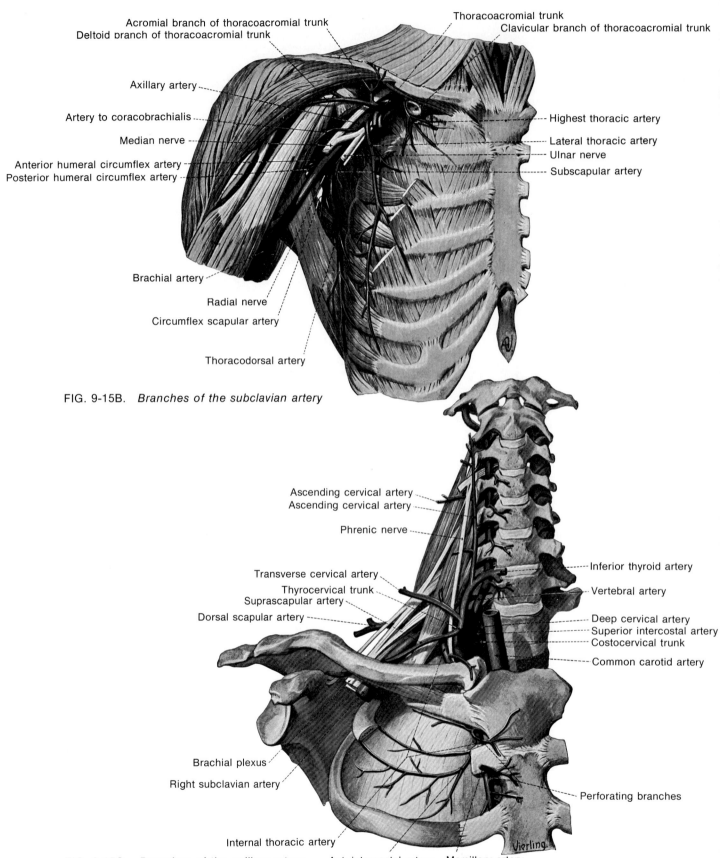

FIG. 9-15B. *Branches of the subclavian artery*

FIG. 9-15C. *Branches of the axillary artery*

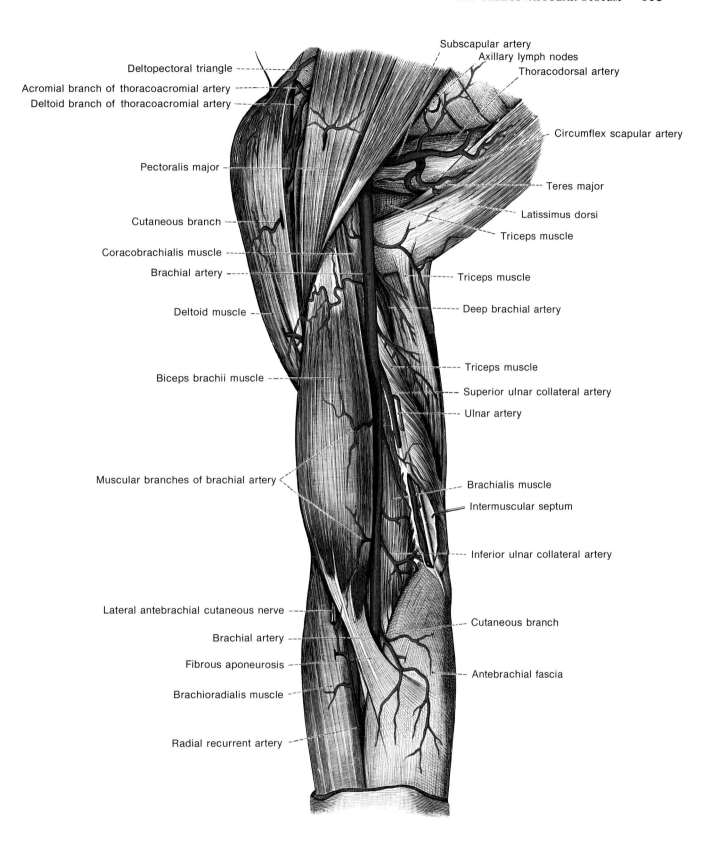

FIG. 9-15D. *Brachial artery and its branches*

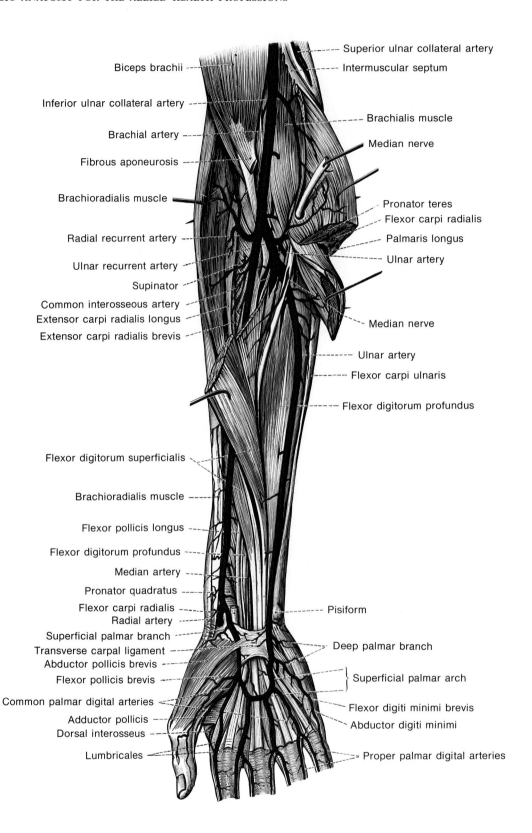

FIG. 9-15E. *Branches of the radial and ulnar arteries*

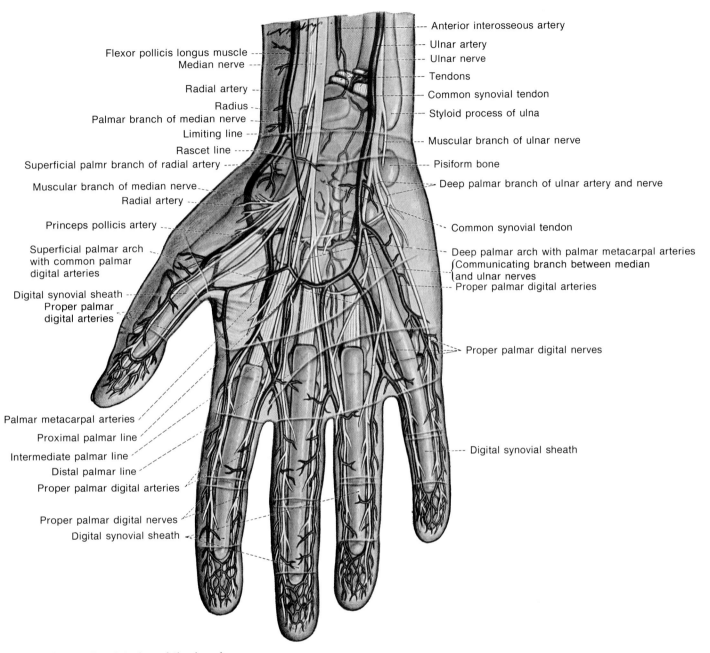

FIG. 9-15F. *Arteries of the hand*

distal to the deep brachial artery and courses with the ulnar nerve.

The *inferior ulnar collateral artery* arises just proximal to the medial epicondyle and supplies blood to the adjacent muscles.

RADIAL AND ULNAR ARTERIES   In the forearm, opposite the neck of the radius, the *brachial artery* usually divides into the *radial and ulnar arteries*, (fig. 9-15*E* and *F*). The *radial artery* gives rise to the *radial recurrent, muscular branches, palmar carpal branch and network,* and a *superficial palmar branch.* The *ulnar artery* gives rise to the *anterior ulnar recurrent, posterior ulnar recurrent, common interosseous arteries, muscular branches, palmar carpal branches* and *dorsal carpal branch.*

The *radial artery* arises at the elbow between the brachioradialis and pronator teres muscles. It traverses the forearm in the interspace between the ventral muscles of the elbow and those of the wrist. At the wrist it winds around the radial side of the carpals

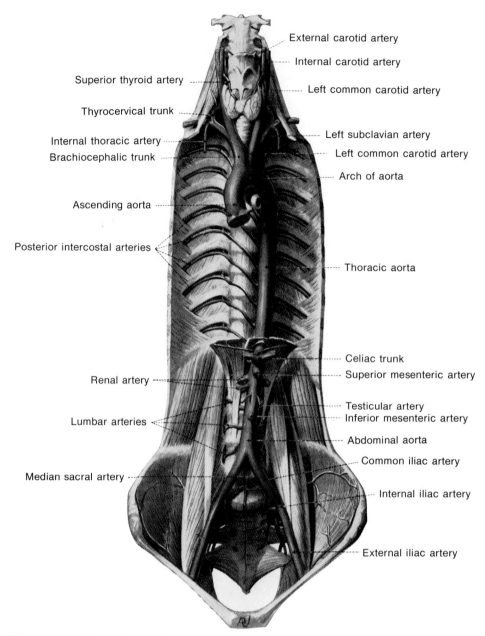

FIG. 9-16A. *Thoracic and abdominal aortas*

to its dorsal aspect, and crosses the palm to form the deep palmar arch. It ends by anastomosing with the ulnar artery. The arch gives off palmar metacarpal arteries, which extend distally to join the common palmar digital, and divide in the first interspace to extend along the first and second digits. In its course the *radial artery* gives off the following branches: the *radial recurrent,* just below the elbow to the adjacent muscles; *numerous rami musculares,* to the forearm musculature; *palmar carpal,* of small size, at the carpus to its palmar aspect; *superficial palmar,* arising at the wrist and coursing deep to the abductor pollicis brevis to join the superficial palmar arch; the *dorsal carpal branch,* extending from the radial artery across the dorsal aspect of the wrist and helping to form the dorsal carpal arch.

The *ulnar artery* begins opposite the neck of the radius. The ulnar artery follows the course of the ulnar nerve and passes deep to the pronator teres muscle and all other superficial flexors. The *branches*

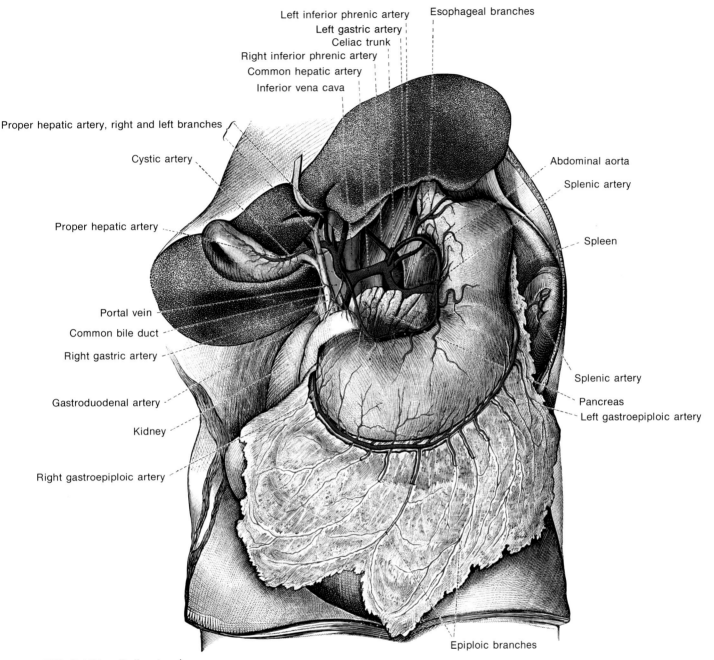

FIG. 9-16B. *Celiac trunk*

of the *ulnar artery* include: the *anterior ulnar recurrent*, which anastomoses with the anterior branches of the superior ulnar collateral and the inferior ulnar collateral arteries; the *posterior ulnar recurrent artery*, which anastomoses with the posterior branches of the superior and inferior ulnar collateral arteries and with the *interosseous recurrent artery*; the *common interosseous*, which divides into *anterior* and *posterior interosseous arteries; muscular branches* of the ulnar artery that supply muscles on the ulnar side of the forearm; a *palmar carpal branch* that arises at the upper border of the flexor retinaculum to unite with the palmar carpal branch of the radial artery; and the *dorsal carpal branch*, which anastomoses with dorsal carpal branch of the radial artery.

Thoracic and Abdominal Aorta

The *thoracic aorta* (fig. 9-16A) gives rise to the following *branches*.

308 ■ BASIC ANATOMY FOR THE ALLIED HEALTH PROFESSIONS

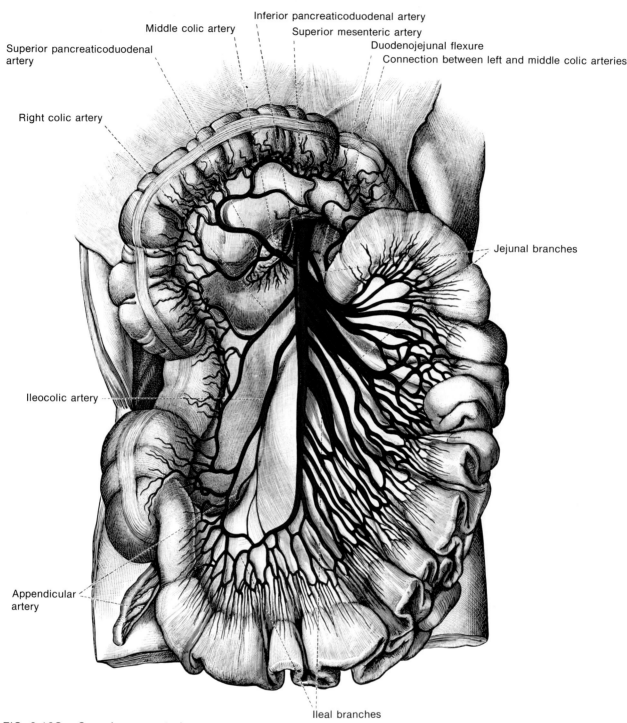

FIG. 9-16C. *Superior mesenteric artery*

The *bronchial arteries* are small branches of the aorta at about the level of the fourth interspace, or at times from the third intercostal artery. They pass to the lungs and supply the lung tissue. Thus the lungs may really have a double blood supply; the pulmonary circulation carries blood mostly through the lungs, and the bronchial arteries carry it to the lung tissues.

The *esophageal arteries* are small branches of the aorta at the different levels, with distribution to the esophagus.

The *pericardial arteries* are minute branches to the more dorsal part of the pericardium.

The *mediastinal arteries* supply the structures in the posterior mediastinum. The more inferior branches of this series are known as the *superior phrenic ar-*

THE CARDIOVASCULAR SYSTEM ▪ 309

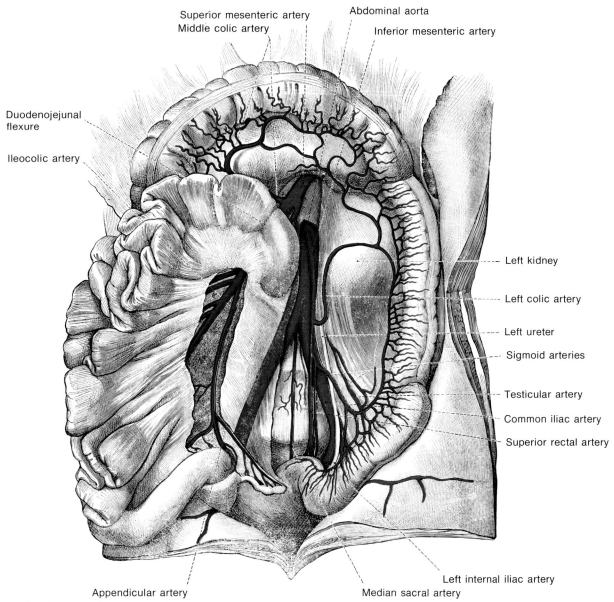

FIG. 9-16D. *Inferior mesenteric artery*

teries, with distribution to the more dorsal part of the diaphragm.

The *posterior intercostals* number 11 pairs and are distributed around the trunk inferior to the 2nd to 12th ribs, the more superior being given off the aorta somewhat inferior to the interspaces in which they course. At the head of the rib each divides into a posterior and anterior ramus. The posterior ramus supplies the spinal canal and its contents, back musculature, and overlying skin; and the anterior ramus follows the intercostal spaces, coursing between the external and internal intercostal muscles, giving off twigs to the adjacent structures and anastomosing ventrally with branches of the internal thoracic.

The *abdominal aorta* (fig. 9-16A) gives rise to the following *branches.*

The *celiac trunk* (fig. 9-16B) is the first branch of the abdominal aorta, at the level of the lesser curvature of the stomach. It divides at once into three branches: the *left gastric,* the *common hepatic,* and the *splenic.* The *left gastric artery* runs in the lesser omentum to the cardiac end of the stomach, and then follows the lesser curvature. At the pyloric end it joins the right gastric branch of the hepatic artery. In its

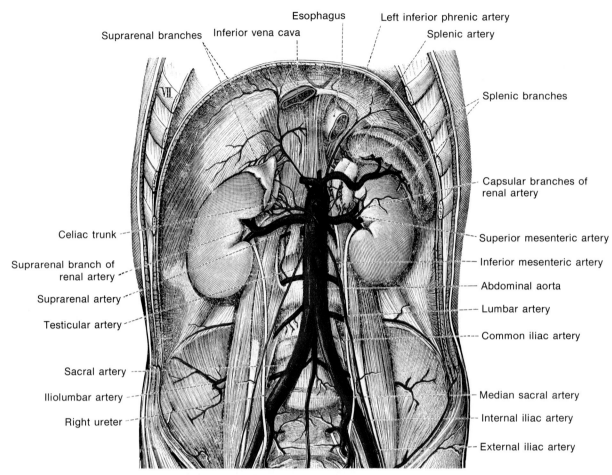

FIG. 9-16E. *Parietal and paired visceral branches of the abdominal aorta*

course it gives off esophageal branches ascending to anastomose with the esophageal branches of the thoracic aorta, and branches to both surfaces of the stomach. The *common hepatic artery* courses in the lesser omentum, with the bile duct and portal vein, toward the right to the portal fissure of the liver, where it divides into branches to the right and left lobes of the liver. The right branch usually gives off the *cystic artery* to the gallbladder. The common hepatic gives off the *gastroduodenal artery,* which in turn divides into the *anterior superior pancreaticoduodenal artery,* to the pancreas and duodenum, and the *right gastroepiploic artery,* along the greater curvature of the stomach from the pyloric end, uniting with the left gastroepiploic artery. It gives off twigs to the stomach and to the greater omentum. The *splenic artery* courses to the left superior to the pancreas and, at first, dorsal to the omental bursa to the hilus of the spleen. In its course it gives off small branches to the pancreas: the *left gastroepiploic artery,* arising close to the spleen and following the greater curvature of the stomach to unite with the right gastroepiploic artery; and branches to the spleen, both surfaces of the stomach, and the greater omentum.

The *superior mesenteric artery* (fig. 9-16C) arises just inferior to the celiac artery, dorsal to the pancreas and splenic vein, and then between the pancreas and inferior part of the duodenum. It enters the mesentery and splits into numerous branches, the chief of which are as follows: the *jejunal* and *ileal branches* are numerous (12–16) and course toward the left to the small intestine, with free anastomoses between neighboring vessels, but with terminal branches that do not anastomose; the *inferior pancreaticoduodenal artery,* the highest branch toward the left, extends to the inferior part of the duodenum and pancreas and anastomoses with the anterior superior pancreaticoduodenal artery; the *middle colic artery,* the next branch toward the left, is large, extends in the mesocolon to the transverse colon, and anastomoses with the right colic artery; the *ileocolic artery,* the most inferior of the right branches, extends retroperitoneally to the juncture of the ileum and colon, continuing to the appendix as the *appendicular artery,* and giving off *right colic* branches to the ascending colon, which anastomose with the middle colic artery.

The *inferior mesenteric artery* (fig. 9-16D) is given off the ventral aspect of the abdominal aorta approximately 3 cm above the origin of the common iliacs. It courses toward the left, retroperitoneally, and is considered to split into three parts: *left colic,* to the descending colon; *sigmoid,* to the sigmoid colon; and *superior rectal,* to the rectum. The three freely anastomose, as also does the superior with the middle rectal arteries.

Additional branches of the abdominal aorta (fig. 9-16E) are as follows.

The *medial sacral artery* is a continuation of the aorta.

The *inferior phrenic artery* arises close to the level of the celiac trunk and is distributed to the inferior aspect of the diaphragm, where it gives off a *superior suprarenal artery* to that gland.

The *middle suprarenal artery* arises between the celiac and renal arteries, and supplies the suprarenal gland.

The *renal artery* arises at the level of the first lumbar vertebra. Near the pelvis of the kidney it splits

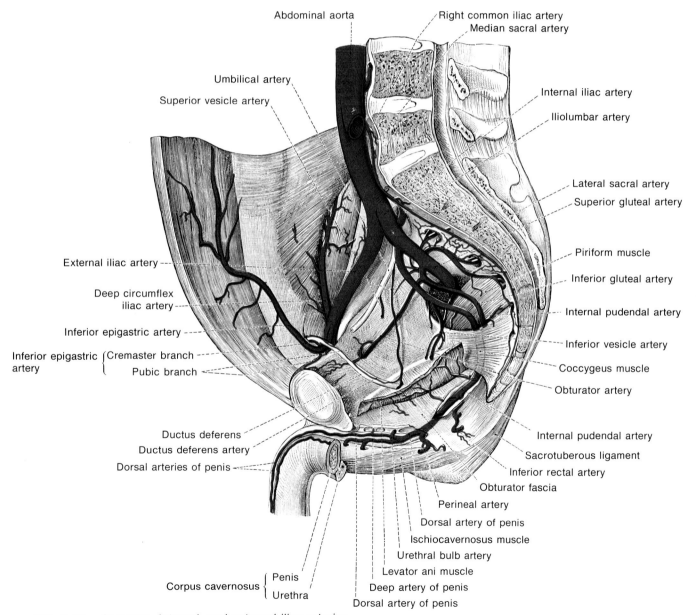

FIG. 9-17. *Common, internal, and external iliac arteries*

into several branches that enter the kidney, and it supplies the *inferior suprarenal artery* to the adrenal gland. Primitively the kidney is lobulated, so it is not unexpected that quite frequently accessory renal arteries are encountered from the aorta to various parts of the kidney.

The *testicular* or *ovarian artery* usually arises an inch or more inferior to the renal artery. The right artery customarily passes ventral to the inferior vena cava. It extends inferolaterally upon the psoas major, and in the male enters the spermatic cord and is distributed to the epididymis and testicle. In the female it descends into the pelvis and passes between the broad ligament to the ovary. In its course it gives off fine twigs to the superficial part of the kidney, to the ureter, to the cremaster in the male, and to the fallopian tube and round ligament in the female.

The *lumbar arteries* are four (pairs) in number, diverging from the aorta segmentally, opposite the first four lumbar vertebrae. They pass dorsal to the psoas major and course between the internal oblique and transversus abdominus around the walls of the abdomen to its ventral aspect.

Iliac Arteries

The *common iliac artery* arises at about the level of the fourth lumbar vertebra, the right one crossing the inferior vena cava upon its ventral surface. At the level of the lumbosacral articulation it divides into the *internal* and *external iliac arteries* (fig. 9-17).

INTERNAL ILIAC ARTERY The *internal iliac artery* enters the true pelvis and divides into two main branches, a *posterior division*, leaving the pelvis as the *superior gluteal artery,* and an *anterior division,* ending in the *inferior gluteal* and *internal pudendal arteries*. In its course the internal iliac artery gives off the following branches (the first five of these may be classed as parietal, and the last five as visceral branches):

The *iliolumbar artery* courses superolaterally from the posterior division, dorsal to the psoas major.

The *lateral sacral artery,* from the posterior division, leaves the pelvis and courses medioinferiorly upon the ventral aspect of the sacrum to the vertebral canal and adjacent structures.

The *superior gluteal artery* leaves the pelvis through the greater sciatic foramen. It divides into a superficial ramus, between the piriformis and the gluteus medius, to supply the gluteal muscles.

The *obturator artery* arises near the origin of the anterior division of the internal iliac artery and runs with the obturator nerve, parallel to the external iliac to the obturator foramen. It passes through this canal and then divides into an anterior and a posterior branch; both go to the structures of the medial thigh and to the hip joint. Within the pelvis are given off vesicular twigs to the bladder; a pubic branch, over the inner aspect of the anterior pelvic wall; and an iliac branch, to the ilium deep to the iliac muscle.

The *inferior gluteal artery,* a terminal part of the anterior division of the internal iliac artery, leaves the pelvis through the inferior part of the greater sciatic foramen.

The *superior vesicle* is usually the highest branch of the anterior division. It extends to the bladder.

In conjunction with the superior vesicular artery is the *lateral umbilical ligament,* representing the fetal umbilical artery, but patent only before birth. It follows the superior vesicular artery for a short space, then courses to the abdominal wall, between the midline and the inferior epigastric, and finally to the umbilicus.

The *deferential artery* arises from the anterior division in relation to one or the other of the vesicular arteries. It extends to the vas deferens and splits into the two branches, one of which accompanies the vas to the epididymis and the other extends to the seminal vesicles.

The *uterine artery* extends to the vicinity of the cervix of the uterus, ascends at the side of the uterus between the layers of the broad ligament, and courses to the uterine tube and ovary, which anastomoses with the ovarian artery. In its course it gives off branches to the vagina and over the surface of the uterus.

The *inferior vesicle* arises from the anterior division and extends to the bladder, supplying blood also to the prostate in males and the vagina in females.

The *middle rectal artery* arises from the internal pudendal artery just before the latter leaves the pelvis. It extends to the rectum and surrounding structures, including the prostate in males and the vagina in females.

The *internal pudendal artery* is one of the terminal branches of the anterior division of the internal iliac artery. It leaves the pelvis through the greater sciatic foramen; it then passes caudally, dorsal to the sacrospinous ligament, ventral to the sacrotuberous ligament, and then through the pudendal canal, in the obturator fascia, to the ischiorectal fossa. In its course it gives off twigs to adjacent muscles and other structures. In the ischiorectal fossa the *inferior rectal* artery passes medially to the rectum. The artery to the perineum passes superficial to the trigonum, and supplies the superficial perineum and scrotum in the male and labia in the female. The artery of the penis enters the trigonum and extends ventrally before leaving it near the arcuate ligament and coursing to the penis as the dorsal penis (clitoridis) artery. In its course it gives off branches to surrounding muscles, urethra, and vestibular structures in the female.

EXTERNAL ILIAC ARTERY The *external iliac artery* courses beneath the inguinal ligament to become the *femoral artery*. Within the pelvis it gives off two branches, both arising just as the artery leaves the pelvis: (1) the *inferior epigastric* courses upon the wall of the ventral abdomen superomedially, first over the falx inguinalis, at the lateral boundary of the inguinal triangle, and then over the dorsal surface of the rectus abdominis, which it penetrates to supply that muscle and to anastomose with the superior epigastric artery; (2) the *deep circumflex iliac artery* courses superolaterally in the angle between the iliac muscle and the abdominal wall, then extends along the iliac crest. It supplies the neighboring structures.

Arteries of the Lower Limbs

Arteries of the lower limbs (fig. 9-18A) include the femoral, popliteal, tibials, and plantars and their branches.

FEMORAL ARTERY The *femoral artery* (fig. 9-18B) is covered in the thigh by the fascia lata and the sartorius muscle. It courses first in the femoral triangle between the iliopsoas and pectineus. In the thigh it passes into the adductor canal, where it is located anterior to the femoral vein, and passes into the popliteal space, posterior to the knee, where it becomes known as the *popliteal artery*. The latter continues between the two heads of the gastrocnemius and deep to the tendinous arch of origin of the soleus muscle, after which it divides into the *anterior* and *posterior tibial arteries*.

BRANCHES OF THE FEMORAL ARTERY The *superficial epigastric artery* arises just inferior to the inguinal ligament, passes through the fascia lata, and extends in the superficial fascia to the region of the umbilicus.

The *superficial circumflex iliac artery* arises from the femoral artery, pierces the fascia lata, and extends in the superficial fascia to the region of the anterior spines of the ilium.

The *external pudendal artery* supplies the external genitalia.

The *deep femoral artery* arises in the femoral triangle and is directed posteriorly. It furnishes the main blood supply of the thigh and courses deep to the adductor longus muscle.

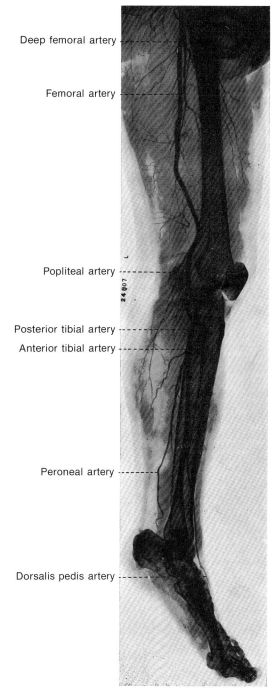

FIG. 9-18A. *Arteries of the lower limb*

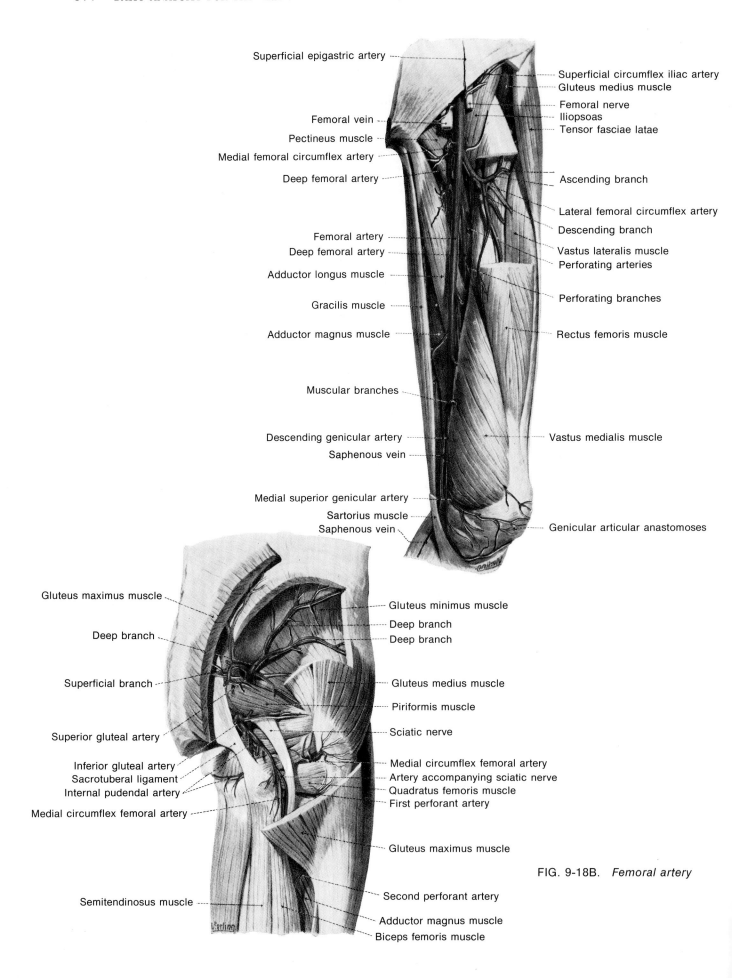

FIG. 9-18B. *Femoral artery*

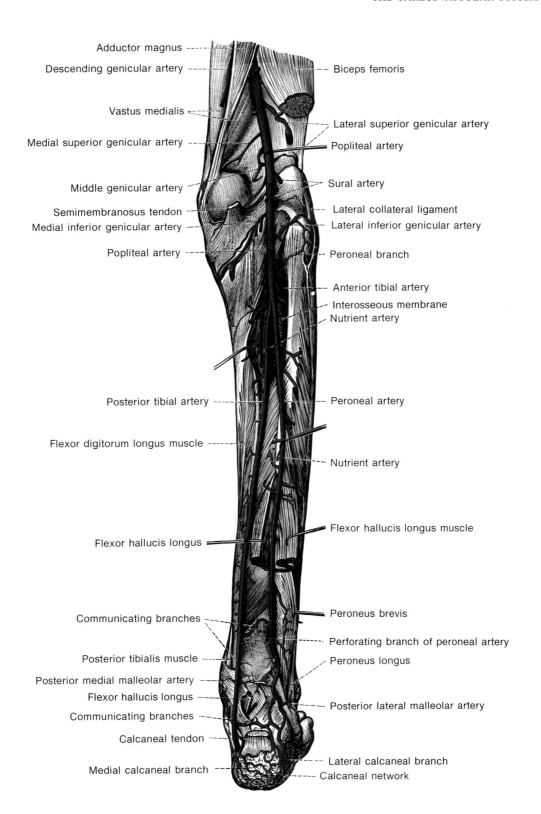

FIG. 9-18C. *Popliteal artery*

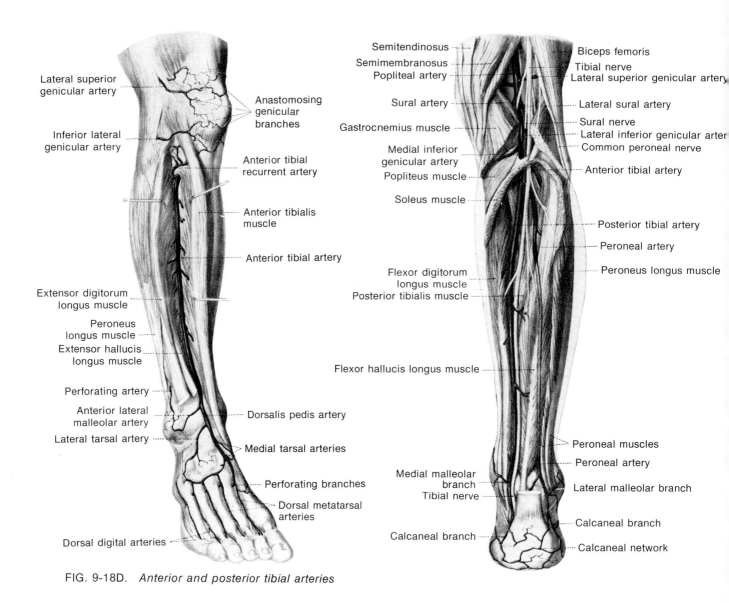

FIG. 9-18D. *Anterior and posterior tibial arteries*

POPLITEAL ARTERY The *popliteal artery* (fig. 9-18C), at the level of the knee and just below, gives off *superior lateral* and *medial*, *inferior lateral* and *medial*, *medial*, *articular*, *patellar*, and *sural branches*, all small, to form the network about the joint. It continues distally as the *posterior tibial artery*, and upon the foot as the *plantar artery*.

BRANCHES OF THE POPLITEAL ARTERY The *posterior tibial artery* (fig. 9-18D) begins in the proximal crus and extends distally toward the medial malleolus, between the superficial group of ventral muscles, to the calcaneus. Near the ankle it passes between the tendons flexor hallucis longus and digitorum longus, and then behind the medial malleolus. It divides into the *medial* and *lateral plantar arteries*. Besides *muscular twigs*, in the proximal crus it gives off the *anterior tibial artery* and the *peroneal artery*, and near the heel, the *calcaneal branches*.

The *anterior tibial artery* (fig. 9-18D) in the proximal crus perforates the interosseous membrane and courses distally upon its anterior surface, between the anterior tibial and extensor digitorum longus. In the crus it gives off a few branches, the *posterior* and *anterior tibial recurrent arteries* near its origin, to the patellae. Upon the foot it continues as the *dorsal pedis artery*. This courses upon the dorsum of the foot between the tendons extensor hallucis longus and digitorum longus. At the level of the tarsometatarsal joints the *dorsal metatarsal arteries* diverge laterally to the interspaces, splitting into the *dorsal digital ar-*

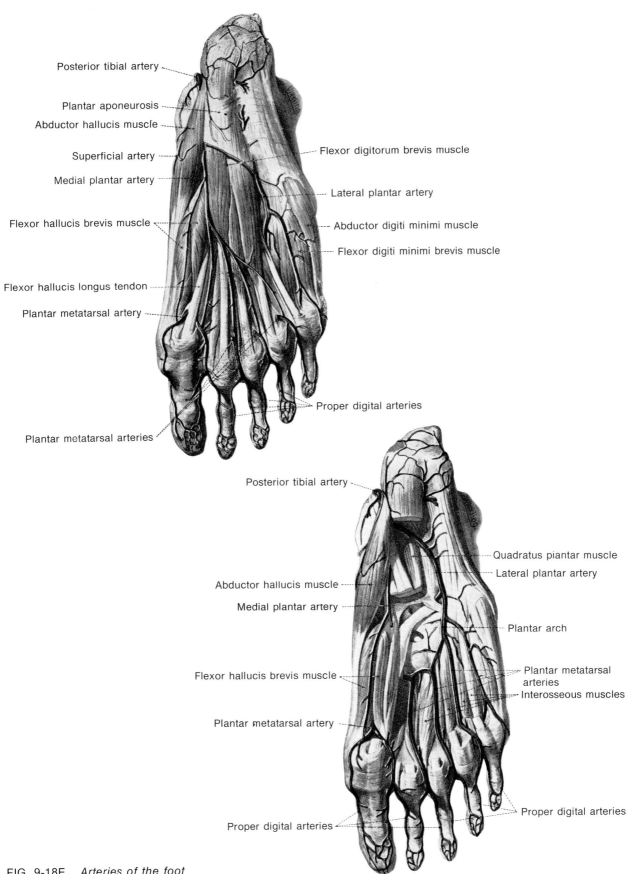

FIG. 9-18E. *Arteries of the foot*

teries to supply contiguous borders of adjacent digits, with perforating branches piercing the interspaces to join the plantar metatarsal arteries. A main branch occurs as the *deep plantar artery,* between the bases of the first and second metatarsals, which joins the plantar arch.

ARTERIES OF THE FOOT The *medial plantar artery* (fig. 9-18*E*) arises from the posterior tibial artery and courses onto the sole of the foot.

The *lateral plantar artery* extends obliquely between the flexor digitorum brevis and the quadratus plantae muscles toward the fourth interspace, where it curves medially to form the *plantar arch,* between the adductor hallucis and interossei. In its course it gives off small twigs to adjacent structures, and from the arch, *plantar metatarsal arteries.* These pass distally in the interspaces, receive perforating branches from the dorsal pedis artery, and split at the bases of the toes to supply the two contiguous digital borders.

VEINS

The venous part of the blood-vascular system may be divided into three portions: (1) afferent vessels carrying oxygenated blood from the lungs to the heart; (2) afferent vessels carrying poorly oxygenated blood from the systemic capillaries to the heart; and (3) vessels carrying blood from the capillaries of the intestines to the capillaries of the liver, comprising the portal system. One may consider that the veins of the heart comprise a fourth division.

The structure of the *veins* (fig. 9-19*A*) is similar to that of the arteries but their walls are considerably thinner, and the elastic and collagenous connective tissue and the smooth muscle fibers are much more scanty than in the arteries. In some veins these fibers are arranged in a circular fashion and in others chiefly in a longitudinal fashion. An empty vein collapses completely, which cannot be said of an empty artery. Veins, unlike arteries, are frequently supplied with *valves,* of semilunar type, derived from folds of inner coat and usually occurring in pairs, with the free border directed toward the heart. At each valve there is a corresponding slight bulge in the wall of the vessel, which may be accentuated in old age or in some vascular disfunctions, thus imparting a knotty conformation to the large veins engorged with clotted blood. The valves function to maintain a one-way flow of blood. The veins are encased in sheaths of areolar tissue, scantier than in the case of the arteries, and these carry the vasa vasorum and autonomic nerve fibers; the nerve fibers are also fewer than the arteries.

As a generality, arteries may be said to course rather deeply, and to supply the surface tissue by smaller branches that run directly to (rather than along) the surface. There are just as many deep veins (fig. 9-19*B*), but in addition, the blood collected by them at the surface is prone to flow along subcutaneous channels (fig. 9-19*C*), particularly in the limbs. Veins accompanying arteries are termed *venae comitantes.* Larger arteries and those of the viscera usually each have one vena comitans, while the smaller arteries in general usually have two each.

Veins of the Heart (Cardiac Veins)

The larger *cardiac veins* (fig. 9-20*A*) and some of the smaller ones open into the coronary sinus, while

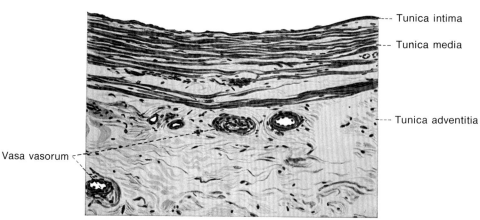

FIG. 9-19A. *Structures of a vein (cross section)*

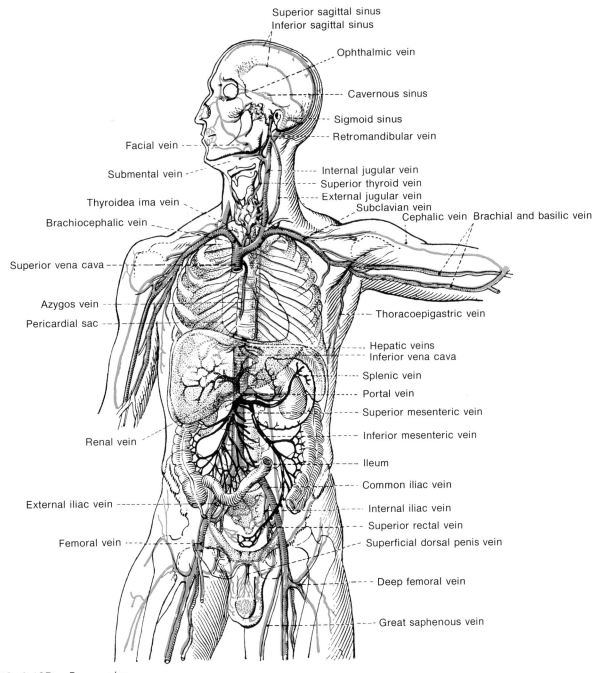

FIG. 9-19B. *Deep veins*

the remainder empty directly into the heart. The first three listed are large vessels and the remainder small. The *great cardiac vein* courses from the apex of the heart to the coronary sinus in the anterior interventricular sulcus. The *middle cardiac vein* extends from the apex to the coronary sinus in the posterior interventricular sulcus. The *small cardiac vein* courses along the inferior or diaphragmatic surface of the heart to the coronary sinus. The *anterior cardiac veins* extend from the surface of the heart to the right atrium, at the coronary sulcus; the *smallest cardiac veins*, from the substance of the heart directly to the atria.

The *coronary sinus* (fig. 9-20B) is located inferior to the base of the heart in the coronary sulcus. It opens into the right atrium just inferior to the valve of the inferior vena cava, and is guarded by a fold forming an incomplete valve.

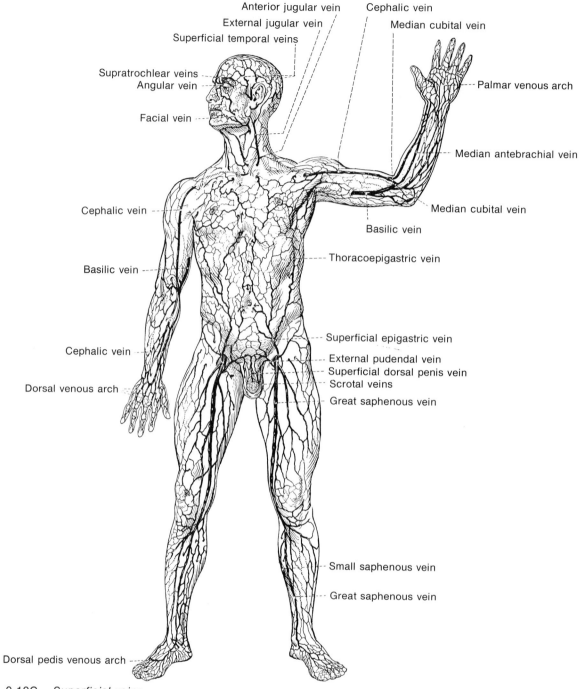

FIG. 9-19C. *Superficial veins*

Pulmonary Veins

The *pulmonary veins* (fig. 9-21A) return oxygenated blood from the lungs to the left atrium of the heart (fig. 9-21B). They are usually four in number, two superior and two inferior, right and left (fig. 9-21C).

Systemic Veins

Because of extensive anastomoses and the diffusion of channels, description of the systemic veins in an orderly manner is more difficult than in the case of the arteries. The systemic veins may be listed in two divisions: *tributaries of the superior vena cava and of the inferior vena cava.*

SUPERIOR VENA CAVA The *superior vena cava* (fig. 9-22) is short, unpaired, and empties into the right atrium. It is formed by the *union of the right and left brachiocephalic veins.*

The *brachiocephalic veins* receive blood draining from the head and neck, arm, and upon the left side, from a considerable part of the thorax. They are formed chiefly by the junction of the internal jugular with the subclavian veins dorsal to the sternal end of the clavicle. The right brachiocephalic vein is short, but the left is long, coursing transversely in respect to the superior vena cava to the arteries of the head and arm. Main tributaries of the brachiocephalic veins are the superficial veins of the head and neck; internal jugular veins; and the vertebral, internal thoracic, and subclavian veins. In addition, the left vein receives the superior intercostal veins, the thyroidea ima vein (when present), usually the accessory hemiazygos vein, and a number of small thoracic visceral branches in addition to those to the superior vena cava and to the azygos system.

The *superficial veins of the head and neck* are considered to comprise chiefly the *facial vein and its tributaries* and the *external jugular vein.*

Facial Vein The *facial vein* (fig. 9-23) extends as a continuation of the *frontal, angular,* and *nasal* branches, between the orbit and nose obliquely over the cheek and the margin of the mandible at the anterior border of the masseter muscle, and empties into the communicating branch joining the facial vein at

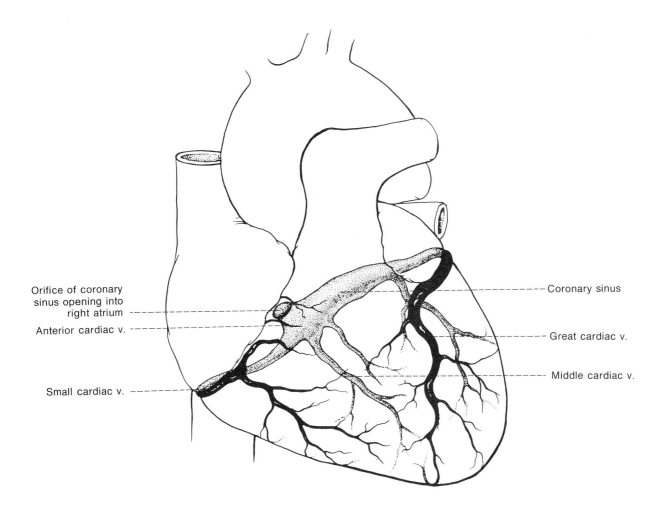

FIG. 9-20A. *Cardiac veins*

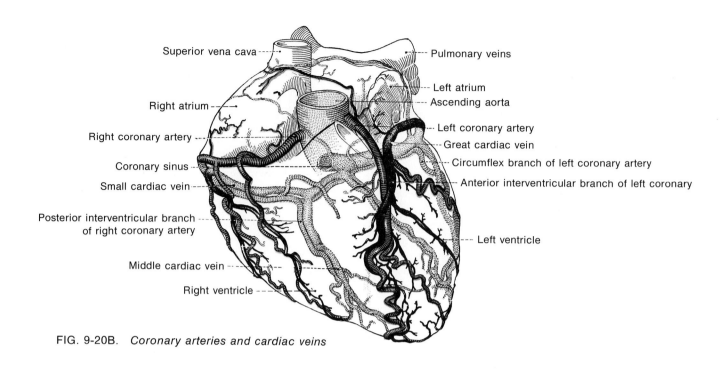

FIG. 9-20B. *Coronary arteries and cardiac veins*

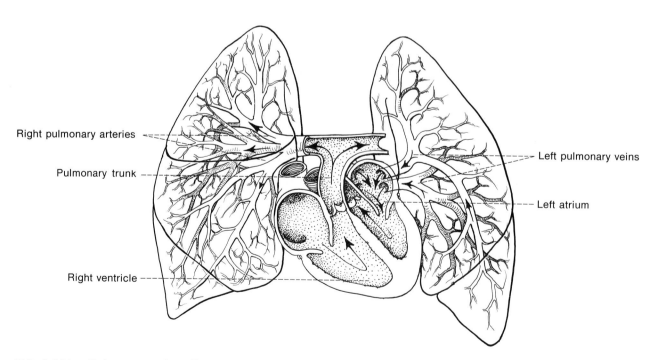

FIG. 9-21A. *Pulmonary veins; diagrammatic view of the lungs and heart*

the internal and external jugulars at the level of the hyoid bone. The *superficial temporal vein* arises anterior to the ear through the union of the larger branches of the superficial temporal veins, draining the superior and anterior parts of the side of the head, with medial temporal vein, which is deeper, from the temporalis. It receives small auricular, parotid, articular, and *transverse facial tributaries.* Below the ear it receives the *maxillary vein,* which drains blood from the *pterygoid plexus* in the pterygopalatine fossa, which receives sphenopalatine, middle meningeal, alveolar, deep temporal, palatine, pterygoid, and other small branches from adjacent structures. It joins the maxillary to form the *retromandibular vein.*

External Jugular Vein The *external jugular vein* (fig. 9-23) is extremely variable. It arises from the *union of the posterior auricular* and *the retromandibular veins* near the angle of the mandible, and courses inferiorly, deep to the platysma muscle, to the subclavian vein. A variable pattern of anterior jugular channels communicates across the midline with the vessel of the opposite side and empties into the subclavian vein.

Internal Jugular Vein The *internal jugular vein* (fig. 9-24*A* and *B*) extends from the jugular foramen to the brachiocephalic vein, coursing in the carotid sheath at first dorsolateral to the internal carotid artery and then ventrolateral to the common carotid artery. It has tributaries that may be divided into the following groups: those of the diploë of the calvaria; the dural sinuses; those of the brain; of the nasal cavity; of the ear; of the orbit; and veins of the neck.

The *diploic veins* are valveless channels of variable pattern in the diploë of the calvaria, with emissary communications both with the veins of the scalp and with the dural sinuses.

The *dural sinuses* are channels, without valves and largely noncollapsible, between the two layers of the cranial dura mater. They receive blood from the diploë, brain, nose, orbit, and ear.

The *superior sagittal sinus* extends sagittally in the basal attachment of the falx cerebri, increasing in size to its junction with the right transverse sinus at the tentorium, where it curves slightly and has an expansion, termed the *confluens of the sinus* or *torcular Herophili.*

The *inferior sagittal sinus* extends sagittally in the free margin of the falx cerebri to its junction with the tentorium and the straight sinus.

The *rectus sinus* extends sagittally in the junction of the falx with the tentorium, from the inferior sagittal sinus to the transverse sinus of the side opposite to that receiving the superior sagittal sinus.

The *intercavernosus sinus* is sometimes considered to comprise two sinuses, an *anterior intercavernosus* and a *posterior intercavernosus.* The intercavernous sinus joins the cavernous sinuses of the two sides, anterior and posterior to the hypophysis.

The *basilar sinus,* or basilar plexus, lies upon the basioccipital bone, posterior to the posterior inter-

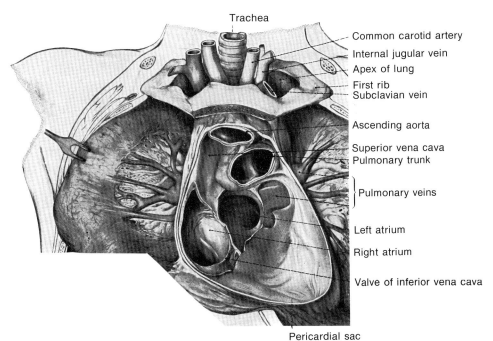

FIG. 9-21B. *Pulmonary veins; lungs, pericardium, and left atrium*

cavernous sinus, and is connected inferiorly with the vertebral plexus of the spinal column.

The foregoing are unpaired sinuses; those that follow, paired.

The *transverse sinus* courses at the basal margin of the tentorium laterally from the confluens sinus, usually being larger upon the right side. It collects most of the blood in the endocranium. In the posterior cranial fossa, before reaching the petrous temporal, it twists inferiorly and medially, and by a further twist, known as the *sigmoid sinus,* approaches the jugular foramen, by which it leaves the cranium as the internal jugular vein.

The *occipital sinus* courses between the sigmoid sinus, skirts the foramen magnum, and extends in the falx cerebelli to the confluens sinus.

The *cavernosus sinus* is a trabeculated space along the side of the hypophysis, covered by the dura stretching between the anterior clinoid process and the dorsum sellae. Through it extend the internal carotid artery, the carotid plexus, and the abducens nerve.

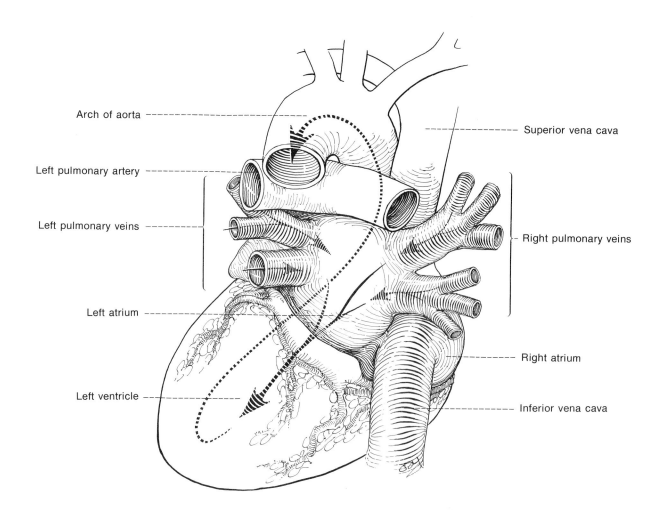

FIG. 9-21C. *Pulmonary veins; diagrammatic view of the left atrium*

The *superior petrosal sinus* extends in the anterolateral margin of the tentorium between the sinus cavernosus and sinus transverse.

The *inferior petrosal sinus* extends medial to the tentorium between the cavernous sinus and the sigmoid portion of the transverse sinus as it enters the jugular foramen.

The *sphenoparietal sinus* occurs in the angle of dura between the anterior and middle cranial fossa, coursing medially to the cavernous sinus.

As already stated, most of these sinuses empty into the internal jugular vein, but they communicate quite freely with the other, more superficial veins, as well as with each other.

Cerebral veins consist of *cerebral* and *cerebellar vessels*.

Veins of the nasal cavity are mostly plexiform and are particularly numerous upon the nasal concha. For the most part they follow the arteries supplying the area, thus communicating with the pterygoid plexus, with the ophthalmic vein in the orbit, the superior labial, the veins of the pharynx, and often with the superior sagittal sinus through the foramen cecum.

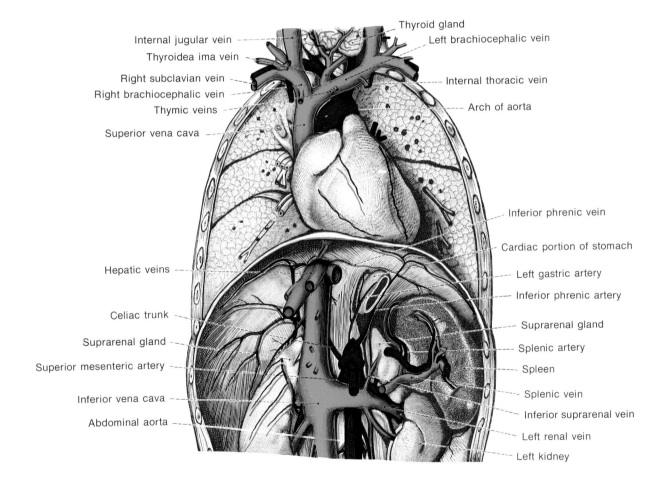

FIG. 9-22. *Superior vena cava*

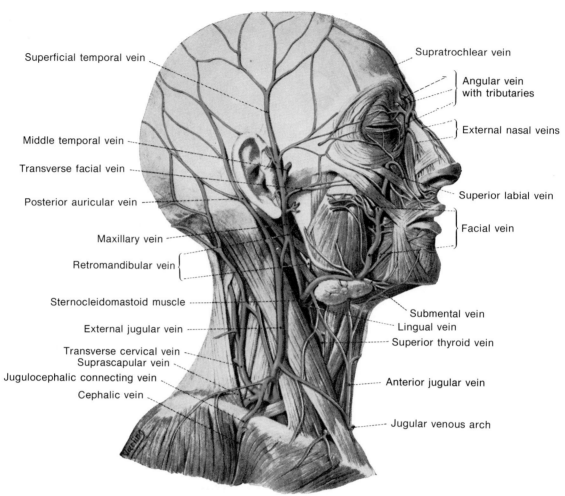

FIG. 9-23. *Facial and external jugular veins*

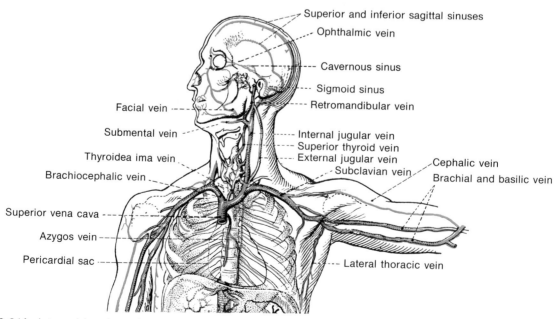

FIG. 9-24A. *Internal jugular veins*

Veins of the ear drain into the superior and inferior petrosal sinuses and the transverse sinus by way of the *internal auditory vein*.

Veins of the orbit consist chiefly of the *ophthalmic*, passing through the superior orbital fissure (not with the artery through the optic foramen) and to the cavernous sinus. Its tributaries, devoid of valves, communicate chiefly with the facial vein through superficial (angular, supraorbital) and deep (infraorbital to the pterygoid plexus) branches, and to a minor extent with the veins of the nasal cavity.

Veins of the neck, in addition to those described with the external jugular vein, comprise the intercommunicating pharyngeal and laryngeal plexuses, which drain chiefly by the superior and inferior laryngeal veins, and the superior thyroid veins to the internal jugular; but the neck veins communicate also with the inferior thyroid and thyroidea ima. The foregoing usually empty into the internal jugular, together or separately, at the level of the thyroid cartilage. The inferior thyroid vein terminates at a lower level.

The *vertebral veins* arise from the posterior vertebral plexus and leave the endocranium with the vertebral arteries, in several channels accompanying the latter down the neck, and passing partially through the foramen of the seventh vertebra as well, terminating in the brachiocephalic veins. They receive small tributaries from the vertebral column and its associated structures.

The *internal thoracic vein* corresponds to the internal thoracic artery.

The *thyroidea ima vein*, unpaired and variable, drains some of the blood from the thyroid region and joins the left brachiocephalic vein near the midline.

Small *esophageal*, *tracheal*, and *mediastinal*, including *thymic*, *branches* join the brachiocephalic and accessory hemiazygos veins.

Subclavian Veins The *subclavian vein* (fig. 9-25A) receives, upon its superior aspect just distal to the internal jugular, the thoracic duct upon the left side, and upon the right, the (usually) three small right lymphatic trunks. The subclavian vein lies ventral to the corresponding artery, and is similarly considered to have adjoining axillary and brachial segments. The veins of the upper limbs are considered to occur in two groups, *superficial* and *deep*.

The *superficial veins of the arm* (fig. 9-25B) occur

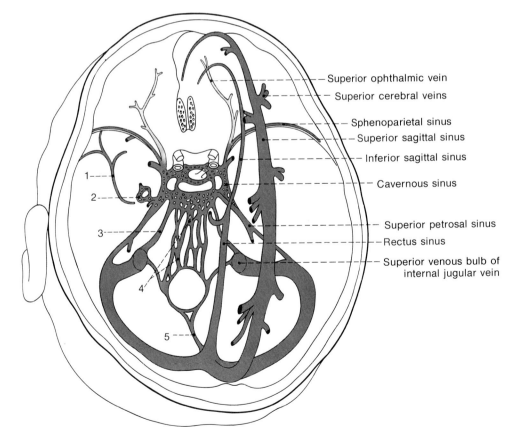

FIG. 9-24B. *Formation of the internal jugular vein*

over the hand and forearm as a venous network within the superficial fascia, containing many longitudinal vessels with valves, and a multitude of communications without valves. As these approach the elbow, or before, they congregate into two main trunks, with two or more large oblique intercommunications. One of these groups becomes the *basilic vein* and the other, the *cephalic vein*.

The *basilic vein* courses proximally along the medial side of the biceps brachii muscle, and at the distal one-third of the brachium pierces the deep fascia, in company with the cutaneous medial antebrachial nerve. The *cephalic vein* courses proximally along the lateral side of the biceps brachii muscle and then in the groove between the pectoralis major and deltoid muscles pierces deep just below the clavicle, and joins the *thoracoacromial vein* and then the *axillary vein*.

The *deep veins of the arm* (fig. 9-25C), distal to the axillary vein, accompany the respective arteries, but there are two venae comitantes of each artery, one on either side, with numerous intercommunications.

Azygos System The *azygos system* (fig. 9-26) comprises a group of veins that receive blood from the intercostal and thoracic branches. The system communicates on either side inferiorly with the ascending lumbar vein, and consists usually of the *accessory hemiazygos vein* on the left side descending to join the ascending *hemiazygos vein*. The latter crosses the vertebral column, usually superior to the middle of the thorax, and joins the *azygos vein*, which ascends in the right posterior mediastinum, first dorsal to and then curving around superior to the root of the right lung, and empties into the dorsal aspect of the superior vena cava. Bronchial branches, draining blood from the lung tissue, join the azygos system, as do tracheal, mediastinal, esophageal, pericardial, and phrenic twigs.

As mentioned, *intercostal veins*, accompanying intercostal arteries, are tributary to the azygos system, as well as to the branches of the internal thoracic vein. In turn, tributary to the intercostal veins are the veins of the spinal column. These consist chiefly of a heavy plexus of (mostly anterior and posterior longitudinal) vessels constituting the *internal vertebral venous plexus* devoid of valves, between the dura mater and the walls of the spinal canal. This plexus communicates freely by intervertebral veins between the vertebrae and accompanying spinal veins from the spinal cord, with the *external vertebral venous plexus* situated upon the external surface of the spinal column.

Superficial Veins of the Trunk The superficial veins form a plexus in the superficial fascia of the trunk (fig. 9-27). At scattered locations there are branches that pierce the deep fascia and the underlying structures to drain into deeper channels, chiefly where the cutaneous branches of the spinal nerve reach the surface, and by these routes the piercing venous branches reach the intercostal veins.

The chief drainage channels from the freely anastomosing veins of the surface, however, comprise the branches of the superficial epigastric vein, emptying into the femoral vein, and the thoracoepigastric vein, a tributary of the lateral thoracic and hence of the axillary veins.

INFERIOR VENA CAVA The *inferior vena cava* (fig. 9-28) may be said to receive all blood drained from the structures inferior to the diaphragm, although this is not strictly true, for the internal thoracic vein receives the inferior epigastric vein from below this level. There is also free communication between some of the tributaries of the superior and inferior vena cava, of importance in establishing a collateral circulation following partial occlusion of the inferior vena cava. Of chief importance to this function are the communications between the internal thoracic vein and the inferior epigastric vein; thoracoepigastric and superficial epigastric veins; the lumbar vein and the azygos system; and between intercostal veins, largely through the vertebral plexuses.

The inferior vena cava extends inferiorly from the right atrium, through the foramen vena cava of the diaphragm to the right of the midline and to the right of the aorta, ending in the two common iliac branches, which diverge dorsal to the right common iliac artery. In its course, it receives the following tributaries: the *inferior phrenic vein, hepatic veins* (two or three), *suprarenal vein, renal vein, testicular* or *ovarian vein, lumbar veins* (four or five pairs), *common iliac vein,* and the *medial sacral vein*. The left suprarenal and left testicular or ovarian veins are tributaries of the left renal vein. There are no tributaries of the vena cava corresponding to the celiac, superior mesenteric, and inferior mesenteric arteries, for the comparable veins are a part of the portal system.

The *inferior phrenic vein* is the first tributary of the inferior vena cava; it corresponds to the inferior phrenic artery.

The *hepatic veins* consist of (usually) two large veins, from the right and left lobes of the liver, a smaller one from the caudate and quadrate lobes, and several small twigs, all entering the inferior vena cava in a cluster just below the diaphragm.

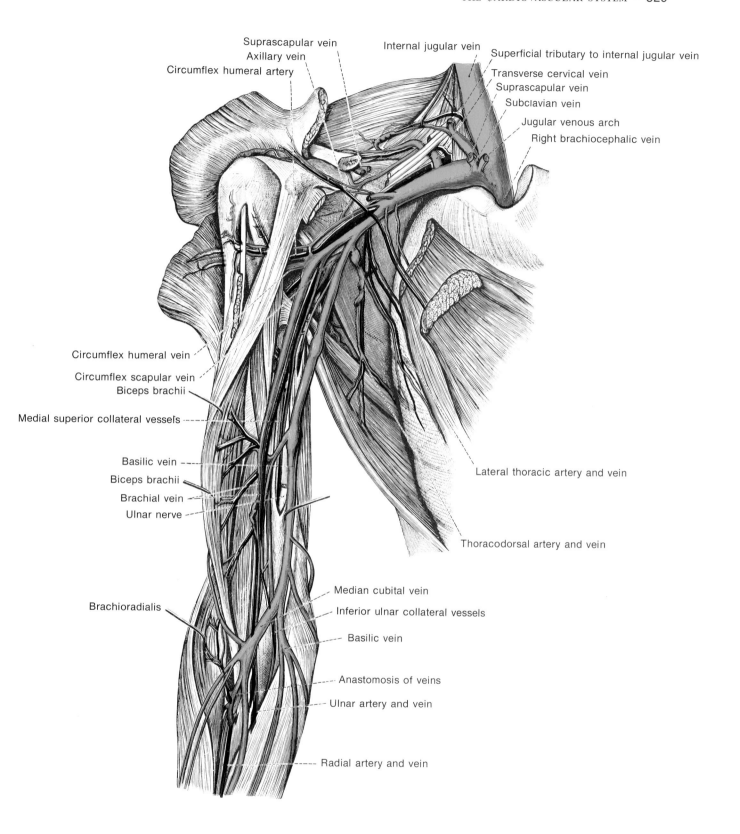

FIG. 9-25A. *Subclavian vein*

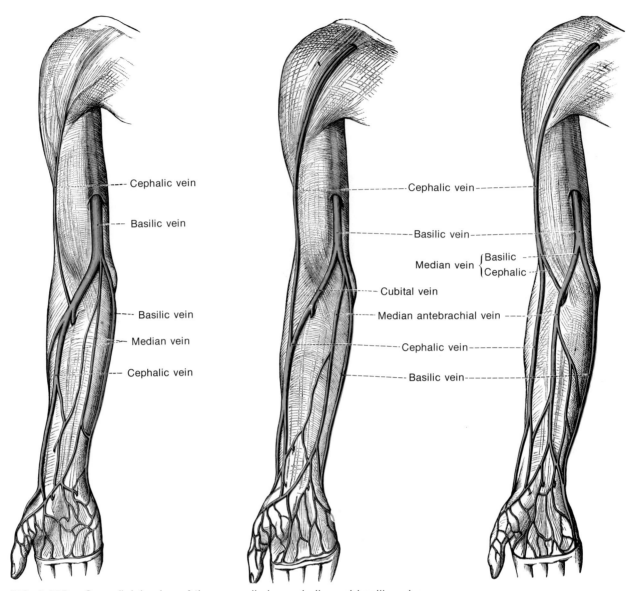

FIG. 9-25B. *Superficial veins of the upper limb; cephalic and basilic veins*

The *suprarenal vein,* from the adrenal gland, enters the adjacent part of the vena cava on the right side and the renal vein on the left.

The *renal vein* is formed of the fusion of several tributaries from the kidney, near the hilus. The left vein is longer, as it must cross (ventral to) the aorta. The left vein receives the suprarenal and testicular or ovarian veins.

On the right side the *testicular (ovarian) vein* joins the inferior vena cava a short distance below the renal vein, while on the left side it is a tributary of the left renal vein to the left of the aorta. It accompanies the corresponding artery, but breaks up into several associated channels and receives small twigs from the tissue about the kidney and the peritoneum. In the male, the distal part of the testicular vein forms the *pampiniform plexus* of veins in the spermatic cord, receiving blood from the testicle and epididymis. In the female, as the ovarian vein, it arises as the *uterovaginal plexus,* forming also a pampiniform plexus, with numerous contributions from the ovary and connections with the uterine branches of the internal iliac vein.

The *lumbar veins* usually comprise four or five pairs, corresponding in the lumbar region to the intercostal veins of the thorax. Adjoining the vertebral body they are interconnected by communications arranged in a longitudinal trunk, emptying inferiorly into the iliolumbar vein, and thus into the common iliac vein, and superiorly into the azygos system.

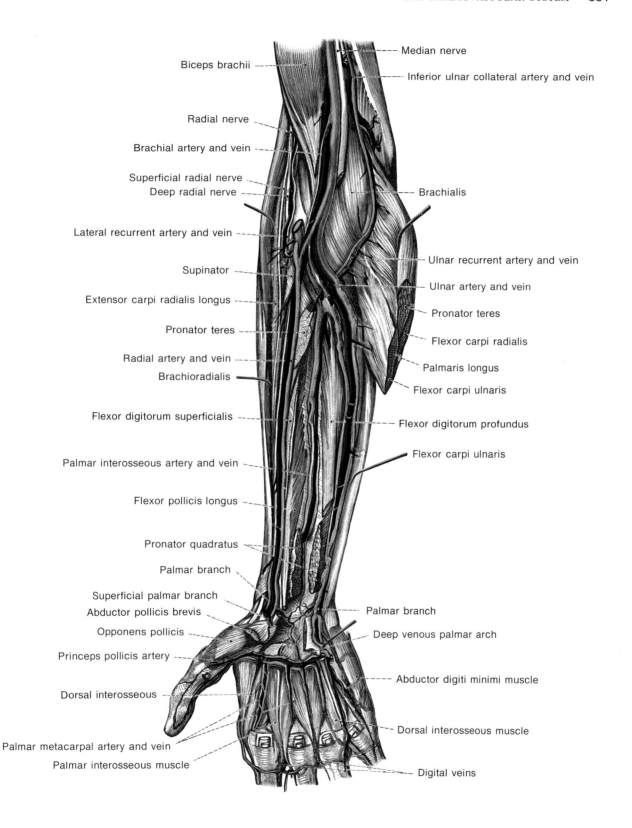

FIG. 9-25C. *Deep veins of the upper limb; tributaries of the subclavian vein*

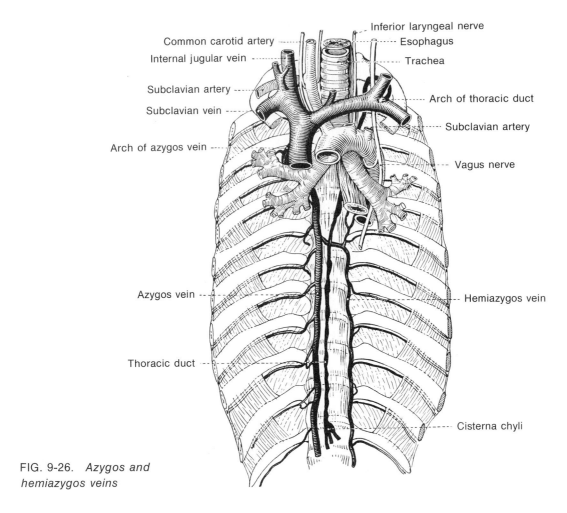

FIG. 9-26. *Azygos and hemiazygos veins*

The *medial sacral vein,* unpaired, represents the caudal part of the inferior vena cava, but in man it is a tributary of the left common iliac vein.

Common Iliac Vein The *common iliac vein* (fig. 9-29) converges on either side from the pelvis to the inferior part of the inferior vena cava, located largely medial to the corresponding artery. It has two main tributaries, the *external and internal iliac veins.*

External Iliac Vein The *external iliac vein* (fig. 9-29) is the intrapelvic continuation of the femoral vein, and is crossed ventrally in its proximal part by the internal iliac artery. Near the abdominal wall it has two tributaries, corresponding to the arteries of the same name, the *inferior epigastric vein* and the *deep circumflex iliac vein.* In addition, it receives a communication from the *obturator vein.*

Internal Iliac Vein The *internal iliac vein* (fig. 9-29) has tributaries that correspond to the branches of the internal ilial artery, except for the remains of the umbilical artery.

Femoral Vein The *tributaries* of the *femoral vein* (fig. 9-30A) drain the lower limb, except the gluteal and obturator regions, and they may be divided into *superficial* and *deep branches.*

The *superficial veins of the lower limb* form a rich plexus of vessels in the superficial fascia that drain to two points: the popliteal fossa and the fossa ovalis.

Femoral and Popliteal Veins The *popliteal vein* (fig. 9-30B) is the proximal continuation of the junction in the superior crus of the two tibial veins. It passes through the tibial space posteromedial and then posterior to the popliteal artery, through the insertional end of the adductor magnus muscle, and there becomes the *femoral vein.* The *femoral vein* ascends in the adductor canal, at first posterior to the femoral artery, but in the femoral triangle medial to it, and passes through the lacuna vasorum, between the femoral artery (laterally) and the anulus femoralis (medially) to become the *external iliac vein* (fig. 9-29). In its course it receives tributaries that correspond in general to the arterial branches, except for the saphenous branches as mentioned. There is much freer communication between the venous than between the arterial branches.

Small and Great Saphenous Veins The *small and great saphenous veins* (fig. 9-31) are superficial veins of the lower limb. The *small saphenous vein* begins as branches over the lateral aspect of the foot, passes behind the lateral malleolus, and courses proximally over the back of the crus, receiving many branches, to the back of the knee where it pierces the deep fascia, enters the popliteal space, and forks, one branch entering the *popliteal vein* (fig. 9-30B) and the other joining the *deep femoral vein* (fig. 9-30B). The *great saphenous vein* begins upon the medial side of the foot, passes anterior to the medial malleolus, and courses proximally, medial to the tibia, the knee, and the thigh, to the fossa ovalis and thus to the *femoral vein* (fig. 9-30A). In the crus it communicates freely with the small saphenous vein and, throughout its

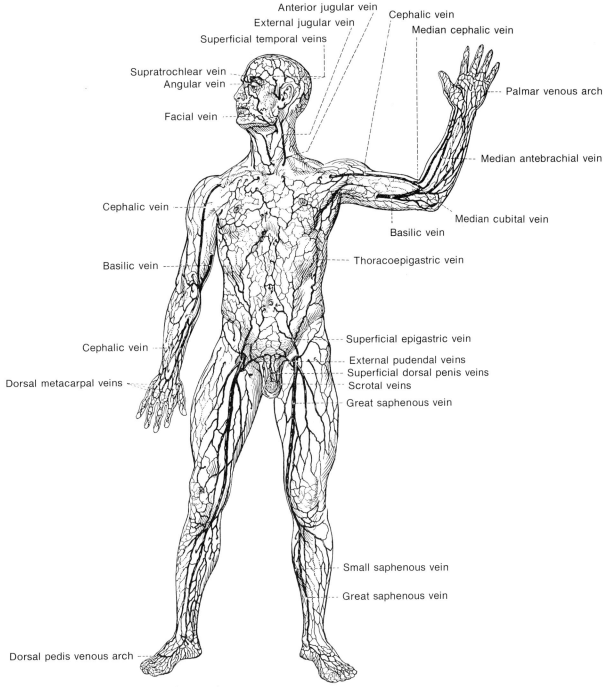

FIG. 9-27. *Superficial veins of the trunk*

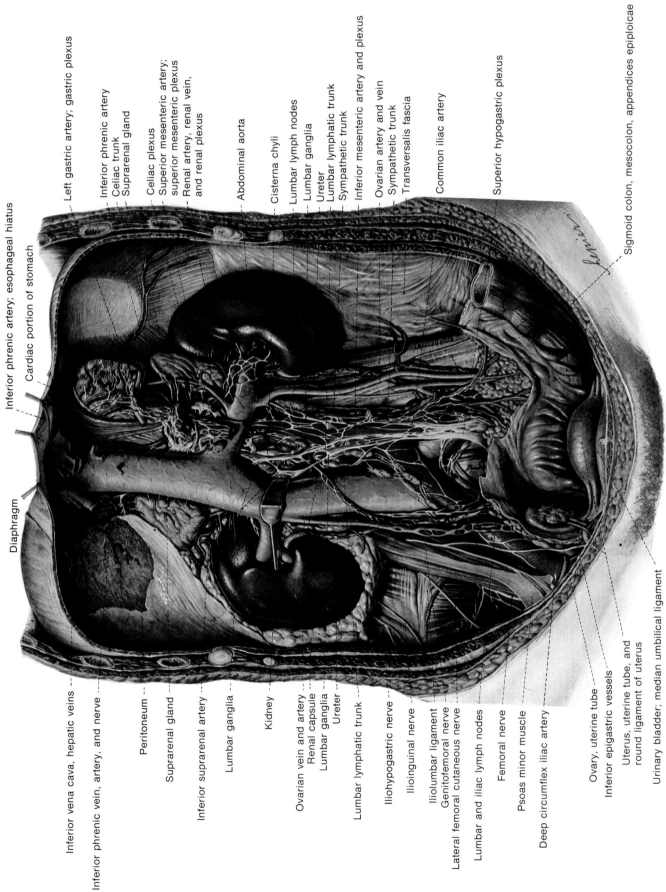

FIG. 9-28. *Inferior vena cava and its tributaries*

course, with the deeper veins, receiving many tributaries. In the fossa ovalis it receives superficial branches from the lower abdomen (*superficial epigastric vein*), from over the crest of the ilium (*superficial circumflex iliac vein*), from the external genitalia and mons pubis [*external pudendal vein*, including the *dorsal penis* (*clitoridis*) and *anterior scrotal* (*labial*) *veins*], and the *lateral* and *medial superficial femoral veins*.

The *deep veins of the lower limb* (fig. 9-30*A*) accompany the corresponding arteries. There are two venae comitantes in all instances except for the popliteal, femoral, and deep femoral veins, which are all single.

Portal System

The *portal system of veins* (fig. 9-32) unlike those of other parts of the body, run between two capillary beds: those of the alimentary canal from the inferior part of the esophagus to the inferior rectum; and of the gallbladder, the pancreas, and the spleen to the capillaries of the liver. It conveys nutrient and other material from the alimentary tract to the liver. The system lacks valves, which predisposes it to relatively excessive pressures of blood in the inferior part, often manifested in the occurrence of hemorrhoids or piles in the internal hemorrhoidal plexus.

PORTAL VEIN The *portal vein* is a trunk about 3 inches in length formed at the level of the pancreas by the *union of the splenic* and *superior mesenteric veins*, whence it passes obliquely superiorly toward the right, between the marginal layers of the lesser omentum as this bridges the epiploic foramen; there it is accompanied by the hepatic artery and common bile duct. Then it enters the porta of the liver and divides into *right* and *left branches*. In its course it receives *splenic, superior mesenteric, coronary, pyloric,* and *cystic branches.*

SPLENIC VEIN The *splenic vein* arises as several branches from the hilus of the spleen, which join to form a large trunk that passes to the right, below the splenic artery and ventral to the aorta, to join the superior mesenteric vein dorsal to the neck of the pancreas. In its course it receives *short gastric veins*, the *left gastroepiploic vein* from the greater curvature of the stomach, *pancreatic branches,* and the *inferior mesenteric vein,* which receives blood from the rectum (by the *superior rectal vein*), the sigmoid colon (by the *sigmoid veins* or inferior left colic veins), and descending colon and left colic flexure (by the *left colic vein*).

SUPERIOR MESENTERIC VEIN The *superior mesenteric vein* arises by intestinal branches from the small intestine and the ascending and transverse colons. These branches are disposed on the left as the *ileal veins* inferiorly, and the *jejunal veins* superiorly; upon the right as the *ileocolic vein* inferiorly, the *right colic,* and the *middle colic veins.* Near its proximal termination the superior mesenteric vein receives also the *right gastroepiploic vein,* the continuation upon the right of the left gastroepiploic vein, and the *pancreaticoduodenal vein,* from the pancreas and duodenum.

CORONARY VEIN The *coronary vein* (right and left gastric veins) extends along the lesser curvature of the stomach, receiving twigs from the inferior part of the esophagus before entering the portal vein.

PYLORIC VEIN The *pyloric vein*, associated with and anastomosing with the coronary vein, arises from the stomach in the vicinity of the pylorus.

CYSTIC VEIN The *cystic vein* brings blood from the gallbladder to the portal vein as the latter enters the liver.

FETAL CIRCULATION

During fetal life food and oxygen are carried to the fetus by the *umbilical vein* (fig. 9-33*A*). This vessel enters the umbilicus and passes in the border of the falciform ligament (a fold of the peritoneum) to the left branch of the portal vein. From this point it continues, as the *ductus venosus,* to the left hepatic vein near the vena cava. When the blood of the inferior vena cava reaches the right atrium it is directed toward the *foramen ovale*, through which some of it passes. The rest enters the right ventricle and so to the pulmonary arterial trunk. However, an insignificant proportion of the blood from the right ventricle reaches the lungs, for these organs do not function to any definite extent before birth, and receive only sufficient blood for their growth. Instead, the blood is diverted from the pulmonary arterial trunk to the aorta through the short *ductus arteriosus,* the remains of the dorsal part of the sixth aortic arch. Little blood is returned by the pulmonary veins to the left atrium, and, therefore, the foramen ovale and ductus arteriosus furnish practically all the blood carried by the aorta. Poorly oxygenated blood is returned to the placenta (fig. 9-33*B*) by the two *umbilical arteries,* which are branches of the internal iliac arteries that course superomedially to the umbilicus over the dorsal aspect of the inferior abdominal wall.

Thus the significant details of the fetal circulation comprise the *umbilical vein, ductus venosus, foramen*

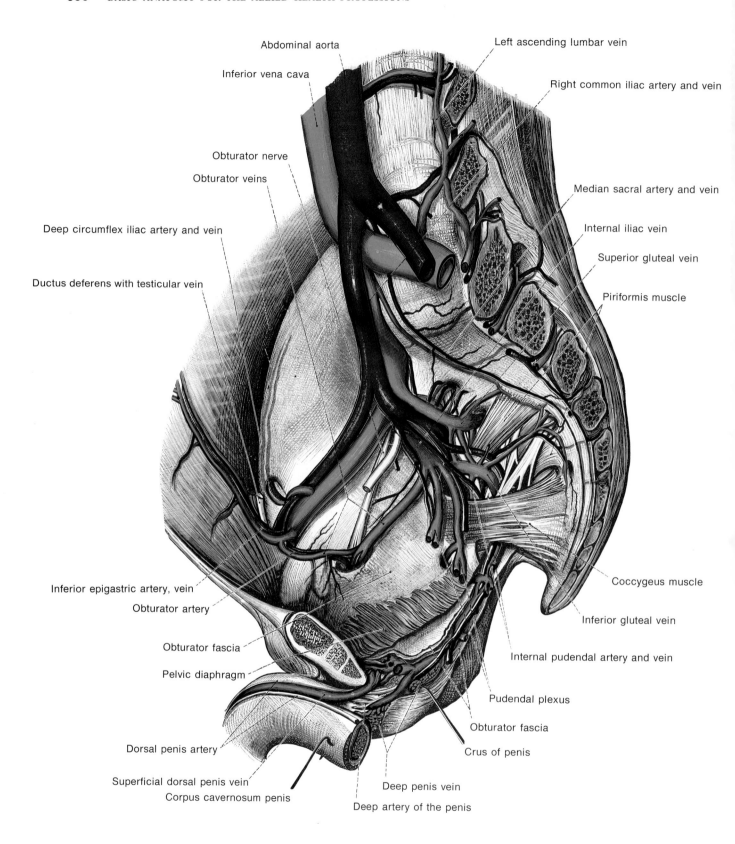

FIG. 9-29. *Iliac veins*

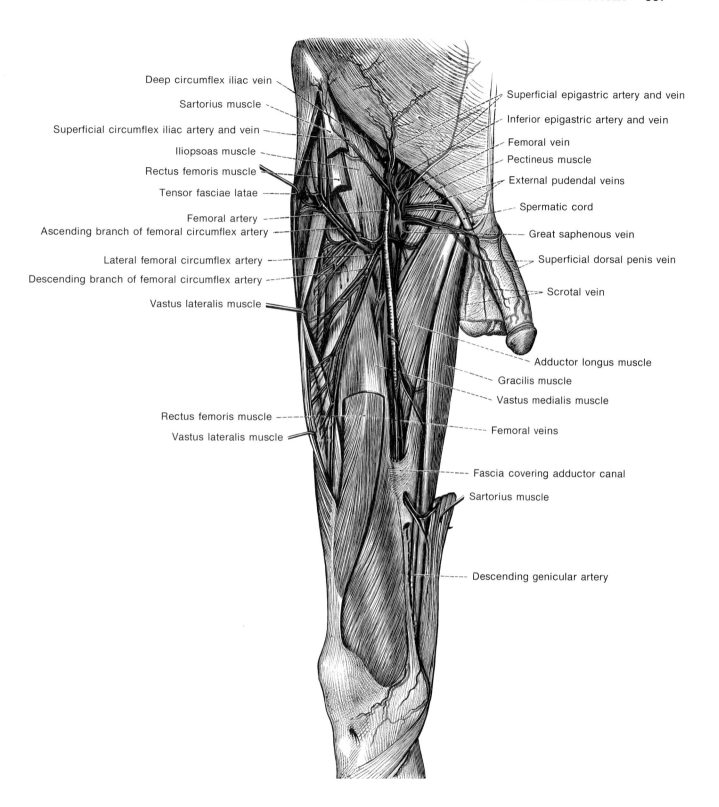

FIG. 9-30A. *Femoral vein*

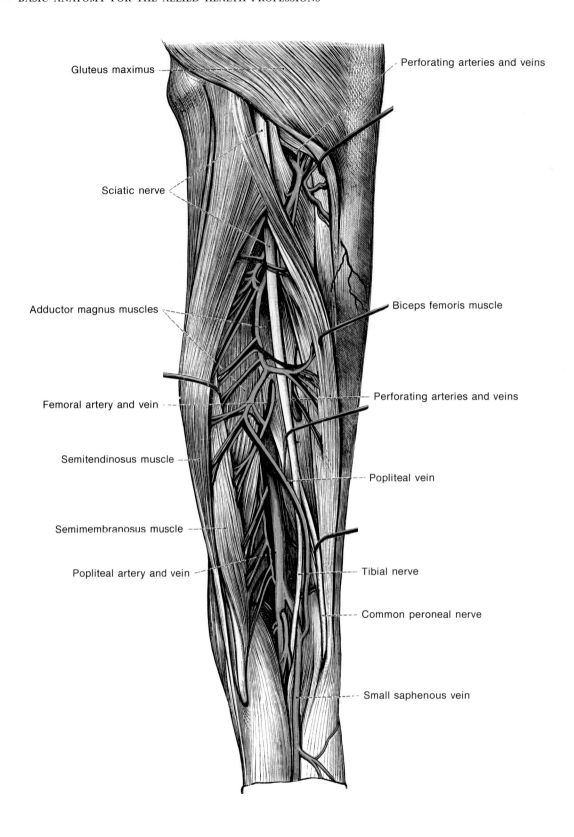

FIG. 9-30B. *Popliteal vein*

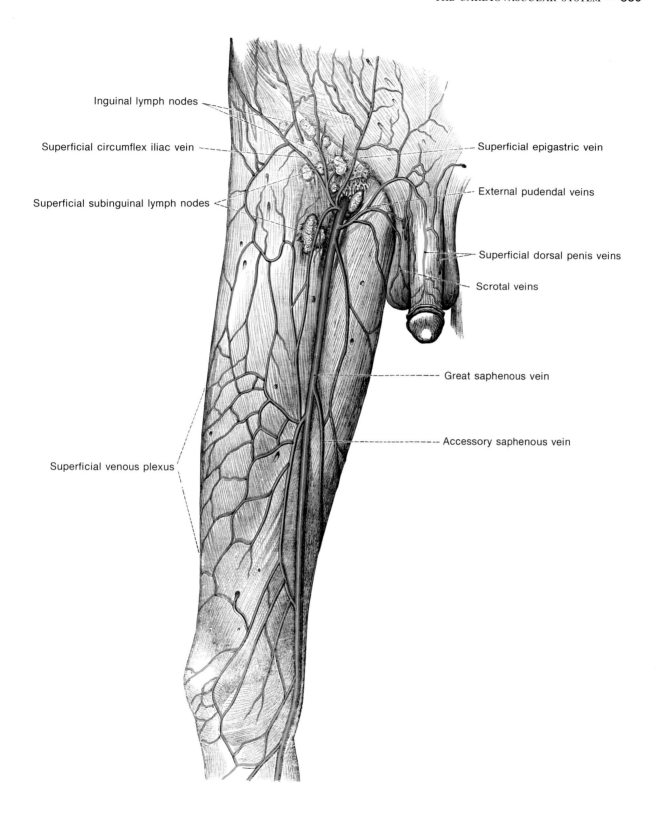

FIG. 9-31A. *Great saphenous vein*

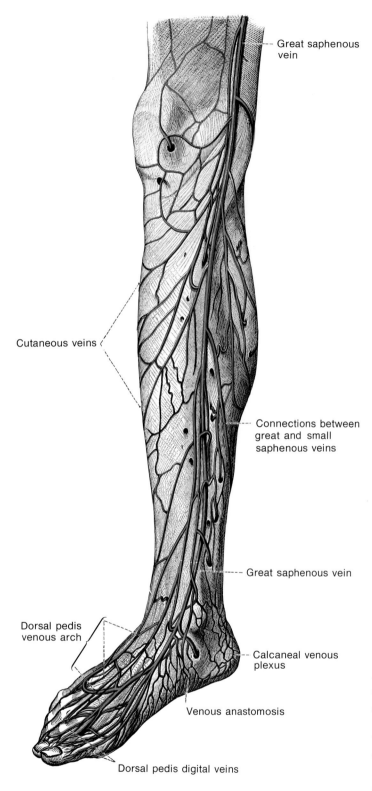

FIG. 9-31B. *Great and small saphenous veins*

ovale, the *relative nonfunctioning of the lungs,* the *ductus arteriosus,* and the *umbilical arteries.*

At birth several drastic changes occur, some at once (such as breathing) and some more gradually. The walls of the exclusively fetal vessels gradually thicken until their lumens are occluded. Thus the *umbilical vein* becomes the *ligamentum teres hepatis;* the *ductus venosus* becomes the *ligamentum venosum;* pressure in the left atrium becomes higher than in the right, so that the flap valve of the foramen ovale is held shut and is then, usually, fused in that position; the *ductus arteriosus* becomes the *ligamentum arteriosum;* and the umbilical arteries become ligamentous, marking the lateral umbilical fold.

CLINICAL APPLICATIONS

Various diseases affecting the arteries are designated as *arteriosclerosis.* A thickening of the tunica intima in large arteries is referred to as *atherosclerosis.* These accumulations are composed of lipids and are known as *atheromas.* Narrowing of the coronary arteries is usually caused by atherosclerosis.

Hypertension, or high blood pressure, may cause a blood vessel to rupture. Current research indicates a genetic defect associated with the sympathetic innervation of the arterioles that causes excessive vasoconstriction. The elasticity of the walls of the arteries is of vital importance in aiding the heart. Partial failure in the functioning of the muscular coat, as in hardening of the arteries, requires the heart to work at overload, and hypertension results because a greater pressure is needed to force the blood through the impaired arteries.

Veins are subject to *phlebitis,* or inflammation, and *varicosities.*

Pericarditis, or inflammation of the pericardium, may cause friction rubs.

The heart may be vulnerable to numerous *arrhythmias* (changes in heart rhythm) such as *tachycardia, atrial flutter, atrial fibrillation,* and *ventricular fibrillation.*

The heart is vulnerable to streptococcal infections, which may lead to *rheumatic heart disease* or *endocarditis.*

Patent ductus arteriosus, pulmonary stenosis, atrial septal defects, interventricular septal defects, and *tetralogy of Fallot* are common examples of *congenital heart diseases.*

Gradual *portal obstruction* results in the establish-

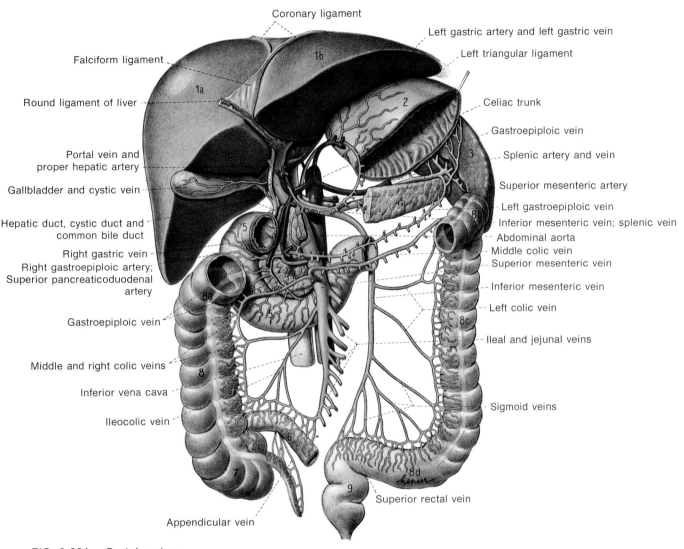

FIG. 9-32A. *Portal system*

ment of a collateral portal circulation by enlargement of the channels indicated as follows: between the left gastric and the esophageal branches of the azygos system; between the superior rectal and the middle rectal branches of the internal iliac vein; between the intrahepatic twigs of the left portal vein and the parumbilical twigs of the falciform ligament flowing to the epigastric veins; between the intrahepatic portal twigs via the bare spot of the liver with the phrenic, internal thoracic, and azygos tributaries; between the portal branches and the inferior vena cava by means of the veins of Retzius, which are undeveloped and grossly undetectable in the normal individual.

Valvular Sounds

The sounds produced by the closure of the valves do not correspond with the position of the valves, but are as follows.

The *bicuspid sound* is heard most distinctly at the apex of the heart as high as the third intercostal space.

The *tricuspid sound* is best heard at the left sternal border between the fifth and sixth costal cartilages.

The *pulmonary sound* is best heard in the second intercostal space to the left of the sternum.

The *aortic sound* is best heard in the second right intercostal space.

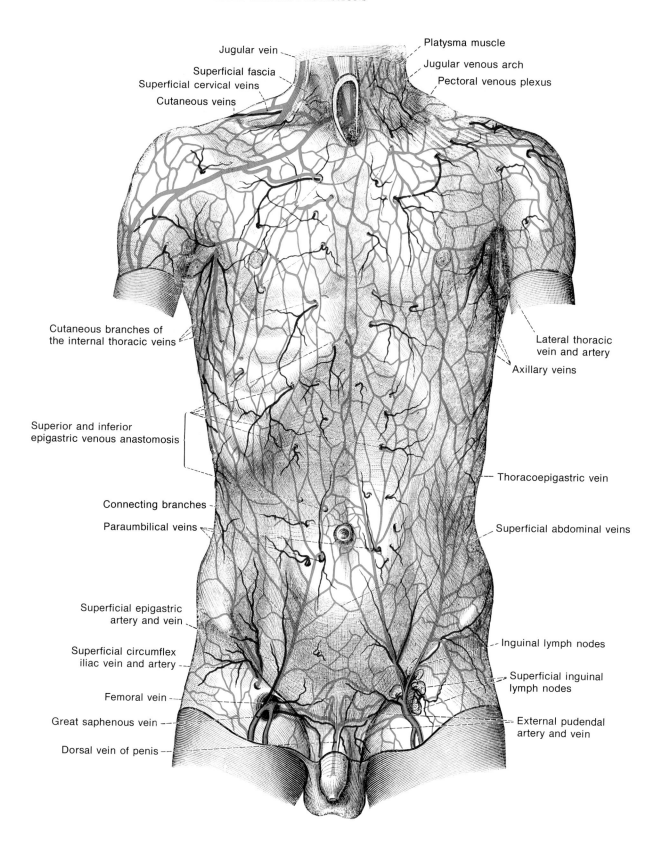

FIG. 9-32B. *Superficial connections to the portal system*

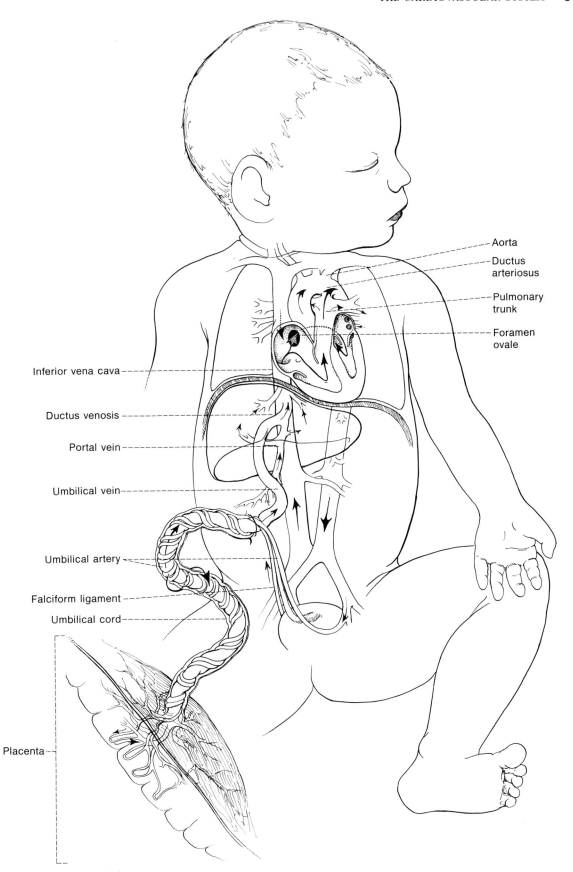

FIG. 9-33A. *Fetal circulation*

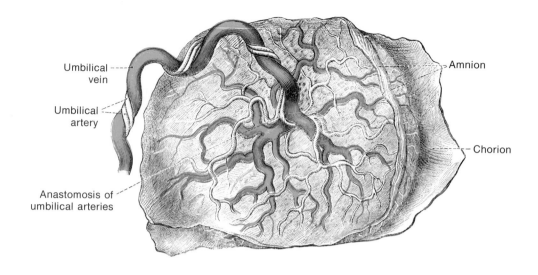

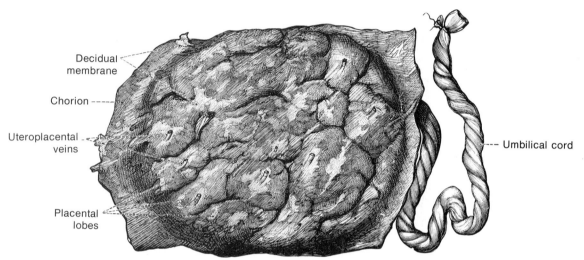

FIG. 9-33B. *Placenta*

REVIEW QUESTIONS:

1. The inferior border or diaphragmatic surface of the heart is located at the level of the _____.
2. The bicuspid valve lies behind the right margin of the sternum opposite the _____ costal cartilage.
3. The pericardial sac consists of two layers referred to as the _____ and _____ pericardium.
4. The heart is a muscular organ, located in the _____ mediastinum.
5. The _____ surface of the heart is formed by the right ventricle.
6. The _____ is the visceral layer of the pericardium.
7. The _____ is the lining of the heart and is continuous with and comparable to the inner coat of the blood vessels.
8. The _____ is the thickened margin surrounding the fossa ovalis.
9. The left atrium exhibits the remains of the foramen ovale upon the interatrial septum, which is called the _____.

10. A groove known as the _____ separates the atria from the ventricles.
11. Internally, ridges and folds of the ventricular walls are called _____.
12. The _____ drains blood from the vessels supplying the heart walls.
13. The right atrioventricular valve is commonly called the _____ valve.
14. The _____ valve is located at the opening between the left ventricle and the aorta.
15. The _____ connects the AV node and the interventricular septum.
16. List the three coats of an artery.
    a.
    b.
    c.
17. The ascending aorta gives rise to the right and left _____ arteries.
18. The arch of the aorta gives rise to the _____, _____, and the _____ arteries.
19. The right common carotid arises from the _____ artery.
20. The external carotid artery usually gives rise to _____ branches.
21. The _____ artery is the terminal branch of the _____.
22. The axillary artery extends from the outer border of the first rib to the lower border of the teres major muscle where it becomes the _____ artery.
23. The _____ is the first branch of the abdominal aorta.
24. At the level of the lumbosacral articulation, the common iliac arteries divide into the _____ and _____ arteries.
25. The _____ artery leaves the pelvis through the greater sciatic foramen.
26. The external iliac artery courses beneath the inguinal ligament to become the _____ artery.
27. The _____ vein courses from the apex of the heart to the coronary sinus.
28. The superior vena cava is formed by the union of the _____ veins.
29. The external jugular vein arises from the union of the _____ and the _____ veins.
30. The _____ sinus extends sagitally in the free margin of the falx cerebri to its junction with the tentorium and the straight sinus.
31. The _____ sinus is a trabeculated space along the side of the hypophysis, covered by the dura stretching between the anterior clinoid process and the dorsum sellae.
32. Veins of the orbit consist chiefly of the _____.
33. The superficial veins of the upper limb include the _____ and _____.
34. The _____ system of veins comprise a group of veins that receive blood from the intercostal and thoracic branches.
35. The _____ and _____ veins are superficial veins of the lower limb.
36. The portal vein is formed by the union of the _____ and _____ veins.
37. During fetal life oxygen is carried to the fetus by the _____ vein.
38. Poorly oxygenated blood is returned to the placenta by the two _____ arteries.
39. The ductus venosus becomes the _____ in the adult.
40. The distal part of the testicular vein forms the _____ of veins in the spermatic cord.

10. The Lymphatic System

> STUDENT OBJECTIVES
> After you have read this chapter, you should be able to:
> 1. Identify the components and functions of the lymph vascular system.
> 2. Describe the function of lymph nodes.
> 3. Identify the locations and functions of the tonsils, spleen, and thymus glands as lymphatic organs.
> 4. Discuss the direction of circulating lymph.
> 5. Compare the structure of veins and lymphatics.

The *lymphatic system* (fig. 10-1A) carries lymph from the fluid spaces of the extracellular tissue of the body to the subclavian veins. In the capillary beds, blood plasma passes through the capillary walls, bathes the tissues, and is collected into *lymphatic capillaries,* which are tributary to the *lymph ducts.* These converge to *lymph nodes,* in the substance of which they break up again into *lymph capillaries,* called *sinuses.* This secondary network in the substance of the lymph node acts as a sieve, from which dead and particulate matter picked up by the lymph is strained out. That is why, in lungs or elsewhere, nodes may be black or gray in color. In the nodes, the capillaries re-form into *efferent lymphatic vessels,* some of which pass to other nodes, and the rest to the lymphatic trunks. The larger trunks are situated in a complicated network upon the dorsal aspect of the trunk cavities, tributary, for the most part, to the thoracic duct. This duct enters the left subclavian vein. Lymph from the right side of the head and neck, right arm, and right shoulder, however, enters the right subclavian vein.

The flow of lymph is much slower than that of the blood and is caused by pressure from the capillaries, the movements of the different parts of the body, and by action of the muscular coat of the lymph vessels. The valves do not, at least in many cases, effect perfect blockage, so that although the flow is in one direction (fig. 10-1B), there are recurring partial reversals of the current, of short duration. The *lymph channels* assist in the removal of waste products from the intercellular spaces, are important in fat absorption from the intestine, in the transport of hormones to the blood stream (thyroid gland), and constitute a means for regulating the equilibrium of the tissue fluids of the body.

Lymph capillaries are abundant in the epithelial and endothelial regions, particularly in the palms of the hands and the soles of the feet, and are present in varying degrees in most other situations. However, they are considered to be absent from the central nervous system, the meninges, eyeball, internal ear, parenchyma of the spleen, lobules of the liver, cartilage, the air cells of the lungs, and the placenta. Some glands have many capillaries and others few. They are particularly numerous in and around the intestines, from which they transport chyle.

Lymphatic vessels are delicate threads in connec-

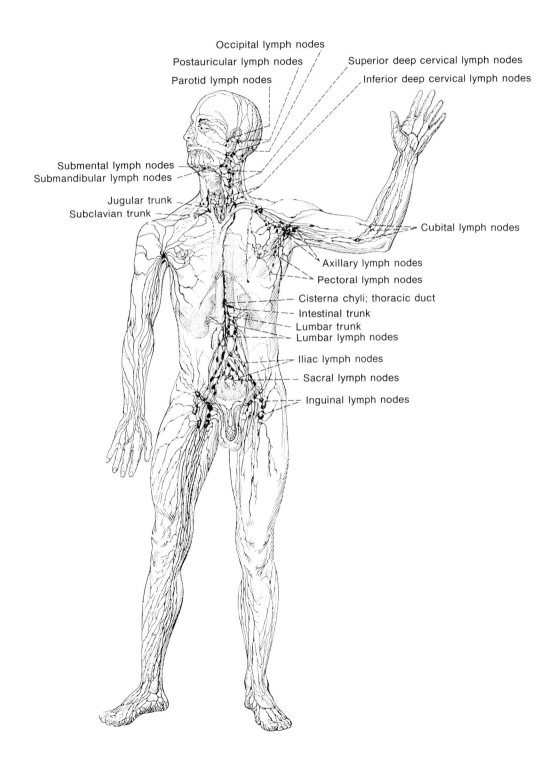

FIG. 10-1A. *Lymphatic system (lymphatic channels and nodes)*

tive tissue coursing to the lymph nodes; some are of great length, as from the foot to the groin. They have numerous slight constrictions, marking positions of valves, and they are present in practically every part of the body, except such nonvascular structures as the nails, cartilage, and lenses of the eyes. The smaller vessels are composed of thin endothelium and connective tissue, but the larger have three coats; the middle layer has transverse and the external has longitudinal fibers of smooth muscle. Lymph vessels are always of relatively small size and plexiform, never joining to form large trunks as in the case of blood vessels, except in the final collecting trunks, upon the dorsal aspect of the trunk cavities, and deep in the right side of the neck. These comprise the *thoracic duct* and the *right lymphatic duct.*

The *thoracic duct* (fig. 10-2) begins in the superior abdominal cavity as a dilated vessel about $\frac{1}{3}$ inch in width and 2 or 3 inches in length, situated between the aorta and the right crus of the diaphragm. This part is termed the *cisterna chyli.* It continues superiorly as the *thoracic duct proper,* of slightly smaller caliber than the cisterna. This duct passes in the posterior mediastinum upon the right of the sagittal plane, but crosses obliquely to the left of this plane at the level of the divergence of the bronchi from the trachea (fifth thoracic vertebra). It then courses dorsal to the subclavian vessels between the vertebral and common carotid arteries, and arches to join the superior aspect of the left subclavian vein just lateral to the internal jugular vein. In its course it receives a number of tributaries.

The *right lymphatic trunk* is variable. It is considered to comprise three vessels that conduct the lymph from the right side of the head and neck, the right arm, and the neighborhood of the right shoulder. These enter the right subclavian vein in the same situation as does the thoracic duct upon the left side. Frequently, however, the three vessels are replaced by a larger number entering the subclavian and internal jugular veins at various points, all being easily overlooked. Less frequently these branches join a fork of the thoracic duct that diverges to join the right subclavian vein.

Lymph nodes are of irregular shapes and normally vary from microscopic, solitary nodules to much larger sizes. In various infections they may far surpass their usual size. Lymph nodes manufacture a part of the white blood corpuscles and act as filters of the lymph; hence the ease with which they become sites for colonies of bacteria, and the fact that they become discolored by arrested particles of carbon and other substances. To and from them pass lymph vessels. On transection of a lymph node the cortex is seen to contain small round masses, 2 mm or less in diameter. These are the *lymph nodules.*

Lymph nodes are situated in the loose connective tissue about the larger blood vessels, but are for the most part within the body contour and within the deep fascia, largely within the body cavities and deep in the neck. A few small ones occur in the limbs, and larger ones in the superficial fascia of the inguinal region. They are particularly associated with the intestinal tract and with the roots of the large vessels to the limbs. The number, sizes, and exact locations of the nodes of any particular area (as in the inguinal region) is individually variable.

FIG. 10-1B. *Lymphatic system (schematic drawing of lymphatic drainage)*

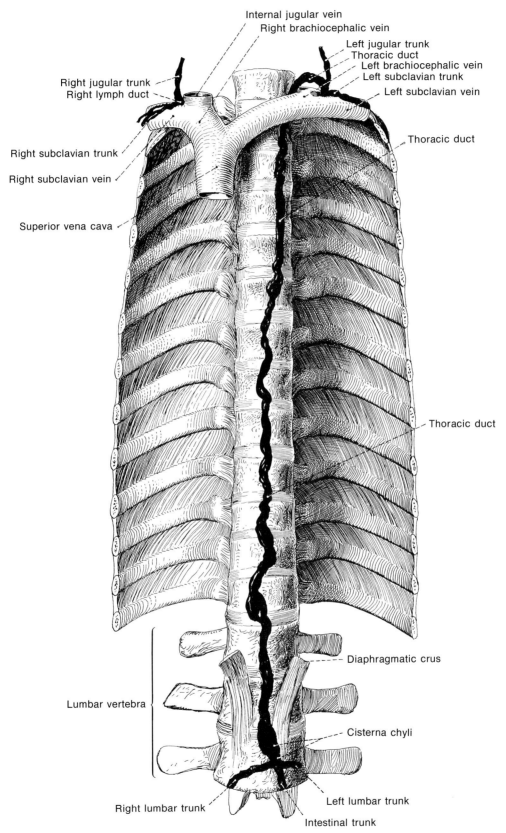

FIG. 10-2. *Thoracic duct*

LYMPHATICS OF THE HEAD AND NECK

The lymphatic nodes (fig. 10-3) in the inferior part of the neck really represent a convergence of a superior part of the axillary and the paratracheal glands of the mediastinum. The chief aggregations are about the internal jugular vein and in the submandibular region, but there are small scattered groups in association with the parotid gland, the external jugular vein, thyrohyoid, laryngeal, and occipital regions.

The lymph vessels of the scalp and forehead pass, for the most part, to the occipital, postauricular and preauricular nodes. These then drain to those of the internal jugular group; those of the eyelid, zygomatic region, auricle, and lacrimal and parotid glands to the preauricular nodes; of the nose, cheek, lips, and chin, and of the submandibular gland to the *submandibular nodes*. Some vessels of all these areas, however, pass directly to the nodes of the *deep jugular chain*. Vessels of the middle ear drain to the preauricular and inter-

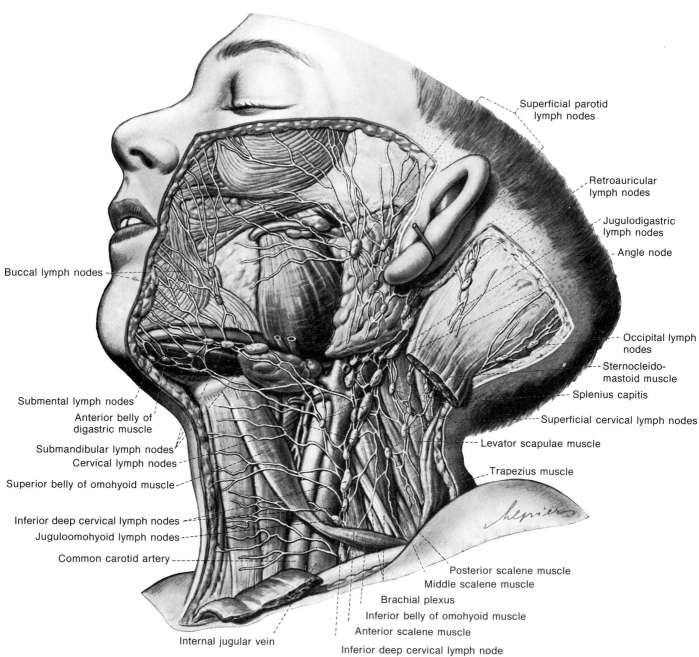

FIG. 10-3A. *Superficial lymphatics of the head and neck*

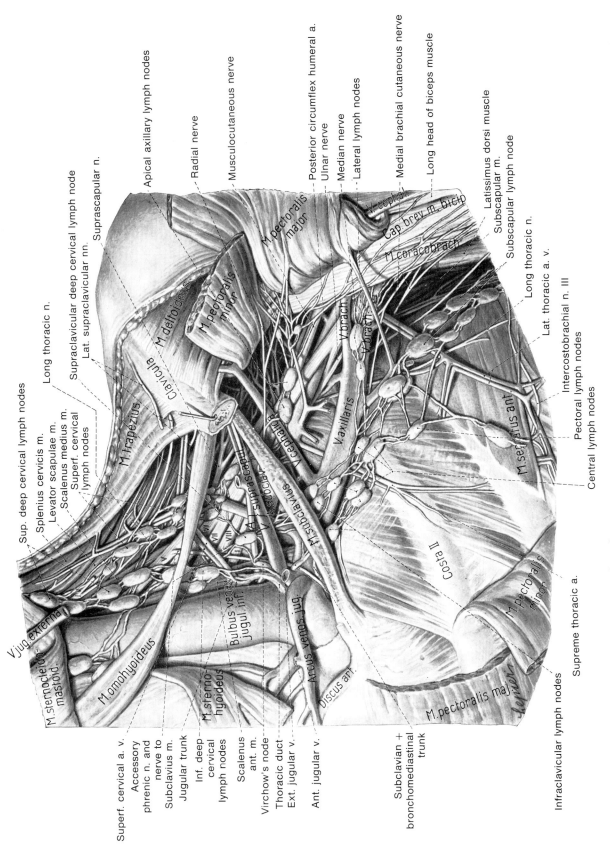

FIG. 10-3B. *Deep lymphatics of the head and neck*

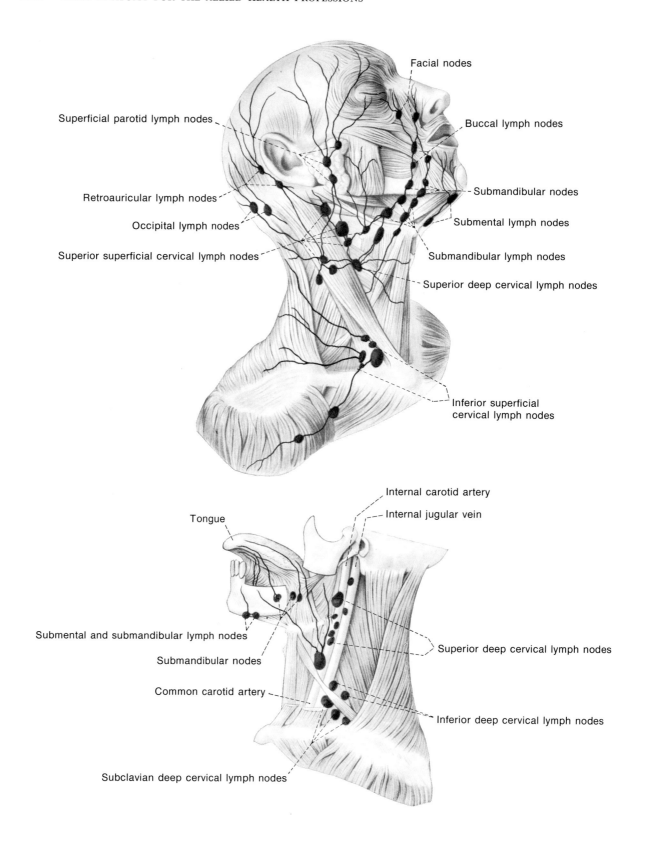

FIG. 10-3C. *Schematic diagram of the lymphatics of the head and neck*

nal jugular nodes; of the nasal cavity to the *submandibular* and *laryngeal nodes;* of the mouth and tongue to the submandibular and *internal jugular nodes;* and of the pharynx to the laryngeal and internal jugular nodes. The lymph vessels of the anterior half of the tongue do not cross the midline, but those of the posterior half do. The vessels of the neck drain, with an obliquely inferior course, toward the nearest nodes.

LYMPHATICS OF THE UPPER LIMBS

Most of the *lymph nodes of the upper limbs* (fig. 10-4) are associated with the axillary and subclavian veins in the axillary fossa and just superomedial thereto, beneath the clavicle. A few may accompany the superior part of the brachial vein. In addition there occur smaller superficial nodes associated with the basilic vein (*cubital* and *supratrochlear*) and the superior part of the cephalic vein (*deltopectoral*). Those of the *axillary group* are numerous and large. Some of them drain directly into the right lymphatic trunk, and others communicate with the more inferior nodes of the internal jugular chain.

The lymph vessels of the upper limb course proximally for the most part in the superficial fascia, a few first to the cubital, supratrochlear, and deltopectoral nodes, but most of them directly to the axillary nodes, which also receive superficial vessels from the side of the thorax. On the medial side of the brachium some of the vessels pierce the deep fascia to accompany the brachial vein.

LYMPHATICS OF THE THORAX

The *lymphatics of the thorax* (fig. 10-5*A* and *B*) may be divided into the *superficial group* and the *deep group*. The former, in the ventral, lateral, and dorsal aspects, drain for the most part into the *axillary nodes*, but in the pectoral region some of them drain into the *subclavian nodes*. There are, however, accessory drainage routes that must be taken into consideration. Some of the flow from the pectoral region is to the deeper nodes associated with the *internal thoracic vein*, and vessels near the midline may drain into the axillary nodes of the opposite side. These facts are of particular importance in neoplasms of the mammary gland.

The deep lymph nodes of the thorax may be said to occur in three groups; the ventral, the inferior, and the dorsal. The *ventral* or *parasternal nodes* occur in association with the internal thoracic vein. The *inferior* or *diaphragmatic group* (fig. 10-5*A*) comprises posterior nodes, which are really the most inferior of the dorsal series; anterior nodes, the most inferior of the ventral series; and middle nodes close to the phrenic nerve. The *dorsal* group is by far the largest and occurs as *paratracheal* (and pulmonary) and posterior *mediastinal nodes,* about the esophagus (fig. 10-5*B*). Those at the hilus of the lung are numerous, and in this area they are often darkened by inspired particulate material. The most inferior nodes of the internal jugular chain (or the most medial of the subclavian nodes) are often included with those of the thorax.

The deeper lymphatic vessels of the thorax are numerous. They comprise vessels of (chiefly the muscular part of) the diaphragm, draining to the nearest of the inferior group of nodes, mostly the middle, and connecting with vessels upon the abdominal surface of the diaphragm and of the liver; of the parietal pleura, to the axillary, subclavian, and thoracic nodes of all three groups; of the visceral pleura and lungs, to the pulmonary nodes of the dorsal group; and of the mediastinal structures, including the pericardium and heart, to the adjoining nodes of the dorsal, ventral, and inferior groups, except that probably no lymph from the heart flows to the nodes of the inferior group. The vessels of the abdominal surface of the diaphragm are of particular importance, for it is chiefly by them that fluid in the peritoneal cavity is absorbed.

LYMPHATICS OF THE ABDOMEN

As in the thorax, the *lymphatics of the abdomen* (fig. 10-6*A*–*C*), including the pelvis, may be divided into a *superficial group* and a *deep group*. The superficial lymph vessels of the superior part of the abdominal wall really belong with those of the thorax, for it must be remembered that the division of the trunk into thorax and abdomen is secondary, and the more fundamental division would be into supra- and infraumbilical portions of the trunk. These vessels empty, for the most part, into *axillary, subclavian,* and *parasternal nodes*. These, however, freely communicate with vessels of the more inferior part of the abdomen, which together with the superficial vessels of the lumbar region and of the external genitalia, discharge into the more *superficial* of the *inguinal nodes*, which in turn communicate with the deeper nodes scattered throughout this region.

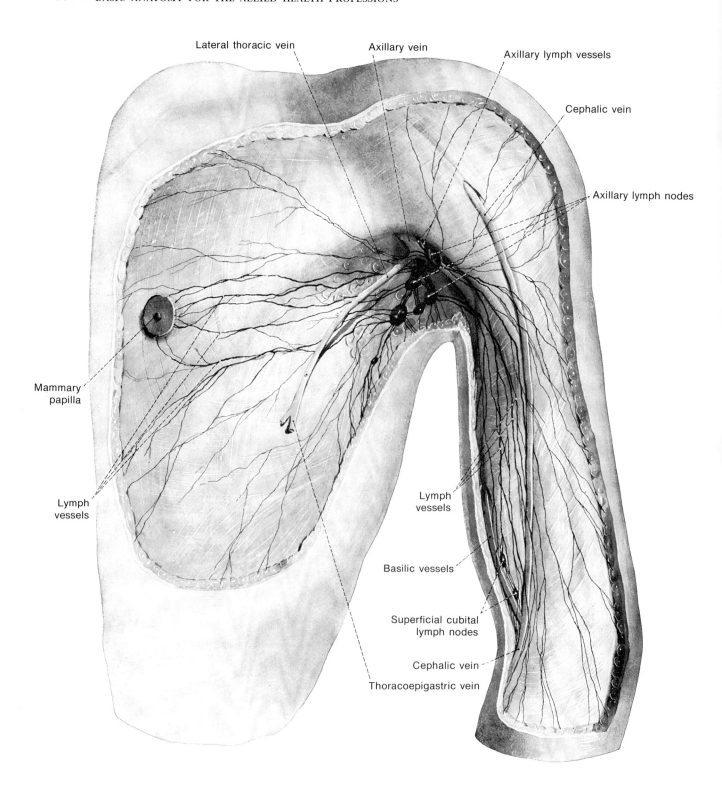

FIG. 10-4. *Lymphatics of the upper limb*

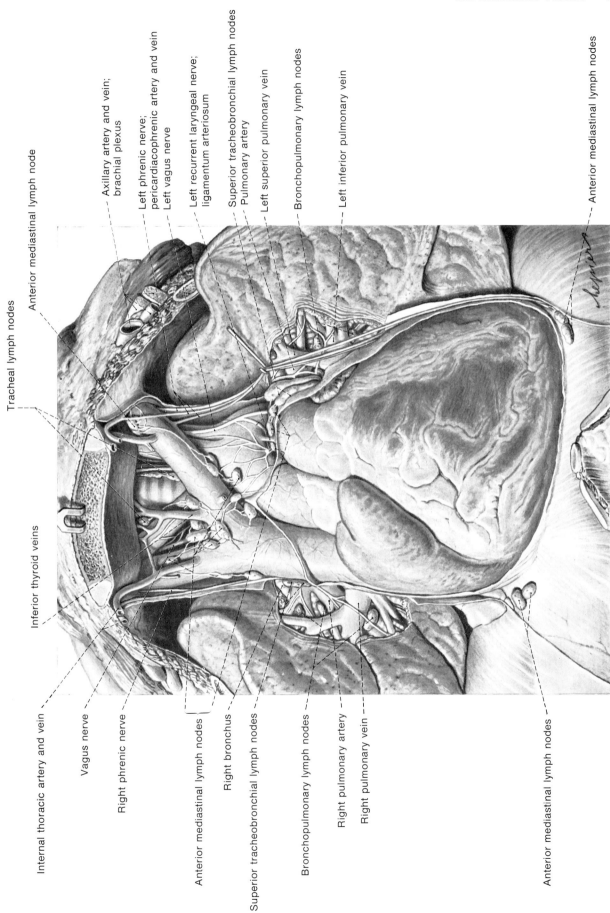

FIG. 10-5A. *Lymphatics of the thorax (anterior view)*

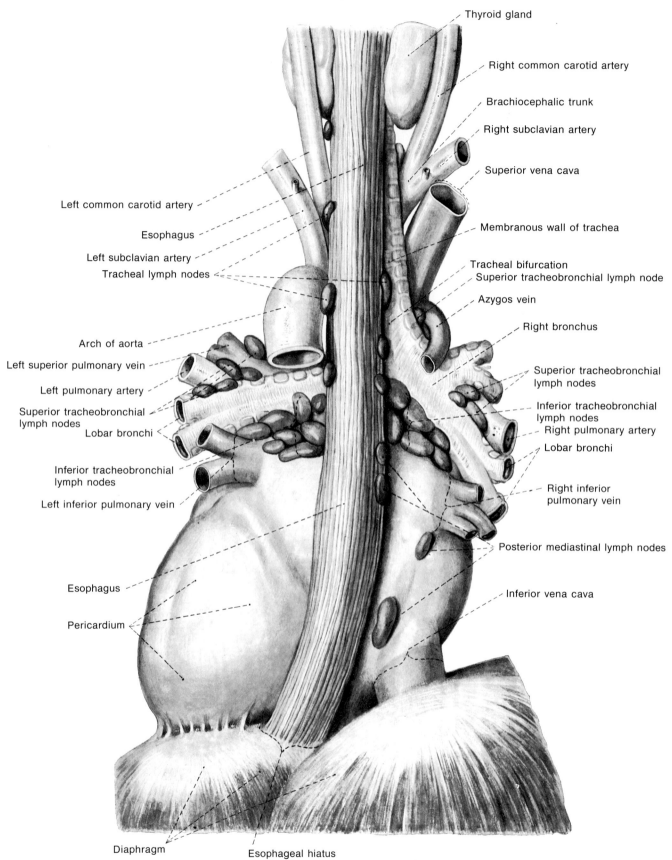

FIG. 10-5B. *Lymphatics of the thorax (posterior view)*

The deep lymph nodes of the abdomen are mostly clustered about the great vessels: the *common iliac* and *aorta* and their larger branches (fig. 10-6A). Some of them are associated particularly with the *celiac, superior mesenteric* and *inferior mesenteric* arteries, with a sparse chain of nodes following these vessels to the liver, stomach, and spleen, and in the mesenteries (fig. 10-6B) to the intestine, sigmoid colon, and rectum. Several nodes also occur in relation to the middle sacral vessels, and between the rectum and bladder (*anorectal* group) (fig. 10-6C).

The deeper lymphatic vessels of the abdomen are particularly numerous about the larger blood vessels, where they converge to the nodes and pass from them to other nodes and to the thoracic duct. The vessels of the various structures drain as follows. Those of the liver drain in part into the vessels of the diaphragm, but chiefly to the nodes about the hepatic artery and bile duct, and then to the celiac axis. The vessels from the pylorus and superior part of the stomach drain to the *cardiac* (the inferior of the esophageal) *nodes* in the lesser omentum, and these to the *celiac* group of nodes; while vessels of the inferior part of the stomach drain to nodes along the greater curvature, in the greater omentum, and to the spleen and then the celiac nodes. Part of the vessels of the pancreaticoduodenal region drain to the *hepatic* (and then to the celiac) and the remainder to the *superior mesenteric nodes*. The lymph vessels of the small intestine (except the duodenum) drain at first to nodes at the periphery of the mesentery, these partly or wholly into intercalated nodes, and these into nodes about the base of the superior mesenteric artery. In the ileocecal region (including the appendix), drainage is into nodes situated in the mesentery superior to the ileocecal junction, and then into the superior mesenteric group. Vessels of the descending and sigmoid colons have a pattern similar to those of the rest of the intestine except that they drain finally into the inferior mesenteric nodes. Vessels of the rectum form a network communicating with the anal vessels and flowing to the inguinal nodes, but drainage is chiefly to the anorectal and then to the middle sacral nodes.

Vessels of the suprarenal gland drain for the most part into adjacent paraaortic nodes, but on the left side they also drain to the vessels of the diaphragm. Those of the kidney drain to the lumbar paraaortic nodes, while those of the ureter have a similar but more extensive drainage, to the iliac nodes as well. Vessels of the anterior part of the bladder drain to a node near the femoral ring, and of the remainder, to deeper nodes of the iliac group. The lymph vessels of the superficial parts of the external genitalia drain to the *inguinal nodes,* but there is some communication with the vessels of the deeper parts of the external and of the internal genitalia, which drain to the iliac nodes.

LYMPHATICS OF THE LOWER LIMBS

The *lymphatic nodes of the lower limb* (fig. 10-7A–C) are mostly clustered about the inguinal region, the majority in the superficial fascia just below the inguinal ligament and a lesser number of smaller nodes medial to the femoral vein, deep to the fascia lata in the femoral triangle. Small, scattered nodes also occur along the courses of the deep vessels of the leg, particularly in the popliteal space.

The lymph vessels of the lower limb are considered to comprise a superficial and a deeper series. The former course from the plexus on the sole (fig. 10-7A) and the lateral aspect of the limb obliquely, superiorly and medially to the vicinity of the great saphenous vein and to the inguinal region, draining to the more *superficial inguinal nodes* (fig. 10-7C). To the inguinal nodes also course the superficial vessels of the buttock, perineum, anus, and those of the superficial parts of the external genitalia, as well as from the inferior abdomen.

The deeper vessels converge to accompany the deeper veins. Some of those of the crus drain into the *popliteal nodes* (fig. 10-7B), and these, as well as most of those of the thigh, empty into the deeper nodes of the femoral triangle. The deep vessels of the buttock and obturator regions, however, accompany the gluteal and obturator veins to terminate in the iliac and hypogastric nodes.

SPLEEN

The *spleen* (fig. 10-8) is located in the upper left quadrant inferior to the diaphragm. Histologically, it resembles a lymph node. The spleen acts as a reservoir for blood during times of stress. It also produces red blood cells in the fetus. It processes lymphocytes and monocytes and destroys microorganisms, platelets, and aging red cells.

THYMUS

The *thymus* (fig. 10-9) is a bilobed pinkish gland situated high in the thoracic region posterior to the sternum. The thymus usually stops growing by the

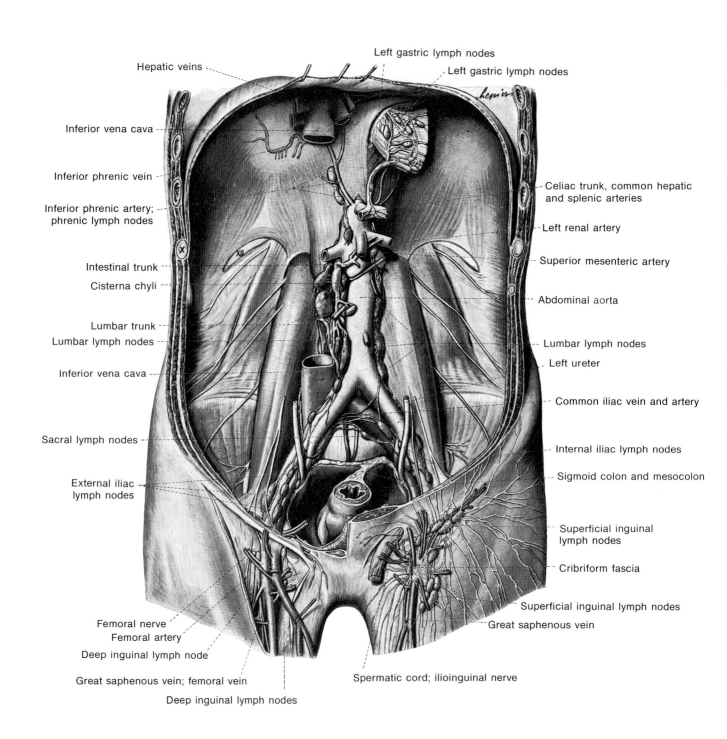

FIG. 10-6A. *Lymphatics of the posterior abdominal wall*

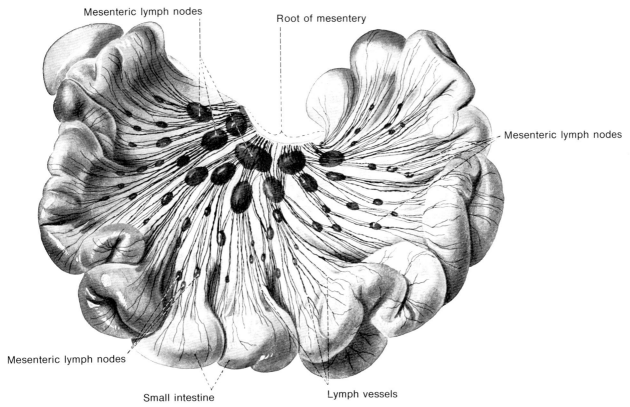

FIG. 10-6B. *Lymphatics of the small intestine*

age of puberty. It is the source of T-cell lymphocytes, the cells concerned with the rejection of skin grafts and delayed hypersensitivity.

TONSILS

The *tonsils* (fig. 10-10) are collections of lymphoid tissue located in the nasal and oral pharynx. The *palatine tonsils* are located in the oral pharynx between the arches of the fauces. The *pharyngeal tonsils* (adenoids) are located in the posterior wall of the nasopharynx. The *lingual tonsils* are situated at the base of the tongue. These masses of lymphoid tissue form a circular defense around the nasal and oral pharynx.

CLINICAL APPLICATIONS

Tonsillitis is usually caused by bacterial invasion of the palatine tonsils. The tonsils become red, swollen, and painful. The tonsillar crypts fill with pus, causing difficulty in swallowing.

The inflammation of lymph nodes is called *lymphadenitis*. The usual agents are staphlococci and streptococci. The nodes may enlarge and become extremely painful.

Lymphangitis is inflammation of the lymphatic vessels. The most common type of pathogens are staphlococci and streptococci, which may enter the blood stream to cause blood poisoning.

Lymphomas are tumors of the lymphatic tissue. The cause is probably related to a virus.

Hodgkin's disease is a lymphoma that commonly begins in a cervical lymph node and spreads to adjacent nodes.

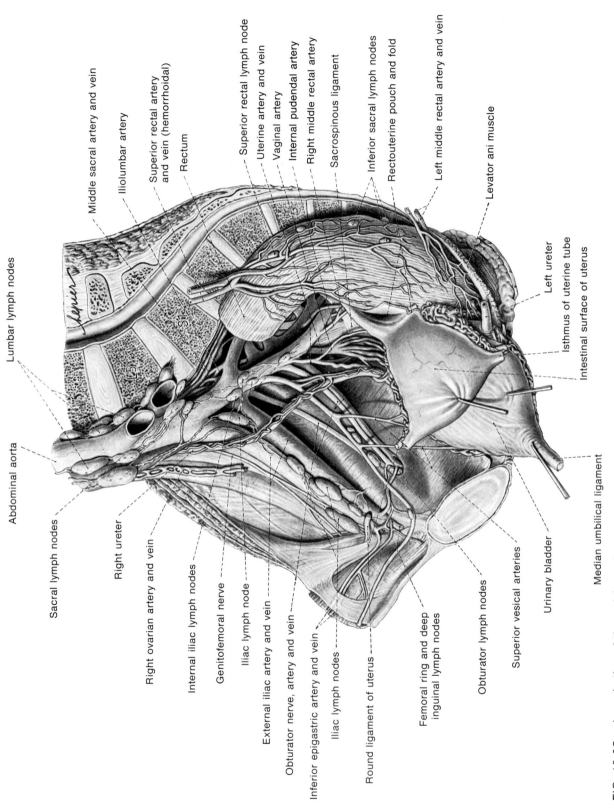

FIG. 10-6C. *Lymphatics of the pelvic cavity*

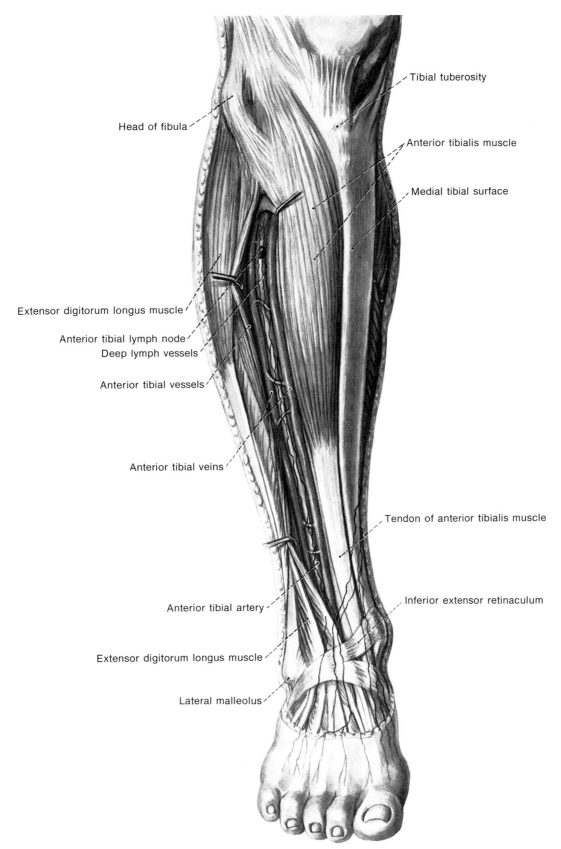

FIG. 10-7A. *Lymphatics of the lower limb*

362 ■ BASIC ANATOMY FOR THE ALLIED HEALTH PROFESSIONS

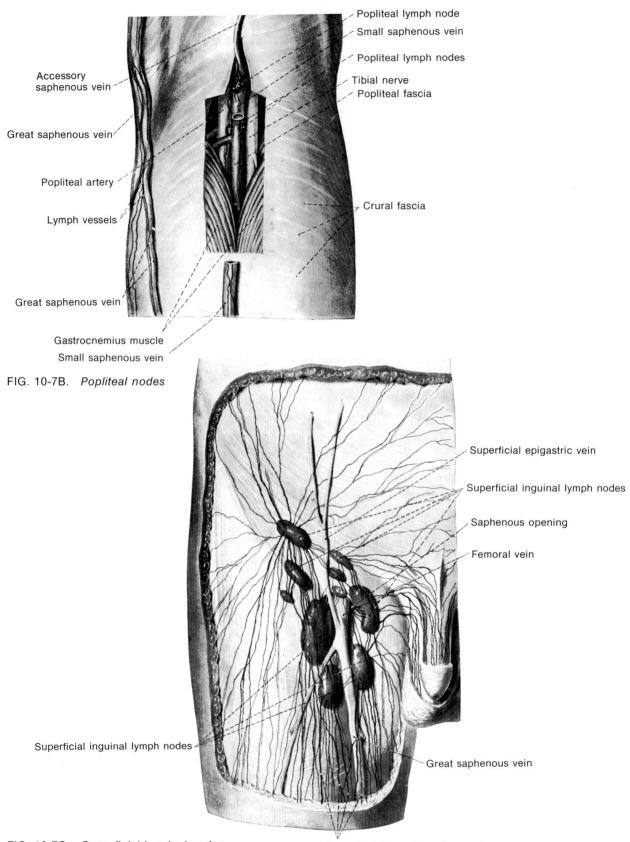

FIG. 10-7B. *Popliteal nodes*

FIG. 10-7C. *Superficial inguinal nodes*

THE LYMPHATIC SYSTEM ■ 363

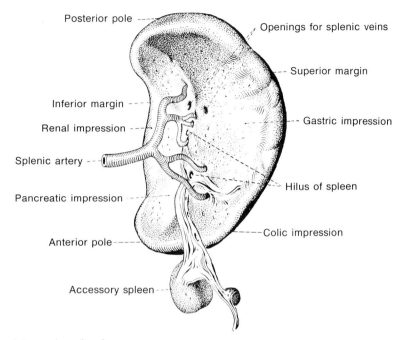

FIG. 10-8A. *Spleen (visceral surface)*

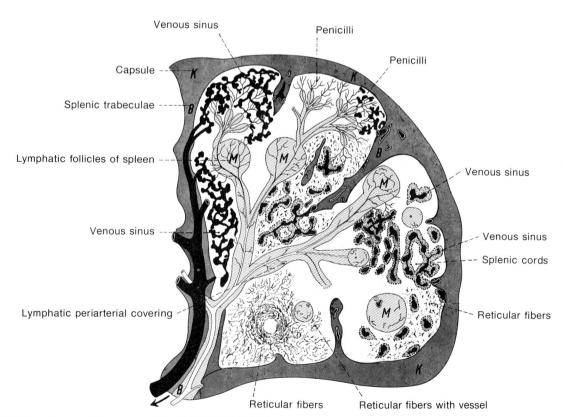

FIG. 10-8B. *Spleen (diagrammatic microscopic view)*

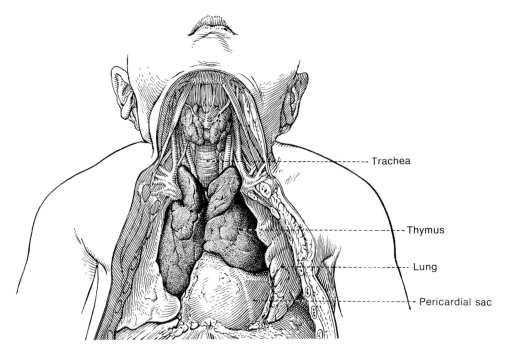

FIG. 10-9A. *Thymus (newborn)*

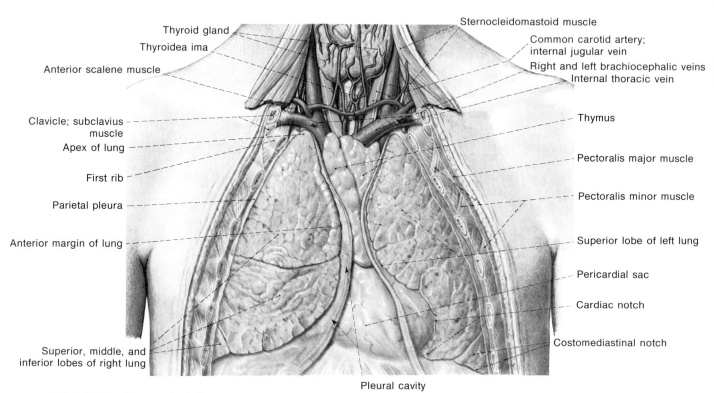

FIG. 10-9B. *Thymus (adult)*

THE LYMPHATIC SYSTEM ■ 365

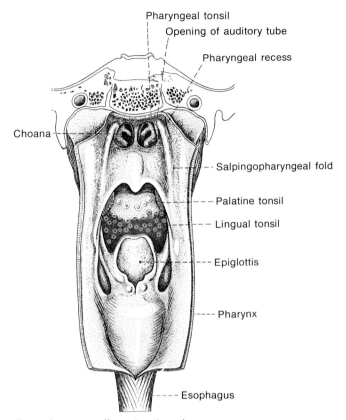

FIG. 10-10A. *Tonsils (tonsils and surrounding structures)*

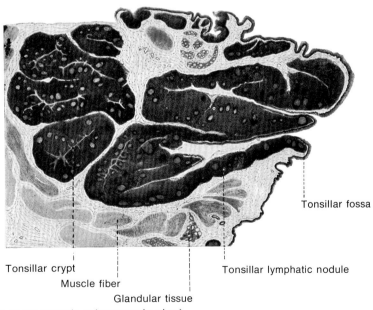

FIG. 10-10B. *Tonsils (diagrammatic microscopic view)*

REVIEW QUESTIONS:

1. The _____ continues superiorly in the posterior mediastinum as the thoracic duct proper.
2. The _____ conduct lymph from the right side of the head and neck, the right arm, and the neighborhood of the right shoulder.
3. The chief aggregations of lymphatics of the head and neck are located about the _____ vein and in the _____ region.
4. The lymphatic channels of the _____ half of the tongue do not cross the midline.
5. The deep lymph nodes of the abdomen are mostly clustered about the _____.
6. Vessels of the suprarenal gland drain for the most part into adjacent _____ nodes.
7. The lymphatic nodes of the lower limb are mostly clustered about the _____.
8. The _____ processes lymphocytes and monocytes, and destroys microorganisms, platelets, and aging red cells.
9. The _____ is a bilobed pinkish gland situated high in the thoracic region posterior to the sternum.
10. The _____ are collections of lymphoid tissue located in the nasal and oral pharynx.
11. List three masses of lymphoid tissue that form a circular defense around the nasal and oral pharynx.
12. The lymph vessels of the external genitalia drain to the _____ nodes.

11. The Respiratory System

> STUDENT OBJECTIVES
> After you have read this chapter, you should be able to:
> 1. Identify the organs of the respiratory system.
> 2. Contrast the three regions of the pharynx and relate their functions in respiration.
> 3. Discuss the anatomical features of the larynx related to sound production.
> 4. Identify the structures that form the bronchial tubes.
> 5. Identify the coverings of the lungs.
> 6. Identify the bronchopulmonary segments.
> 7. Discuss the gross anatomical features of the lung.
> 8. Discuss the role of alveoli in respiration.

The *respiratory system* consists of the *nasal cavities, larynx, trachea, bronchi,* and *lungs.* The lungs are the essential element, the larynx chiefly the mechanism whereby food and liquids are excluded from the lungs, while the remainder are air passages. The *pharynx* is fundamentally a part of the digestive system that is shared by the respiratory system to connect the nasal cavities with the larynx.

The nose is formed by nasal cartilage and bone (fig. 11-1*A*).

NASAL CAVITIES

The *nasal cavity* (fig. 11-1*B* and *C*) is situated between the cranial cavity and the mouth, and between the external and the internal nares or choanae, by which it communicates with the nasopharynx. It is divided by the *nasal septum* into two *nasal fossae*, which are rarely bilaterally symmetrical in the adult. The septum is composed of the *ethmoid* and *vomer* bones and the *septal cartilage.* Each nasal fossa is

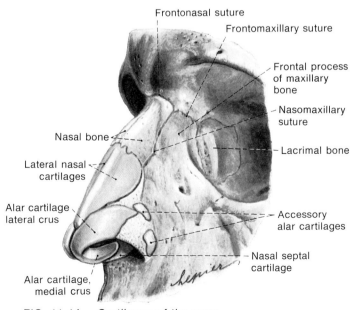

FIG. 11-1A. *Cartilages of the nose*

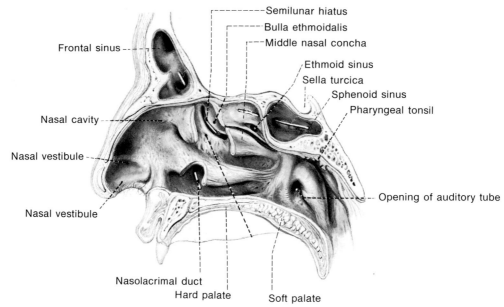

FIG. 11-1B. *Lateral wall of the nasal cavity*

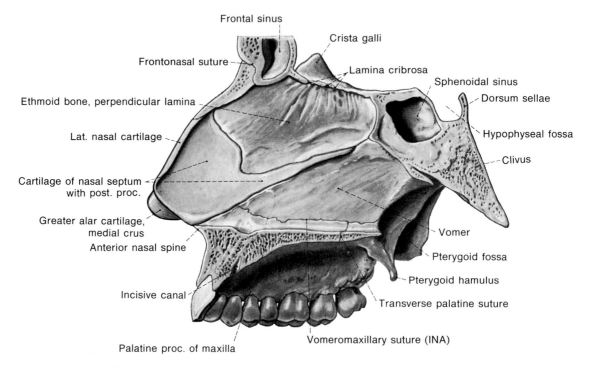

FIG. 11-1C. *Nasal septum*

lined with mucous membrane (except for the vestibule), beneath which is the periosteum (or perichondrium). The lining of the superior part of the cavity, both laterally and medially, constitutes the olfactory area, covered with thin *olfactory epithelium,* while the remainder is lined with *respiratory epithelium,* rich in mucous glands. Particularly over the inferior and middle conchae this is thick and extremely vascular and subject to much swelling during infections (common colds).

Projecting from the lateral wall of the nasal fossa are the nasal conchae, of variable size; a large and long *inferior,* a shorter *middle,* and a very short *superior concha,* which is at times crowded by an unusually large middle concha until it is almost obliterated. The free space inferior to the inferior concha is the *inferior nasal meatus,* while the *middle meatus* is inferior to the middle, and the *superior meatus* inferior to the superior concha.

Just within each external aperture is a superior dilatation, the *nasal vestibule,* which supports a growth of hairs in older individuals, particularly males. The *sphenoethmoidal recess* adjoins the aperture of the sphenoidal sinus; the *posterior nasal sulcus* is posterior to the inferior and middle conchae. The orifice of the *auditory tube* is in the lateral wall of the posterior nasal cavity. It is sickle-shaped and the posterior margin overlaps the anterior.

At the anterior margin is a prominence caused by the levator veli palatine muscle, and the posterior margin projects slightly because of the underlying cartilage of the auditory tube. This projection is the *torus tubarius,* and the raised margin continues inferiorly as the *salpingopharyngeal fold.* The nasopharynx continues posteroinferior to this level.

The *paranasal sinuses* (fig. 11-1*D*) comprise *frontal, maxillary, sphenoidal,* and variable *ethmoid sinuses* considered to occur in two groups, anterior and posterior. Their passages open into the nasal cavity mostly beneath the conchae, as does the nasolacrimal duct. Beneath the inferior concha is the *nasolacrimal duct,* usually rather close to the attachment of the concha. Beneath the middle concha is the *semilunar hiatus,* which is variable, but is typically a linear recess into which usually opens the duct of the frontal sinus. Into this, or close by, opens the duct (or ducts) of the anterior ethmoidal cells. The prominence posteriorly bordering the infundibulum is the *bulla ethmoidalis,* and that anteriorly, the *uncinate process.* The duct of the maxillary sinus also opens beneath the middle concha, either close to the attachment of the middle concha or, often, more inferiorly. The ducts of the posterior ethmoid cells usually open into the superior meatus. The duct of the sphenoidal sinus opens into or near the sphenoethmoidal recess, in the angle between the two bones indicated by that name.

The ducts of the sinuses are of such size that the cilia of their epithelial lining effect their drainage, this action being assisted by gravity in some cases. The duct of the sphenoidal sinus, and particularly of the maxillary sinus, is not at the lowest point of the cavity, and hence exudates are more prone to accumulate in the maxillary than in the frontal sinus.

The nerve supply of the nasal cavity comprises special sensory innervation by the olfactory nerve to the superior part of the cavity, and somatic sensibility reaching the cavity by two pathways: nasociliary twigs of the ophthalmic nerve from the orbit, and branches of the maxillary nerve by way of the pterygopalatine ganglion, comprising nasopalatine and palatine twigs. Parasympathetic fibers, from the facial nerve, reach the mucosa via the pterygopalatine ganglion. Sympathetic fibers follow the course of the blood vessels.

The arterial supply of the nasal cavity is chiefly by the sphenopalatine branch of the maxillary artery, supplying most of the concha, the septum, and the anterior sinuses (but the maxillary sinus is also supplied by infraorbital and superior alveolar twigs). The descending palatine branch of the maxillary also supplies the floor of the fossa, while the superior part is supplied by the anterior and posterior ethmoidal branches of the ophthalmic artery, and the anterior part by nasal and superior labial twigs of the facial

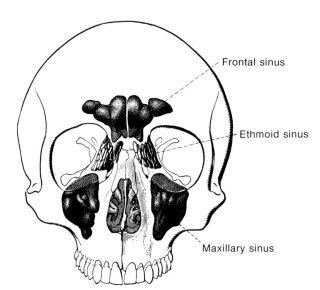

FIG. 11-1D. *Paranasal sinuses*

artery. The essentials of the venous drainage correspond in plan to that of the arteries.

LARYNX

The *larynx* (table 11-1) has the primary function of excluding food and liquids from the trachea, and the secondary function of producing sounds. In both functions it may be said that practically the entire laryngeal mechanism participates. The larynx is a tubular structure composed of cartilage, membrane, and muscle, communicating below with the trachea and above with the pharynx by the *laryngeal aperture,* bounded ventrally by the *epiglottis,* prolonged superiorly, and laterally by the *aryepiglottic folds.*

Cartilages of the larynx (fig. 11-2A) comprise three unpaired (*epiglottic, thyroid, cricoid*) and three paired (*arytenoid, corniculate, cuneiform*) units, the last two of which are small. The *epiglottic cartilage* is situated anterior to the laryngeal aperture, at the root of the tongue and superior to the thyroid cartilage. It is leaf-shaped and to its margins are attached the *aryepiglottic folds* (fig. 11-2B). The *thyroid cartilage* is the largest of the group, broad and folded, the fold being in the midline and constituting the laryngeal prominence, which is better developed in adult males and in them is known as the Adam's apple. It has a *superior thyroid notch* in the midline, and paired *lateral laminae,* each with an *oblique line* marking the attachment of muscles. The dorsal border of each lamina continues as a *superior cornu,* attached by a ligament to the greater cornu of the hyoid bone, and an *inferior cornu,* articulating with the cricoid cartilage. The *cricoid cartilage,* situated inferior to the thyroid and superior to the first tracheal ring, surrounds the tracheal tube. It is narrow ventrally and broad dorsally. It articulates with the thyroid and arytenoid cartilages. The *arytenoid cartilages* are situated, one on either side, beneath (dorsal to) the thyroid cartilage and are concerned with the vocal apparatus. Each has a *base* and a superior curved apex, terminating in a *vocal process,* for the attachment of the *vocal ligament.* Articulating with the apex are the small *corniculate* and *cuneiform cartilages,* which are responsible for the tubercles of the same name on the aryepiglottic folds.

The laryngeal cartilages are bound together by ligaments, membranes, and muscles (fig. 11-2C–E). The chief membranes are the *thyrohyoid,* between the thyroid cartilage and hyoid bone, and the *conus elasticus,* which is attached to the superior margin of the cricoid cartilage inferiorly and to the vocal ligaments superi-

TABLE 11-1: Muscles of the Larynx

MUSCLE	ORIGIN	INSERTION	INNERVATION	ACTION
Cricothyroid	Cricoid cartilage	Thyroid cartilage	Superior laryngeal	Tenses the vocal cords
Posterior cricoarytenoid	Cricoid cartilage	Arytenoid cartilage	Inferior laryngeal	Dilates the glottis
Lateral cricoarytenoid	Cricoid cartilage	Arytenoid cartilage	Inferior laryngeal	Narrows the glottis
Transverse arytenoid	Arytenoid cartilage	To the same part on the opposite side	Inferior laryngeal	A sphincter muscle
Oblique arytenoid	Arytenoid cartilage	Opposite arytenoid cartilage	Inferior laryngeal	A sphincter muscle
Aryepiglottic	Arytenoid muscle in aryepiglottic folds	Oblique	Inferior laryngeal	A sphincter muscle
Thyroarytenoid	Thyroid cartilage	Arytenoid cartilage	Inferior laryngeal	Regulates the tension in the Vocalis muscle.
Vocalis	From medial fibers of thyroarytenoid muscle	Along vocal ligament	Inferior laryngeal	Regulates tension in the Vocalis muscle
Thyroepiglottic	Continuation of thyroarytenoid muscle		Inferior laryngeal	A sphincter muscle

THE RESPIRATORY SYSTEM ■ 371

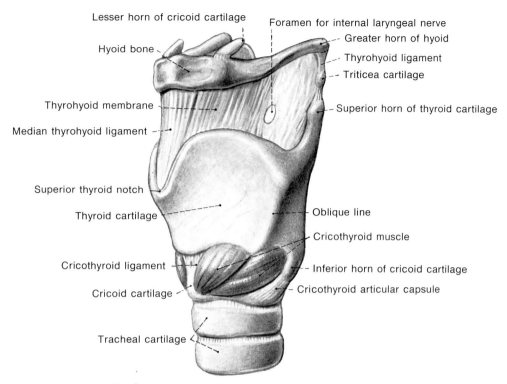

FIG. 11-2A. *External view of the larynx*

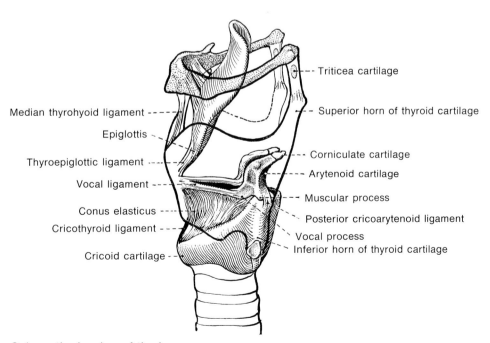

FIG. 11-2B. *Schematic drawing of the larynx*

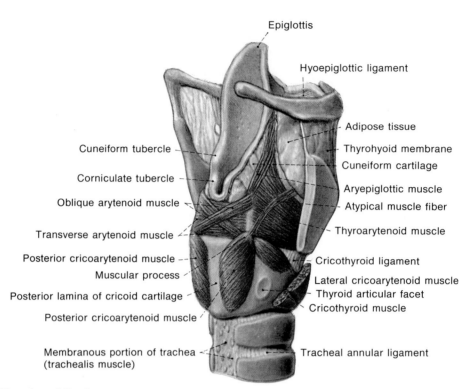

FIG. 11-2C. *Muscles of the larynx*

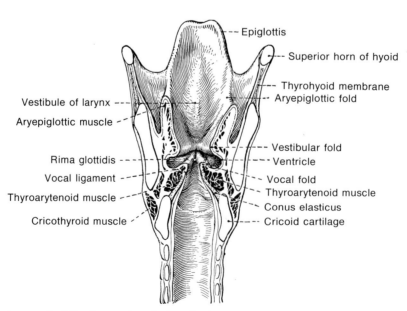

FIG. 11-2D. *Internal aspect of the larynx (posterior view)*

THE RESPIRATORY SYSTEM ▪ 373

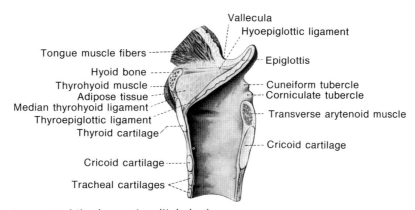

FIG. 11-2E. *Internal aspect of the larynx (sagittal view)*

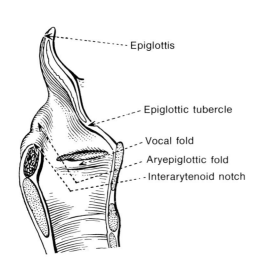

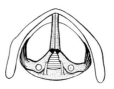
Vocal folds, closed

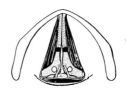
Vocal folds, closed; lateral cricoarytenoid muscle, contracted

Vocal folds, abducted; posterior cricoarytenoid muscle, contracted

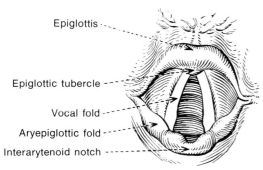

FIG. 11-2G. *Abducted vocal folds*

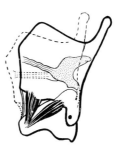
Cricothyroid muscle shortened, pulling thyroid cartilage forward to increase tension on vocal cords

FIG. 11-2F. *Vocal folds*

orly (fig. 11-2F). These ligaments constitute the thickened border of the conus elasticus, extending from the superior thyroid notch of the thyroid cartilage, diverging to the vocal processes of the arytenoid cartilages, thus forming, with the investing mucous membrane, the vocal cords, medial to the vocalis muscle.

The cavity of the larynx, lined with mucous membrane, has a slightly expanded *vestibule* between the laryngeal aperture and the constriction of the *vestibular folds*. The latter constitute the *false vocal cords*, superior to the true vocal cords and separated from them by another, but short, transverse expansion caused by the *ventricles*, one on either side. These vary in depth. The two vocal cords, with the intervening triangular space (*rima glottidis*) constitute the *glottis*. Inferior to the glottis the larynx becomes tubular as it joins the trachea.

Action of the larynx is of two sorts. In vocalization the vocal cords are tensed and approximated to different degrees by the *cricoarytenoid* and *thyroarytenoid* muscles. The vocal cords of man, however, are not essentially different from those of most mammals; speech is made possible by the better nervous control of the vocal cords, vestibule, the tongue, and the more mobile lips. The swallowing action is complicated. Preparation for deglutition is voluntary; its accomplishment involuntary. Foods or liquids are pushed

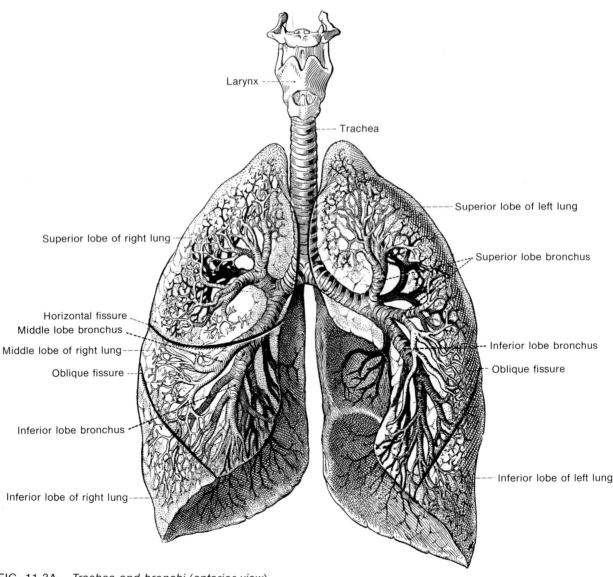

FIG. 11-3A. *Trachea and bronchi (anterior view)*

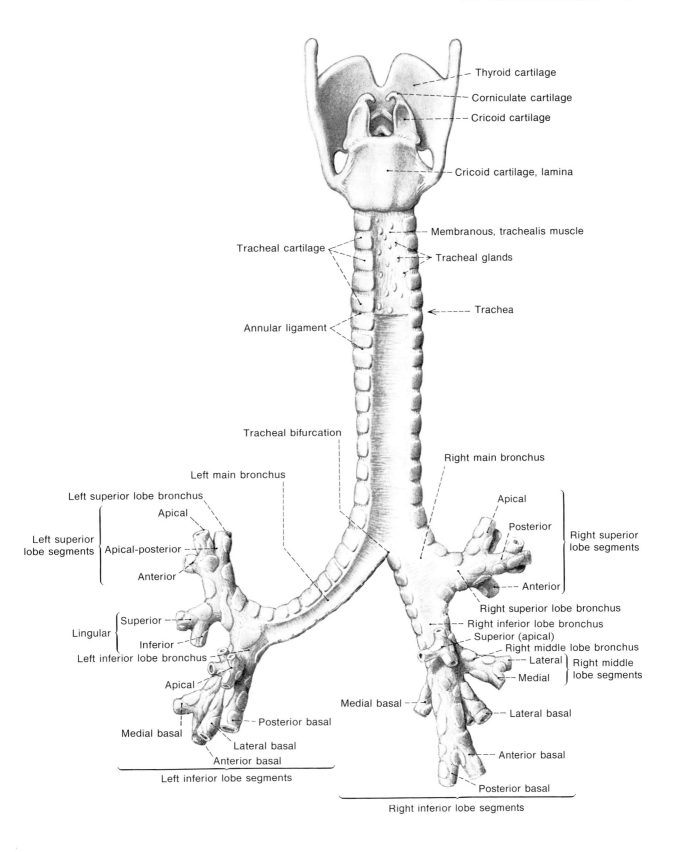

FIG. 11-3B. *Trachea and bronchi (posterior view)*

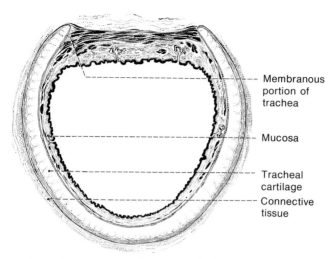

FIG. 11-3C. *Trachea* (cross-section)

by the tongue through the pillars of the fauces into the pharynx, at which point the action becomes involuntary. The choanae are occluded by the elevation of the soft palate, largely by the levator veli palatini muscle, and this assists in raising the larynx. The part played by the epiglottis is not precisely understood. It seems certain that it does not fold down over the laryngeal aperture as one might expect. The closure of the larynx is effected by the crowding action of the base of the tongue coupled with the elevation of the larynx, obliteration of the vestibule by constriction of its walls, and approximation of both the vestibular folds and the vocal folds. The food thus passes by the laryngeal aperture into the esophagus, through which it is forced by peristaltic action. Passage into the stomach may or may not be delayed for a short space by the action of the musculature of the distal esophagus. Vomiting is accomplished by violent contraction of the stomach and reversal in direction of the peristaltic action of the esophagus.

Arterial supply of the larynx is by the *superior laryngeal* and by the *cricothyroid* branch of the *superior thyroid artery*, both from the *external carotid artery;* and by the *inferior laryngeal* branch of the *inferior thyroid artery*. Venous drainage is by the laryngeal branches of the *superior* and *inferior thyroid veins*. Innervation is by the autonomic system, both sympathetic and parasympathetic. The latter is carried by the *vagus nerve,* involving taste (from the epiglottis), visceral sensibility, and branchiomeric motor by the *inferior* (recurrent) *laryngeal nerve*. The cricothyroid muscle is innervated by the *superior laryngeal* branches of the vagus nerve.

TRACHEA AND BRONCHI

The *trachea* (fig. 11-3A and B) is a cylindrical tube, slightly flattened dorsally, that extends as a continuation of the larynx into the mediastinum, where it bifurcates to form the *bronchi*. It lies in the median plane, ventral to the esophagus, from the sixth cervical to about the fifth thoracic vertebra. Bordering it in the neck are the carotid sheaths, and more inferiorly the recurrent laryngeal nerves, while the thyroid gland rests upon it (ventral to it). In its inferior part the arch of the aorta lies ventral and then to the left of it, while the three great vessels of the arch have a similar relationship to it. To the right is the superior vena cava and the right brachiocephalic vein.

The *trachea* is composed of a series (16–20) of incomplete cartilaginous rings, describing some two-thirds of a circle, with the hiatus directed dorsally. The bronchi have a similar structure. The rings are held slightly apart from each other by an investing fibroelastic membrane, and the whole is lined with mucous membrane. Transverse smooth muscle fibers join the ends of the rings (fig. 11-3C).

The two main *bronchi* extend inferolaterally from the trachea, the right being slightly shorter, of larger caliber, and more vertical, so that it is usually the one in which foreign bodies may lodge. The bifurcation lies ventral to the esophagus, the aorta arches over the left bronchus and the left pulmonary artery crosses central to it, while the azygos vein arches over the right bronchus. All of the left bronchus is crossed by the left pulmonary artery, but the right artery crosses only the main part of the right bronchus, as the eparterial bronchus branch is above it.

THORACIC CAVITY

The *thoracic cavity* (fig. 11-4A) is the part of the body cavity superior to the diaphragm. It is enclosed by the ribs, vertebral column, and sternum, and contains the lungs and the mediastinal contents.

The *mediastinum* (fig. 11-4B) is the space in the thoracic cavity between the pleura, bounded sagittally by the vertebral column and the sternum. For convenience it is customary to regard the space as divisible into superior and inferior portions, of which the inferior is again subdivided into anterior, middle, and posterior parts. The *superior mediastinum,* superior to the level of the pericardium, contains the great vessels of the region (including intercostal vessels and parts of the azygos system), thoracic duct, thymus gland, trachea, esophagus, phrenic, vagus, left recurrent

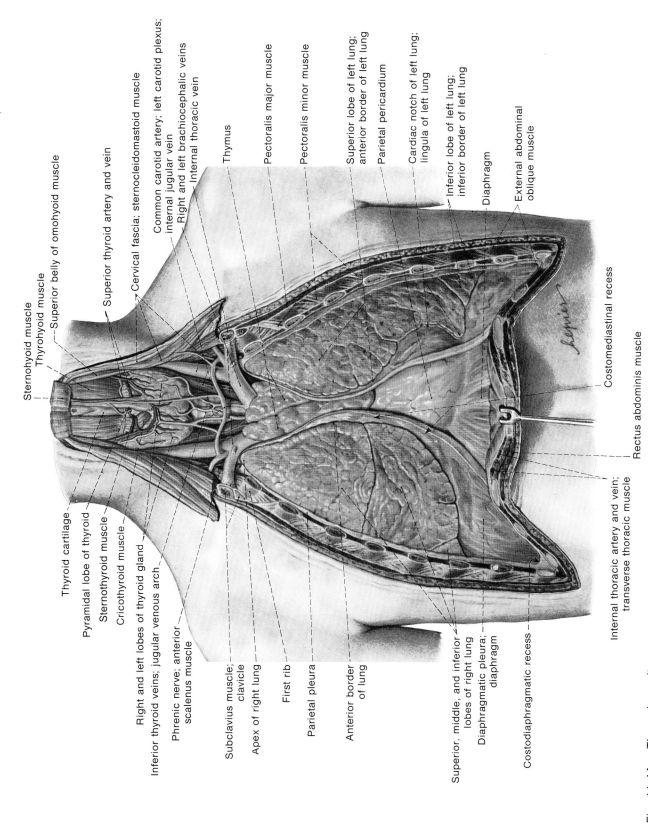

Fig. 11-4A. *Thoracic cavity*

378 ■ BASIC ANATOMY FOR THE ALLIED HEALTH PROFESSIONS

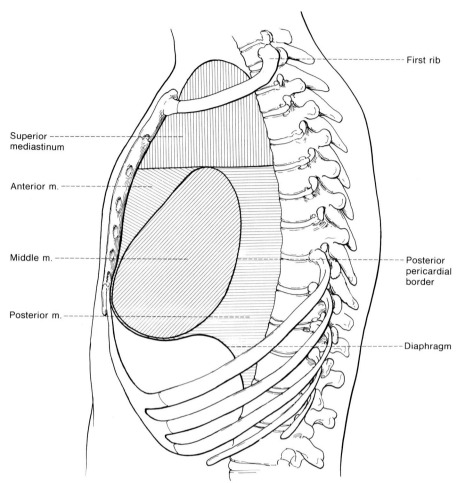

FIG. 11-4B. *Mediastinum*

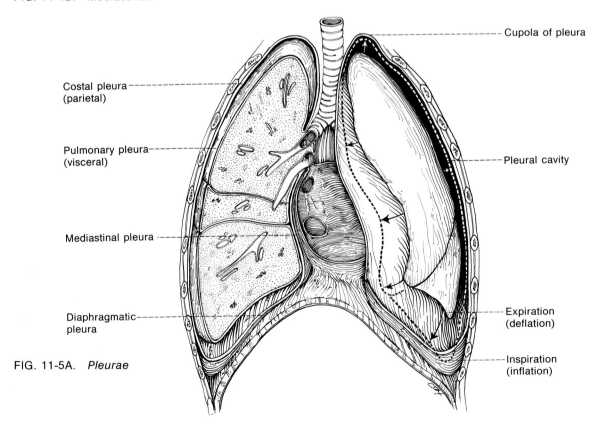

FIG. 11-5A. *Pleurae*

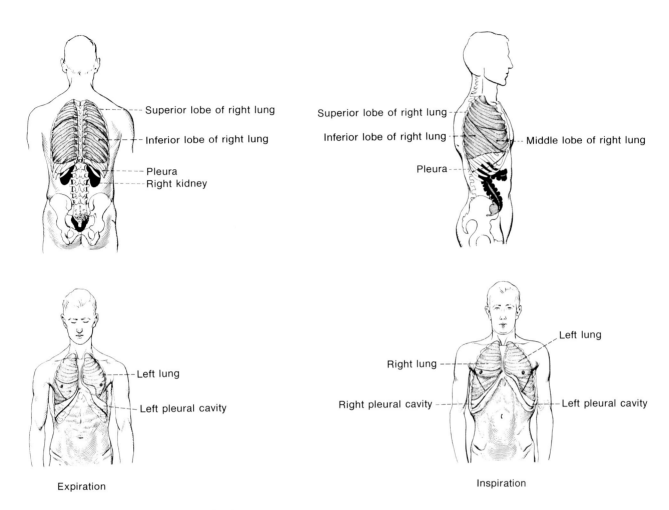

FIG. 11-5B. *Pleurae during respiration*

laryngeal, cardiac nerve branches, and lymph nodes. The *anterior mediastinum*, ventral to the pericardium, contains connective tissue, internal thoracic vessels, a few lymph nodes, and the ventromedial folds of the pleura. The *middle mediastinum* contains the pericardium and its contents and the phrenic nerves. The *posterior mediastinum,* dorsal to the pericardium, shares many of the structures occurring in the superior mediastinum. It contains the esophagus, descending aorta and its local branches, parts of the azygos system, lymph nodes, thoracic duct, vagus nerves, and local sympathetic ganglia and nerves.

PLEURAE

The *pleurae* (fig. 11-5A) are a pair of closed serous sacs formed by the walls of the pericardial and pleuroperitoneal parts of the celomic cavity, which the developing lungs invade. Thus each lung is covered, except at the hilus, by a closely adherent visceral pleura, which is reflected at the hilus as the parietal pleura, lining the pleural cavity. Except when lesions are present the two layers are free of each other (fig. 11-5B).

The *parietal pleura* is attached by loose connective tissue (often toughened by lesions) to the costae, the diaphragm, and the pericardium. Accordingly it is said to be divided into these corresponding regions. An additional area is often recognized—the *cervical pleura*, elevated by the apex of the lung through the arc of the first rib. The *costal pleura* covers the inner surface of the ribs and usually of the sternum, particularly upon the right side. The *diaphragmatic pleura* covers the lateral part of the diaphragm not occupied by the pericardium, not, however, completely filling the fissure between the diaphragm and thoracic wall. It is here reflected upon the costal pleura at an acute angle. The *mediastinal pleura,* except at the hilus of the lung, is the medial boundary of the mediastinum,

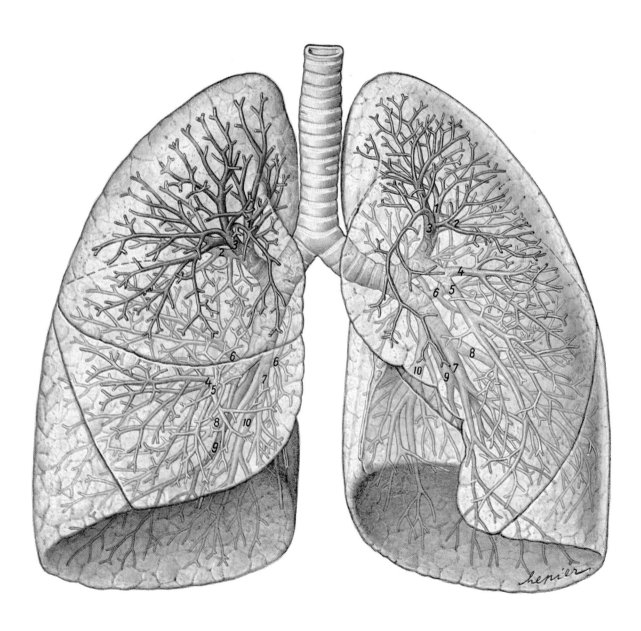

FIG. 11-6A. *Right and left lungs. The ten segmented bronchi on each side are designated with different colors and numbered 1–10.*

conforming to the curved surface of the pericardium. It is continuous with the costal pleura, uninterrupted dorsolaterally, and over a sharp angle ventrally.

The *visceral* or *pulmonary pleura* is so closely adherent to the substance of the lungs that even small areas of it cannot be removed without tearing. It continues into the fissures, thus dividing the lung into lobes. The visceral pleura is reflected around the hilus or root of the lung into the parietal pleura. There is thus a bare area at the hilus (potential rather than actual), through which the pulmonary vessels and bronchus pass, comparable to the bare spot of the liver. This area is not circular but is elongated superoinferiorly, with a narrow prolongation inferiorly, so that here the dorsal and ventral reflected borders approach contact. This narrow feature is known as the *pulmonary ligament*.

LUNGS

Grossly, the *lungs* (fig. 11-6) are masses of highly elastic tissue, spongy because of contained air spaces (*alveoli*), that communicate by *bronchioles* with the bronchi. Through the substance of the lungs blood vessels ramify, in addition to the bronchial subdivisions. The lungs comprise two units, lying on either side of the mediastinum and within the pleura, occupying the pleural cavities. The right lung is slightly larger and wider, but shorter, than the left, partly because of the position of the pericardium and partly because the right dome of the diaphragm is higher than the left. The *apex of the lung* reaches the level of the neck of the first rib, and is therefore superior to the clavicle, while the lung extends inferiorly to the approximate level, dorsally, of the neck of the 11th

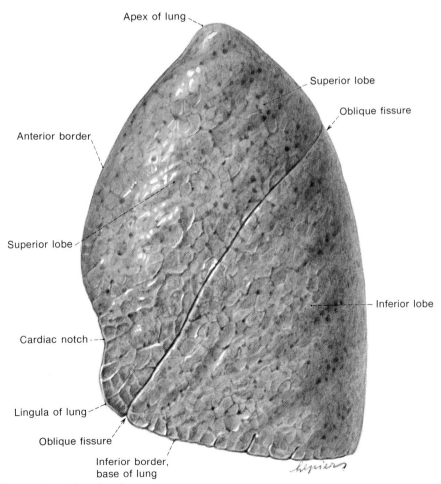

FIG. 11-6B. *Left lung (costal surface)*

rib. The anterior and *diaphragmatic margins* of the lungs, however, do not usually extend to the limits of the adjoining reflected pleural margins. Upon the diaphragm, each lung has a base that tapers irregularly to a superior apex, in addition to a *costal* and a *mediastinal surface;* the latter is concave, chiefly to conform to the shape of the pericardium. Upon the right side this surface has a groove that curves around the superior border of the hilus, for the azygos vein; another dorsally for the esophagus; and a third upon the superior part for the superior vena cava. Upon the mediastinal surface of the left lung the aortic groove curves around the superior part of the hilus, with a groove for the subclavian artery superiorly, and an esophageal groove inferiorly. Each lung has a deep interlobular *oblique fissure* extending from near the apex dorsally, obliquely to the lateroventral part of the diaphragmatic surface, and penetrating almost to the hilus. The part dorsal and lateral to the fissure is the *inferior lobe.* In the left lung the entire part ventrosuperior to the fissure is the *superior lobe,* but in the right this portion is further subdivided by another interlobular, more transverse, *horizontal fissure,* into a *superior* and a *middle lobe.* Thus the left lung has two lobes and the right lung, three.

The *root* or *hilus* of the lung is the pedicle by which the bronchus and vessels enter and leave the lung. There is thus at this point a *bare area,* devoid of pleura. It is shaped slightly differently upon the right and left sides. On the right the bronchial branches are situated dorsally, the venous branches ventrally and inferiorly, and the arterial branches superiorly, between bronchi and artery. Upon the left the artery is situated superiorly, the upper vein ventrally, and the lower vein inferiorly, while the bronchial branches are between them.

Within the lung the bronchi branch and redivide in treelike fashion, in intimate association with a similar

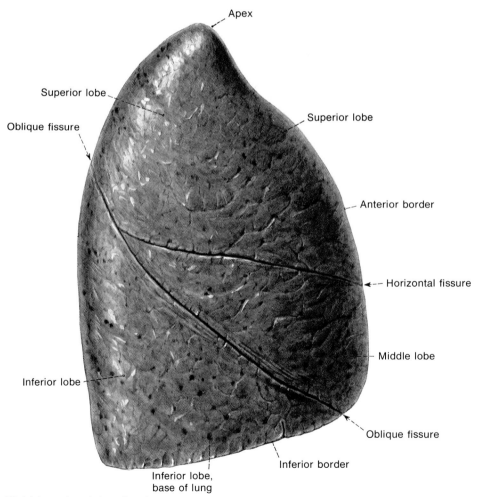

FIG. 11-6C. *Right lung (costal surface)*

arrangement of arterial and venous branches. The highest branch of the *right bronchus* arises superior to the point at which the bronchus is crossed by the pulmonary artery, and this is termed the *eparterial bronchus*. All other branches occur below this level, upon both right and left sides.

The lung tissue sinks in water before birth and floats after birth, because of imprisoned air. The lung cannot be completely emptied of air. It is said that only the peripheral 3 mm of the lung tissue is used in quiet breathing, and 30 mm in forced breathing. In breathing during rest only about 300 cc of air is involved. The amount varies individually, but a fair average in a male may be said to involve some 1600 or 1700 cc additional in forced expiration, and a like amount in forced inspiration, so that the vital capacity is in the neighborhood of 3600 cc. But the lungs contain another 1600 cc or so of air, the residual capacity, that one cannot expel. It is the loss of a part of this that causes discomfort when one has "the wind knocked out" of him. The lungs themselves are, of course, entirely passive agents. It is a partial vacuum within the thorax that causes inspiration, and pressure within the thorax that results in expiration. If air is admitted through the thoracic wall the adjacent lung will promptly collapse.

Bronchial arteries from the aorta or intercostal arteries, two for the left lung and one for the right, supply the pulmonary pleura, lung tissue, and bronchi, while the pulmonary arteries bring blood to the lungs for oxygen and to the alveoli. The pulmonary veins carry away oxygenated blood from the alveoli and alveolar ducts, while the bronchial veins drain to the brachiocephalic veins. The intercostal, azygos, hemiazygos, and pulmonary veins carry oxygenated blood from the rest of the lung tissue and from the pulmonary pleura. The smooth musculature of vessels, bronchi, and the alveolar sphincters of the lungs

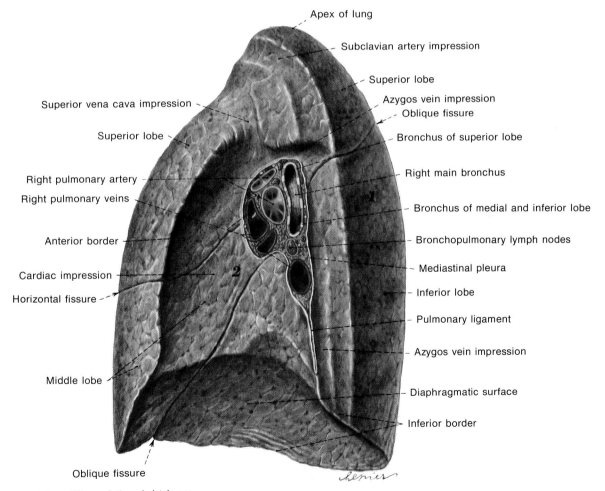

FIG. 11-6D. *Hilus of the right lung*

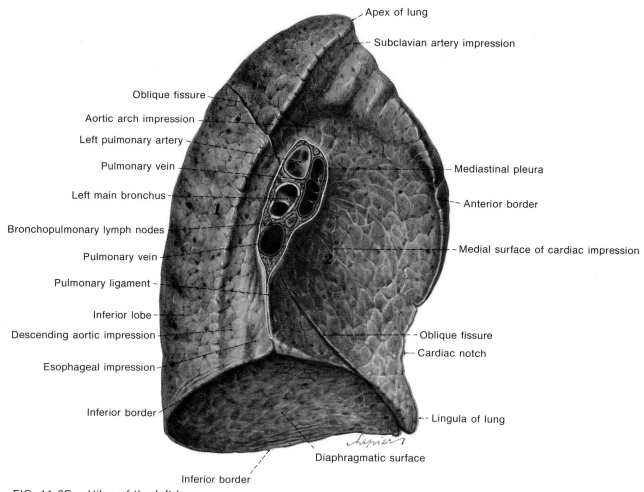

FIG. 11-6E. *Hilus of the left lung*

are supplied by sympathetic (of the second, third, and fourth thoracic segments) and pulmonary branches of the vagus, through the pulmonary plexuses. The lungs, pulmonary pleura, and serous coat of the parietal pleura are devoid of sensibility. The external surface of the parietal pleura is supplied with sensory fibers by intercostal nerves, and inferiorly, by the phrenic nerve.

CLINICAL APPLICATIONS

Hypoxia is associated with a reduction of oxygen in the body tissues.

Pneumonia is an inflammation of the bronchioles and the alveoli. Pneumonia is often caused by an encapsulated microorganism known as *Diplococcus pneumoniae*.

Abnormal breathing may be described by any one of several terms. *Dyspnea* is caused by obstruction in the air passages. *Tachypnea* is rapid breathing and *orthopnea* is the inability to breathe in the horizontal position.

The *common cold* is caused by numerous types of viruses and may involve the upper respiratory tract or extend to the larynx and trachea.

Influenza is caused by several types of viruses that cause inflammation of the respiratory mucous.

Numerous diseases can be directly related to the respiratory system, such as *pulmonary tuberculosis, diphtheria, psittacosis, emphysema, bronchial asthma,* and *cancer.*

THE RESPIRATORY SYSTEM ■ 385

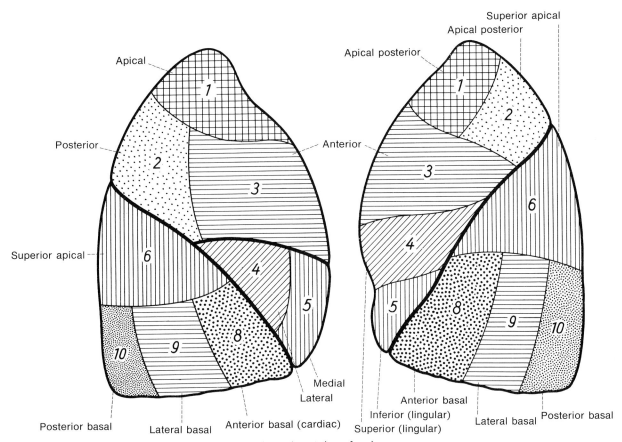

Fig. 11-6F. *Bronchopulmonary segments of the lung (costal surface)*

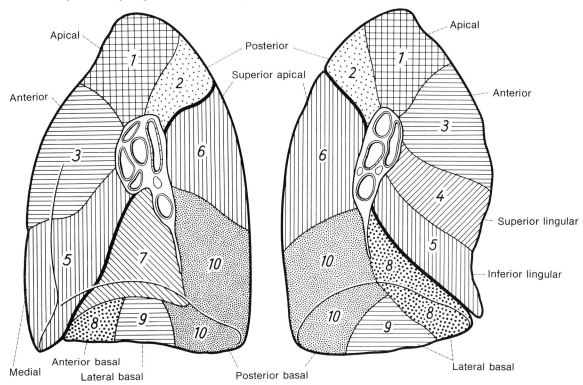

FIG. 11-6G. *Bronchopulmonary segments of the lung (mediastinal view)*

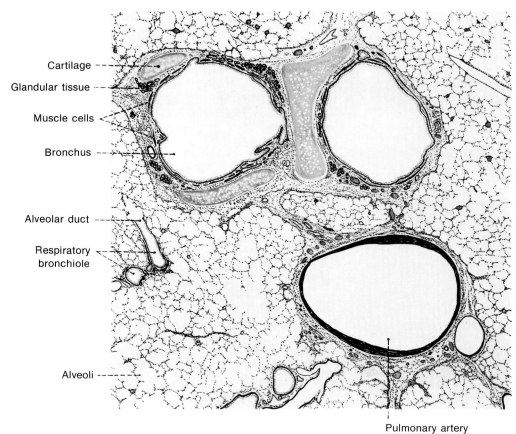

FIG. 11-6H. *Microscopic view of the adult lung*

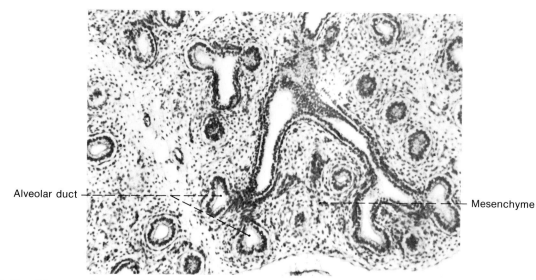

FIG. 11-6I. *Microscopic view of the fetal lung*

REVIEW QUESTIONS:

1. The nasal septum is composed of the _____ and _____ bones and the _____ cartilage.
2. The free space inferior to the inferior concha is the _____.
3. The paranasal sinuses comprise _____, _____, _____, and _____.
4. The frontal sinus usually opens into the _____.
5. Cartilages of the larynx comprise _____ unpaired and _____ paired.
6. The _____ cartilage is situated anterior to the laryngeal aperture, at the root of the tongue and superior to the thyroid cartilage.
7. The two vocal cords, with the intervening triangular space, _____, constitute the _____.
8. The _____ is a cylindrical tube that extends as a continuation of the larynx into the mediastinum.
9. The two main bronchi extend inferolaterally from the trachea, the _____ being slightly shorter, of larger caliber, and _____.
10. The _____ is the space in the thoracic cavity between the pleura.
11. The apex of the lung reaches the level of the neck of the _____ rib.
12. Each lung has a deep interlobular _____ fissure.
13. The highest branch of the right bronchus arises superior to the point at which the bronchus is crossed by the pulmonary artery, and this is termed the _____ bronchus.
14. The bronchial arteries may arise from the _____ or _____ arteries.
15. The _____ of the lung is the pedicle by which the bronchus and vessels enter and leave the lung.
16. List the two serous layers of pleura.
17. The _____ mediastinum contains the pericardium and its contents and the phrenic nerve.
18. The vestibular folds constitute the _____ vocal cords.
19. The _____ is attached to the superior margin of the cricoid cartilage inferiorly and to the vocal ligaments superiorly.
20. Articulating with the apex of the arytenoid cartilages are the small _____ and _____ cartilages.

12. The Urinary System

> **STUDENT OBJECTIVES**
> After you have read this chapter, you should be able to:
> 1. Identify the external and internal morphology of the kidneys.
> 2. Identify the anatomical features of a nephron.
> 3. Trace blood through the kidney.
> 4. Identify the collecting tubules of the kidney.
> 5. Describe the location of the kidneys, ureters, bladder, and urethra.
> 6. Describe the internal aspect of the bladder and urethra.

The *urinary system* (fig. 12-1) consists of *paired kidneys,* for the production of urine, and *paired ureters,* an *unpaired bladder,* and the *urethra* for its transport and rejection by the body.

KIDNEYS

The *kidney* is the chief organ assisting in the maintenance of a constant composition of the inorganic and organic constituents of the plasma and of the tissue fluids. It is located in the dorsal part of the abdominal cavity, dorsal to the peritoneum, with the left unit usually slightly superior to the right, and the inferior border 1–2 inches superior to the highest point of the iliac crest. It is not firmly anchored but is held in place largely by surrounding structures, most of which are themselves yielding. The kidney is capped by the adrenal gland, is surrounded by a loose fascial envelope in which there is some fat, and is adjoined by much soft *perirenal fat.* Dorsally the kidney adjoins the diaphragm, psoas major, and quadratus lumborum muscles. Ventrally the right kidney is adjoined by the liver superiorly, and the right flexure of the colon and second part of the duodenum inferiorly; the left kidney is adjoined by the stomach, pancreas, and to a slight extent by the spleen.

The kidney (fig. 12-2A) is a bean-shaped structure of smooth texture about $4\frac{1}{2}$ inches in length, 2 inches in breadth, 1 inch in thickness, and $\frac{1}{2}$ lb in weight. The lateral border is convex, the ends rounded, and the medial border concave and indented. It is enclosed in a tough, glistening, translucent, fibrous capsule, which continues around the medial border to the interior, through the depression known as the *renal sinus.* At the sinus, the ureter and other vessels enter and leave, hence the margin of the sinus constitutes the *hilus* of the kidney. In longitudinal section it is seen that the structure of the kidney consists of a peripheral *cortex* and a deeper *medulla,* which contains the *pyramids.* Adjacent pyramids are separated from each other by *renal columns.* The *base* of each pyramid is toward the capsule of the kidney, and the *apex* of each ends in the *papilla,* projecting into the sinus of the kidney.

The *nephron* is the functional unit of the kidney. It is estimated that approximately one million nephrons exist in each kidney. The chief features include the *renal corpuscle* and the *renal tubule.* The corpuscle includes a capillary tuft called the *glomerulus,* which

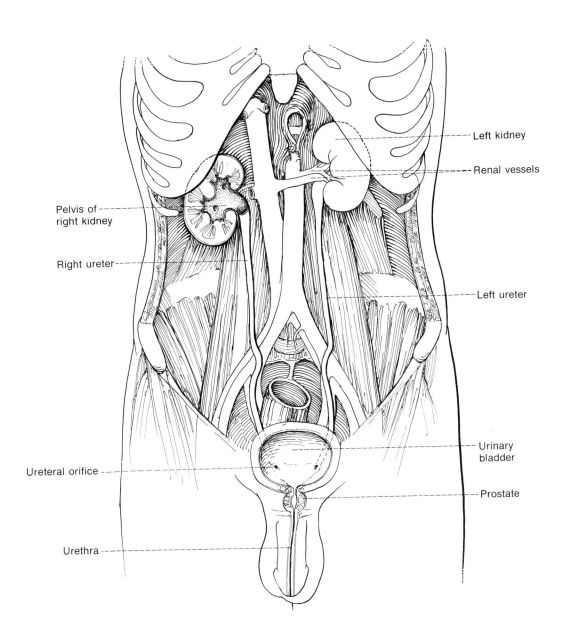

FIG. 12-1A. *Urinary system (diagrammatic view)*

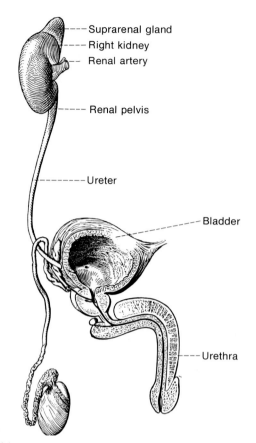

FIG. 12-1B. *Urinary system (male)*

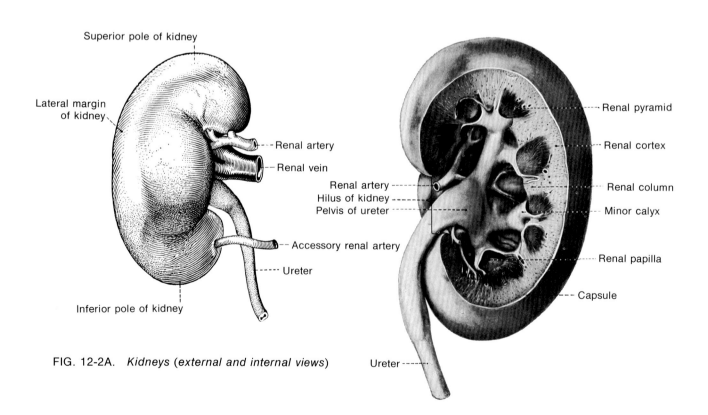

FIG. 12-2A. *Kidneys (external and internal views)*

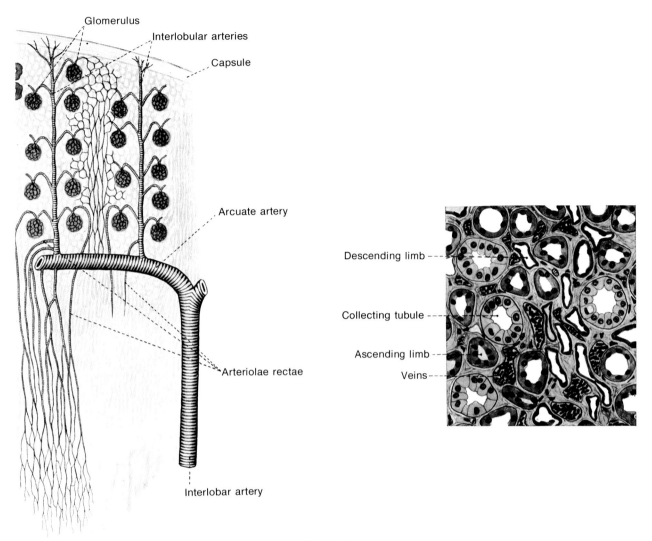

FIG. 12-2B. Kidneys (diagrammatic and microscopic view)

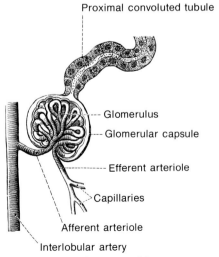

FIG. 12-2C. Kidneys (diagrammatic view of a renal corpuscle)

is enclosed within a double-walled membrane called *Bowman's capsule*. This capsule is the dilated and invaginated blind end of the *proximal convoluted tubule*. The *distal convoluted tubule* enters a *collecting tubule*, which joins other collecting tubules to form the renal pyramids (fig. 12-2*B* and *C*).

Blood reaches the nephron from the *renal artery* via the *interlobar artery, arcuate artery, interlobular artery,* and the *afferent arteriole.* The *efferent arteriole* carries the blood away from the glomerulus to the capillary networks that surround the tubules. The blood then returns to the renal vein via the *interlobular vein,* the *arcuate vein,* and the *interlobar vein.*

URETERS

The *ureter* (fig. 12-3) is the excretory duct from kidney to bladder. Its most proximal part constitutes the *minor calices,* attached to the fibrous capsule of the kidney around the papillae, at the summit of which open the excretory tubules of the kidney. The minor open into two to three *major calices,* which in turn converge to form the pelvis of the ureter. The part of the pelvis outside the sinus is located dorsal to the renal vessels and decreases in diameter as it curves to follow an inferior course as it becomes the ureter, ventral to the psoas major muscle and dorsal to the peritoneum. It is crossed by the testicular or ovarian vessels, crosses the common iliac vessels and enters the pelvic cavity, where it lies ventral to the internal iliac artery and is crossed by the ductus deferens (in males), and then enters the lateral side of the inferior border of the bladder.

The arterial supply of the kidney is chiefly by the renal artery. Venous drainage is by the renal vein to the vena cava. The proximal part of the ureter is supplied by the renal and testicular or ovarian arteries, and the distal part by vesicular and middle rectal branches of the internal iliac artery, with corresponding venous drainage. Innervation of these structures is entirely autonomic. Sympathetic fibers to the kidney are derived from the lesser and least splanchnic nerves via the celiac and renal plexuses, and to the ureter by the renal and inferior mesenteric plexuses.

URINARY BLADDER

The *urinary bladder* is a distensible vesicle situated dorsal to the symphysis pubis and ventral to the rectum in the male, and the uterus in the female. Superiorly it is covered by peritoneum and folds of the small intestine. The relations, as well as the shape, change with the distention of the bladder. There is a shallow fold of peritoneum, the *rectovesical pouch,* situated superiorly between the rectum and bladder in the male, and a deep, slitlike *rectouterine pouch,* between the rectum and uterus in the female.

The bladder (fig. 12-4*A* and *B*) has a *base* or *fundus* dorsally and a *vertex* or *apex* ventrally, with a *body* between, and a *neck* inferiorly, where diverges the

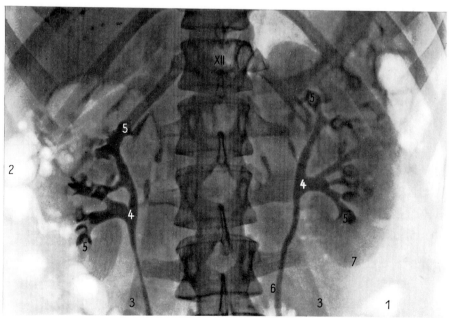

1—Descending colon
2—Ascending colon
3—Psoas major muscle
4—Pelvis of ureter
5—Renal papillae
6—Ureter
7—Kidney

FIG. 12-3A. *Kidneys (radiogram)*

THE URINARY SYSTEM ■ 393

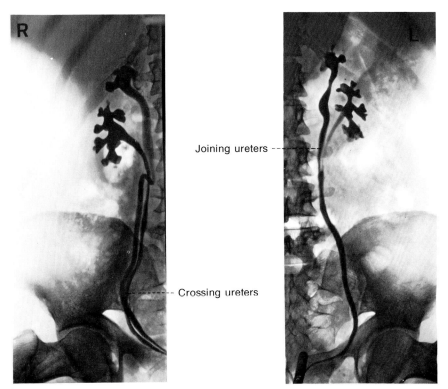

FIG. 12-3B. *Bifid ureters*

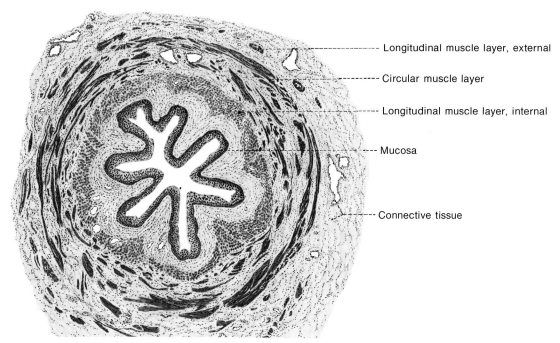

FIG. 12-3C. *Microscopic cross section of the ureter*

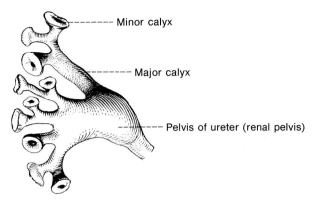

FIG. 12-3D. *Renal calices*

urethra. The *urachus* or *median umbilical ligament* extends from vertex to umbilicus, sometimes having a short mesentery. The ureters join the bladder obliquely upon either side of the base.

Beneath the peritoneal or *serous coat* of the bladder is a thick *tunica muscularis* of smooth muscle fibers disposed meshlike, and successively a *submucous* and a *mucous coat*, of which the latter is highly rugose in the empty bladder, but smooth when it is full. At the fundus is the *vesical trigone*, a triangular area with the orifices of the ureters at the lateral angles of the triangle and of the urethra at the ventral angles. Toward each urethral orifice there is an *interureteric fold*, and toward the urethra, a midline elevation termed the *uvula*. The walls of the bladder continue into the walls of the urethra in the female, and into the prostate in the male.

Arterial supply of the bladder is by vesicular branches of the internal iliac artery. The venous drainage is similar. Innervation (fig. 12-4C) is autonomic, the sympathetic fibers issuing from the lumbar nerves via the hypogastric plexus, and parasympathetic by the sacral outflow.

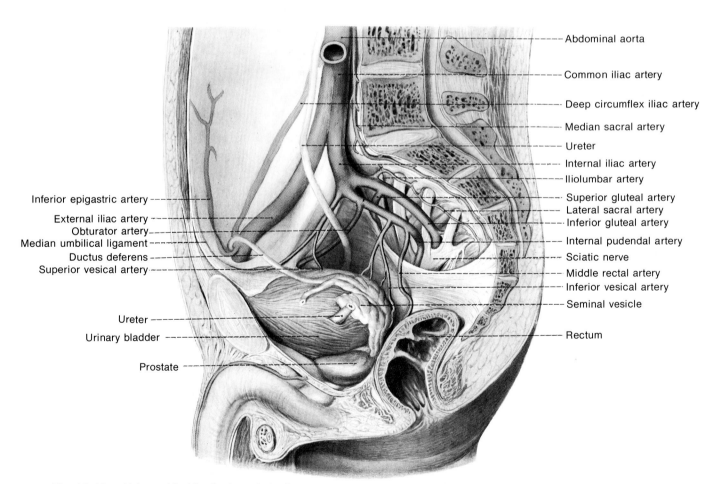

Fig. 12-4A. *Urinary bladder (external view)*

URETHRA

The *male urethra* is divided into three parts: the *prostatic*, which runs through the prostate gland; the *membranous*, which passes through the urogenital diaphragm; and the *cavernous*, which extends the length of the penis.

The *female urethra* extends from the bladder to the external urethral orifice situated between the clitoris and the vaginal opening.

CLINICAL APPLICATIONS

Infection is the most common disease of the urinary system. *Inflammation* caused by infections of the kidney, ureter, bladder, and urethra are referred to as *pyelonephritis, ureteritis, cystitis,* and *urethritis,* respectively.

Glomerulonephritis involves damage to the glomeruli and is usually associated with previous infection with streptococcus.

Acute renal failure is usually associated with tubular necrosis.

Congenital anomalies of the kidney and ureter include *horseshoe kidney, fetal lobulation, bifid ureters, ectopic kidney,* and *bifid pelves.*

Urinary calculi are formed in the urinary tract by the precipitation of crystalloids.

Many types of *cancer* occur in the kidneys and bladder.

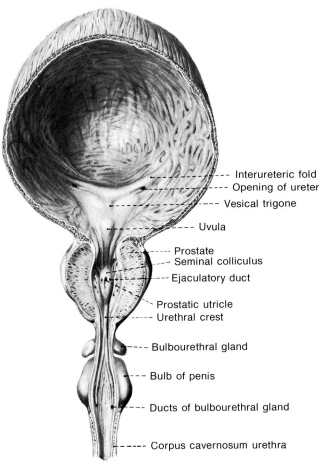

FIG. 12-4B. *Bladder and urethra (internal view)*

REVIEW QUESTIONS:

1. Adjacent pyramids are separated from each other by _____.
2. The apex of each renal pyramid ends in _____ projecting into the sinus of the kidney.
3. The _____ is the functional unit of the kidney.
4. The renal corpuscle includes a capillary tuft called the _____.
5. The _____ is the excretory duct from the kidney to the bladder.
6. The minor calices open into _____.
7. The _____ or _____ extends from the vertex of the bladder to the umbilicus.
8. Arterial supply of the bladder is by vascular branches of the _____ artery.
9. The glomerulus is inclosed within a double-walled membrane called _____.
10. The _____ carries the blood away from the glomerulus to the capillary networks that surround the tubules.
11. In longitudinal section it is seen that the structure of the kidney consists of a peripheral _____ and a deeper _____.
12. The blood returns from the efferent arteriole to the renal vein via the _____ vein, the _____ vein, and the _____ vein.

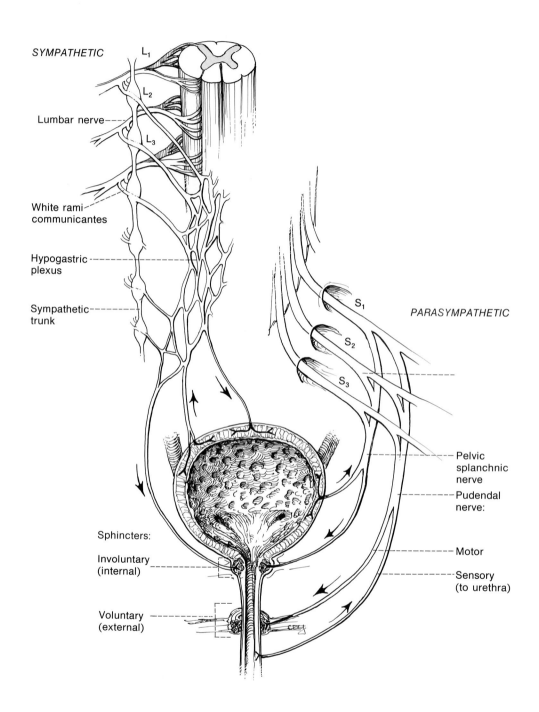

FIG. 12-4C. *Innervation of the urinary bladder and urethra*

13. The Reproductive System

> **STUDENT OBJECTIVES**
> After you have read this chapter, you should be able to:
> 1. Discuss the morphology of the testes.
> 2. Describe the location, structure and function of the epididymis.
> 3. Discuss the location of the ductus deferens.
> 4. Identify the location of the ejaculatory duct.
> 5. Identify the subdivisions of the male urethra.
> 6. Explain the location and functions of the seminal vesicles, prostate gland, and bulbourethral glands.
> 7. Explain the morphology of the penis.
> 8. Identify the location and morphology of the ovaries.
> 9. Describe the anatomical features of the uterus.
> 10. Locate the round ligament, uterine tubes, and ovarian ligaments.
> 11. List the components of the vulva.
> 12. Describe the anatomical features of the female perineum.
> 13. Describe the normal position of the uterus in relationship to the vagina.

MALE REPRODUCTIVE SYSTEM

The essential features of the *male reproductive system* (fig. 13-1) comprise the *testes, epididymis, ductus deferens, prostate, urethra,* and *penis.* The testes descend from the abdominal cavity into the scrotum shortly before birth; the processus vaginalis of the peritoneum introduces into the scrotal sac representative layers of the abdominal wall structures, within which the testes accordingly come to be situated.

Scrotum

The *scrotum* (fig. 13-2A) is a pouch of skin, fascia, and muscle containing the two testes, each with its epididymis, vessels, and nerves. The skin is thin and is supplied after puberty with a scanty covering of coarse hairs. Because of the smooth musculature beneath, it is subject to much wrinkling, particularly when cold; when warmed the skin becomes smooth. The left testis usually hangs slightly lower than the right. Along the midline the position of the *scrotal raphe* is indicated by a slight ridge, and this is accentuated dorsally near the perineum.

Beneath the skin is the *tunica dartos,* the scrotal representative of the superficial fascia. It is devoid of fat but is supplied with smooth muscle fibers. Part of its connective tissue fibers continue over the midline to interlace with those of the opposite side, while a part of them continue in the midline to form the *scrotal septum,* by which the scrotum is divided into two chambers. This is the only layer of fascia continuous across the midline of the scrotum. Succeeding fascias invest each testis and ductus deferens separately. The second layer of fascia beneath the skin is the *external spermatic fascia.* This is the scrotal representative of the fascia of the external abdominal oblique muscle. The third layer beneath the skin of the scrotum is the *cremasteric fascia* and *muscle,* derived chiefly from the muscle and fascia of the internal abdominal oblique muscle. The fourth and final layer of fascia is the *internal spermatic fascia,* derived from the transversalis fascia of the abdomen. Between this final layer of fascia and the testis is the double layer of serous membrane, the *tunica vaginalis testis,* derived from the processus vaginalis of the peritoneum. As with all other serous membrane this is a closed sac, with a parietal and a visceral layer, one reflected onto the other at the border of the hilus of the testis.

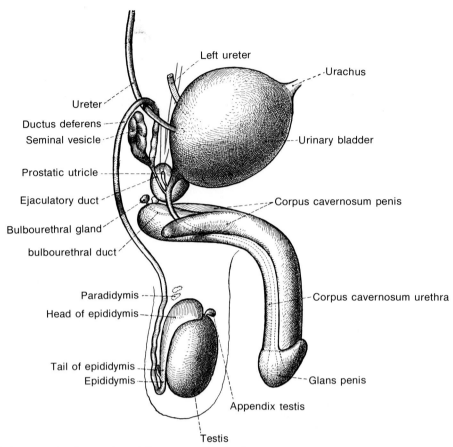

FIG. 13-1A. *Male reproductive system (diagrammatic view)*

Anterior Abdominal Wall and
Scrotal Homologues (fig. 13-2B)

Abdomen	Scrotum
Skin	Skin
Superficial fascia	Tunica dartos
External abdominal oblique muscle	External spermatic fascia
Internal abdominal oblique muscle	Cremasteric fascia and muscle
Transversalis fascia	Internal spermatic fascia
Peritoneum	Tunica vaginalis testis

The arterial supply of the scrotum and the nongenital part of its contents is by scrotal branches of the internal and external pudendal arteries. The veins correspond to the arteries. In scrotal innervation, sensibility is carried by the genital branch of the genitofemoral nerve, the anterior scrotal branch of the ilioinguinal nerve, and posterior scrotal branches of the internal pudendal and the posterior cutaneous femoral nerves; motor fibers to the cremaster muscle by the genital branch of the genitofemoral nerve; and sympathetic fibers to the smooth musculature of the tunica dartos by the hypogastric plexus.

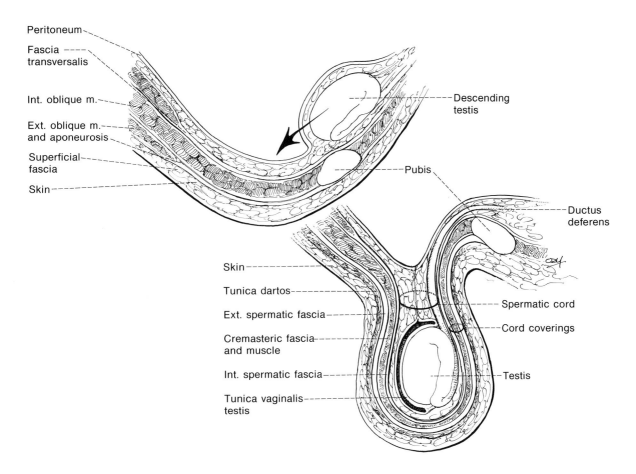

FIG. 13-2B. *Anterior abdominal wall and scrotal homologues*

the deep inguinal ring. It crosses dorsal to the obliterated umbilical artery and the ureter, and becomes somewhat expanded into an *ampulla.* Near the prostate it receives the slender duct of the *seminal vesicle,* and thereafter is known as the *ejaculatory duct.* This duct passes through the prostate and opens by a minute aperture upon the colliculus of the urethral crest, close to the utricle.

Seminal Vesicles

The *seminal vesicles* are a pair of irregularly shaped sacs that are diverticula of the ductus deferens, lateral to the ductus and partly between the bladder and rectum. Each has a length of about 2 inches and tapers toward its duct, by which it communicates with the ductus deferens. It contains smooth muscle fibers by which it is constricted during ejaculation, and the lining epithelium secretes a fluid that is considerably more viscous than is the seminal fluid as it is finally discharged.

The arterial supply of the testis is by the testicular artery, and of the epididymis by the epididymal branches of the testicular artery. The ductus deferens and the terminal part of the epididymis is supplied by the deferential branch of the internal iliac artery, and the seminal vesicles are supplied partly by this and partly by the neighboring inferior vesicular branch. The veins correspond to the arteries but are more

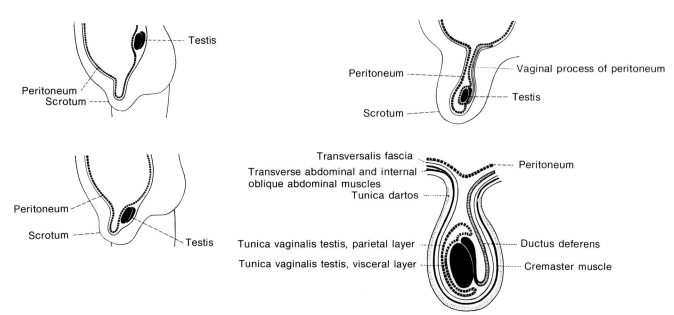

FIG. 13-3A. *Descent of the testes*

plexiform, particularly the testicular vein, whose network is known as the *pampiniform plexus*. The innervation is autonomic, through the hypogastric plexus.

Penis

The *penis* (fig. 13-5*A–C*) is designed for both the introduction of spermatozoa into the female, and for the excretion of urine. The common duct for both functions is the *urethra;* but to accomplish the former act the penis is elongated and stiffened by the cavernous tissue of the two erectile bodies, the *corpora cavernosa penis*. The urethra is also surrounded by erectile, cavernous tissue, the *corpus cavernosum urethrae*.

The skin of the penis is thin and highly mobile, with a median raphe upon the undersurface continuous with both the raphe of the scrotum and with the frenulum penis. It covers the glans, as the *prepuce*, and then is reflected proximally and inward to the base of the glans, where it is firmly attached. It is once more reflected distally, fused with the glans, to the urethral aperture. At the inferior part of the glans it is folded to form the midline *frenulum*.

The *corpora cavernosa penis* are bilateral, each arising by dense fibrous tissue from a firm anchorage upon the medial margin of the ramus of the ischium. This part is termed the *crus penis,* and is surrounded by the ischiocavernosus muscle. The two crura converge, as they pass ventrally, increase in size, and gradually develop cavernous tissue; each one is then surrounded by a dense, fibrous *tunica albuginea*. Near the symphysis pubis they contact one another and their coats fuse to form a septum that is not entirely complete. The septum continues to the distal end, which tapers to some extent and is capped by the *glans penis*. Trabeculae continue inward from the tunica, dividing the cavernous tissue into chambers which, upon the proper stimuli, become engorged with blood and result in erection. Upon the ventral or inferior surface of the junction of the two corpora there is a longitudinal groove, into which fits the *corpus cavernosum urethrae*.

The *male urethra,* extending from bladder to glans, is divisible into *prostatic, membranous,* and *spongy portions*. The first of these, with an almost vertical direction, pierces the prostate gland ventral to its center and is dilated. Its dorsal surface is ridged by the *urethral crest,* with a slight *papilla* upon which opens the blind duct of the *prostatic utricle,* and the two *ejaculatory ducts* somewhat laterally. On either side of the urethral crest are the minute openings of the ducts of the prostate.

The *membranous urethra* pierces the urogenital diaphragm and is surrounded by the sphincter urethrae membranaceae muscle. It is constricted and relatively inelastic, and its course is ventroinferior.

The *spongy urethra* extends from the urogenital diaphragm to the glans. As it leaves the urogenital diaphragm it enters the posterior aspect of the bulb of the penis and then bends to extend longitudinally in the penis. As it enters the bulb the urethra increases slightly in size and here it receives the two minute ducts of the bulbourethral glands. Just proximal to

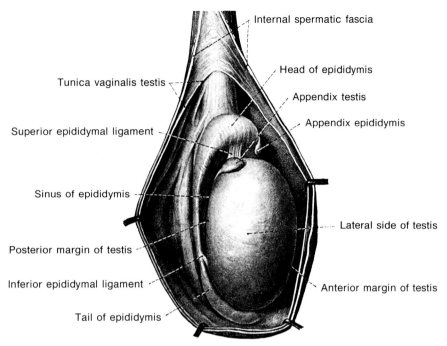

FIG. 13-3B. *Testis, epididymis, and spermatic cord*

the glans the urethra is slightly constricted once more, and within the glans it is again expanded rather abruptly to form the *fossa navicularis*, which in form is slitlike in the sagittal plane.

The spongy urethra, surrounded by the *corpus cavernosum urethrae*, is not of equal thickness throughout. Proximally there is an expanded *bulbus penis*, surrounded by the bulbocavernosus muscle, that extends dorsal or posterior to the urethra and is inferior to the urogenital diaphragm, between the crura of the corpora cavernosa penis. Distal to the bulb the corpus cavernosum urethrae occupies the groove inferior to and between the two corpora cavernosa penis, and it terminates in the expansion termed the *glans penis*, which fits caplike upon the tapered ends of the corpora. The free surface of the glans is covered by extremely thin skin, the prepuce, continuous with that which encircles the penis except at the frenulum of the prepuce.

The arterial supply of the superficial, chiefly the proximal, part of the penis is by external pudendal branches of the femoral artery, coursing in the superficial fascia. At a deeper level dorsally are the dorsalis penis branches and pudendal and perineal branches of the internal pudendal artery, which supply all three corpora cavernosa. The tunica intima of the ar-

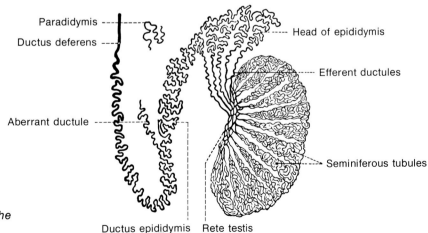

FIG. 13-3C. *Tubular network of the testis and epididymis*

teries to the erectile tissue is peculiarly thickened, and this coat is capable of an unusual degree of dilatation, under erectile stimuli, with corresponding increase in the blood supply. Venous drainage largely conforms to arterial supply, except for a more plexiform arrangement whereby drainage during erection is retarded. Innervation by somatic sensory fibers involving skin and fascia is by the anterior scrotal branch of the ilioinguinal nerve and chiefly by the dorsalis penis branch of the pudendal nerve; somatic motor innervation to the striated musculature is by the perineal branches of the pudendal nerve; and the autonomic supply is by sympathetic and (sacral) parasympathetic fibers from the hypogastric plexus. Those from S3 and S4 are sometimes termed nervi erigens, as their stimulation results in erection.

The accessory sexual glands of the male comprise the *prostate, bulbourethral glands,* and the *seminal vesicles.*

Prostate Gland

The *prostate gland* (fig. 13-5D) is composed partly of glandular and partly of smooth muscle tissue. It is a fig-shaped structure surrounding the proximal part

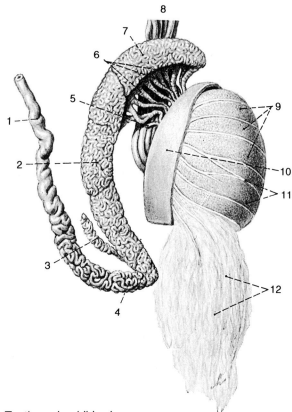

Testis and epididymis

1 Ductus deferens
2 Body of epididymis
3 Aberrant ductule
4 Tail of epididymis
5 Aberrant ductule
6 Lobules of epididymis
7 Head of epididymis
8 Pampiniform plexus
9 Lobules of testis
10 Tunica albuginea
11 Septula testis
12 Seminiferous tubules

FIG. 13-3D. *Transected tubule of the testis showing various stages of spermatogenesis*

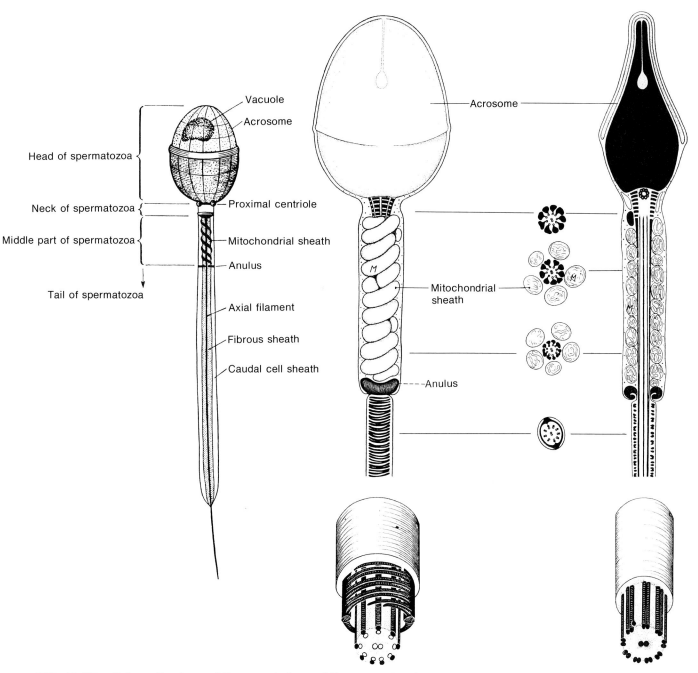

FIG. 13-3E. *Schematic views of the morphology of the spermatocyte*

of the urethra, between the bladder and urogenital diaphragm, that is a specialization of the walls of the urethra. It has a base against the bladder and an apex against the urogenital diaphragm, fitting between the levator ani muscles of either side. Its individual glands are variable in number, with small ducts opening upon the floor of the prostatic urethra on either side of the urethral crest. The prostate is also pierced obliquely by the ejaculatory ducts and contains the blind duct of the prostatic utricle. The part of the prostate superior to the urethra constitutes the *middle lobe,* which is the portion usually affected in hypertrophy of the gland, while the *lateral lobes,* involving largely the more glandular tissue, comprise the inferolateral parts. The *posterior lobe* is situated below the ejaculatory duct. The whole is enclosed in

406 ■ BASIC ANATOMY FOR THE ALLIED HEALTH PROFESSIONS

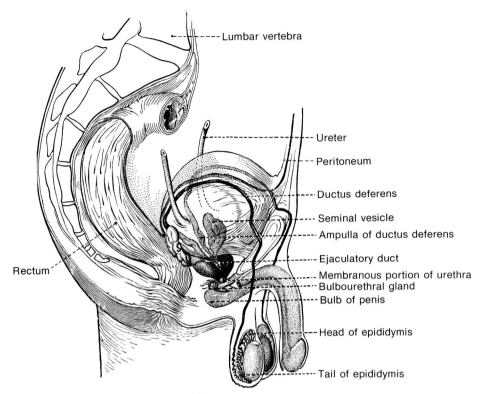

FIG. 13-4A. *Ductus deferens and seminal vesicles*

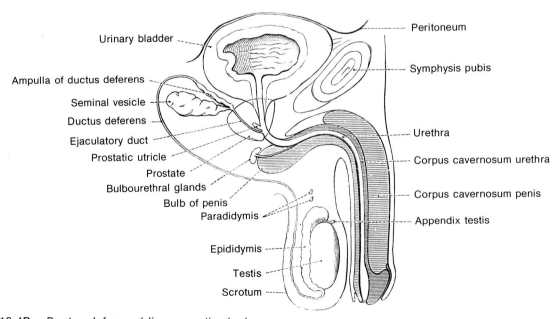

FIG. 13-4B. *Ductus deferens (diagrammatic view)*

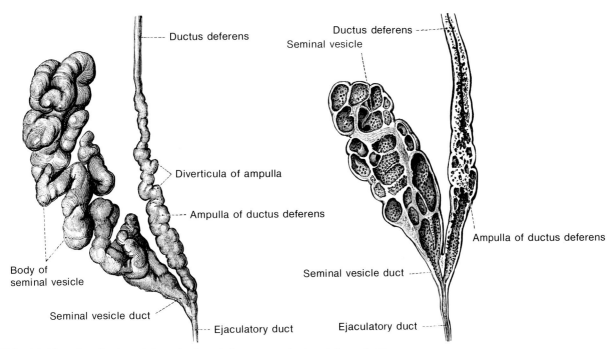

FIG. 13-4C. *Seminal vesicles, ductus deferens, and ejaculatory duct*

the capsule of the prostate. The prostate furnishes most of the seminal fluid, but its secretion is thinner than that of the final product. It has an alkaline reaction and has the effect of mobilizing the spermatozoa.

The arterial supply of the prostate is by the middle rectal and inferior vesicular branches of the internal iliac artery. Venous drainage is by the prostatic plexus of the internal iliac vein. Innervation is autonomic by way of the hypogastric plexus.

The *bulbourethral* or *Cowper's glands* in the male are two whitish bodies, each the size of a small pea, embedded among the fibers of the sphincter urethrae membranaceae muscle between the two fascial layers of the urogenital diaphragm, dorsal to and on either side of the urethra. The duct of each gland courses obliquely distally and opens at an angle into the urethra by a minute orifice just distal to the bulb. The alkaline mucoid secretion is discharged during sexual excitement with the alleged purpose of neutralizing any trace of acid urine in preparation for the passage of the spermatozoa.

FEMALE REPRODUCTIVE SYSTEM

The essential features of the *female reproductive system* (fig. 13-6) comprise the *ovaries,* the *uterine tubes, uterus,* and *vagina.*

Ovaries

In the fetal position the *ovaries* (fig. 13-6) are located more inferiorly than are the testes in the male, but unlike the latter, the ovaries experience no further alteration in position. They are a pair of oval, flattened gonads, usually an inch or more in length but variable in size and considerably smaller in old age. They are subject to moderate displacement, caused by crowding of neighboring structures, but most often they are situated upon the lateral wall of the pelvic cavity midway between the anterior superior spine of the ilium and the pubic tubercle of the opposite side, with the long axis almost vertical. The superior end is termed the *tubal end,* and the inferior, the *uterine end.* The ovary lies within a peritoneal fold of the *broad ligament* with a short mesentery, the *mesovarium.* Its free border is directed somewhat dorsally and the mesovarial border ventrally, this being the border that receives nerves and vessels; accordingly it is sometimes termed the hilus of the ovary.

The *ovary* is ductless; the ova rupture from its surface through the germinal epithelium into the peritoneal cavity, but pass at once, because of the overlying fimbriae of the uterine tube and the action of its cilia, into that tube.

Associated with the ovary are the *epoophoron* and *paroophoron,* which are the respective female homo-

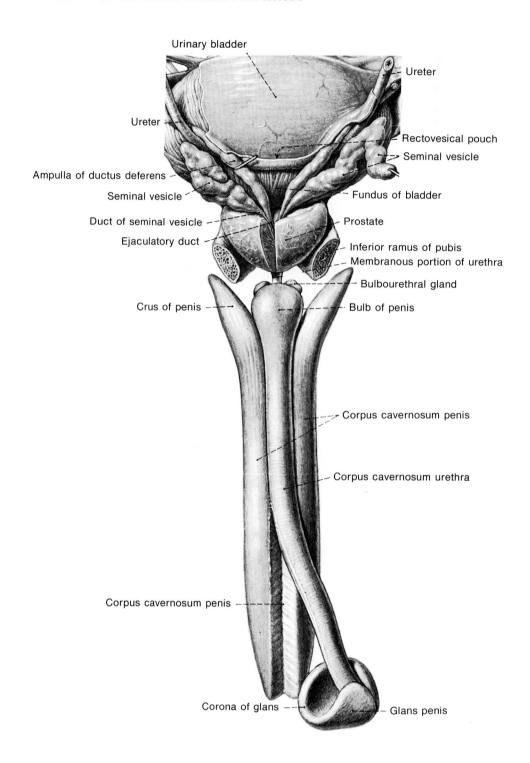

FIG. 13-5A. *Penis, urethra, and prostate*

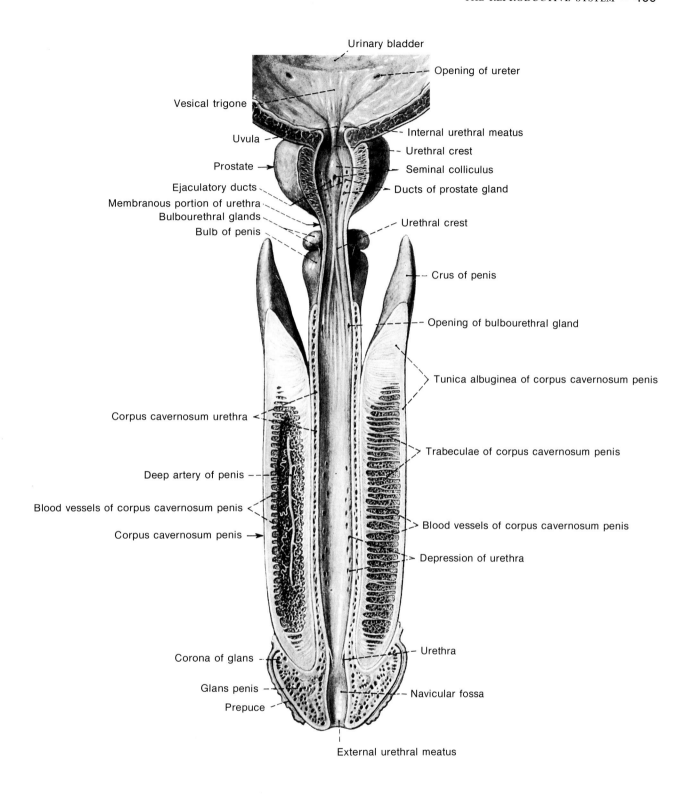

FIG. 13-5A. *Diagrammatic view of bladder, prostate, penis and urethra*

logues of the male epididymis and paradidymis, but in the female both are vestigial structures without function. They occur, but are rarely well marked, as connected or segregated blind tubules in the broad ligament between the ovarian ligament and the uterine tube.

The arterial supply of the ovary is by the ovarian artery, the female representative of the testicular artery in the male, which passes to the gland in the ovarian ligament, and by ovarian branches of the uterine artery. Venous drainage is comparable, with a *pampiniform plexus* as in the male. Innervation is autonomic, from the ovarian plexus, which follows the artery inferiorly from the renal plexus, and from the hypogastric plexus.

Uterine Tubes

The *uterine tube* (*fallopian tube*) or *oviduct* (fig. 13-7*A, B,* and *E*) is a tube some 4 inches long in the superior border of the round ligament that opens into the peritoneal cavity near the ovary at one end and into the uterus at the other. The medial end is essentially transverse, but the lateral end curves inferiorly. The part near the uterine end is termed the *isthmus*, which expands into the *ampulla* near the ovarian end, which in turn enlarges funnelwise as the *infundibulum*. The opening, known as the *ostium abdominale*, is surrounded by irregular *fimbriae*, one of which is usually somewhat longer than the others. The fimbriae and ostium clasp the ovary, and receive the ova that rupture through its walls. The part of the broad ligament adjacent to the tube is termed the *mesosalpinx*.

The arterial supply of the uterine tube is by tubal branches of the ovarian and uterine arteries, with a similar venous drainage. Innervation is the same as for the ovary and uterus.

Uterus

The *uterus* (fig. 13-7*C* and *D*) is an unpaired hollow organ with thick walls containing smooth musculature. It lies mostly between the two layers of peritoneum forming the broad ligament. The inferior part is situated between the bladder and rectum, while the remainder is dorsosuperior to the collapsed bladder. It is roughly thumb-shaped, with a slight central constriction, the *isthmus*. The central cavity, lined by *endometrium*, communicates at the superior end with the uterine tubes, and opens inferiorly into the vagina. The muscle layer is called the *myometrium* and the serous layer is called the *epimetrium*. An average length is about 3 inches, although it increases in size during pregnancy to enclose the fetus. The part above the isthmus is the *body,* ending superiorly in the rounded *fundus.* This is the broadest part of the uterus, and is slightly projected laterally to receive the uterine tubes at the *lateral angle* of the uterus. The part below the isthmus is the *cervix,* which is smaller than the uterine body after puberty and has a *supravaginal* and a *vaginal part;* the latter projects into the vagina. The nature of the tunica mucosa of the cervix differs, however, from that of the body; in the former it produces a fluid rich in mucous. The vaginal part of the uterus is hemispherical, and upon the prominence is the orifice termed the *external os,* round or oval before childbirth, and slitlike (transversely) thereafter.

The cavity of the uterus is mostly collapsed, but it has two potential expansions: one, the *canal of the cervix,* is between the external os and the *internal os* (which is a contraction at the level of the isthmus); and the other, the *canal of the body,* is superior to the isthmus. This has a triangular form, with the uterine tubes and the internal os at the three angles. The part of the broad ligament adjacent to the uterus is termed the *mesometrium.*

Arterial supply of the uterus is chiefly by the uterine artery (from the internal iliac artery), which anastomoses with the ovarian artery. The venous drainage is by the uterine plexus by veins that correspond to the arteries, but communicate also with surrounding venous plexuses. Innervation is autonomic, from the uterovaginal plexus, on either side of the uterus; these are continuations of the hypogastric plexus, with parasympathetic fibers from S2, S3, and S4.

Ligaments of Internal Genital System

Ligaments associated with the internal genital system (fig. 13-7*F*) of the female comprise the following.

The *broad ligament* of the uterus is a transverse fold of peritoneum directed superiorly between the pelvic walls laterally and situated sagittally between the bladder and rectum. Between its two layers are enclosed the uterus, uterine tubes, and ovaries, with their vessels and nerves. At the base of the broad ligament the peritoneum is reflected anteriorly upon the bladder and laterally forms the *vesicouterine pouch;* and posteriorly upon the rectum and more dorsal pelvic floor, forming the *rectouterine pouch.* The superior border of the broad ligament is free and encloses the fundus of the uterus and the uterine

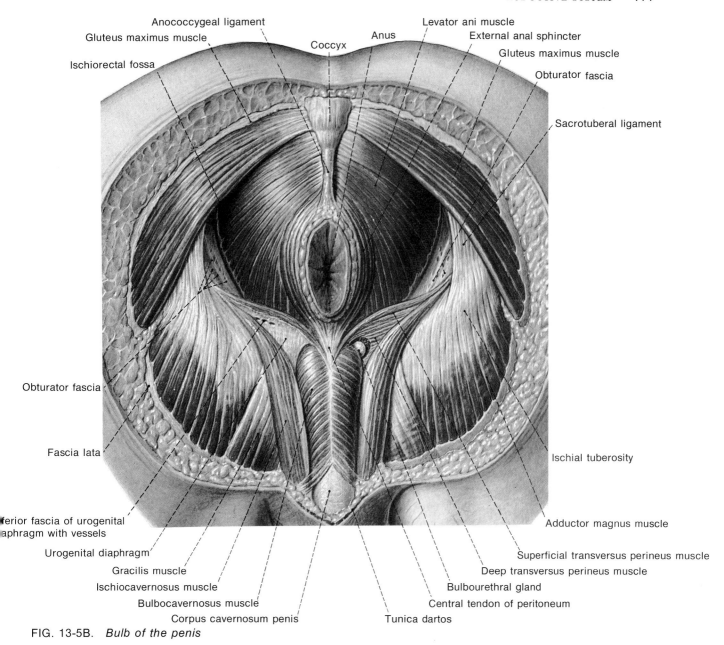

FIG. 13-5B. *Bulb of the penis*

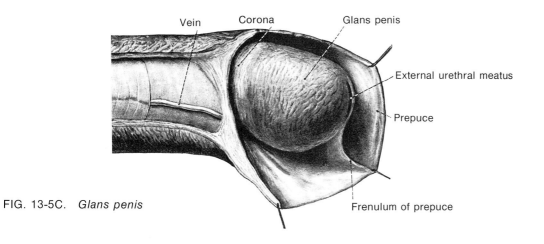

FIG. 13-5C. *Glans penis*

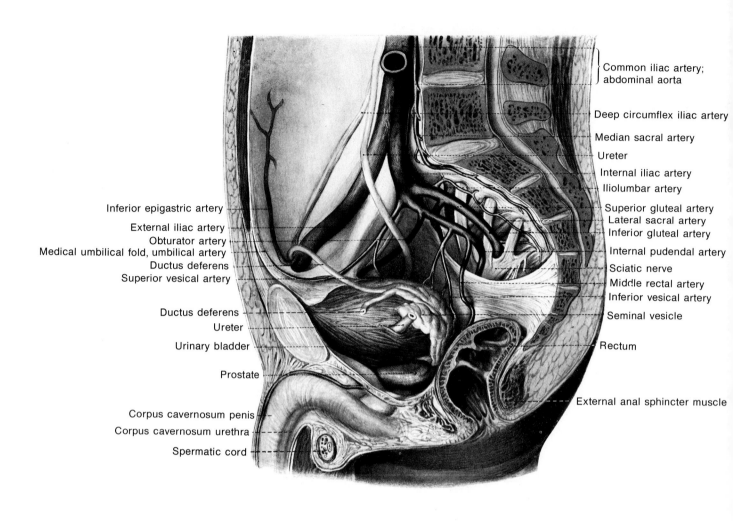
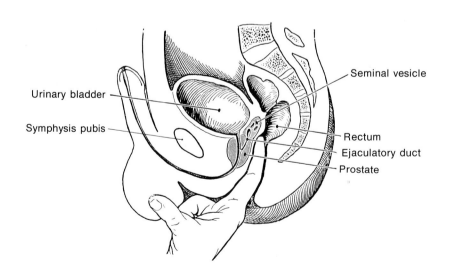

FIG. 13-5D. *Rectal examination of the prostate and seminal vesicles*

THE REPRODUCTIVE SYSTEM ■ 413

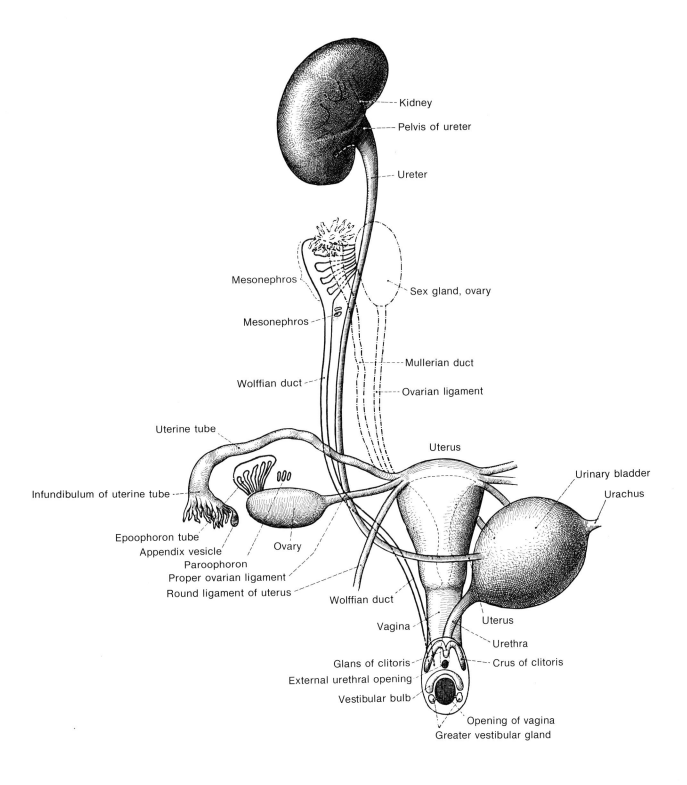

FIG. 13-6A. *Female reproduction system (diagrammatic view)*

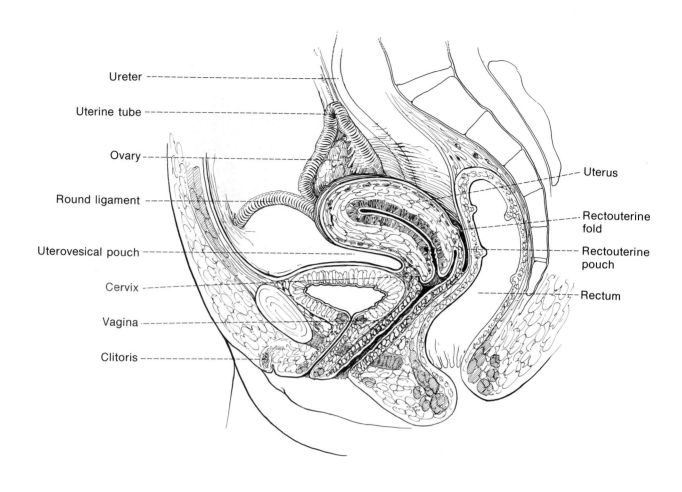

FIG. 13-6B. *Female reproductive system (sagittal view)*

tubes. Projecting from the ventral wall of the broad ligament on either side of the uterus is a small additional fold of mesentery supporting the ovaries. This is the *mesovarium*. The part of the broad ligament superior to this, concerned with the uterine tubes, is the *mesosalpinx,* and the part inferior to it, lateral to the uterus, is the *mesometrium*.

The *suspensory ligament of the ovary* is the name employed for the free fold of the lateral termination of the broad ligament, which extends superiorly between the most lateral part of the uterine tube and the pelvic wall, and encloses the ovarian vessels.

The *ovarian ligament* is a fibrous band at the base of the mesovarium connecting the uterine end of the ovary to the lateral angle of the uterus, dorsal to the uterine tube.

The *ligamentum teres* or *round ligament* is a slightly flattened fibrous band which, near the uterus, contains smooth muscle fibers. Ostensibly its abdominal end is upon the uterus just inferior to the end of the

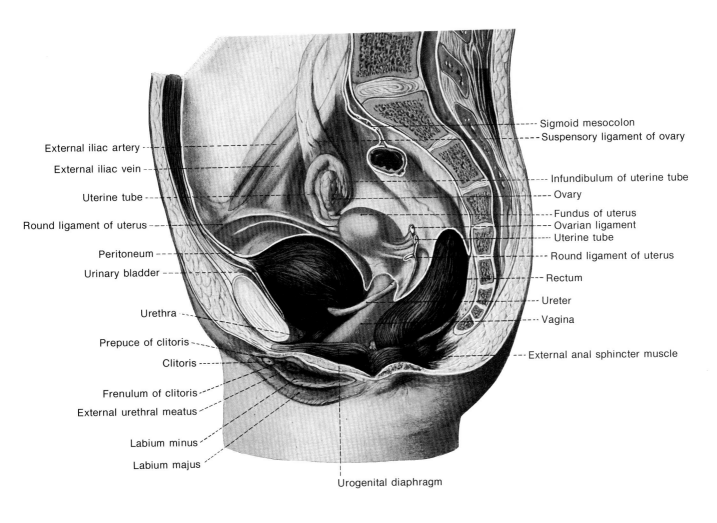
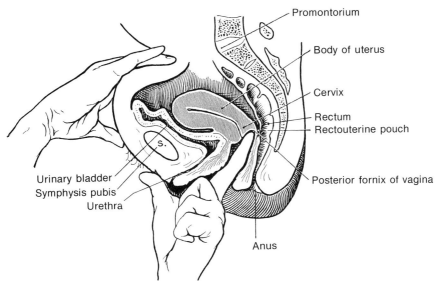

FIG. 13-7A. *Uterine tubes, uterus, and ovaries (parasagittal view)*

FIG. 13-7B. *Female pelvis*

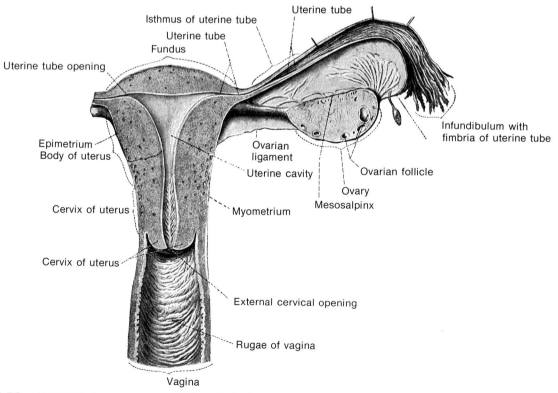

FIG. 13-7C. *Uterine tubes and uterus (internal view)*

uterine tube, but as already stated it is merely anchored to the uterus at this point and continues to the ovary. It passes in the broad ligament inferolaterally to the anterior pelvic wall, then between the wall and the peritoneum ventrally and slightly superiorly, across the obliterated umbilical artery to the abdominal inguinal ring, through the canal and out the superficial inguinal ring. This part of its course is thus entirely comparable with that of the ductus deferens in the male. The ligament becomes lost in the superficial fascia of the labium majus.

The *uterosacral ligament* is a fibrous band containing some smooth muscle fibers that passes on either side from the supravaginal part of the cervix to the lateral part of the sacrum. Its tension causes the slight ridge of the *rectouterine fold* along the ventral margin of the rectouterine pouch.

The *urethra* in the female is concerned solely with urination, but as in the male, it is included in the genital system. In the female it corresponds only to the prostatic and membranous portions of the male urethra and is slightly more than an inch in length. It courses inferiorly with a slight ventral inclination and opens in the vestibule at the external urethral orifice, between the clitoris and the vaginal orifice.

Vagina and External Genitalia

The *vagina* (fig. 13-8A and B) is a dilatable canal that opens inferiorly into the genital vestibule and then continues superodorsally, ending blindly in the *fornix of the vagina*. The tunica mucosa of the vagina and character of the vaginal contents change in conformity with the menstrual alterations of the uterus. The vagina is usually slightly more than 3 inches in length, and near the fornix there projects into it from the ventrosuperior aspect the vaginal part of the cervix uteri. Accordingly, the external orifice of the uterine canal opens into the vagina. It is lined with mucous membrane and its walls are supplied with two layers of smooth musculature: the inner of circular fibers less well defined, and the outer of longitudinal fibers. The inferior portion of the vagina passes between the two halves of the urogenital diaphragm, this part offering more resistance to dilatation than the rest. At the junction of the vagina with the vestibule there is a thin membranous diaphragm with a central hiatus of variable extent. This is the *hymen*. After rupture by coition or by other means its remains may often be identified as small flaps, the *carunculae hymenales*. Occasionally the hymen is imperforate.

The *external genitalia* or *vulvae* (fig. 13-8A and B) in the female consist of a sagittally linear depression containing the vestibule of the vagina, into which projects the clitoris and opens the urethra, bordered by the labia minora and the labia majora. The *labia majora* are thick folds of skin containing fat, areolar and glandular tissue, covered laterally with hair after puberty. They are homologous with the scrotum in the male and continue from the *mons pubis*, superficial to the symphysis pubis on either side, decrease in size to the vicinity of the perineal body, and are there joined across the midline in a slight transverse fold, the *posterior commissure*.

The *labia minora* (fig. 13-8C and D) are usually much smaller as well as shorter than the majora, between which they lie. They have thin skin grossly resembling mucous membrane. Anteriorly they converge and join the glans as the *frenulum clitoridis*, and also join one another anterior to the glans in a fold that is termed the *prepuce of the clitoris*. Posterior to the vestibule they join across the midline in a low fold.

The *vestibule* of the vagina is the space between the labia minora external to the hymen or its remains. The glans clitoridis is located within the vestibule, into which opens the external urethral orifice upon a slight elevation between the clitoris and vagina.

The *clitoris*, about 1 inch in length and somewhat flattened transversely, is the miniature homologue of the penis in the male, but without the urethra. It is partly erectile and is composed of two adjacent *corpora cavernosa clitoridis* that form the body, and is continuous with the two converging *crura*, which arise from the rami of the pubis; each crura is covered by the ischiocavernosus muscle. The clitoris is capped by a minute *glans clitoridis*, connected with the bulbs of the vestibule by a thread of erectile tissue. The clitoris also has a *suspensory ligament*, which is frequently poorly defined, corresponding to the structure of the same name in the male.

The bilateral *vestibular bulbs* contain erectile tissue and are located upon each side of the vaginal orifice, inferior to the urogenital diaphragm. The bulbs are covered by the bulbocavernosus muscle. It is thick posteriorly and tapers anteriorly, where it joins its fellow of the opposite side in the *commissure of the bulb*, which continues thinly to the glans. It is the homologue of the male corpus cavernosum urethrae.

The *greater vestibular* or *bulbourethral glands* are bilateral and located at the base of the bulb lateral to the posterior part of the vaginal orifice. The duct opens into the vestibule in the angle between the hymen and labium minora. It discharges an alkaline

418 ■ BASIC ANATOMY FOR THE ALLIED HEALTH PROFESSIONS

FIG. 13-7D. *Ovary (cross section)*

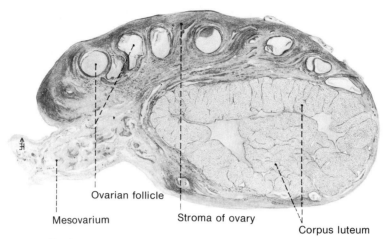

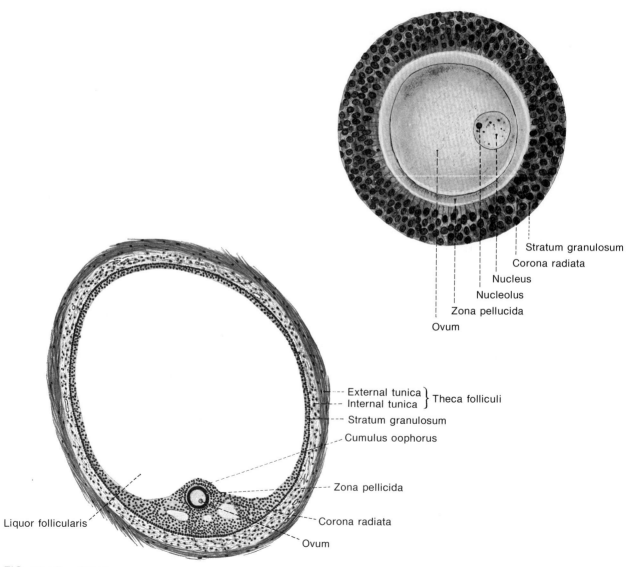

FIG. 13-7E. *Ovum*

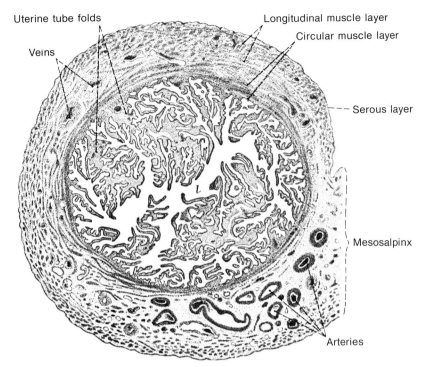

FIG. 13-7F. *Uterine tube (microscopic section)*

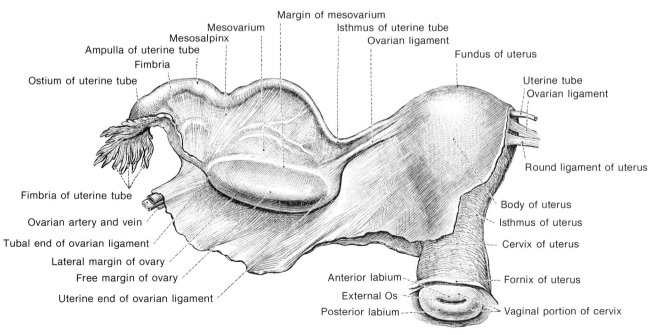

FIG. 13-7G. *Uterus (external view)*

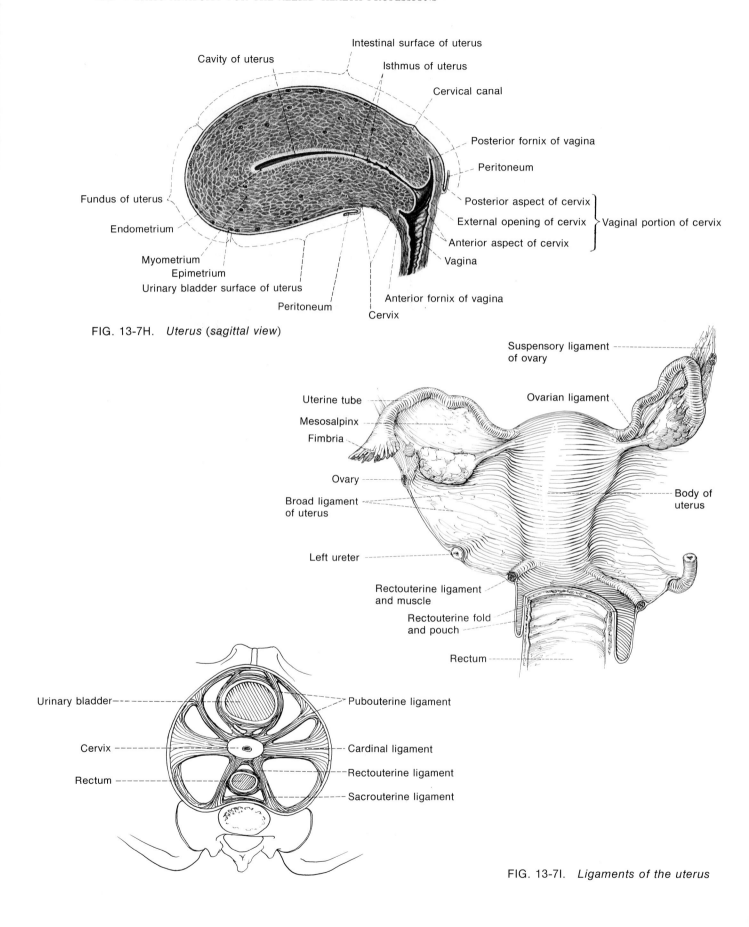

FIG. 13-7H. *Uterus (sagittal view)*

FIG. 13-7I. *Ligaments of the uterus*

mucoid secretion during sexual excitement that is partly for lubrication and partly for the correction of acidity, in preparation for the reception of the spermatozoa. The *lesser vestibular glands* are grossly indistinguishable mucous glands that open in the anterior part of the vestibule.

The arterial supply of the vagina and external genitalia is by vaginal branches of the uterine, inferior vesical, medial rectal, and internal pudendal arteries, as well as the genital branches of the external pudendal artery. Venous drainage corresponds. Innervation is partly autonomic, by the hypogastric plexus with parasympathetic fibers from S3 and S4; somatic sensory by the ilioinguinal nerve, the genital branch of the genitofemoral nerve, the labial branches of the pudendal and posterior cutaneous femoral nerves; and somatic motor to the striated musculature by branches of the pudendal nerve.

CLINICAL APPLICATIONS

A *varicocele*, commonly located on the left side, involves the elongation and dilation of the veins of the pampiniform plexus.

The descent of the testes may be involved in a number of congenital anomalies, such as *incomplete descent, maldescent,* and *improper rotation.*

Because of the bend in the urethra as it enters the bulb, the constriction of the membranous portion, and the tendency for the particular involvement of the bulbourethral ducts during infections, this portion of the urethra is particularly subject to *mechanical injury* during the introduction of various surgical instruments.

The middle lobe of the prostate is often involved in *benign hypertrophy.* The posterior lobe, however, is commonly involved with *cancer,* which frequently shows *skeletal metastases* in the lower vertebral column.

The processus vaginalis may be associated with several congenital anomalies, such as *hydroceles* or hernial sacs for *indirect inguinal herniae.*

Ectopic pregnancies involve implantation of the fertilized ovum outside the uterus. The ovary may release its ova into the peritoneal cavity rather than the uterine tube and hence an *abdominal pregnancy* occurs. Implantation of a fertilized ovum may also occur within the uterine tube.

The position of the uterus is related to several factors, such as: (1) muscle tone of the levator ani muscle; (2) vessels and nerves (vaginal, uterine, and hypogastric) associated with the uterus; and (3) associated ligaments (pubocervical, sacrocervical, transverse cervical, and broad). If the fundus and body are bent backwards on the vagina, the uterus is said to be *retroverted.* If the body of the uterus is, in addition, bent backwards on the cervix, it is said to be *retroflexed.*

REVIEW QUESTIONS:

1. The internal spermatic fascia is derived from the _____ of the abdomen.
2. The tunica vaginalis testis is derived from the _____ of the peritoneum.
3. The scrotal representative of the fascia of the external abdominal oblique is the _____ fascia.
4. The surface of the testis is invested, except where covered by the epididymis, by the closely adherent visceral layer of the _____.
5. The _____ forms the capsule of the testis.
6. The _____ is a group of aberrant tubules above the head of the epididymis.
7. The _____ is the continuation of the tail of the epididymis to the prostatic urethra.
8. Near the prostate the _____ joins the slender duct of the seminal vesicle to form the _____ duct.
9. The crus penis is surrounded by the _____ muscle.
10. The male urethra is divisible into _____, _____ and _____ portions.
11. The _____ urethra pierces the urogenital diaphragm and is surrounded by the _____ muscle.
12. The _____ urethra extends from the urogenital diaphragm to the glans.
13. The bulbus penis is surrounded by the _____ muscle.

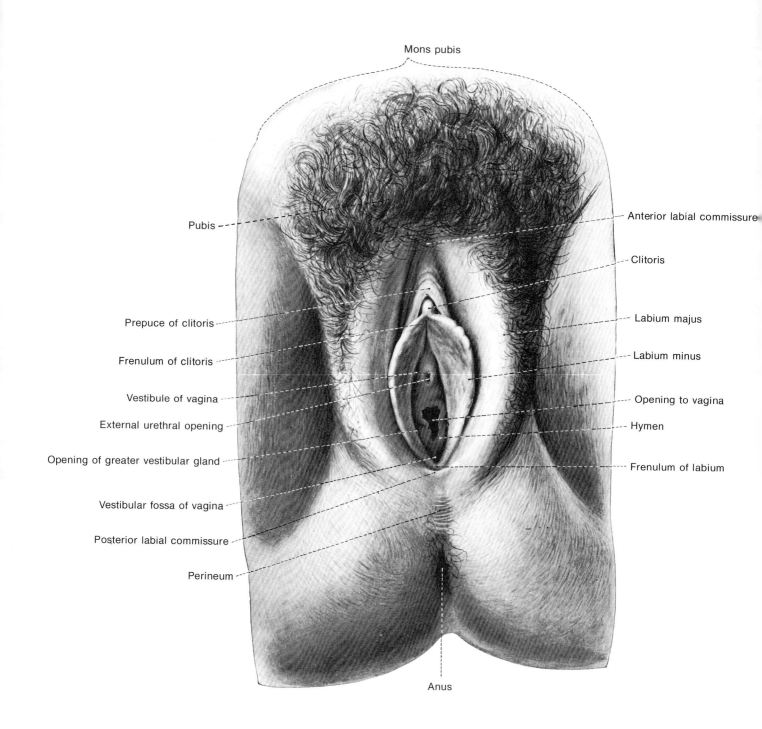

FIG. 13-8A. *Female external genitalia*

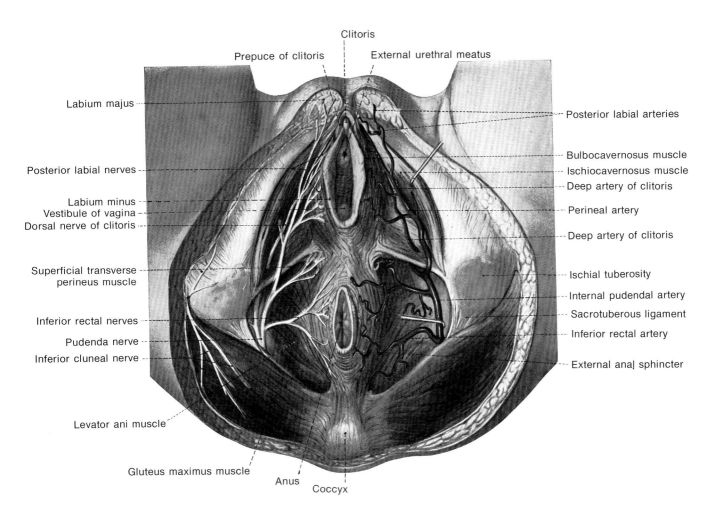

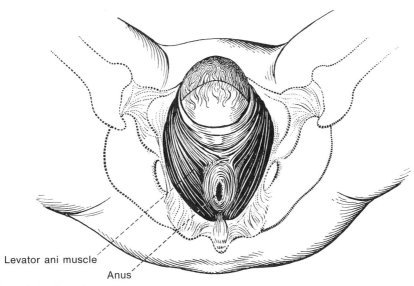

Fig. 13-8B. *Muscles of the female perineum*

424 ■ BASIC ANATOMY FOR THE ALLIED HEALTH PROFESSIONS

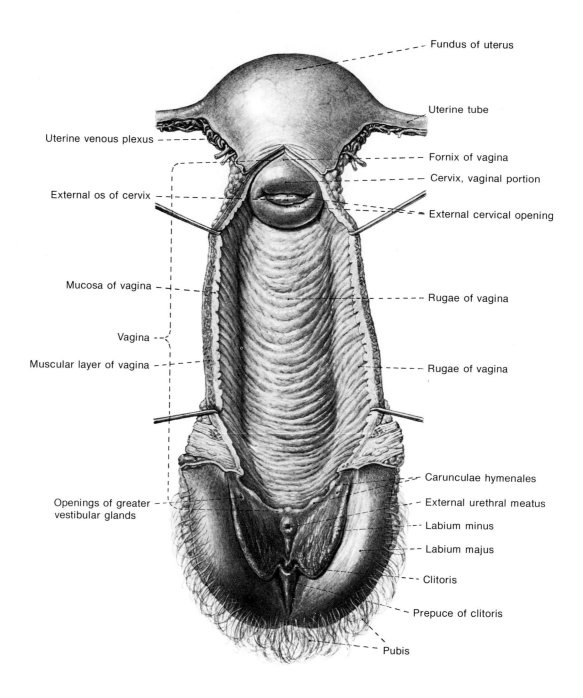

Fig. 13-8C. *Vagina and cervix (internal view)*

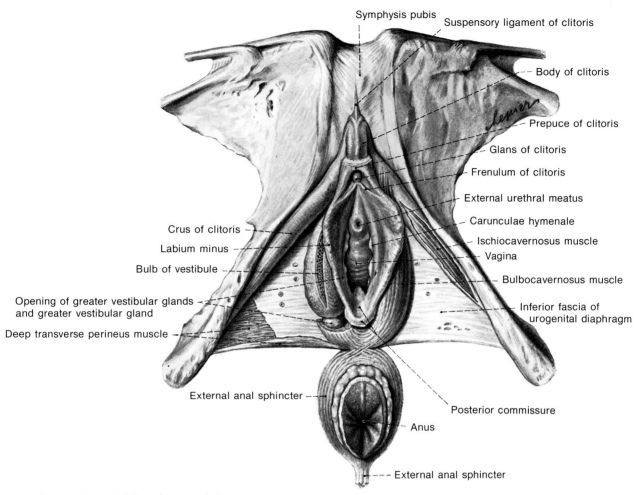

Fig. 13-8D. *Superficial perineum of the female*

14. The accessory sexual glands of the male comprise the _____, _____ and the _____.
15. The _____ is located between the bladder and the urogenital diaphragm.
16. The part of the broad ligament adjacent to the uterine tube is termed the _____.
17. The _____ lines the uterine cavity.
18. The muscle layer of the uterus is called the _____.
19. The broadest portion of the uterus is the _____.
20. The part of the broad ligament adjacent to the uterus is termed the _____.
21. The ducts of the greater vestibular glands open into the _____.
22. The vestibular bulbs are covered by the _____ muscle.
23. The _____ or _____ in the female consist of a sagittally linear depression containing the vestibule of the vagina, clitoris, urethra, labia minora and labia majora.
24. The _____ is a fibrous band at the base of the mesovarium connecting the uterine end of the ovary to the uterus.
25. The canal of the cervix is located between the _____ and the _____.

14. The Endocrine Glands

> **STUDENT OBJECTIVES**
> After you have read this chapter, you should be able to:
> 1. Define an endocrine gland.
> 2. Describe the location and function of the thyroid gland.
> 3. Describe the location and morphology of the suprarenal glands.
> 4. Distinguish between endocrine and exocrine functions of the pancreas.
> 5. Identify the location and function of the pineal gland.
> 6. Identify the location, structure, and functions of the thymus gland.
> 7. Discuss the endocrine aspects of the testes and ovaries.
> 8. Discuss the location, morphology and functions of the pituitary gland.

The *endocrine glands* (fig. 14-1), preponderantly macroscopic in character, are physiologically related to each other and to other parts of the body by the circulating blood, into which they discharge their products directly. They are distinguished from glands with ducts by the terms ductless and endocrine, in contrast with the term exocrine.

The functions of the endocrine glands (*thyroid, parathyroid, thymus, pituitary, pineal, suprarenal, ovaries, testes,* and *pancreas*) concern the general metabolism and survival of the animal. They control the carbohydrate, protein, fat and water metabolism, growth, and sexual behavior. They do not function independently of each other, but influence and modulate one another in an interrelated manner through the blood stream, largely by neuroendocrine control.

THYROID GLAND

The *thyroid gland* (fig. 14-2*A–C*) is usually a somewhat U-shaped, highly vascular structure clasping the trachea inferior to the thyroid cartilage, deep to the infrahyoid musculature, and surrounded by the pretracheal layer of cervical fascia. It has *bilateral lobes,* each directed superiorly upon the side of the trachea,

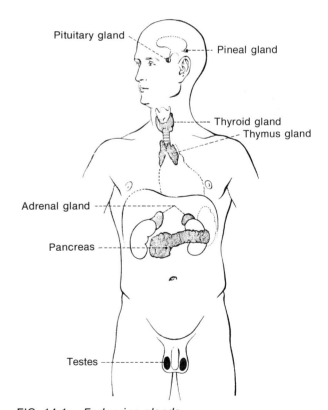

FIG. 14-1. *Endocrine glands*

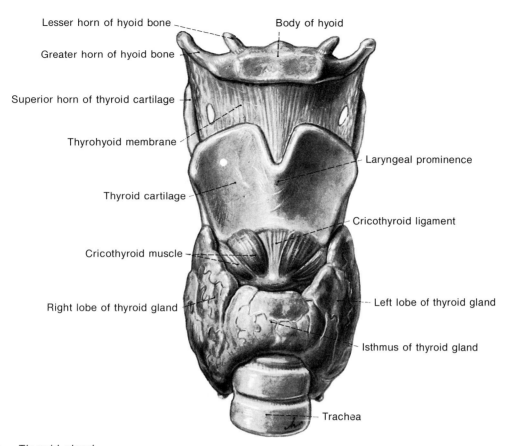

FIG. 14-2A. *Thyroid gland*

and connected across the midline inferiorly by the *isthmus,* commonly at about the level of the third tracheal ring. Often from one lobe or the other there is a short, superiorly-directed, *pyramidal lobe* near the midline. Frequently when this is present, and at times when it is not, there extends from it or the comparable part of the gland to the thyroid cartilage or hyoid a band that is either fibrous, the *median suspensory ligament,* or muscular, the *levator glandulae thyroideae muscle.* This levator muscle merely represents a detached portion of the infrahyoid musculature.

Arterial supply of the thyroid is by the superior and inferior thyroid arteries, and occasionally by the unpaired thyroidea ima artery from the brachiocephalic artery. Venous drainage is from a rich plexus to the superior, medial, and inferior thyroid (internal jugular), and quite frequently the thyroidea ima vein, to the left innominate vein. Innervation is autonomic, mainly sympathetic from the middle and inferior cervical ganglia, and also with parasympathetic vagal fibers from the laryngeal nerves.

PARATHYROID GLANDS

The *parathyroids* (fig. 14-2C and D) are two pairs of glands usually about the size of a pen embedded in the dorsal surface of the thyroid and within its fascia. An average position of the inferior unit is close to the trachea near the inferior border of the thyroid, and of the superior, close to the inferior larynx at about the middle of the lobe. There is much individual variation, however, both in the size of the parathyroids and in their positions, which may be below, beside, or above the thyroid; and even in their number, alleged at times to be less than 4 or as many (but smaller) units as 16. The color varies from pink to yellowish or brownish.

The parathyroid glands control the calcium metabolism, and hence the deposition of calcium in the body. The arterial supply and venous drainage is by adjacent branches of the thyroid vessels, and innervation is grossly in common with that of the thyroid gland.

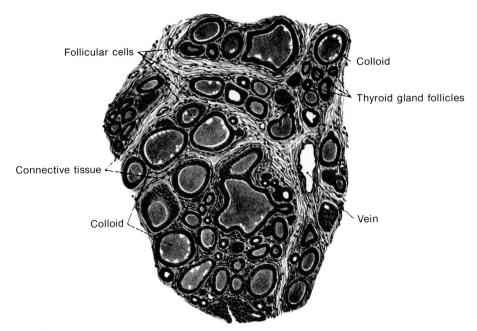

FIG. 14-2B. *Thyroid gland (microscopic view)*

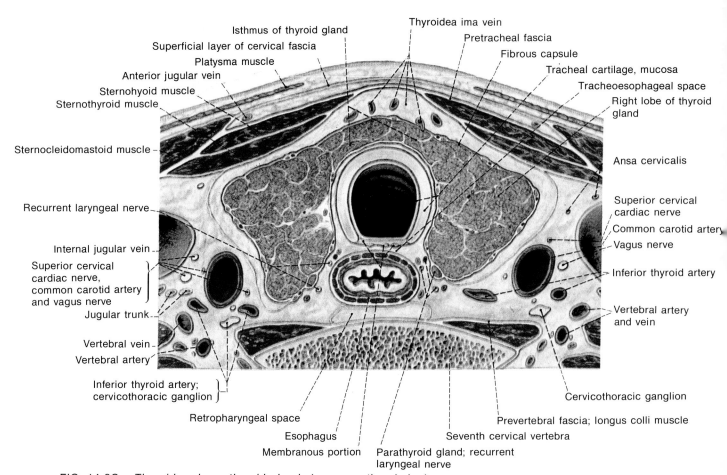

FIG. 14-2C. *Thyroid and parathyroid glands (cross sectional view)*

THYMUS GLAND

The *thymus gland* is an originally bilateral mass of soft, finely lobulated glandular tissue in the ventral part of the superior mediastinum between the thoracic wall on one side, and the pleura, pericardium, and great vessels on the other. It consists of a right and a left gland, termed *lobes,* joined by connective tissue, and it is of an irregular, flattened shape. At birth (fig. 14-3*A*) it is relatively larger than later, its absolute size being largest at puberty; between these times it grows only slowly. After puberty (fig. 14-3*B*) it usually, although not invariably, decreases in size until in middle life it no longer can be recognized as a gland, but is merely a mass of fibrous tissue in the superior mediastinum, at times with a fibrous connection to the thyroid gland.

The arterial supply is by branches of the internal thoracic artery; venous drainage is by the internal thoracic vein and left subclavian vein; and innervation by sympathetic and vagal parasympathetic fibers.

PITUITARY GLAND (HYPOPHYSIS)

The *pituitary gland* or *hypophysis* (fig. 14-4*A*) is a small mass consisting of an *anterior* and a *posterior lobe,* connected with the hypothalamus by a short stalk, the *infundibulum,* which joins the posterior lobe. The posterior lobe constitutes 20–25% of the whole. The gland is enclosed in a pocket of dura mater constricted around the infundibulum, and is situated in the hypophyseal fossa of the endocranium between the tuberculum sellae and the dorsum sellae. The pituitary gland is surrounded by the cavernous and intercavernous sinuses.

The arterial supply of the pituitary gland (fig. 14-4*B*) is by the hypophyseal twigs of the internal carotid artery and from several parts of the circle of Willis; venous drainage is to the cavernous and intercavernous sinuses. Innervation of at least a portion of the gland is by way of the supraopticohypophyseal tract. There may also be sympathetic innervation via the arteries.

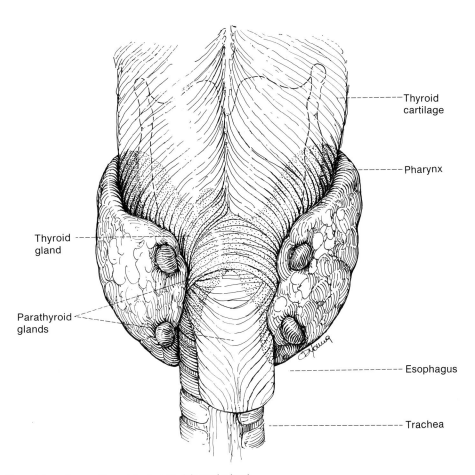

FIG. 14-2D. *Thyroid and parathyroid glands (dorsal view)*

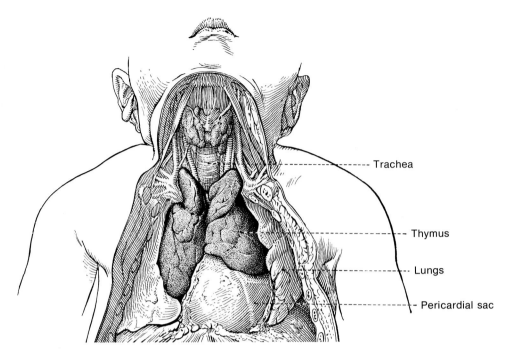

FIG. 14-3A. *Thymus gland (newborn)*

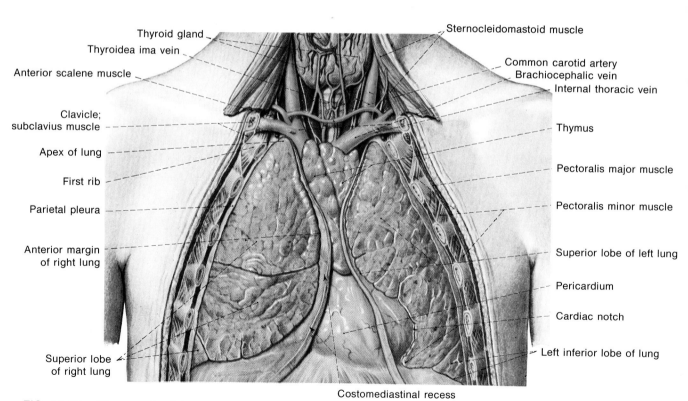

FIG. 14-3B. *Thymus gland (young adult)*

PINEAL BODY (EPIPHYSIS)

The *pineal body* (*epiphysis*) (Fig. 14-5) develops in the midline from the epithalamic part of the diencephalon as an evagination of the roof of the third ventricle. It is a small, oval structure with a hollow *stalk,* located between the superior colliculi of the mesencephalon.

SUPRARENAL GLANDS

The *suprarenal glands* (*adrenal glands*) (fig. 14-6) are paired structures that cap the superior end of the kidneys. Each has a thick *cortex* surrounding a much smaller *medulla.* The cortical part arises from celomic mesoderm, and parts of it may become separated from the main mass to form accessory cortical bodies, which are at times carried by adjacent shifting organs that carry them to distant points, as in the broad ligament of the uterus or into the scrotum. In contrast, the medulla is a part of the chromaffin system that has become associated with the cortex; thus the suprarenal gland in part belongs to this system. In the fetus the cortex is relatively bulky, but shrinks toward term, and at birth two cortical portions are recognizable: a peripheral *true cortex* overlying a *fetal cortex.* The latter experiences a relative degeneration during early infancy, but the gland as a whole has an accelerated growth during late childhood.

Each suprarenal gland is an irregular mass of extremely vascular tissue, with a fibrous capsule surrounded by areolar tissue within the renal fascia. Its shape is largely influenced by the pressure of surrounding structures and it is often associated in old age with considerable soft fat. The right suprarenal is subtriangular and the left is more semilunar and slightly larger. In relation to each gland is the kidney inferomedially and the diaphragm dorsally. In addition, the right gland is in relation to the vena cava medially, the liver superiorly, and at times the duodenum ventrally; the left gland with the omental bursa ventrally and the stomach, splenic vessels and pancreas, and sometimes the spleen laterally. The hilus of the gland medially is marked by the entry of the vessels of the gland.

The arterial supply of the suprarenal gland is by three sources: the superior suprarenal artery, from the inferior phrenic artery; the middle suprarenal from the aorta; and the inferior suprarenal from the renal artery. Venous drainage is by the single suprarenal vein, the right emptying into the inferior vena cava and the left into the left renal vein. There is minor and accessory drainage of the capsule to the inferior phrenic vein or to the small veins of the surrounding adipose tissue. Innervation of the medulla is exclusively sympathetic, through the greater splanchic nerve and the suprarenal plexus.

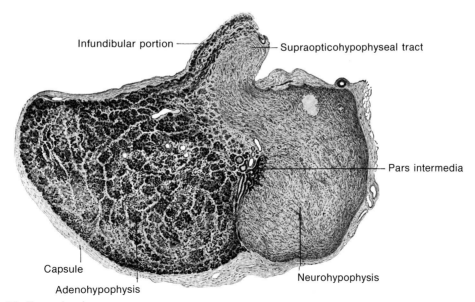

FIG. 14-4A. *Pituitary gland*

CHROMAFFIN SYSTEM

The *chromaffin system* (fig. 14-7) is composed of numerous small or minute bodies serially arranged along the aorta or its continuations, comprising paraganglia, carotid, aortic, and coccygeal bodies. The system gets its name from the affinity for chromium salts, effecting a brownish stain, and the units arise in common with, and from the same ectodermal tissue of the ganglionic crest as, the sympathetic ganglia. Accordingly most of them are in intimate association with the sympathetic ganglia, which is why they are termed *paraganglia*.

The paraganglia of the sympathetic trunk are minute bodies within or close to the sheaths of the sympathetic trunk ganglia, while other collateral paraganglia are scattered throughout the sympathetic ganglia associated with the viscera of the trunk cavities.

Chromaffin bodies in direct relationship with the aorta or its continuations are considered to number three. The *carotid body* is a small oval structure, sometimes double, in the vicinity of the bifurcation of the common carotid artery. It has many fine blood vessels and, in addition, nerve filaments to the carotid plexus and a longer one—the *carotid sinus nerve*—which passes with the pharyngeal branch of the glos-

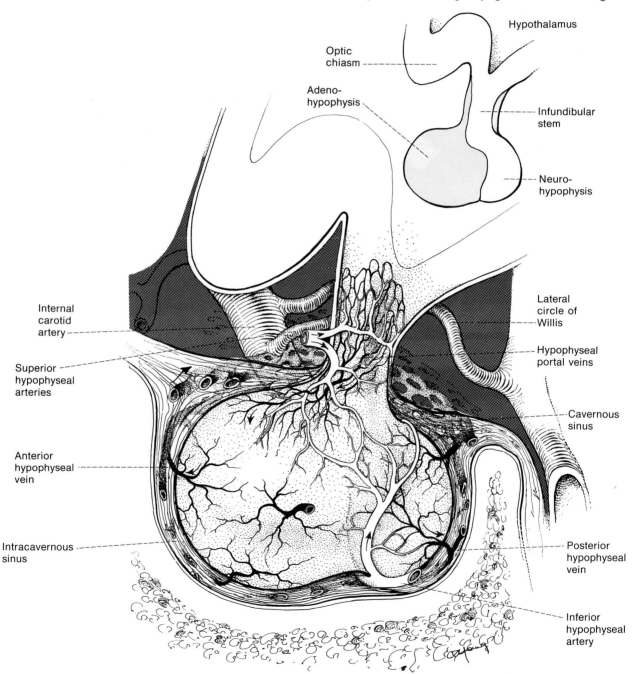

FIG. 14-4B. *Blood supply to the hypophysis: pituitary gland*

sopharyngeal nerve to join both the latter nerve and the ganglion nodosum of the vagus nerve. The *aortic bodies* (*paraganglia aortica*) are a pair of bilateral structures associated with the collateral paraganglia in the vicinity of the origin of the inferior mesenteric artery. They tend to shrink after infancy, and disappear (grossly) by puberty. The *coccygeal body* is a small, unpaired structure ventral to the tip of the coccyx in relationship to the terminal part of the middle sacral artery.

OVARIES AND TESTES

The *ovaries* (fig. 14-8*A*) produce female sex hormones (estrogen and progesterone) that are responsible for the development and maintenance of the female sexual characteristics.

The *testes* (fig. 14-8*B*) produce the male sex hormones (testosterone) that stimulate the development and maintenance of the male sexual characteristics.

PANCREAS

The *pancreas* (fig. 14-9) may be classified as both endocrine and exocrine. The *endocrine* portion of the pancreas consists of cells clustered together to form the *islets of Langerhans*. The beta cells secrete insulin and the alpha cells secrete glucagon. The *exocrine* portion is concerned with digestive juices to the digestive system.

CLINICAL APPLICATIONS

Disorders of the pituitary may include *growth dysfunctions* (*dwarfism, giantism,* and *acromegaly*).

Disorders of the thyroid may cause *cretinism* or *goiters.*

Disorders of the parathyroids may cause *tetany* or *osteitis fibrosa cystica.*

Disorders of the suprarenal glands may cause *Addison's disease* (hyposecretion of glucocorticoids) or *Cushing's syndrome* (hypersecretion of glucocorticoids).

Disorders of the pancreas may cause *diabetes mellitus* (hyposecretion of insulin).

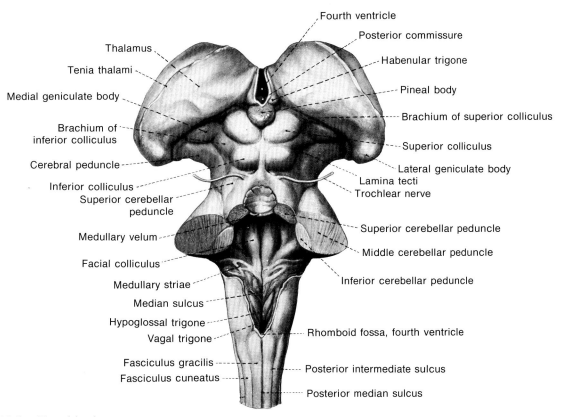

FIG. 14-5. *Pineal body*

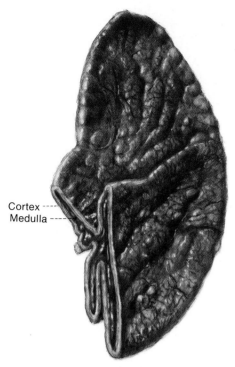

FIG. 14-6A. *Suprarenal gland*

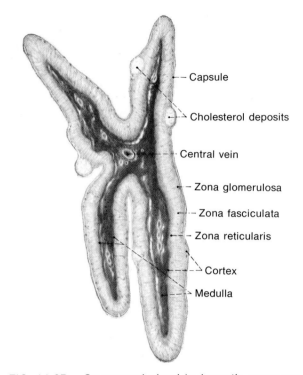

FIG. 14-6B. *Suprarenal gland (schematic cross-section)*

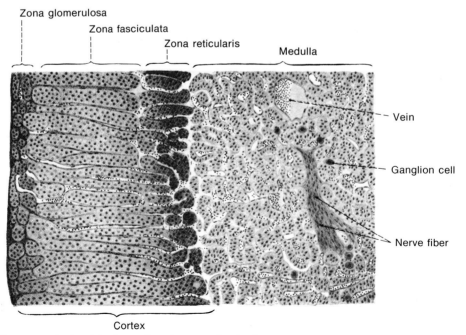

FIG. 14-6C. *Suprarenal gland (schematic microscopic view)*

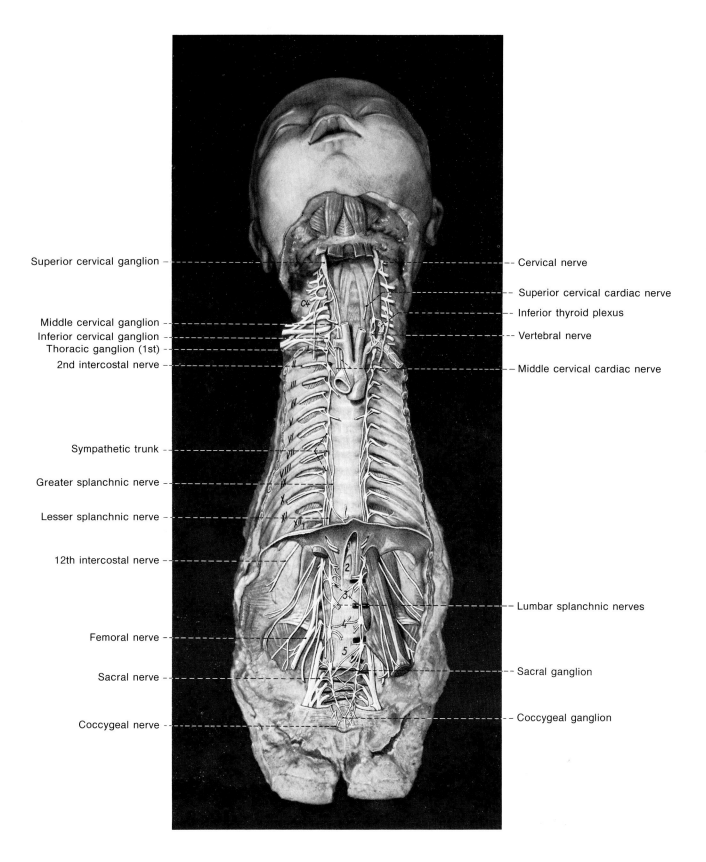

FIG. 14-7A. *Chromaffin system*

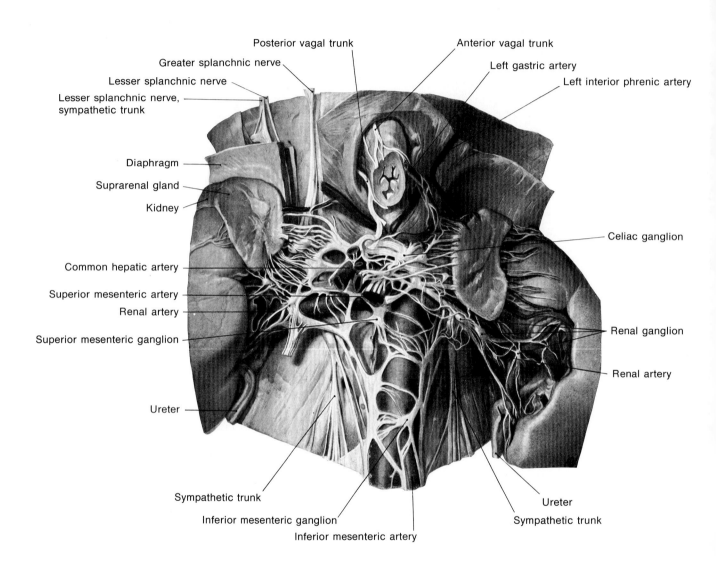

FIG. 14-7B. *Ganglia of the chromaffin system*

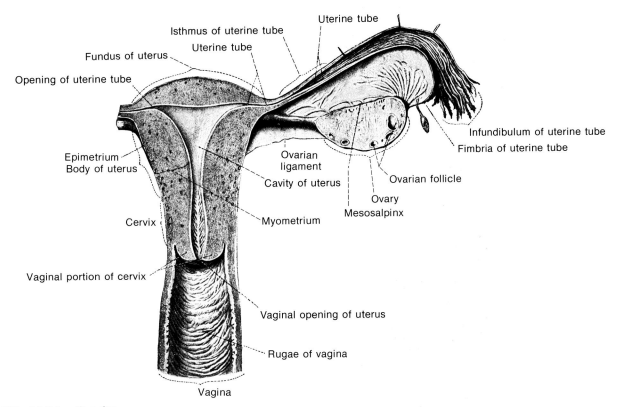

FIG. 14-8A. *Ovaries*

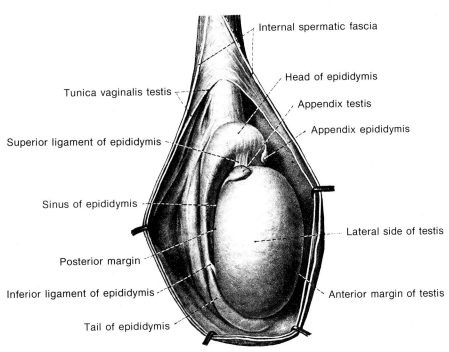

FIG. 14-8B. *Testes*

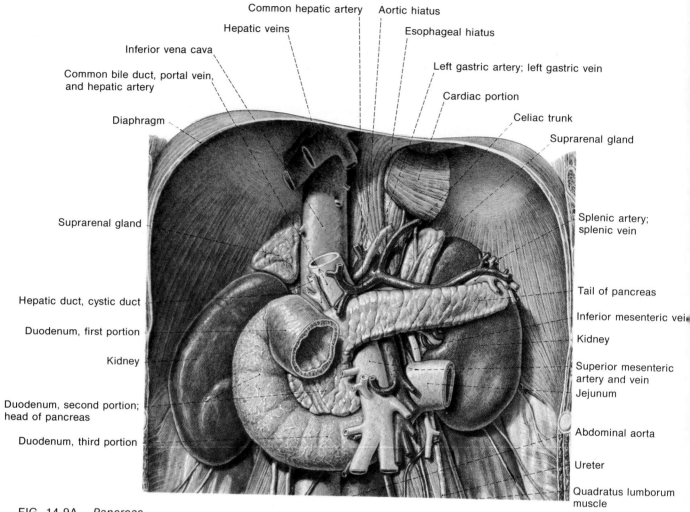

FIG. 14-9A. Pancreas

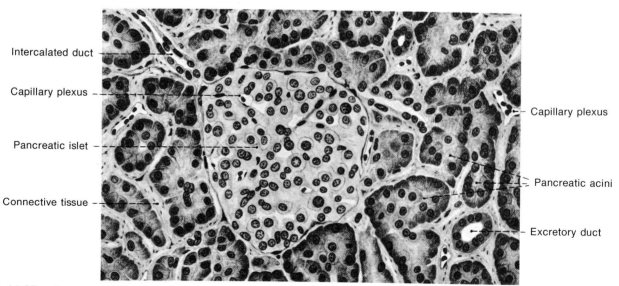

FIG. 14-9B. Pancreas (*microscopic view*)

REVIEW QUESTIONS:

1. The _____ is a small mass of tissue consisting of an anterior and a posterior lobe, connected with the hypothalamus by the _____.
2. The _____ develops in the midline from the epithalamic part of the diencephalon as an evagination of the roof of the third ventricle.
3. The _____ cap the superior end of the kidneys.
4. The _____ is composed of numerous small bodies serially arranged along the aorta or its continuations, comprising paraganglia, and carotid, aortic, and coccygeal bodies.
5. The ovaries produce _____ and _____.
6. The testes produce _____.
7. The endocrine portion of the pancreas consists of cells clustered together to form the _____.
8. The _____ gland is located in the ventral part of the superior mediastinum.
9. The _____ influence calcium metabolism.
10. Arterial supply of the thyroid is by the _____ and _____ arteries.

Answers

CHAPTER 1
1. structure
2. forward
3. plantar
4. distal
5. lateral
6. right
7. sagittal, coronal
8. chemical
9. tissue
10. tissues
11. regional, systemic
12. systemic

CHAPTER 2
1. skin, integument
2. epidermis, dermis
3. lucidum
4. corneum
5. basale
6. papillae
7. crescent
8. arrector pili
9. sudoriferous
10. ciliary, ceruminous, mammary
11. superficial
12. third

CHAPTER 3
1. axial, appendicular
2. optic, superior orbital
3. zygomatic arch
4. maxillae, nasal
5. mental
6. lambdoid
7. pterygopalatine
8. mandibular
9. Eustachian
10. sagittal
11. anterior, middle, posterior
12. dorsum sellae, posterior clinoid process
13. arcuate
14. foramen magnum
15. optic canal
16. superior orbital fissure
17. maxillary
18. ethmoid
19. hiatus semilunaris
20. maxillary
21. glossopharyngeal, vagus, spinal accessory
22. intervertebral foramina
23. transverse foramen
24. axis
25. thoracic
26. anulus fibrosus
27. true ribs
28. costal groove
29. manubrium, body, xiphoid process
30. supraspinous, subscapular, infraspinous
31. acromion
32. manubrium, acromial
33. olecranon
34. styloid
35. scaphoid, lunate, triangular, pisiform
36. pelvic girdle
37. ischium, ilium, pubic
38. acetabulum
39. medial, femur
40. patella
41. medial malleolus
42. calcaneus
43. radial
44. coronoid
45. two vertebrae
46. atlas
47. 7th cervical
48. pterygopalatine, nasal
49. middle cranial, pterygopalatine
50. pterygoid

CHAPTER 4
1. synarthrosis—immovable joints amphiarthrosis—slightly movable joints diarthrosis—freely movable joints
2. hyaline
3. sphenomandibular
   stylomandibular
   lateral temporomandibular
4. atlantooccipital
5. anulus fibrosus
6. posterior longitudinal ligament
7. temporomandibular and sternoclavicular
8. conoid and trapezoid
9. medial malleolus, calcaneus, navicular
10. anterior, posterior

CHAPTER 5
1. facial
2. depresses
3. trigeminal
4. abducens

5. oculomotor
6. trochlear
7. hypoglossal
8. genioglossus
9. accessory
10. a. sternocleidomastoid
    b. semispinalis capitis
    c. splenius capitis
    d. longissimus capitis
11. cervical
12. scapula
13. hyoid
14. facial
15. mandible, 1st cervical
16. a. iliocostalis
    b. longissimus
    c. spinalis
17. a. external abdominal oblique
    b. internal abdominal oblique
    c. transverse abdominal
    d. rectus abdominal
18. diaphragm, phrenic
19. coracoid
20. a. levator scapulae
    b. rhomboideus major
    c. rhomboideus minor
21. long thoracic
22. thoracodorsal
23. abducts, axillary
24. greater tubercle, humerus
25. olecranon, ulna
26. musculocutaneous
27. median
28. ulnar
29. radial
30. median
31. radial
32. obturator
33. femoral
34. flexes
35. plantar, inverts

CHAPTER 6

1. central peripheral
2. fissures, sulci
3. longitudinal
4. lateral
5. orbital, triangular, opercular
6. insula
7. fusiform
8. splenium
9. cuneus
10. cingulate
11. septum pellucidum
12. putamen, globus pallidus
13. claustrum
14. internal capsule
15. cerebral aqueduct
16. epithalamus, thalamus, subthalamus, hypothalamus
17. interventricular foramen
18. pineal body
19. thalamus
20. tectum
21. oculomotor
22. basal, tegmental
23. glossopharyngeal, vagus, accessory
24. pons
25. vermis
26. conus medullaris
27. dorsolateral
28. cauda equina
29. leptomenix
30. denticulate
31. filum terminale
32. vertebral, radicular
33. cavernous sinus
34. trochlear
35. mandibular
36. abducens
37. glossopharyngeal
38. hypoglossal
39. C_2, C_4
40. autonomic
41. ciliary, pterygopalatine, submandibular, otic
42. ventricular
43. fimbria, hippocampus
44. thalamus
45. trochlear
46. external
47. lamina terminalis
48. infundibulum
49. uncus
50. lateral geniculate body

CHAPTER 7

1. mitral, tufted, granule
2. vorticose
3. vitreous body
4. circular, meridional
5. oculomotor
6. optic
7. pterygoid
8. palpebral conjunctiva
9. palpebral aperture
10. gustatory
11. facial
12. glossopharyngeal
13. cochlear
14. antihelix
15. external acoustic meatus
16. tensor tympani
17. malleus, incus, stapes
18. facial
19. petrotympanic
20. membranous, bony
21. membranous, perilymphatic
22. basilar, vestibular
23. glossopharyngeal
24. auditory tube
25. three

CHAPTER 8

1. vestibule
2. palate
3. sublingual
4. foramen cecum
5. parotid, submandibular, sublingual
6. parotid
7. oral
8. cricoid
9. parietal
10. falciform
11. hepatogastric, hepatoduodenal
12. bile duct, hepatic artery, portal vein
13. epiploic
14. rugae
15. duodenum, jejunum, ileum
16. common bile duct, main pancreatic duct
17. accessory pancreatic duct
18. appendices epiploicae
19. teniae coli
20. hepatic
21. anus
22. liver
23. porta hepatis
24. cystic duct
25. pancreas

CHAPTER 9

1. xiphisternal junction
2. 4th
3. fibrous, serous
4. middle
5. diaphragmatic
6. epicardium
7. endocardium
8. limbus fossae ovalis
9. valvulae foramina ovalis
10. coronary sulcus
11. trabeculae carneae
12. coronary sinus
13. tricuspid
14. aortic semilunar
15. atrioventricular bundle
16. tunica media, tunica intima, tunica adventitia
17. coronary
18. brachiocephalic, left common carotid, left subclavian
19. brachiocephalic
20. 8
21. external carotid
22. brachial
23. celiac trunk
24. internal, external iliac
25. superior gluteal
26. femoral
27. great cardiac
28. brachiocephalic
29. posterior auricular, retromandibular
30. inferior sagittal sinus
31. cavernosus sinus
32. ophthalmic
33. cephalic, basilic
34. azygos
35. small, great saphenous
36. splenic, superior mesenteric
37. umbilical
38. umbilical
39. ligamentum venosum
40. pampiniform plexus

CHAPTER 10

1. cisterna chyli
2. right lymphatic trunk
3. internal jugular, submandibular
4. anterior
5. great vessels
6. paraaortic
7. inguinal region
8. spleen
9. thymus
10. tonsils
11. palatine tonsils, pharyngeal tonsils, lingual tonsils
12. inguinal nodes

CHAPTER 11

1. ethmoid, vomer, septal
2. inferior nasal meatus
3. frontal, maxillary, sphenoidal, ethmoid
4. semilunar hiatus
5. 3, 3
6. epiglottic
7. rima glottidis, glottis
8. trachea
9. right
10. mediastinum
11. first
12. oblique
13. eparterial
14. aorta, intercostal
15. hilus
16. parietal, visceral
17. middle
18. false
19. conus elasticus
20. corniculate, cuneiform

CHAPTER 12

1. renal columns
2. papilla
3. nephron
4. glomerulus
5. ureter
6. major calices
7. urachus, median umbilical ligament
8. internal iliac
9. glomerular (Bowman's) capsule
10. efferent arteriole
11. cortex, medulla
12. interlobular, arcuate, interlobar

CHAPTER 13

1. transversalis fascia
2. processus vaginalis
3. external spermatic
4. tunica vaginalis
5. tunica albuginea
6. paradidymis
7. ductus deferens
8. ductus deferens, ejaculatory duct
9. ischiocavernosus
10. prostatic, membranous, spongy
11. membranous, sphincter urethrae membranaceae
12. spongy
13. bulbocavernosus
14. prostate, bulbourethral glands, seminal vesicles
15. prostate gland
16. mesosalpinx
17. endometrium
18. myometrium
19. fundus
20. mesometrium
21. vestibule
22. bulbocavernosus
23. external genitalia, vulva
24. ovarian ligament
25. external os, internal os

CHAPTER 14

1. pituitary (hypophysis), infundibulum
2. pineal body (epiphysis)
3. suprarenal glands (adrenal glands)
4. chromaffin system
5. estrogen, progesterone
6. testosterone
7. islets of Langerhans
8. thymus
9. parathyroids
10. superior, inferior thyroid

Index

Abdominal aorta 282, 309, *291*, 306*, 310**
abdominal cavity 252, 256
abdominal pregnancy 421
abducens nerve 137, 173, 179, 222, 223
abscess 274
accessory hemiazygos vein 328
accessory nerves 137, *181*, 182**
accessory pancreatic duct 272
accessory plantar ligament 81
acetabulum 62
acromegaly 433
acromial artery 299
acromial process 57
acromioclavicular ligament 73
acromioclavicular articulation *73–75**
acute glaucoma 237
acute mastoiditis 237
acute renal failure 395
Addison's disease 433
adductor brevis 117
adductor longus 117
adductor magnus 117
adductor tubercle 64
aditus 232
adrenal glands 431
afferent arteriole 392
ala 31
alveolar border 33
alveolar process 33
alveoli 31, 33, 239, 381
amphiarthrosis 70
ampulla 233, 235, 267, 401, 410
  of Vater 272
amygdala 129
amygdaloid nucleus 130
anal sinuses 267
anatomical position 3, *4**
anatomy 3
  applied 4
anconeus 100
angle 33, 39, 49, 225
angular branch 284, 285
angular gyrus 124
angular vein 220
ankle 60
annular ligament of the radius 76
anorectal node 357
ansa cervicalis 187
ansa subclavia 210
antebrachial cutaneous nerve 191
anterior 3, 35, 90, 150, 292, 299, 307
anterior abdominal wall
  anteriolateral view *95**
  anterior view *96**
  inguinal region *96**
  lateral view *95**
  posterior view *97**
  and scrotal homologues *398*, 401**
  transverse section *96**
anterior atlantooccipital membrane 72
anterior band 75
anterior branch 195
anterior cardiac veins 319
anterior cerebral artery 151, 157
anterior circumflex humeral artery 299
anterior clinoid process 36
anterior commissure 128, 130
anterior communicating arteries 151, 157

anterior compartment 101, 118
anterior cruciate ligament 79
anterior cutaneous branches 195
anterior deep temporal vein 288
anterior division 312
anterior ethmoid canal 36
anterior ethmoid foramina 41
anterior ethmoidal sinuses 39
anterior femoral compartment 101
anterior gray column 146
anterior horn 129
anterior inferior cerebellar artery 153, 296
anterior intercavernous sinus 323
anterior interventricular branch 282–83
anterior interventricular sulci 278
anterior limb of the internal capsule 129
anterior lobe 429
anterior longitudinal ligament 73
anterior median fissure 135, 145
anterior mediastinum 379
anterior medullary velum 132, 140
anterior muscular compartments 117
anterior nasal aperture 33, 37
anterior nasal spine 33
anterior nodes 353
anterior perforated substance 127, 128
anterior pole 214
anterior process 230
anterior ramus 201
anterior sacroiliac ligament 76
anterior scrotal veins 335
anterior spinal artery 150, 153, 296
anterior superior iliac spine 62
anterior superior pancreaticoduodenal artery 310
anterior talofibular ligament 79
anterior tibial arteries 313, *316**
anterior tibial recurrent artery 316
anterior tympanic artery 285
anterior ulnar recurrent artery 305, 307
anterior white commissure 146
anterolateral sulcus 135
antihelix 227
antitragus 227
anulus fibrosus 45, 73
anus *266*, 267**
aorta 280, 281, 357, *291*, 292*, 293**
  abdominal 282, *306**, 309, *310**
  descending 282
  isthmus 282
  parietal and paired visceral branches of 310
  thoracic *306**, 307
aortic bodies 433
aortic semilunar valve 280
aortic sinuses 281
aortic sound 341
aortic spindle 282
aorticorenal plexus 211
apex 275, 277, 388, 392
apex of the lung 381
appendages of the skin 26
appendices epiploicae 264
appendicitis 272
appendicular artery 311
appendix 266
aqueous humor 218
arachnoid membrane 146
arachnoid granulations 148

arch of the aorta 281, *293**
arcuate artery 392
arcuate vein 392
areola area 28
arm 55
  muscles 100
arrector pili muscle 26
arrhythmias 340
arterial circle of Willis *161**, 292
Arteries *16*, 276*,* 280, 289, *290**
  anterior cerebral 151, 157
  anterior circumflex humeral 299
  anterior communicating 151, 157
  anterior inferior cerebellar 153, 296
  anterior spinal 150, 153, 296
  anterior superior pancreaticoduodenal 310
  anterior tibial 313, *316**
    recurrent *316**
  anterior tympanic 285
  anterior ulnar recurrent 305, 307
  appendicular 311
  arcuate 392
  ascending cervical 297
  ascending palatine 284
  ascending pharyngeal 284
  axillary 298, 299, *302**
  basilar 135, 151, 153, 295, 296
  brachial 295, 298, 299, *303**, 305
  brachiocephalic 283
  bronchial 308, 383
  caroticotympanic 292
  carotid 151
  celiac 357
  central 217, *220**
  central, of retina 292
  cerebal *300**
  choroid 158, 292
  circumflex scapular 299
  common hepatic 310
  common iliac 312
  common interosseous 305
  coronary *292*, 322**
  cortical 151
  cricothyroid 376
  cystic 310
  deep auricular 285
  deep brachial 299
  deep cervical 298
  deep circumflex iliac 313
  deep plantar 318
  deferential 312
  dorsal carpal branch 305–7
  dorsal digital 316–318
  dorsal metatarsal 316
  dorsal pedis 316
  dorsal scapular 295, 298
  esophageal 308
  external carotid 283, *295**, 376
  external iliac *311**, 312, 313
    facial 284
  femoral 313, *314**
    of the foot *317**
    of the hand *305**
    highest intercostal 298
    highest thoracic 298
  ileocolic 310
  iliolumbar 312
  inferior epigastric 297, 313

443

Arteries
　inferior gluteal 312
　inferior lateral 316
　inferior mesenteric *309**, 311, 357
　inferior pancreaticoduodenal 310
　inferior phrenic 311
　inferior rectal 312
　inferior suprarenal 312
　inferior thyroid 297, 376
　inferior ulnar collateral 305
　infraorbital 288
　interlobular 392
　internal auditory 295
　internal carotid 151, 158, 283, 288
　internal iliac *311**, 312
　internal pudendal 312
　internal thoracic 297
　interosseous 307
　jejunal 310
　lacrimal 292
　lateral plantar 316, 318
　lateral sacral 312
　lateral striate 158
　left common carotid 282, 283
　left coronary 281, 282
　left gastric 309
　left gastroepiploic 310
　left pulmonary 280
　left subclavian 282
　lingual 284
　of the lower limb *313**
　lumbar 312
　main arterial channel in upper limb 295
　maxillary 284, 285
　　branches *296**
　medial 316
　medial cerebral 157
　medial femoral 101
　medial plantar 318
　medial sacral 311
　medial striate 158
　mediastinal 308
　middle cerebral 151
　middle colic 310
　middle meningeal 285
　middle rectal 312
　middle suprarenal 311
　musculophrenic 297
　obturator 312
　occipital 284, 285
　ophthalmic 220, *222**, 292, *299**
　ovarian 312
　pericardial 308
　peroneal 316
　plantar 316
　plantar metatarsal 318
　popliteal 313, 315, 316
　posterior auricular 284, 285
　posterior cebellar 153, 296
　posterior cerebral 151, 295, 296
　posterior circumflex humeral 299
　posterior communicating 151, 158, 292
　posterior interosseous 307
　posterior radicular 150
　posterior spinal 150, 153, 296
　posterior tibial 313, 316
　posterior tibial recurrent 316
　posterior ulnar recurrent 305, 307
　radial 305, 306
　radial recurrent 305, 306
　radicular 150
　renal 311, 392
　right colic 311
　right common carotid 283
　right coronary 281, 282
　right gastroepiploic 310
　right subclavian 283
　sigmoid 311
　sphenopalatine 288
　splenic 310

Arteries
　superficial circumflex iliac 313
　superficial epigastric 313
　superficial temporal 284, 288
　superior cerebellar 153, 296
　superior epigastric 297
　superior gluteal 312
　superior labial 168
　superior mesenteric *308**, 310, 357
　superior phrenic 308–309
　superior rectal 311
　superior suprarenal 311
　superior thyroid 284, 376
　superior ulnar collateral 299
　supraorbital 292
　suprascapular 297
　supraspinous 292, 353
　testicular 312
　transverse facial 288
　ulnar 305, 306
　umbilical 335, 340
　upper limb 295, *301**
　uterine 312
　vertebral 150, 151, 153, 295
arterioles 281
arteriosclerosis 340
articular 4, 316
articular disk 70
articular eminence 35
articular facets 49
articular process 39
articular tubercle 72
articularis genus 113
articulation 70
aryepiglottic fold 251, 370
arytenoid cartilages 370
ascending aorta 281, *292**
ascending cervical artery 297
ascending colon 254, 256, 266
ascending palatine branch 284
ascending pharyngeal artery 284
atheromas 340
atherosclerosis 340
atlantooccipital articulation *72**, *73**
atlas 45, *49**
atrial fibrillation 340
atrial flutter 340
atrial septal defects 340
atrioventricular bundle 280
atrioventricular (AV) node 280
auditory tube *231**, 232 369
auricle *227**
auriculotemporal nerve 172
auricular branch 177
autonomic nervous system 122, *199**, 207
axillary artery 295, 298, 299, *302**
axillary group 353
axillary nerve 192
axillary nodes 353
axillary vein 328
axis 45, *49**
azygos vein 328, *332**

Bacterial 212
bare area 271, 382
basal cell 30
basal portion 135
basal vein 158
basilar artery 135, 151, 153, 295, 296
basilar membrane 235
basilar sinus 323
basilar venous plexus 159
basilic vein 328, *330**
benign hypertrophy 421
biceps 100
biceps femoris 113
bicuspid 280
bicuspid sound 341

bifid pelves 395
bifid ureters *393**, 395
bilateral lobes 426
bladder and urethra (internal view) *395**
blade 62
blind spot 217
blood flow, path of *285**
Bones *10**
　calcaneus 64
　capitate 60
　carpals 60, *61**
　coxal 53, 60, *63**
　cuboid 64
　cuneiform 64
　eminence of petrous portion of temporal
　　bone 36
　ethmoid 37, 367
　facial 42, *46**
　fibula 64 *66**
　of the forearm *60**
　frontal 33, 35, 37
　hamate 60
　lacrimal 37
　of the lower limbs 53
　lunate 60
　malleus 230
　mandible 39, 42
　manubruim 53
　maxilla 33, 37
　medial malleolus 64
　metacarpals 60, *61**
　metatarsals 64, *67**
　middle nasal conchae 37
　nasal 33
　occipital 33, 35
　palatine 37
　parietal 33, 35
　patella 64, *65**, 79
　phalanges:
　　mani- 60, *61**
　　pedi- 64, *67**
　pisiform 60
　sacral vertebrae 45, *52**
　sacroiliac 76
　scaphoid 60
　scapula 53, 55, *58**
　sphenoid 33, 37
　sternum 53, *56**
　talus 64
　tarsals 64, *67**
　tarsus 225
　temporal 33, 36, 123
　tibia 64, *66**, 79
　trochanter 31, 64
　of the upper limbs 53
　vomer 33, 37, 367
　zygomatic 33, 37
bony orbit 219, *221**
bony pelvis 60
Bowman's capsule 392
brachialis 100
brachial artery 295, 298, 299, *303**, 305
brachial plexus 187, *189**
brachiocephalic artery 282, 283
brachiocephalic veins 321
brachium of the inferior colliculus 132
brachium of the superior colliculus 132
brain, blood supply of: *158**, *159**, *160**,
　*161**
brain stem 122, 130, *132**, *133**, *134**, *135**,
branchiomeric motor (special visceral motor)
　162
breast cancer 30
broad ligament 407, 410
bronchi 367, *374**, *375**, 376
bronchial arteries 308, 383
bronchial asthma 384
bronchioles 381
bronchopulmonary segments of the lung *385**
buccal artery 288

buccal nerve 170, 174
bulb 26
bulbourethral glands 404, 407, 417
bulbus penis 403, *411**
bulla ethmoidolis 37, 369
burns 30
bursae 70

Calamus scriptorius 140
calcaneal branches 316
calcaneofibular ligaments 79
calcaneovalgocavus 120
calcaneus 64
calcarine fissure 128
calculi 274
calvaria 35
canal 31
　of the body 410
　central 145
　of the cervix 410
　of the skull 39
cancer 274, 384, 395, 421
　breast 30
　liver 272
　skin 30
canines 244
canthi 225
capitate 60
capitulum 57, 230
capsule of Glisson 270
cardia 260
cardiac branches 179
cardiac nodes 357
cardiac plexuses 211
cardiac veins 318, *321*, 322**
cardiovascular system *276*, 277**
caroticotympanic artery 292
carotid artery 151
carotid body 432
carotid canal 35, 36, 41
carotid groove 36
carotid sinus 283
carotid sinus nerve 432
carpals 60, *61**
carpometacarpal articulations 76, *77**
cartilages 370
　of the nose *367**
carunculae hymenalis 417
cauda equina 141
caudal 3
caudate 129
caudate lobe 270
caudate nucleus 129, 130
cavernous artery 211, 295
cavernous sinus 159, 220, 325
cavity of the nose 37
cecum 266
celiac artery 357
celiac nodes 357
celiac trunk *307*, 309*
cellular level 4
cementum 244
central artery 217, *220**
central artery of the retina 292
central canal 145
central gelatinous substance 146
central lobule 141
central nervous system *122**
central pulp cavity 244
central sulcus 123
cephalic vein 328, *330**
cerebellum 122, 140, *142*, 143**
cerebral aqueduct 130
cerebral arteries *300**
cerebral cortex of the insula 129
cerebral fossae 35
cerebral palsy 212
cerebral peduncles 135
cerebral veins 158, *163*, 164*, 325

cerebral vessels 325
cerebrovascular accidents 212
cerebrum 122
ceruminous glands 25, 227
cervical branch 174
cervical enlargements 144
cervical nerves 141, 146
　ventral ramus 187
cervical pleura 379
cervical plexus 187, *188*, 211*
cervical region 183
cervical sympathetic trunk and ganglia *207**
cervical vertebrae 42, *48**
cervix 410, *424**
chemical level 4
chorda tympani nerve 173, 225, 232, 237
chordae tendineae 280
choroid artery 158, 292
choroid layer 216
choroid plexus 129, 130
　of the fourth ventricle 140
chromaffin system 432, *435**
cilia 225
ciliary 162, 292
ciliary body 216
ciliary ganglion 207
ciliary glands 28, 225
ciliary muscle 217, *218*, 219*
ciliary processes 216
cingulate gyrus 128
cingulate sulcus 128
circular fibers 219
circular folds 263
circular sinus 160
circulus arteriosus 151
circumflex branch 283
circumflex scapular artery 299
circumvallate papillae 248
cirrhosis 272
cisterna chyli 348
claustrum 129
clava 137
clavicles 53, 55, 57, *59**
clavicular branches 299
clavicular facets 53
claw 119
clitoris 335, 417
coccygeal body 433
coccygeal ligament 147
coccygeal nerve 141, 146
coccygeal plexus 205
coccyx 45
cochlea 227, 232, *234**
cochlear aqueduct 233
cochlear duct 235
cochlear recess 235
cochleariform process 230
collateral fissure 127
collateral ligament 76, 81
collecting tubule 392
comminuted fractures 66
commissure of the bulb 417
common annular tendon 219
common bile duct 270, *271**
common carotids 294
common cold 384
common hepatic artery 307, 310
common hepatic duct 270
common iliac lymph nodes 357
common iliac artery *311*, 312*
common iliac vein 328, 332
common interosseous arteries 305, 307
compartments 101
compound fractures 64
compression fractures 64
concha 227
conduction system 280, *287*, 288*
condylar fossa 35, 41
condylar process 39
condyle 31, 33

confluens of the sinus 159, 323
congenital anomalies 395
congenital heart disease 340
conjoined tendon 94
conjunctival fornix 225
conoid ligament 75
conus elasticus 370
conus medullaris 141, 145
coracobrachialis muscle 100
coracoclavicular ligament 75
coracoid process 57, 75
cords 187
cornea 216
corneal segment 214
corniculate cartilages 370
coronal plane 4
coronal suture 33, 35
coronary arteries *292*, 322**
coronary branch 335
coronary ligaments 271
coronary sinus 278, 319
coronary sulcus 278
coronary vein 335
coronoid fossa 57
coronoid process 33, 39, 60
corpora cavernosa clitoridis 417
corpora cavernosa penis 402
corpus callosum 128, 129
corpus cavernosum urethrae 402, 403
cortex 129, 388, 431
cortical artery 151
costal cartilages 49
costal groove 49
costal pleura 379
costal surface 382
costocervical trunk 295, 298
costosternal articulation *74**
Cowper's gland 407
coxal bones 53, 60, *63**
cranial 3
cranial nerves 160, *169*, 170**
cranial sensory ganglia 162
cranium 35, *36*, 42*
cremaster muscle 94
cremasteric fascia 397
crescent 26
crest 31
cretinism 433
cribiform plates 35
cricoarytenoid articulation 370
cricoid cartilage 370
cricothyroid artery 376
crista galli 36
crown 239
crura 230, 417
crus of the helix 227
crus penis 402
cubital node 353
cuboid bone 64
culmen 141
cuneate tubercle 137
cuneiform bones 64
cuneiform cartilage 370
cuneus 128
cupula 233
Cushing's syndrome 433
cutaneous branches of the cervical plexus 187
cutaneous innervation
　of the inferior limb 205
　of the upper limb *191*, 193**
cystic artery 310
cystic branch 335
cystic duct 270
cystic vein 335
cystitis 395
cysts 274

Deciduous teeth 244, *246*, 247**
declive 141

decussation of the pyramids 135, 137
deep auricular artery 285
deep brachial artery 299
deep cardiac plexus 211
deep cervical artery 298
deep circumflex iliac artery 313
deep circumflex iliac vein 332
deep femoral vein 333
deep group 353
deep inguinal ring 94
deep jugular chain 350
deep lymphatics of the head and neck *351**
deep middle cerebral vein 158
deep muscular and dorsal penis branches 205
deep plantar artery 318
deep posterior compartments 117, 118
deep radial nerve 193
deep veins *319**
   of the arm 328
   of the lower limbs 335
   of the upper limbs *331**
deferential artery 312
deformities of the spinal column 64
deltoid ligament 79
deltoid muscle 299
deltoid tuberosity 57
deltopectoral node 353
dental branches 171
dental caries 272
dental pulp 244
denticulate ligaments 148
dentine 244
dermal papillae *24**
dermatome chart of spinal nerves:
   cutaneous distribution *184**
cutaneous innervation *152**
dermatome chart representing cutaneous
   innervation of upper limb *191**
dermis 23, *24**
descending aorta 282
descending colon 254, 257, 267
descent of the testes *402**
diabetes mellitus 433
diaphragmatic group 353
diaphragmatic margins 382
diaphragmatic pleura 379
diaphragmatic surface 275, 277
diaphysis 31
diathrosis 70
diencephalon 122, 130, *135**, *136**
digestive system 4, 239, *240**, *241**, *242**, *243**, *244**
digits, 60, 64, *106**
dilator pupillae 219
diphtheria 384
diplococcus pneumoniae 384
diploe 35
diploic veins 323
diplopia 237
direct inguinal hernia 120
directional terms *5**
distal 3
distal convoluted tubule 392
diverticula 272
diverticulitis 272
divisions 187
dorsal 3, 146, 211, 353
dorsal carpal branch 305-7
dorsal digital arteries 316-18
dorsal metatarsal arteries 316
dorsal pedis artery 316
dorsal penis veins 335
dorsal rami of the spinal nerves 183, *185**
dorsal root 141
dorsal root ganglion 141
dorsal scapular artery 295, 298
dorsal scapular nerve 192
dorsal sensory root 183
dorsal surface 123
dorsum sellae 36

ductus arteriosus 335, 340
ductus choledochus 270
ductus deferens 397, 400, *406**, *407**
ductus reuniens 235
ductus venosus 335, 340
duodenal recesses 257
duodenum 263
dura mater 146
dural sinuses 165, 323
dwarfism 433
dyspnea 384

Ectopic kidney 395
ectopic pregnancy 421
efferent arteriole 392
efferent lymphatic vessels 346
efferent tubules 399
ejaculatory ducts 401, 402, *407**
elbow articulation 75, *76**
eminence of the petrous portion of the
   temporal bone 36
eminences 33
emphysema 384
enamel 244
encephalitis 212
endocarditis 340
endocardium 278
endocrine 4, 433
endocrine glands *426**
endolymphatic system 233, 235
endometrium 410
endosteum 32
eparterial bronchus 383
epicardium 278
epicondyle 31
epidermis 23, *24**
epididymis 397, 400, *403**
epidural space 146
epigastric hernia 120
epiglottis 370
epilepsy 212
epimetrium 410
epiphysis 31, 431
epiploic foramen 257
epithalamus 130, 132
epitympanic recess 230
epoophoron 407
equator 214
equilibrium 214
equinovarus 120
esophageal vein 327
esophageal arteries 308
esophagus 239, 251, *253**, *254**
ethmoid bone 37, 367
ethmoid sinus 39, 369
ethmoidal anterior nerve 167
eustachian tube 251
exocrine 433
extensor muscles 118
external 3, 179, 283
external acoustic meatus 33, 227, *228**
external capsule 129
external carotid artery 283, *295**, 376
external genitalia 417
external iliac artery *311**, 312, 313
external iliac vein 332
external jugular vein 321, 323, *326**
external nasal branches 168
external oblique muscle 94
external occipital protuberance 35
external os 410
external pudendal vein 335
external spermatic fascia 94, 397
external vertebral venous plexus 328
extreme capsule 129
eyeball 214, *216**, *217**
   muscles 84, *89**, *90**
eyelids 223, 224

Facial artery 284
facial bones *46**
facial expression—muscles 84, *87**, *88**
facial nerve 173, *180**, *201**, 207, 225, 236
   palsy 237
facial vein 321, *326**
falciform ligament 253, 271
fallopian tube 410
false pelvis 60-62
false vocal cords 374
falx cerebri 148
fascia 94
fasciculus 137, 145
female external genitalia *422**
female pelvis *416**
female perineal muscles 423
female reproductive system 407, *413**, *414**
female urethra 395
femoral artery 313, *314**
femoral branch 197
femoral nerves 200
femoral vein 332, 333, *337**
femur 64, *65**, 79
   muscles *110**, *111**
fenestra cochleae 230
fenestra vestibuli 230
fetal circulation *343**
fetal cortex 431
fetal lobulation 395
fibrous 277
fibula 64, *66**
fibular 3
fibular collateral ligament 79
fila radicularia 141
filiform 248
filum terminale 148
fimbria 130, 410
fimbriated fold 248
fissures 123, 140
fixed (sacral and coccygeal) vertebrae 60
flexor muscles 118
floor of the cranial cavity 35, *36**
folium 141
fontanelles 42, *46**
foot 60, 64
   muscles *112**, *113**, *114**, *115**,
foot-drop 120
Foramen 31, 33, 39, *43**, *44**
   anterior ethmoid 41
   cecum 36, 39, 135, 160, 248
   epiploic 257
   greater palatine 33, 41
   greater sciatic 62
   incisive 33, 41
   infraorbital 33, 41
   interventricular 128-30
   intervertebral 42
   jugular 35, 37, 41
   lacerum, 35, 36, 41
   lesser palatine 33, 41
   lesser sciatic 62
   of Luschka 140, 148
   of Magendie 140, 148
   magnum 35, 36, 135
   mandibular 41
   mental 33, 41
   optic 33, 41
   ovale 35, 36, 41, 335
   palatine *44**
   posterior ethmoid 41
   root 244
   rotundum 36, 41
   sphenopalatine 37, 41
   spinosum 35, 36, 42
   stylomastoid 35, 42
   transverse 42
   vertebral 42
   of Winslow 257
   zygmaticofacial 33

forearm 55
    muscles 101, *102*, 103*, 104*, 105*, 109*
fornix 128, 129
    of the vagina 417
fossa 31, 57
    cerebral 35
    condylar 35, 41
    coronoid 57
    gluteal 62
    ileac 62
    incisive 33
    infraspinous 57
    infratemporal 33
    interpeduncular 135
    jugular 35
    lacrimal 33, 225
    mandibular 35
    middle cranial 66
    nasal 367
    navicularis 403
    ovalis 278
    pterygoid 35
    pterygopalatine 41
    radial 57
    scaphoid 227
    triangular 227
    trochlear 222, 223
fourth ventricle 137
fovea centralis 217
fracture 64, 66
    of the cranium 66
frenulum 248, 402
    clitoridis 417
frontal 32, 321
frontal bone 33, 35, 37
frontal lobe 123
frontal nerve 167
frontal region 33
frontal process 33
frontal sinus 39, 369
frontal suture 33
fundus 225, 233, 260, 392, 410
fungal infections 30
fungiform papillae 248
funiculi 145
fusiform gyrus 127

Gallbladder 253, *268*, 270, *271*
gallstones (radiogram) *271*
ganglion 162
    of the autonomic nervous system 160
    cervical sympathetic *207*
    of the chromaffin system *436*
    ciliary 207
    cranial sensory 162
    dorsal root 141
    Gasserian 166
    geniculate 162, 173, 225, 236
    inferior cervical 210
    jugular 176
    middle cervical 210
    nodose 176
    pterygopalatine 218
    root 162
    splanchnic 210
    submandibular 162, 207, 208
    superior 175
    superior cervical 210
    trunk 162
    vestibular 175
gastrocnemius muscle 118
gastrocolic ligament 254
gastroduodenal artery 310
gastrolienal ligament 254
gastrophrenic ligament 254
general somatic sensory nerve 162
general visceral sensory nerve 162
geniculate ganglion 162, 173, 225, 236

genital branch 198
genitofemoral nerve 197
genu 128, 129
giantism 433
glabella 33
glans clitoridis 417
glans penis 402, 403, *411*
glaucoma 237
glenoid cavity 55
glenoid labrum 78
glioma 212
globus pallidus 129
glomerulonephritis 395
glomerulus 388
glossoepiglottic fold 251
glossopharyngeal nerve 137, 162, 175, *181*,
    182, *201*, 209, 225
glottis 370
gluteal fossa 62
gluteal nerve 203
gluteus maximus 113
gluteus medius 113
gluteus minimus 113, 117
goiters 433
gracilis 117
granular pits 35
granule 214
gray commissure 146
gray ramus communicans 209
gray substance 145
great cardiac vein 319
great cerebral vein (of Galen) 159
great saphenous vein 333, *339*, *340*
greater auricular nerve 187
greater curvature 260
greater horn 53
greater occipital nerve 183
greater omentum 253, 254
greater palatine foramen 33, 41
greater palatine nerve 169
greater sac 259
greater sciatic foramina 62
greater sciatic notch 62
greater splanchnic nerve 210
greater trochanter 64
greater tubercle 57
greater vestibular bulb 417
green stick fracture 66
growth dysfunctions 433
gustatory organs (taste buds) 225, *226*
gyrus 122, *125*, *126*
    angular 124
    cingulate 128
    fusiform 127
    inferior frontal 124
    inferior temporal 124, 127
    middle temporal 124
    parahippocampal 127, 128
    postcentral 124
    precentral 123, 124
    rectus 127
    subcallosal 128
    superior 123, 124
    superior frontal 128
    superior temporal 124
    supermarginal 124

Habenula 132
hair *26*
hallux 64
hamate bone 60
hand 55
    muscles:
        palmar *105*
        dorsal *106*
handle 230
hard palate 33, 244
haustra 266
Haversian systems 32

head, muscles *91*, *92*
heart 277, *281*, *282*, *283*, *286*, *322*
    chambers *284*
    valves 280, *286*
helical network *28*
helix 227
hemianopsia 237
hemiazygos vein 328, *332*
hemorrhage 272
hemorrhoids 272
hepatic flexure 266
hepatic nodes 357
hepatic veins 238, 328
hepatoduodenal ligament 257
hepatogastric ligament 257
hiatus semilunaris 37
highest intercostal artery 298
highest thoracic artery 298
hilus 382, 388
    of the left lung *384*
    of the right lung *383*
hip:
    articulation 76, *78*, *79*
    muscles 101
hippocampal fissure 127
hippocampus 130
Hodgkin's disease 359
horizontal 140
horizontal fissure 382
horseshoe kidney 395
humerus 57, *59*
    muscles *101*, *102*
hyaloid membrane 218
hydroceles 421
hyoid 53, *56*
    muscles *92*, *93*
hypertension 335
hypogastric plexus 211
hypoglossal canal 35, 37, 41
hypoglossal nerve 137, 181, *182*, *183*
hypoglossal trigone 140
hypophysis 429
hypothalamic groove 129
hypothalamus sulcus 130
hypothalamus 130, 132
hypoxia 384

Ileal branches 310
ileal veins 335
ileocolic artery 310
ileocolic recesses 257
ileocolic vein 335
ileum 264
iliac crest 62
iliac fossa 62
iliac veins *336*
iliofemoral ligament 78
iliohypogastric nerve 196, 197
iliolumbar artery 312
ilium 62
improper rotation 421
incisive foramen 33, 41
incisive fossa 33
incisive papilla 245
incomplete descent 421
incus 230
indirect inguinal hernia 120, 421
infection 212, 395
inferior 3
inferior alveolar nerve 171, 285
inferior angle 57
inferior articular process 42
inferior border 275
inferior cardiac branches 179
inferior cardiac nerve 210
inferior cerebellar peduncle 137, 140
inferior cerebral veins 158
inferior cervical ganglion 210
inferior colliculi 132

inferior concha 369
inferior cornu 370
inferior epigastric artery 297, 313
inferior epigastric vein 332
inferior flexure 263
inferior frontal gyrus 124
inferior ganglion 175
inferior gemelli 113
inferior glenohumeral ligaments 75
inferior gluteal artery 312
inferior group 353
inferior gyrus 123, 124
inferior horns 129
inferior labial branch 284
inferior labial frenula 239
inferior laryngeal nerve 376
inferior lateral artery 316
inferior lateral cluneal rami 203
inferior lobe 382
inferior mesenteric artery 309*, 311, 357
inferior mesenteric plexus 211
inferior mesenteric vein 335
inferior nasal concha 37
inferior nasal meatus 37, 369
inferior ophthalmic veins 220
inferior orbital fissure 33, 37, 41, 222, 223
inferior palpebral 168
inferior pancreaticoduodenal artery 310
inferior petrosal sinus 325
inferior phrenic artery 311
inferior phrenic vein 328
inferior ramus 165
inferior ramus of the pubis 62
inferior recess 259
inferior rectal artery 312
inferior rectal nerve 205
inferior sagittal sinus 159, 323
inferior suprarenal artery 312
inferior temporal gyrus 124, 127
inferior temporal line 33
inferior temporal sulcus 124, 127
inferior thyroid artery 297, 376
inferior thyroid veins 376
inferior ulnar collateral artery 305
inferior vena cava 278, 321, 328
inferior tributaries 334
inferior vesicle 312
inflammation 395
inflammation reactions 212
influenza 384
infraglenoid tubercle 57
infrahyoid muscles 90
infraorbital artery 288
infraorbital canal 33
infraorbital foramen 33, 41
infraorbital groove 33
infraorbital nerve 168
infraspinous fossa 57
infratemporal fossa 33
infratrochlear nerve 167
infundibular stalk 130
infundibulum 127, 132, 410, 429
inguinal hernia 97*
inguinal ligament 94
inguinal nodes 357
inner ear 232
inner layer 35
inner sheath 26
innervation:
    of the foot 198*
    of the hand 192*
    of the lower limb 196*
    of the upper limb 190*
    of the urinary bladder and urethra 396*
insula 129
insular 123
integument 4, 23
interatrial septum 278
intercarpal articulations 76, 77*
intercavernous sinus 323

intercondylar area 64
intercondylar eminence 64
intercondylar notch 64
intercostal veins 328
intercostobrachial nerves 196
intercrural fibers 94
interfascial space 214
interlobar vein 392
interlobular artery 392
intermediate branches 187
intermedius nerve 173
internal 3, 179
internal acoustic meatus 39, 233
internal aspects of the cerebral hemispheres:
    coronal section 129*
    transverse section 130*
internal auditory artery 295
internal auditory vein 327
internal capsule 129
internal carotid artery 151, 158, 283, 288, 297*, 298*
internal ear 232*
internal iliac artery 311*, 312
internal iliac veins 332
internal jugular nodes 353
internal jugular veins 321, 323, 326*, 327*
internal oblique muscle 94
internal os 410
internal pudendal arteries 312
internal pudendal vein 335
internal spermatic fascia 94, 397
internal strabismus 237
internal thoracic artery 297
internal thoracic trunks 295
internal thoracic vein 327, 353
internal vertebral venous plexus 328
interosseous artery 307
    recurrent 307
interparietal sulcus 124
interpeduncular fossa 135
interphalangeal articulation 81*, 82*
interpubic disk 78
intersigmoid recess 257
intertarsal articulation 79, 81*, 82*
interthalamic adhesion 128
intertubercular groove 57
interureteric fold 394
interventricular foramen 128-30
interventricular septal defects 340
interventricular septum 278
intervertebral articulation 73*, 74*
intervertebral disks 42, 45, 53*, 73
intervertebral foramina 42
intraarticular ligament 73
intraperitoneal 254
intrapetrous part of the facial nerve 236
iris 217, 219*
ischiofemoral ligament 78
ischium 62
islets of Langerhans 433
isthmus 128, 227, 410, 427
isthmus of the aorta 282

Jejunal artery 310
jejunal veins 335
jejunum 264
joints of the shoulder 73
jugular nerve 162
jugular foramen 35, 37, 41
jugular fossa 35
jugular ganglion 176
jugular notch 53

Kidney 388, 390*, 391*, 392*
knee 78, 80*
    muscles 110*
kyphosis 64
    articulation

Labia majora 417
labia minora 417
labial vein 335
labial branches 205
labial tubercle 239
labyrinthine vein 153
lacrimal artery 292
lacrimal bone 37
lacrimal caruncle 224*, 225
lacrimal duct 225
lacrimal fossa 33, 225
lacrimal gland 220, 223*
lacrimal nerve 167
lacrimal papillae 225
lacrimal punctum 225
lacrimal sac 225
lactiferous ducts 28
lacunar ligament 94
lambdoid 33
lambdoid suture 35
lamina 31, 42, 370
    terminalis 130
large intestine 239, 263*, 264*, 265*, 266*
laryngeal aperture 370
laryngeal nodes 352
laryngeal plexuses 211
laryngeal portion of the pharynx 251
larynx 367, 370, 371*
    muscles 372*
lateral 3, 57, 187, 225, 233
lateral angle 410
lateral antebrachial cutaneous nerve 188
lateral branches 185, 187, 195
lateral compartment 118
lateral condyle 64
lateral crura 94
lateral cutaneous branches 195
lateral dorsal cutaneous pedis nerve 204
lateral epicondyle 57, 64
lateral facets 53
lateral femoral muscle 101
lateral femoral compartment 113
lateral femoral cutaneous nerve 200
lateral fibular notch 64
lateral fissure 123
lateral geniculate body 132
lateral incisors 244
lateral laminae 370
lateral ligament 79
lateral lobes 405
lateral malleolus 64
lateral menisci 79
lateral muscular compartments 117
lateral nasal branch 285
lateral olfactory striae 127, 214
lateral pectoral nerves 188
lateral plantar artery 316, 318
lateral plantar nerve 204
lateral process 230
lateral pterygoid 72
lateral pterygoid plates 33, 35
lateral raphe 225
lateral recess 140
lateral sacral artery 312
lateral striate arteries 158
lateral superficial femoral veins 335
lateral supracondylar line 64
lateral surface 123
lateral temporomandibular ligament 72
lateral thoracic branches 299
lateral umbilical ligament 253, 312
lateral ventricles 129
lateral wall 37
least splanchnic nerve 211
left atrioventricular aperture 280
left atrioventricular valve 280
left atrium 278, 323*, 324*
left border 275
left branch 335
left branch bundle 280

left colic vein 311, 335
left common carotid artery 282, 283
left coronary artery 281, 282
left gastric artery 309
left gastroepiploic artery 310
left lobe 270
left pulmonary arteries 280
left subclavian arteries 282
leg 60, 64
lens 218, *221**
lentiform nucleus 129
leptomeninges 146
lesser curvature 260
lesser horn 53
lesser occipital nerve 187
lesser omentum 257
lesser palatine foramen 33, 41
lesser palatine nerve 169
levator palpebrae superioris 219
lesser petrosal nerve 175, 230
lesser sac 257
lesser sciatic foramina 62
lesser sciatic notch 62
lesser trochanter 64
lesser tubercle 57
levator glandulae thyroideae muscle 427
ligament *11**, 70, 410
   accessory plantar 81
   annular 76
   anterior cruciate 79
   anterior longitudinal 73
   anterior sacroiliac 76
   anterior talofibular 79
   arterial 340
   broad 407, 410
   calcaneofibular 79
   coccygeal 147
   collateral 76, 81
   conoid 75
   coracoclavicular 75
   coronary 271
   deltoid 79
   denticulate 148
   falciform 253, 271
   fibular collateral 79
   gastrocolic 254
   gastrolienal 254
   gastrophrenic 254
   hepatoduodenal 257
   hepatogastric 257
   iliofemoral 78
   inferior glenohumeral 75
   inguinal 94
   intraarticular 73
   ischiofemoral 78
   lacunar 94
   lateral 79
   lateral temporomandibular 72
   lateral umbilical 253, 312
   long plantar 81
   long posterior sacroiliac 76
   medial 79
   medial palpebral 225
   median suspensory 427
   middle glenohumeral 75
   ovarian 414
   patellar 316
   phrenicocolic 254
   phrenicolienal 254
   posterior cruciate 79
   posterior deep 285
   posterior interosseous 76
   posterior longitudinal 73
   posterior talofibular 79
   pubofemoral 78
   pulmonary 381
   radial collateral 75
   reflected inguinal 94
   round 414
   sacrospinous 78

ligament
   sacrotuberous 76
   short posterior sacroiliac 76
   sphenomandibular 72
   stylomandibular 72
   superior glenohumeral 75
   supraspinous 57
   suspensory, of the ovary 414
   teres 414
   teres hepatis 253, 340
   tibial collateral 79
   transverse 79
   transverse acetabular 78
   transverse metatarsal 81
   trapezius 75
   triangular 271
   ulnar collateral 75
   uterosacral 417
   venosum 340
   vocal 370
limbic system 128
limbus fossae ovalis 278
linea 31
linea alba 94
linea aspera 64
limiting sulcus 140
lingual nerve 170
lingual artery 284
lingula 141
liver 239, 253, 267, *268**, *269**, *270**
   cancers 272
lobes 429
lobule 227, 399
long ciliary nerve 167
long plantar ligament 81
long posterior sacroiliac ligaments 76
long thoracic nerve 192
longitudinal fissure 123
longitudinal fold 263
lordosis 64
lower crus 227, 230
lower limbs 4, 60, *62**
lower nerve 192
lumbar arteries 312
lumbar enlargements 144
lumbar nerves 141, 146, 196, *208**
lumbar plexus 198, 200
lumbar region 185
lumbar splanchnic nerves 211
lumbar veins 328, 330
lumbar vertebrae 45, *51**
lumbosacral plexus *194**
lumens 28
lunate bone 60
lungs *322**, *323**, 367, *381**, *386**
lunula 26
lymphadenitis 359
lymphangitis 359
lymphatic 4
lymphatic system 346, *347**, *348**, *352**, 353, 359
lymphomas 359

Main arterial channel in the upper limb 295
main pancreatic duct 272
major calices 392
major duodenal papilla 264
major papilla 272
malabsorption syndromes 272
maldescent 421
male reproductive system 397, *398**, *399**
male urethra 395, 402
malleus 230
mammary glands *28**, *29**
mammillary bodies 127, 132
mandible 39, 42
mandibular 32
mandibular division of the trigeminal nerve 169

mandibular foramen 41
mandibular fossa 35
mandibular nerve 174
mandibular notch 33
mandibular region 33
manubrium 53
marginal 282
masseteric 285
mastoid process 33
mastoiditis 237
maxilla 32, 33, 37
   associated foramina *44**
maxillary artery 284, 285
   branches *296**
maxillary division of the trigeminal nerve 168
maxillary process 33
maxillary region 33
maxillary sinus 37, 369
maxillary tuberosity 33
maxillary vein 323
meatus 31
mechanical injury 421
Meckel's diverticulum 264, 272
medial 3, 57, 187, 316
medial artery 316
medial branches 185, 187, 195
medial cerebral artery 157
medial condyle 64
medial crura 94
medial eminence 140
medial epicondyle 57, 64
medial femoral muscle 101
medial femoral compartment 117
medial geniculate body 132
medial incisors 244
medial ligament 79
medial longitudinal stria 128
medial malleolus 64
medial menisci 79
medial olfactory striae 127, 214
medial palpebral ligament 225
medial pectoral nerves 188
medial plantar artery 318
medial plantaris muscle 204
medial pterygoid plates 35
medial sacral artery 311
medial sacral vein 328, 330
medial striate arteries 158
medial superficial femoral veins 335
medial supracondylar line 64
medial sural cutaneous nerve 204
medial surface 123
medial surface of brain *128**
medial wall 37
median artery 316
median nerve 190
median palatine raphe 245
median sulcus 135, 140
median suspensory ligament 427
median umbilical ligament 253, 394
mediastinal arteries 308
mediastinal nodes 353
mediastinal pleurae 379
mediastinal surface 382
mediastinal vein 327
mediastinum 376, 378
mediastinum testis 399
medulla 130, 388, 431
medulla oblongata 122, 135, *140**, *141**
melanoma 30
membranous 395
membranous labyrinth 233
membranous urethra 402
meningeal branch 168
meningeal vessels 35
meninges 146, *153**, *154**, *155**, *156**
meningitis 212, 237
menisci 79
mental foramen 33, 41

mental nerve 171
mental protuberance 33
mental tubercle 33
mentolabial sulcus 239
meridional fibers 219
mesencephalon 122, 130, 132, *137**, *138**
mesenteries 254, 256
mesoappendix 256
mesometrium 410, 414
mesosalpinx 410, 414
mesovarium 407, 414
metacarpals 60
metatarsals 64, *67**
metatarsophalangeal articulation *81**, *82**
metopic suture 33
midbrain 132
middle 35, 192
middle concha 369
middle cardiac nerve 210
middle cardiac vein 319
middle cerebellar peduncle 135, 137, 141
middle cerebral artery 151, 292
middle cerebral veins 158
middle cervical ganglion 210
middle colic artery 310
middle colic veins 335
middle cranial fossa 66
middle ear 227, 230
middle glenohumeral ligaments 75
middle gyrus 123, 124
middle lobe 382, 405
middle meatus 37, 369
middle mediastinum 379
middle meningeal artery 285
middle nasal conchae 37
middle rectal artery 312
middle rectal vesicular 211
middle suprarenal artery 311
middle temporal gyrus 124
middle temporal sulcus 124, 127
minor calices 392
minor duodenal papilla 264
minor papilla 272
mitral 214
mitral valve 280
modiolus 233
mons pubis 417
motor root 166, 173
mouth 239
mucous coat 394
multiple sclerosis 212
Muscles *12**, *85**, *86**, 397
  arm 100
  arrector pili 26
  ciliary 217, *218**, 219
  coracobrachialis 100
  cremaster 94
  deltoid 299
  extensor 118
  external oblique 94
  facial expression 84, *87**, *88**
  female perineum 423
  femur *110**, *111**
  flexor 118
  foot and toes *112**, *113**, *114**, *115**
  forearm 101, *102**, *103**, *104**, *105**, *109**
  gastrocnemius 118
  hand (dorsal) *106**
    (palmar) *105**
  head *91**, *92**
  hip 101
  humerus *101**, *102**
  hyoid *92**, *93**
  infrahyoid 90
  internal oblique 94
  knee joint *110**
  larynx 370, *372**
  lateral femoral 101
  levator palpebrae superioris 219
  of mastication 84, 87, 88

Muscles
  medial femoral 101
  medial plantaris 204
  oblique 219
  pectinate 278
  pectineus 117
  pectoral 299
  peroneus brevis 118
  peroneus longus 118
  peroneus profundus 203
  peroneus tertius 118
  piriformis 117
  plantaris 118
  popliteus 118
  postaxial 100–101
  preaxial 100–101
  quadratus femoris 113
  quadriceps femoris 113
  rectus femoris 113
  respiration *98**, *100**
  sartorius 113
  semimembranosus 113
  semitendinosus 113
  soleus 118
  stapedius 230
  subclavian 188
  suprahyoid 90
  tensor tympani 230
  thigh 60, 101
  thyroarytenoid 370
  thyroideae (levator glandular) 427
  tibialis anterior 118
  tibialis posterior 118
  tongue *91**
  trapezius 60
  triceps 100
  vastus intermedius 113
  vastus lateralis 113
  vastus medialis 113
  vertebral column 90, *93**, *94**
muscular 4
muscular arteries 292
muscular atrophy 119
muscular branches 305, 307
muscular dystrophies 118–119
muscular twigs 316
musculocutaneous nerve 188
musculophrenic artery 297
myenteric plexus (of Auerbach) 211
mylohyoid nerve 171
myocardium 278
myometrium 410
myopathy 118

Nails *26**
nasal 32, 321
nasal bones 33
nasal cavity *39**, *40**, 367
nasal conchae 33, 37
nasal fossae 367
nasal region 33
nasal septum 33, 39, 367, *368**
nasal vestibule 369
nasociliary nerve 167
nasolabial sulcus 239
nasolacrimal duct 225, 369
navicular bone 64
neck 2, 39, 49, 55, 57, 64, 230, 239, 272, 392
necrosis 274
nephron 388
Nerves *13**, *14**, *122**, *199**
  abducens 137, 173, 179, 222, 223
  accessory 137, *181**, *182**
  antebrachial cutaneous 191
  auriculotemporal 172
  autonomic 122, *199**, 207
  axillary 192
  carotid sinus 432

Nerves
  cervical 141, 146
    ventral ramus 187
  chorda tympani 173, 225, 232, 237
  coccygeal 141, 146
  cranial 160, *169**, *170**
  deep radial 193
  dorsal scapular 192
  ethmoidal anterior 167
  facial 173, *180**, *201**, 207, 225
    geniculate ganglion 236
    intrapetrous part 236
    palsy 237
  femoral 200
  frontal 167
  general somatic sensory 162
  general visceral sensory 162
  genitofemoral 197
  glossopharyngeal 137, 162, 175, *181**, *182**, *201**, 209, 225
  gluteal 203
  greater auricular 187
  greater occipital 183
  greater palatine 169
  greater splanchnic 210
  hypoglossal 137, 181, *182**, *183**
  iliohypogastric 196, 197
  inferior alveolar 171, 285
  inferior cardiac 210
  inferior laryngeal 376
  inferior rectal 205
  infraorbital 168
  infratrochlear 167
  intercostobrachial 196
  intermedius 173
  lacrimal 167
  lateral antebrachial cutaneous 188
  lateral dorsal cutaneous pedis 204
  lateral femoral cutaneous 200
  lateral pectoral 188
  lateral plantar 204
  least splanchnic 211
  lesser occipital 187
  lesser palatine 169
  lesser petrosal 175, 230
  lingual 170
  long ciliary
  long thoracic 192
  lower 192
  lumbar 141, 146, 196, *208**
  lumbar splanchnic 211
  mandibular division of trigeminal 169, *176**, *177**
  maxillary division of trigeminal 168, *178**
  medial pectoral 188
  medial sural cutaneous 204
  median 190
  mental 171
  middle cardiac 210
  middle cervical ganglion 210
  musculocutaneous 188
  myenteric plexus (of Auervach) 211
  mylohyoid 171
  nasiociliary 167
  nasociliary 167
  obturator 201
  oculomotor 135, 165, *172**, 207
  olfactory 162, *171**, 214, *215**
  optic 127, 164, *172**
  peroneal 202
  phrenic 187
  piriform 202
  prepelvic 198
  proprioceptive sensory 162
  pterygoid plexus 220, 323
  pudendal *195**
  pudendal plexus 205
  radial 193
  sacral 141, 146, 209
  sacral plexus 198, 202
  saphenous 201

Nerves
  short ciliary 219
  spinal *149*, 150*, 151*, 152**, 160, 182, *184**
  spinal accessory 179, *183**
  spinal cord 122, 141, *144*, 145*, 146*, 147*, 148*, 150*, 185*, 186**
  splanchnic *204**, 210
  splanchnic ganglion 210
  submandibular ganglion 162, 207, 208
  submucosal plexus (of Meissmer) 211
  subscapular 192
  superficial cardiac plexus 211
  superior cervical ganglion 210
  superior cluneal 186
  superior ganglion 175
  superior gluteal 202
  supraclavicular 187
  supraorbital 167
  suprascapular 188
  supratrochlear 167
  sympathetic trunk *208**, 209
  thoracic 141, 146, 195, *208**
  thoracic spinal 193
  thoracodorsal 192
  thoracolumbar sympathetic trunk 209
  thyrocervical trunk 296
  trigeminal 135, 166, 168, 169, *174*, 175*, 176*, 177*, 178**
  ulnar 190
  vagus 137, 162, 176, *181*, 182*, 202**, 376
  vestibulocochlear 174, *180**
  zygomatic 169
  zygomaticofacial 169
  zygomaticotemporal 169
nervous 4
network 305
nodes 348
nodose 162
nodose ganglion 176
nodulus 141
nucleus gracilis 137
nucleus pulposus 45, 73

Occipital bone 33, 35
olfactory nerve 162, *171**, 214, *215**
olfactory pathways *216**
olfactory sulcus 127
olfactory tract 127, 214
olfactory trigone 127
omental bursa 254
opercular fold 124
ophthalmic vein 327
ophthalmic artery 220, *222**, 292, *299**
ophthalmic division of the trigeminal nerve 166
optic canal 37
optic chiasma 127
optic disk 217
optic foramen 33, 41
optic groove 36
optic nerve 127, 164, *172**
optic tract 127, 135
ora serrata 217
oral cavity *245**
  cavity proper 244
orbit 32, 37
orbital cavity *37*, 38**
orbital part 224
orbital region 33
orbital surface 127
organ levels 4
organs:
  of Corti 175
  of the special senses 214
orthopnea 384
osteitis fibrosa cystica 433
osteomalacia 64
osteoporosis 64

ostium abdominale 410
ostium ganglion 162, 207, 209
otitis medial 237
outer layer 35
outer sheath 26
ovarian artery 312
ovarian ligament 414
ovarian vein 328, 330
ovaries 407, *415*, 418**, 426, 433, *437**
oviduct 410
ovum *418**

Pacemaker 280
paired kidneys 388
paired ureters 388
palate 244, 249
palatine bone 37
  associated foramina *44**
palatine branches 288
palatine tonsil 251, 359
palatoglossal arch 245
palatopharyngeal arch 245, 251
palmar 3
palmar carpal branch 305–307
palpebral aperture 225
palpebral conjunctiva 225
pampiniform plexus 330, 402, 410
pancreas 239, 272, 273, 426, 433, *438**
pancreatic branches 335
pancreatic ducts *273**
pancreaticoduodenal vein 335
papella 388, 402
papillae 23
papilledema 237
paradidymis 400
parahippocampal gyrus 127, 128
paraganglia 432, 433
paranasal sinuses *40*, 41*, 369**
parasternal nodes 353
parasympathetic motor 162
parasympathetic system *200*, 201*, 202*, 204**, 207–9
parathyroid 426, 427, *428*, 429**
paratracheal nodes 353
parietal 33, 123
parietal bone 33, 35
parietal emissary 160
parietal layer 277
parietal lobe 124
parietal pleura 379
parietooccipital fissure 128
Parkinsonism 212
parolfactory area 128
paroophoron 407
parotid gland 249
pars tensa 227
patella 64, *65**, 79
patellar ligament 316
patent ductus arteriosus 340
path of blood flow *285**
pattern of cristae 23, *25**
pectinate muscle 278
pectineus muscle 117
pectoral muscle 299
pedicle 42
pelvic brim 60
pelvic girdle 62
pelvic plexuses 211
penis 397, 402, 408
pericardial arteries 308
pericardial cavity 277
pericardial fluid 277
pericardial sac (heart removed) 275, *279*, 280**
pericarditis 340
pericardium *323**
perilymphatic duct 233, 235
perilymphatic fluid 233
perineal rami 203, 205

peripheral nervous system 122, 160
  anterior view *123**
  autonomic *168**
  cranial nerves *166**
  posterior view *124**
  spinal nerves *167**
perirenal fat 388
peritoneum 253, 254
peritonitis 272
perivascular space 148
permanent teeth 244, *246*, 247**
peroneal artery 316
peroneal nerve 202
peroneus brevis 118
peroneus longus 118
peroneus profundus 203
peroneus tertius 118
petrosal 162
Peyer's patches 263
phalanges 60, 64, *67**
pharyngeal aperture 251
pharyngeal branch 177, 168–69
pharyngeal tonsils 359
pharyngeal recess 251
pharynx 239, 251, *252**, 367
philtrum 239
phlebitis 340
phrenic nerve 187
phrenicocolic ligament 254
phrenicolienal ligament 254
pia mater 146, 148
pineal body 132, 426, 431, *433**
pineal recess 130
piriform nerves 202
piriformis muscle 117
pisiform bone 60
pituitary gland 426, 429, *431**
placenta *344**
plantar 3
plantar arch 318
plantar artery 316
plantar metatarsal artery 318
plantaris muscle 118
pleurae 378, *379**
plicae tunicae mucosae 270
pneumonia 384
pons 122, 130, 135, *139**
pontine 153, 296
popliteal artery 313, 315, 316
popliteal nodes 357, 362
popliteal vein 333, 338
popliteus 118
porta hepatis 270
portal obstruction 340
portal system 341
portal vein 335
postaxial muscles (extensors) 100, 101
postcentral gyrus 124
postcentral sulcus 124
posterior 3, 35, 90, 233
posterior auricular artery 284, 285
posterior atlantooccipital membrane 72
posterior band 75
posterior branch 195
posterior branches of the brachial plexus 191
posterior cerebellar arteries 153, 296
posterior cerebral arteries 151, 153, 295
posterior chambers of the eye 218
posterior circumflex humeral arteries 299
posterior clinoid process 36
posterior commissure 130, 132, 417
posterior communicating artery 151, 158, 292
posterior compartment 101
posterior cruciate ligament 79
posterior deep ligament 285
posterior division 312
posterior elastic membrane 216
posterior ethmoid canals 36
posterior ethmoid foramina 41
posterior ethmoidal branches 168

posterior ethmoidal sinuses 39
posterior femoral 101
posterior femoral compartment 113
posterior femoral cutaneous 203
posterior gray column 146
posterior gray commissure 146
posterior horn 129
posterior inferior spine 62
posterior intercavernous sinus 323
posterior intercostals 309
posterior interosseous artery 307
posterior interosseous ligament 76
posterior interventricular branches 282
posterior interventricular sulci 278
posterior limb of the internal capsule 129
posterior lobe 405, 429
posterior longitudinal ligament 73
posterior median sulcus 135, 145
posterior mediastinum 379
posterior nasal apertures 33, 37
posterior nasal sulcus 369
posterior pole 214
posterior radicular arteries 150
posterior ramus 201
posterior spinal artery 150, 153, 296
posterior superior iliac spine 62
posterior talofibular ligament 79
posterior tibial artery 313, 316
posterior tibial recurrent artery 316
posterior trunks 187
posterior two-thirds of the tongue 225
posterior ulnar recurrent artery 305, 307
posterointermediate sulcus 145
posterolateral sulcus 145
postganglion neuron 207
postganglion sympathetic motor cell bodies 209
postganglion sympathetic cells and fibers *206**
postganglion tubercle 35
postpelvic group of nerves 198
practical anatomy 4
preaxial muscles (flexors) 100, 101
precentral gyrus 123, 124
preganglionic neuron 207
preganglionic sympathetic motor cell bodies 209
preganglionic sympathetic cells and fibers *206**
prepelvic group of nerves 198
prepuce 402
  of the clitoris 417
primary regions 4
process 31
profundus ramus 204
proprioceptive sensory nerves 162
prostate gland 397, 404, 408
  rectal examination of *411**
prostatic 211, 295, 402
prostatic utricle 402
proximal 3
proximal convoluted tubule 392
psittacosis 384
pterygoid canal 35, 41, 288
pterygoid canal branches 288
pterygoid fossa 35
pterygoid hamulus 35
pterygoid plexus 220, 323
pterygopalatine 162, 207
pterygopalatine fossa 41
  associated foramina *43**
pterygopalatine ganglion 208
ptosis 237
pubic arch 62
pubis 62
pubofemoral ligaments 78
pudendal canal 205
pudendal nerve *195**
pudendal plexus 205
pulmonary branches 210
pulmonary ligament 381

pulmonary pleura 281
pulmonary semilunar valves 280
pulmonary sound 341
pulmonary stenosis 340
pulmonary trunk 280
pulmonary tuberculosis 384
pulmonary veins 280, 320, *322**, *323**, *324**
pulvinar 132
putamen 129
pyelonephritis 395
pyloric branch 335
pyloric valve 260
pyloric vein 335
pylorus 260
pyramid 137, 141, 388
pyramidal eminence 230
pyramidal lobe 427

Quadrate lobe 270
quadratus femoris 113
quadriceps femoris 113
quadrigeminal plate 132

Radial 3
radial artery 305, 306
radial collateral ligament 75
radial fossa 57
radial nerve 193
radial notch 60
radial recurrent artery 305, 306
radicular arteries 150
radiocarpal articulation 76
radius 57
ramus 33, 39
  of the ischium 62
  musculares 306
rectal columns 267
rectal examination of the prostate and seminal vesicles *411**
recti 219
rectouterine fold 47
rectouterine pouch 392, 410
rectum 267
rectus femoris 113
rectus sinus 159, 323
recurrent laryngeal branch 179
recurrent meningeal branch 177
recurrent meningeal filament 183
red marrow 32
reflected inguinal ligament 94
refracting media 218
renal artery 311, 392
renal calcices *394**
renal columns 388
renal corpuscle 388
renal failure 395
renal sinus 388
renal tubule 388
renal vein 328, 330
reproductive 4
respiration muscles *98**, *100**
respiratory epithelium 369
respiratory system 4, 367
rete testes 399
retina 217, *219**, *220**
retroflexed uterus 421
retromandibular vein 323
retroperitoneal 254
retroverted uterus 421
rheumatic heart disease 340
rhinal fissure 127
ribs 49, *54**, *55**
rickets 64
right atrioventricular aperture 280
right atrioventricular valve 280
right atrium 278
right border 275
right branch 335

right branch bundle 280
right bronchus 383
right colic artery 311
right colic vein 335
right common carotid arteries 283
right coronary arteries 281, 282
right gastroepiploic artery 310
right gastroepiploic vein 335
right lobe 270
right lymphatic duct 348
right lymphatic trunk 348
right subclavian artery 283
rima glottidis 370
root 187, 239, 382
root canal 244
root foramen 244
root ganglion 162
rostrum 128
round ligament 414
rugae 260

SA node 280
sacculus 235
sacral 141
sacral coccygeal regions 186
sacral nerves 141, 146, 209
sacral plexus 198, 202
sacral vertebrae 45, *52**
sacroiliac 76, *78**, *79**
sacroiliac articulation 76, *78**, *79**
sacrospinous ligament 78
sacrotuberous ligament 76
sagittal plane 4
sagittal suture 35
salivary glands 239, 249, 251
salpingopharyngeal fold 369
saphenous nerve 201
saphenous veins 333, *339**, *340**
sartorius muscle 113
scala tympani 230, 233
scala vestibuli 233
scalene tubercle 53
scaphoid bone 60
scaphoid fossa 227
scapula 53, 55, *58**
  winged 119
sclera 216
sclera segment 214
scoliosis 64
scrotal (labial) branches 205
scrotum 397, *400**
sebaceous glands *27**, 28
second molars 244
second premolars 244
sella turcica 36
semicircular canals 233, 235
semilunar 162, 166, 225
semilunar fold *224**
semilunar hiatus 369
semilunar valve 281
semimembranosus 113
semitendinosus 113
seminal vesicles 401, 404, *406**, *407**
  rectal examination of *411**
seminiferous tubules 399
sensory root 166
septal cartilage 367
septula testis 399
septum pellucidum 128
serous coat 394
serous pericardium 277
shaft 49, 57, 64
short ciliary nerves 167, 219
short crus 230
short gastric veins 319, 335
short posterior sacroiliac ligaments 76
shoulder 55
  articulation 73, 75
  girdle *99**, *100**

sight 214
sigmoid arteries 311
sigmoid colon 267
sigmoid groove 37
sigmoid mesocolon 257
sigmoid sinus 159, 325
sigmoid veins 335
simple fractures 64
sinoatrial node 280
sinus 31, 346
  anal 267
  aortic 281
  basilar 323
  carotid 283
  cavernosus 159, 220, 325
  circular 160
  confluens of 159, 323
  coronary 278, 319
  dural 165, 323
  ethmoid 39, 369
  frontal 39, 369
  inferior petrosal 325
  inferior sagittal 159, 323
  intercavernous 323
  maxillary 37, 369
  rectus 159, 323
  renal 388
  sigmoid 311
  small unpaired occipital 159
  sphenoidal 39, 369
  sphenoparietal 160, 325
  superior petrosal 325
  superior sagittal 159, 323
  superior sagittal venous 35
  tonsillar 251
  transverse 159, 325
  transverse venous 37
  tympanic 230
  venosus sclerae 216
skeletal 4
skeletal metastases 421
skeleton of the appendages 53
skin 23
  cancer 30
skull *32**, *46**
small cardiac vein 319, 333
small intestine 239, 253-54, *260**, *261**, *262**, 263
small unpaired occipital sinus 159
small saphenous vein 333
smallest cardiac veins 319
smell 214
soft palate 244
soleus muscle 118
somatic motor nerves 162
somatic sensory nerves 162
special senses 4
special sensory nerves 162
spermatic cord *403**
sphenoethmoidal 39
sphenoethmoidal recess 369
sphenoid bone 33, 37
  associated foramina *44**, *45**
sphenoidal sinus 39, 369
sphenomandibular ligament 72
sphenopalatine artery 288
sphenopalatine foramen 37, 41
sphenoparietal sinus 160, 325
sphincter pupillae 219
spinal accessory nerve 179, *183**
spinal cord 122, 141, *150**
  dorsal rami of *185**
  external aspect, dorsal view *144**
  external aspect, ventral view *145**
  internal aspect, diagrammatic view of reflex arc *146**
  internal aspect, diagrammatic view of sensory and motor pathways *147**
  transverse sections of various levels of spinal cord *148**

spinal cord
  transverse section of vertebral column, spinal cord and its covering *147**
  ventral rami of *186**
spinal nerves 160, 182, 184
  dermatome chart of cutaneous distribution *184**
  dermatome chart for cutaneous innervation *152**
  dorsal and ventral roots *151**
  dorsal view of vertebral canal, spinal cord and spinal nerves *150**
  spinal cord segments *149**
  transverse section *151**
spine 31
spine of the scapula 57
spinous process 42
spiral lamina 233
spiral organ (organ of Corti) 235, *236**
spiral valve 270
splanchnic ganglion 210
splanchnic nerve 210
spleen 357, *363**
splenic artery 310
splenic branch 335
splenic flexure 267
splenic vein 335
splenium 128
spongy urethra 402
squamous cell 30
squamous portion of the temporal bone 33
stalk 431
stapedius 230
stapes 230
stenosis 272
sternal angle 53
sternoclavicular articulation *74**
sternum 53, *56**
stomach 239, 253, *257**, *258**, *259**
stratum basale 23
stratum corneum 23
stratum granulosum 23
stratum lucidum 23
stratum spinosum 23
structural organization *8**
structure 3
styloid process 33, 35, 57, 60
stylomandibular ligament 72
stylomastoid foramen 35, 42
subarachnoid cisterns 148
subarachnoid space 146, 148
subcallosal gyrus 128
subclavian 295
subclavian nodes 353
subclavian portion 295
subclavian vein 327, 329, *331**
subclavian muscle 188
subdural space 146
sublingual gland 251
sublingual papilla 248, 249
submandibular ganglion 162, 207, 208
submandibular gland 249
submandibular nodes 350, 353
submental branch 284
submucosal plexus (of Meissner) 211
submucous coat 394
subscapular 57, 299
subscapular nerves 192
subthalamus 130
sudoriferous glands *27**, 28
Sulcus 31, 123, *125**, *126**, 150
  anterior intraventricular 278
  anterolateral 135
  central 123
  cingulate 128
  coronary 278
  of the corpus callosum 128
  hypothalamic 130
  inferior temporal 124, 127
  interparietal 124

Sulcus
  interventricular 278
  limiting 140
  median 135, 140
  mentolabial 239
  middle temporal 124, 127
  nasolabial 239
  olfactory 127
  postcentral 124
  posterior interventricular 278
  posterior median 135, 145
  posterior nasal 369
  posterointermediate 145
  posterolateral 145
  superior temporal 124
  terminalis 248
superciliary ridges 33
superficial 3, 151, 158, 327, 353
superficial cardiac plexus 211
superficial circumflex iliac artery 313
superficial circumflex iliac vein 335
superficial connections to the portal system *342**
superficial cutaneous branch 193
superficial epigastric artery 313
superficial epigastric vein 335
superficial fascia 28
superficial group 353
superficial inguinal nodes 357
superficial inguinal ring 94
superficial lymphatics of the head and neck *350**
superficial palmar branch 305, 306
superficial perineum of the female *425**
superficial posterior compartment 117, 118
superficial temporal artery 284, 288
superficial temporal vein 323
superficial veins *320**
  of the arm 327
  of the head and neck 321
  of the trunk 328, *333**
  of the upper limb *330**
superior 3, 162, 233
superior alveolar arch 33
superior avleolar branches 169, 288
superior angle 57
superior articular process 42
superior borders 57
superior branches 179
superior cerebellar artery 153, 296
superior cerebellar peduncle 132, 141
superior cerebral veins 158
superior cervical ganglion 210
superior cluneal nerve 186
superior colliculi 132
superior concha 369
superior cornu 370
superior epigastric artery 297
superior flexure 263
superior frontal gyrus 128
superior ganglion 175
superior gemelli 113
superior glenohumeral ligaments 75
superior gluteal artery 312
superior gluteal nerve 202
superior gyrus 123, 124
superior labial artery 168
superior labial branch 284
superior labial frenula 239
superior laryngeal vein 376
superior laryngeal branch 177
superior lateral artery 316
superior lobe 382
superior meatus 37, 369
superior mediastinum 376
superior mesenteric artery *308**, 310, 357
superior mesenteric branch 335
superior mesenteric node 357
superior mesenteric vein 335
superior nasal conchae 37

superior nuchal line 35
superior ophthalmic veins 220
superior orbital fissure 33, 36, 37, 41, 220, 222, 223
superior palpebral fold 225
superior petrosal sinus 325
superior phrenic arteries 308–09
superior radioulnar articulation 76
superior ramus 165
    of the pubis 64
superior recess 257
superior rectal artery 311
superior rectal vein 335
superior sagittal sinus 159, 323
superior sagittal venous sinus 35
superior suprarenal artery 311
superior tarsus 224
superior temporal gyrus 124
superior temporal line 33
superior temporal sulcus 124
superior thyroid artery 284, 376
superior thyroid notch 370
superior thyroid veins 376
superior ulnar collateral artery 299
superior vena cava 278, 321, 325
superior vesicle 312
supermarginal gyrus 124
supraclavicular nerve 187
supraglenoid tubercle 57
suprahyoid muscles 90
supraorbital artery 292
supraorbital margins 33
supraorbital nerve 167
supraorbital notch 33, 42
suprarenal gland 426, 431, *434**
suprarenal vein 328
suprascapular artery 297
suprascapular nerve 188
suprascapular notch 57
supraspinous ligament 57
supratrochlear artery 292, 353
supratrochlear nerve 167
supravaginal 410
sural branches 316
surface projections of the heart
    (diagrammatic views) *278**
surgical neck 57
suspensory ligament 218, 407
suspensory ligament of the ovary 414
sustentaculum tali 64
suture 31
sweat glands 28
sympathetic division of the autonomic
    nervous system *205*, 206*, 207*, 208**, 209
symphysis 31
symphysis menti 33
symphysis pubis 64, 76
    articulation *78*, 79**
synarthrosis 70
synovial sheaths 70
system level 4
systemic anatomy
    articular system *11**
    cardiovascular system *16*, 17**
    digestive system *15**
    endocrine system *21**
    integumentary system (skin) *9**
    ligamental system *11**
    lymphatic system *18**
    muscular system *12**
    nervous system *13**
    nervous system:
        neonatal view *14**
        posterior view *14**
    reproductive system *20**
    respiratory system *19**
    skeletal system *10**
    special senses *15**
    urinary system *19**

Tachycardia 340
tachypnea 384
tail 130, 272
talipes 120
talocrural articulation 79
talus 64
tarsal glands 225
tarsal part 224
tarsals 64, *67**
tarsometatarsal articulation *81*, 82**
tarsus 225
taste 214
teeth 239, 244, *246*, 247*, 248**
tegmen tympani 230
tegmental portion 135
temporal bone 33, 36, 123
    associated foramina *43**
temporal branch 174
temporal lobe 123, 124
temporal process 33
temporomandibular articulation *71**
teniae coli 264
tensor fasciae latae 117
tensor tympani muscle 230
tentorial incisure 148
tentorium cerebelli 37, 148
testes 397, 399, *403*, 404**, 426, 433, 437
testicular artery 312
testicular vein 328, 330
tetanus 212
tetany 433
tetralogy of Fallot 340
thalamus 128, 129, 130, 131, 132
thickened periosteum 235
thigh muscles 60, 101
thin plate of bone 236
third molars 244
third portion of the axillary artery 299
third ventricle 130
thoracic aorta *291*, 306**, 307
thoracic cavity 376, 377
thoracic duct 348, 349
thoracic nerves 141, 146, 195, *208**
thoracic region 185
thoracic spinal nerve 193
thoracic vertebrae 45, *50**
thoracoacromial artery 299
thoracoacromial vein 328
thoracodorsal artery 299
thoracodorsal nerve 192
thoracolumbar sympathetic trunk 209
thymic vein 327
thymus 357, 426, 429
    adult *359*, 430**
    newborn *364*, 430**
thyroarytenoid muscle 370
thyrocervical trunk 295, 296
thyrohyoid muscle 370
thyroid cartilage 370
thyroid gland 426, *427*, 428*, 429**
thyroidea ima vein 327
tibia 64, *66*, 79*
tibial 3
tibial collateral ligaments 79
tibialis anterior muscle 118
tibialis posterior muscle 118
tissue damage 30
tissue level 4
tongue 246, *250**
tongue
    muscles *91**
tonsillar branch 284
tonsillar sinus 251
tonsillitis 359
tonsils 359, 365
torcular herophili 323
torus tubarius 369
traveculae carneae 278
trachea 367, *374*, 375*, 376**

tracheal vein 327
tragus 227
transversalis fascia 94
transverse 94
transverse acetabular ligament 78
transverse band 75
transverse cervical nerve 187
transverse cervical artery 297
transverse colon 253, 267
transverse crista 233
transverse facial artery 288
transverse facial tributaries 323
transverse foramen 42
transverse groove 37
transverse ligament 79
transverse mesocolon 256
transverse metatarsal ligament 81
transverse palatine plicae 245
transverse plane 4
transverse process 42
transverse sinus 159, 325
transverse venous sinus 37
trapezius 60
    ligament 75
trench mouth 272
triangular 60, 124
triangular fossa 227
triangular ligaments 271
triceps 100
tricuspid sound 341
tricuspid valve 280
trigeminal nerve 135, 166
    central connections *178**
    cutaneous distribution *178**
    diagrammatic view *175**
    mandibular division 169, *176*, 177**
    maxillary division 168, *178**
    lateral view *174**
trochanter 31, 64
trochlea 31, 57, 166
trochlear fossa 222, 223
trochlear nerve 132, 166, 173
trochlear notch 57
true cortex 431
true pelvis 62
trunk 4
trunk ganglion 162
trunks 187
tubal end 407
tuber 141
tuber cinereum 127, 130, 132
tubercle 31, 49
tubercular facet 49
tuberculum cinereum 137
tuberculum sellae 36
tuberosity 57
tubular network of the testis and epididymis *403**
tufted 214
tumors 212
tunica:
    adventitia 281
    albuginea 399, 400, 402
    dartos 397
    intima 281
    media 281
    muscularis 394
    vaginalis 399
    vaginalis testis 397
tympanic antrum 232
tympanic branch 175
tympanic cavity 227
tympanic membrane 227, 229
tympanic nerve 230
tympanic notch 227
tympanic plexus 175, 230
tympanic promontory 230
tympanic sinus 230
typical thoracic spinal nerves *193**

Ulcers 272
ulna 57
ulnar 3
ulnar artery 305, 306
ulnar collateral ligament 75
ulnar nerve 190
umbilical arteries 335, 340
umbilical hernia 120
umbilical vein 335, 340
umbo 227
uncinate process 369
uncus 127
unpaired bladder 388
upper 192
upper crus 227
upper limb 4, 55, *57**
urachus 253, 394
ureter 392, *393**
urethra 388, *395**, 397, 402, 408, 417
urethral crest 402
urethritis 395
urinary 4
urinary bladder 392, *394**
urinary calculi 395
urinary system 388, *389**, *390**
uterine artery 312
uterine end 407
uterine tubes 407, 410, *415**, *416**, *419**
uterosacral ligament 417
uterovaginal plexus 330
uterus 407, 410, *415**, *416**, *419**, *420**
    retroflexed 421
    retroverted 421
utriculus 233
uvula 141, 245, 394

Vagina 407, 417, *424**
vaginal 211
vaginal part 410
vagus nerve 137, 162, 176, *181**, *182**, *202**, 376
vallecula 251
valves 318
valvulae foramina ovalis 278
varicocele 421
varicosities 340
vasa casorum 281
vascular 4
vascular grooves 35
vastus intermedius 113
    lateralis 113
    medialis 113
Veins *17**, *27**, 158, *318**
    accessory hemiazygos 328
    angular 220
    anterior cardiac 319
    anterior deep temporal 288
    anterior scrotal 335
    arcuate 392
    axillary 328
    azygos 328, *332**
    basal 158
    basilic 328, *330**
    brachiocephalic 321
    cardiac 318, *321**, *322**
    cephalic 328, *330**
    cerebral 158, *163**, *164**, 325
    common iliac 328, 332
    coronary 335
    cystic 335
    deep *319**
        of the arm 328
        of the lower limbs 335
        of the upper limbs *331**
    deep circumflex iliac 332
    deep femoral 333
    deep middle cerebral 158
    diploic 323

Veins
    dorsal penis 335
    of the ear 327
    external jugular 321, 323, *326**
    facial 321, *326**
    femoral 332, 333, *337**
    great cardiac 319
    cerebral (of Galen) 159
    saphenous 333, *339**, *340**
    hemiazygos 328, *332**
    hepatic 238, 328
    ileal 335
    ileocolic 335
    iliac *336**
    inferior epigastric 332
    inferior mesenteric 335
    inferior ophthalmic 220
    inferior phrenic 328
    inferior thyroid 376
    inferior vena cava 278, 321, 328, 334
    intercostal 328
    interlobar 392
    internal auditory 327
    internal iliac 332
    internal jugular 321, 323, *326**, *327**
    internal pudendal 335
    internal thoracic 327, 353
    jejunal 335
    labial 335
    labyrinthine 153
    lateral superficial femoral 335
    left colic 311, 335
    lumbar 328, 330
    maxillary 323
    medial sacral 328, 330
    medial superficial femoral 335
    mediastinal 327
    middle cardiac 319
    middle cerebral 158
    middle colic 335
    of the nasal cavity 325
    of the neck 327
    ophthalmic 327
    of the orbit 327
    ovarian 328, 330
    pancreaticoduodenal 335
    popliteal 333, 338
    portal 335
    pulmonary 280, 320, *322**, *323**, *324**
    pyloric 335
    renal 328, 330
    retromandibular 323
    right colic 335
    right gastroepiploic 335
    short gastric 319, 335
    small cardiac 319, 333
    smallest cardiac 319
    splenic 335
    subclavian 327, 329, *331**
    superficial circumflex iliac 335
    superficial epigastric 335
    superficial temporal 323
    superior cerebral 158
    superior laryngeal 376
    superior mesenteric 335
    superior ophthalmic 220
    superior rectal 335
    superior thyroid 376
    superior vena cava 278, 321, 325
    suprarenal 328
    testicular 328, 330
    thoracoacromial 328
    thymic 327
    thyroidea ima 327
    tracheal 327
    transverse venous sinus 37
    umbilical 335, 340
    vertebral 327
    vorticose 216

velum 245
venae comitantes 318
venous distribution 150
venous lacunae 159
ventral 3, 146
ventral motor root 183
ventral pulmonary plexuses 211
ventral rami of the cervical nerves 187
    of the spinal nerves *186**
ventral ramus 183, 186, 195
ventral root 141
ventral surface 123
ventral surface of the brain *127**
ventricles 370
ventricular fibrillation 340
ventricular system of the cerebral
    hemispheres *131**
vermiform process 266
vermis 141
vertebra, typical *47**
vertebra prominens 45
vertebral 295
vertebral artery 150, 151, 153, 295
vertebral column, muscles 90, *93**, *94**
vertebral foramen 42
vertebral veins 327
vertex 392
vertigo 237
vesical trigone 394
vesicouterine pouch 410
vestibular 227
vestibular aqueduct 233
vestibular bulbs 417
vestibular folds 374
vestibular ganglion 175
vestibular labyrinth 230
vestibular membrane 235
vestibule 232, *233**, 239, 257, 374, 417
vestibulocochlear nerve 174, *180**
villi *261**, *262**, 263
viral 212
viral diseases 30
visceral 381
visceral layer 277
visceral sensory nerve 162
visceral pathways *225**
vitreous body 218
vocal folds *373**
vocal ligament 370
vocal process 370
volar 3
vomer bone 33, 37, 367
vorticose veins 216
vulvae 417

Walls of the heart 278
white ramus communicans 209
white substance 145
winged scapula 119
wrist 55
wrist-drop 119

Xiphoid process 53

Yellow bone marrow 32

Zonula ciliaris 219
zygomatic arch 32, 33
zygomatic bone 33, 37
zygomatic branch 174
zygomatic nerve 169
zygomatic process 33
zygomaticofacial foramen 33
zygomaticofacial nerves 169
zygomaticotemporal nerves 169